CRC
Handbook
of
Atomic
Absorption
Analysis

Volume II

Author

Asha Varma, Ph.D., F.A.I.C.
Naval Air Development Center
Warminster, Pennsylvania

CRC Press, Inc.
Boca Raton, Florida

Library of Congress Cataloging in Publication Data

Varma, Asha, 1942—
    Handbook of atomic absorption analysis.

    Bibliography: p.
    Includes indexes.
    1. Atomic absorption spectroscopy--Handbooks,
manuals, etc. I. Title.
ZD96.A8V37 1985   543′.0858   83-18925
ISBN 0-8493-2985-X (v. 1)
ISBN 0-8493-2986-8 (v. 2)

Direct all inquiries to CRC Press, Inc., 2000 Corporate Blvd., N.W., Boca Raton, Florida, 33431.

© 1984 by CRC Press, Inc.

International Standard Book Number 0-8493-2985-X (v. 1)
International Standard Book Number 0-8493-2986-8 (v. 2)

Library of Congress Card Number 83-18925
Printed in the United States

PREFACE

Atomic absorption spectroscopy, since its introduction by Walsh, has witnessed a more rapid growth than any other analytical technique. Over the period, automation, optimization, and elaborate computer systems for handling large numbers of data, have made this technique one of the best and most widely used analytical methods for the determination of major and minor constituents in agriculture, biology, geology, mining and metallurgy, pollution, etc. It seems to be replacing the traditional wet chemical methods used for analysis.

Several introductory and text books have been written from time to time on the theory, instrumentation, and applications of atomic absorption spectroscopy. Although some stabilization has taken place in the fundamental development of instrumentation and methodology, applications of this technique continue to extend at a fast pace.

A present day analyst is confronted with an increasing demand for greater sensitivity, reliability, and speed to analyze complex materials. Modern technology has provided new reagents, procedures, and instruments to deal with this growing demand. The data required for selection of a technique is mostly scattered in abstracts, journals, monographs, periodicals, and textbooks. Without an easy access to the applicable literature pertinent to the analysis desired, an analyst can generally overlook the most reliable and sensitive method. A desk reference handbook could be a valuable asset to the analyst.

To assemble all the possibly available literature on atomic absorption spectroscopy, its development of theory, instrumentation, and applications, and present it in a ready-reference form is a tedious and demanding task, but it is necessary to the continued growth of analytical chemistry.

In this handbook, an attempt has been made to provide the practicing analyst or research scientist a concise, convenient, and critical reference guide through the vast literature of atomic absorption spectroscopy in a simplified manner. The material has been arranged to develop a knowledge and practical use of the technique.

The first chapter deals with the introduction to atomic absorption spectroscopy. The general principles are now well established. Theory has been treated from a nonmathematical viewpoint. A theoretical spectroscopist can obtain detailed information from various standard textbooks listed in Appendix IV. The practical aspects of the technique, care of the instrument, precautions to be taken for handling the instrument, preparation of the standard and sample solutions, and calculations are discussed in detail. Lists of manufacturers of atomic absorption spectrophotometers and accessories (Appendix VIA), hollow cathode lamps (Appendix VA), and chemical suppliers (Appendix VIC) are provided.

The reference data for general instrumentation, development and improvements of the atomic absorption spectroscopy are presented in Chapter II. A separate application index is included here, although all the subsequent chapters are an application of this technique.

The periodic table arrangement is used, where one group of elements is assigned to one section. Group VIII has been divided in two sections; Fe, Co, and Ni in one and Rh, R, Pd, Ir, Os and Pt as 'platinum group metals' in the other.

Elements that can not be analyzed with the atomic absorption method are completely ignored and no attempt is made to give an alternate suitable method for analysis. However, an indirect use of atomic absorption method to analyze certain elements is mentioned.

In each chapter, a little introduction to the element, possible dissolution method of the sample, interferences, standard solution preparation, and instrumental parameters for flame atomic absorption method are described for the element. References begin with the earliest attempt to analyze the element and end with the latest one. References are followed by author and subject indexes. This simple arrangement is maintained throughout all the chapters, which makes data search complete and convenient for each and every element ever analyzed with the use of the atomic absorption method.

Since every chapter and element is complete in itself, there are repetitions of references to confine the pertinent data to that particular element. A separate index for the analysis of biological products is provided in the chapters on calcium, copper, zinc, and lead. The appendixes include lists of abstracts and reviews in Appendix II and general references in Appendix III, as well as a glossary for the terms and definitions used in atomic absorption spectroscopy.

This handbook, a record of references (up to December 1982), is dedicated to the scientists, engineers, and technicians who designed, developed, and applied atomic absorption spectroscopy to elemental analysis and revolutionized the field of analytical chemistry.

I am pleased to express my appreciation to the Perkin-Elmer Corporation for allowing the use of the following material as appeared in the "Analytical Methods for Atomic Absorption Spectroscopy": (1) instrumental parameters for lutetium, (2) instrumental parameters for technetium, (3) table for absorbance vs. percent absorption values, and (4) calculations given on page 31.

Thanks are also due to my professional acquaintances and the advisory board members, who have encouraged me to bring this handbook to its present state.

At last but not least, I will like to acknowledge the help, encouragement, patience, and love of my family, without which I would have never attempted to take this task and complete it to share it with fellow chemists.

The errors, omissions, and limitations are my sole responsibility. I will be grateful for any comments to improve subsequently.

Asha
March 12, 1982

THE AUTHOR

Asha Varma, Ph.D., is affiliated with the Instrumentation Laboratories at the Naval Air Development Center, Warminster, Pennsylvania.

Dr. Varma received her B.Sc. and M.Sc. degrees in 1958 and 1960 from Bareilly College, Bareilly, India, and obtained her Ph.D. in Analytical Chemistry from Banaras Hindu University, Varanasi, India in 1963. She was appointed Senior Research Fellow (1963 to 1966) by the Council of Scientific & Industrial Research, Government of India. She served at the Banaras Hindu University and National Chemical Laboratory, Poona and established Protein Chemistry Laboratory. She became the first woman in 1966 to be selected by the Madhya Pradesh Government as an Assistant Director of the Forensic Science Laboratory, Sagar. She held a postdoctoral fellowship at the Chemistry Department (1967) and a research associateship at the Institute of Metallurgy, University of Connecticut, Storrs, from 1973 to 1975. She has served as the Supervisor of Analytical Laboratories at the Laboratory for Research on the Structure of Matter, University of Pennsylvania, Philadelphia from 1977 to 1982.

Dr. Varma is a Certified Professional Chemist. She is a fellow of the American Institute of Chemists; member of American Chemical Society and Coblentz Society and associate member of National Wildlife Federation and American Museum of Natural History. Dr. Varma is listed in American Men & Women of Science.

Dr. Varma has thirty publications. She is involved with the identification of metals in lubricants, alloys, test specimens, eutectic materials, corrosion products, additives, protective coatings, resins, epoxys, and polymers. Her current research interests are in developing new methods for trace contaminants in organic and inorganic materials.

To My Husband with Love

TABLE OF CONTENTS

Volume I

GROUP IVB ELEMENTS

GROUP VB ELEMENTS

GROUP VIB ELEMENTS

GROUP VIIB ELEMENTS

GROUP VIII ELEMENTS (Fe, Co, Ni)

GROUP VIII ELEMENTS (Platinum Group Metals)

Volume II

GROUP IB ELEMENTS (Cu, Ag, Au)

GROUP IIB ELEMENTS (Zn, Cd, Hg)

GROUP IIIA ELEMENTS

GROUP VIA ELEMENTS

GROUP VIIA ELEMENTS (F, Cl, Br, I)

ZERO GROUP ELEMENTS (He, Ne, Ar, Kr, Xe)

APPENDIXES

Group IB Elements — Cu, Ag, and Au

COPPER, Cu (ATOMIC WEIGHT 63.546)

Copper is one of the earliest known, recognized, and used metals by man. It occurs widely in nature as a free element as well as in the combined form in sulfide ores and minerals. It is also found in the sun, meteorites, rocks, soil, mineral and ocean water, plants, and animals. Copper is used in the electrical industry, agriculture industry, and in manufacturing other alloys (brass, bronze, monel, etc.). Its compounds are used as poisons, algicides for water purification, and in analytical chemistry as chemical reagents. Copper determination is required in ores, minerals, rocks, copper-base alloys, nonferrous alloys, and other metallurgical products, as a additive or an impurity in steels and cast iron, paints, lubricants, water, waste, industrial waste, plants and biological products, etc.

For atomic absorption analysis, sample dissolution depends on the nature of the sample. Some of the samples could be water soluble, while others such as refractory materials may need fusion. Dissolution of metallic copper usually requires nitric acid or hot perchloric acid. Hydrochloric acid or sulfuric acid with 30% hydrogen peroxide can also be used as oxidizing agents. Samples containing organic material should be heated with a combination of nitric, sulfuric, and perchloric acids.

No spectral or chemical interferences have been reported for copper analysis. However, depression of the copper signal has been observed at high zinc/copper ratios, which can be minimized by the use of a lean air-acetylene flame or a nitrous oxide-acetylene flame.

Standard Solution

To prepare 1000 mg/ℓ copper solution, dissolve 1.000 g of metallic copper in 50 mℓ of 1:1 nitric acid and dilute to 1 ℓ with 1% (v/v) nitric acid.

Instrumental Parameters

Wavelength	324.7 nm
Slit width	0.7 nm
Light source	Hollow cathode lamp
Flame	Air-acetylene, oxidizing (lean, blue)
Sensitivity	0.03—0.07 mg/ℓ
Detection limit	0.002 mg/ℓ
Optimum range	5 mg/ℓ

Secondary wavelengths and their sensitivities

327.4 nm	0.1—0.2 mg/ℓ
217.9 nm	0.3 mg/ℓ
218.2 nm	0.4 mg/ℓ
216.5 nm	0.47—0.6 mg/ℓ
222.6 nm	1.1—1.5 mg/ℓ
249.2 nm	5.0—5.8 mg/ℓ
244.2 nm	11.0—14.0 mg/ℓ
224.4 nm	15.7—24 mg/ℓ

REFERENCES

1. **Strasheim, A., Strelow, F. W. E., and Butler, L. R. P.,** The determination of copper by means of atomic absorption spectroscopy, *J. S. Afr. Chem. Inst.*, 13, 73, 1960.
2. **Allan, J. E.,** The determination of copper by atomic absorption spectrophotometry, *Spectrochim. Acta*, 17, 459, 1961.
3. **Strasheim, A. and Verster, F.,** The determination of copper and zinc in human tissues using atomic absorption spectroscopy, *Tydskr. Natuurwet.*, 1, 197, 1961 (in Afrikaans).
4. **Elwell, W. T. and Gidley, J. A. F.,** *Atomic Absorption Spectrophotometry*, Pergamon Press, New York, 1961.
5. **Fabricand, B. P., Sawyer, R. R., Ungar, S. G., and Adler, S.,** Trace metal concentrations in the ocean by atomic absorption spectroscopy, *Geochim. Cosmochim. Acta*, 26, 1023, 1962.
6. **Barras, R. C.,** Application of atomic absorption to the petroleum industry, *Jarrell Ash Newslett.*, June 1962.
7. **Schuler, V. L. O., Jansen, A. V., and James, G. S.,** The development of atomic absorption methods for the determination of silver, copper, iron, lead and zinc in high purity gold and the role of organic additives, *J. S. Afr. Inst. Min. Metall.*, 62, 807, 1962.
8. **Strasheim, A., Butler, L. R. P., and Maskew, E. C.,** Determination of certain impurities in gold by atomic absorption spectroscopy, *J. S. Afr. Inst. Min. Metall.*, 62, 796, 1962.
9. **Zeeman, P. B. and Butler, L. R. P.,** The determination of lead, copper and zinc in wines by atomic absorption spectroscopy, *Appl. Spectrosc.*, 16, 120, 1962.
10. **Butler, L. R. P. and Brink, D.,** The determination of magnesium, calcium, potassium, sodium, copper and iron in water samples by atomic absorption spectroscopy, *S. Afr. Ind. Chem.*, 17, 152, 1963.
11. **Wallace, F. J.,** Determination of copper in metallurgical materials by atomic absorption, *Hilger J.*, 7, 65, 1963.
12. **Sprague, S. and Slavin, W.,** Determination of the metal content of lubricating oils by atomic absorption spectrophotometry, *At. Absorpt. Newsl.*, No. 12, April 1963.
13. **Herrman, R. and Lang, W.,** Determination of serum copper by atomic absorption flame photometry, *Z. Klin. Chem.*, 1, 182, 1963 (in German).
14. **Belt, C. B., Jr.,** Atomic absorption spectrophotometry and the analysis of silicate rocks for copper and zinc, *Econ. Geol.*, 59, 240, 1964.
15. **Belt, C. B., Jr.,** The determination of copper and zinc in Bayer process liquor, *At. Absorpt. Newsl.*, No. 19, March 1964.
16. **Frey, S. W.,** The determination of copper, iron, calcium, sodium and potassium in beer by atomic absorption spectrophotometry, *At. Absorpt. Newsl.*, 3, 127, 1964.
17. **Kinson, K. and Belcher, C. B.,** Determination of minor amounts of copper in iron and steel by atomic absorption spectrophotometry, *Anal. Chim. Acta*, 31, 180, 1964.
18. **Morgan, M. E.,** Determination of copper in milk by atomic absorption spectroscopy, *At. Absorpt. Newsl.*, No. 22, July 1964.
19. **Slavin, W. and Sprague, S.,** The determination of trace metals in blood and urine by atomic absorption spectrophotometry, *At. Absorpt. Newsl.*, No. 17, January 1964.
20. **Sprague, S. and Slavin, W.,** Determination of very small amounts of copper and lead in potassium chloride by organic extraction and atomic absorption spectrophotometry, *At. Absorpt. Newsl.*, No. 20, May 1964.
21. **Trent, D. J. and Slavin, W.,** Determination of various metals in silicate samples by atomic absorption spectrophotometry, *At. Absorpt. Newsl.*, 3, 118, 1964.
22. **Willis, J. B.,** Determination of copper in butter and butterfoil by atomic absorption spectroscopy, *Aust. J. Dairy Technol.*, 70, 1964.
23. **Novoselov, V. A. and Aidarov, T. K.,** Spectrographic determination of the trace elements Ag, Cu, Pb, Cd and Al in solutions by using a hollow cathode source, *Tr. Pokhim. Khim. Technol.*, 108, 1964.
24. **Berman, E.,** Application of atomic absorption spectrometry to the determination of copper in serum, urine and tissue, *At. Absorpt. Newsl.*, 4, 296, 1965.
25. **Means, E. A. and Ratcliffe, D.,** Determination of wear metals in lubricating oils by atomic absorption spectroscopy, *At. Absorpt. Newsl.*, 4, 174, 1965.
26. **Beyer, M.,** Determination of manganese, copper, chromium, nickel and magnesium in cast iron and steel, *At. Absorpt. Newsl.*, 4, 212, 1965.
27. **Biechler, D. G.,** Determination of trace copper, lead, zinc, cadmium, nickel and iron in industrial waste waters by atomic absorption spectrometry after ion exchange concentration on DOWEX A-I, *Anal. Chem.*, 37, 1054, 1965.
28. **Bordonali, C., Biancifiori, M. A., and Besazza, G.,** Determination of traces of metals in sodium by atomic absorption (spectrometry). Determination of copper, manganese, iron, lead, nickel and zinc, *Chim. Ind. (Milan)*, 47, 397, 1965.

29. **Farrar, B.,** Determination of copper and zinc in ore samples and lead-based alloys, *At. Absorpt. Newsl.,* 4, 325, 1965.
30. **Khalifa, H., Svehla, G., and Erdey, L.,** Precision of the determination of copper and gold by atomic absorption spectrophotometry, *Talanta,* 12, 703, 1965.
31. **Magee, R. J. and Rahman, A. K. M.,** Determination of copper in sea water by atomic absorption spectroscopy, *Talanta,* 12, 409, 1965.
32. **Rousselet, F. and Girard, M. L.,** Use of atomic absorption for micro determination of copper and zinc in biological media, *C. R.,* 260, 3780, 1965 (in French).
33. **Sprague, S. and Slavin, W.,** Determination of iron, copper and zinc in blood serum by an atomic absorption method requiring only dilution, *At. Absorpt. Newsl.,* 4, 228, 1965.
34. **Bradfield, E. G. and Spencer, D.,** Leaf analysis as a guide to the nutrition of fruit crops. VI. Determination of magnesium, zinc and copper by atomic absorption spectroscopy, *J. Sci. Food Agric.,* 16, 33, 1965.
35. **Bradfield, E. G. and Osborne, M.,** Application of atomic absorption spectroscopy to the determination of Zn, Fe, Cu, Mn and Pb in bottled ciders, *Long Ashton Agric. Hort. Res. Stn. Univ. Bristol, Annu. Rep.,* 157, 1965.
36. **Dumanski, J.,** Adaptation and modification in determining copper, zinc and manganese in plant material by atomic absorption spectroscopy, *Rocz. Nauk Roln. Ser. A,* 90, 431, 1961 (in Polish).
37. **Ostroumenko, P. P.,** Some aspects of the excitation of copper lines in hollow cathode discharge, *Zh. Prikl. Spektrosk.,* 5, 581, 1966.
38. **Adda, J., Rousselet, F., and Mocquat, Ct.,** A solution to the problem of copper in the butter industry, *Rev. Laitiere Francaise,* 231, 227, 1966 (in French).
39. **Bell, G. F.,** The analysis of aluminum alloys by means of atomic absorption spectrophotometry, *At. Absorpt. Newsl.,* 5, 73, 1966.
40. **Cheek, D. B., Powell, G. K., Reba, R., and Feldman, M.,** Manganese, copper and zinc in rat muscle and liver cells and in thyroid and pituitary insufficiency, *Bull. John Hopkins Hosp.,* 118, 338, 1966.
41. **Fishman, M. J.,** The use of atomic absorption for analysis of natural waters, *At. Absorpt. Newsl.,* 5, 102, 1966.
42. **Gaumer, M. W., Sprague, S., and Slavin, W.,** An automated procedure for the determination of trace metals by atomic absorption spectroscopy, *At. Absorpt. Newsl.,* 5, 58, 1966.
43. **Heneage, P.,** Five element lamp, *At. Absorpt. Newsl.,* 5, 67, 1966.
44. **Kirkbright, G. F., Peters, M. K., and West, T. S.,** Determination of trace amounts of copper in niobium and tantalum by atomic absorption spectroscopy, *Analyst (London),* 114, 1966.
45. **Passmore, W. O. and Adams, P. B.,** The determination of copper in glass by flame emission spectroscopy and atomic absorption spectrometry, *At. Absorpt. Newsl.,* 5, 77, 1966.
46. **Sattur, T. W.,** Routine atomic absorption analysis on non-ferrous alloys and plant intermediates, *At. Absorpt. Newsl.,* 5, 37, 1966.
47. **Slavin, S. and Slavin, W.,** Fully automatic analysis of used aircraft oils, *At. Absorpt. Newsl.,* 5, 106, 1966.
48. **Simpson, G. R. and Blay, R. A.,** Rapid method for determination of the metals copper, zinc, tin, iron and calcium in foodstuffs by atomic absorption spectroscopy, *Food Trade Review,* August 1966.
49. **Atsuya, I.,** Determination of copper in aluminum metals, aluminum alloys and steel by atomic absorption spectroscopy, *Bunseki Kagaku,* 15, 247, 1966.
50. **Holtzman, N. A., Elliott, D. A., and Heller, R. H.,** Copper intoxication, *N. Engl. J. Med.,* 275, 347, 1966.
51. **Spitzer, H.,** Atomic absorption spectrophotometry for Cu, Pb, Fe, Mn, Ni, and Ca in zinc oxide, *Z. Erzbergbau Metallhuttenwes.,* 19, 567, 1966 (in German).
52. **Kanabrocki, E. L., Case, L. F., Graham, L., Fields, T., Miller, E. B., Oester, Y. T., and Kaplan, E.,** Non-dialyzable manganese and copper levels in serum of patients with various diseases, *J. Nucl. Med.,* 8, 166, 1967.
53. **Scott, T. C., Roberts, E. D., and Cain, D. A.,** Determination of minor constituents in ferrous materials by atomic absorption spectrophotometry, *At. Absorpt. Newsl.,* 6, 1, 1967.
54. **Bohmer, M., Auer, E., and Bartels, H.,** Determination of iron and copper by the atomic absorption method, *Arztl. Lab.,* 13, 258, 1967 (in German).
55. **Brooks, R. R., Presley, B. J., and Kaplan, I. R.,** Determination of copper in saline waters by atomic absorption spectrophotometry combined with APDC-MIBK extraction, *Anal. Chim. Acta,* 38, 321, 1967.
56. **Girard, M. L. and Rousselet, F.,** Atomic absorption applied to biology. V. Progress made in the micro-determination of copper and zinc, *Ann. Pharm. Fr.,* 25, 353, 1967.
57. **Hanig, R. C. and Aprison, M. H.,** Determination of calcium, copper, iron, magnesium, manganese, potassium, sodium, zinc and chloride concentrations in several brain areas, *Anal. Biochem.,* 21, 169, 1967.
58. **Ostroumenko, P. P.,** Determining the absolute copper atom concentration in an acetylene air flame by the linear absorbance method, *Zh. Prikl. Spektrosk.,* 7, 752, 1967.

59. **Marshall, J. C. and Blanchard, D. P.**, The determination of copper (II) iodate by atomic absorption spectrophotometry, *At. Absorpt. Newsl.*, 6, 109, 1967.

60. **Ramakrishna, T. V., Robinson, J. W., and West, P. W.**, Determination of copper, cadmium and zinc by atomic absorption spectroscopy, *Anal. Chim. Acta*, 37, 20, 1967.

61. **Strunk, D. H. and Andreasen, A. A.**, Collaborative study using atomic absorption spectrophotometry for the determination of copper in alcoholic products, *J. Assoc. Off. Anal. Chem.*, 50, 339, 1967; reprinted in *At. Absorpt. Newsl.*, 6, 111, 1967.

62. **Sunderman, F. W., Jr. and Roszel, N. O.**, Measurement of copper in biologic materials by atomic absorption spectrometry, *Am. J. Clin. Pathol.*, 48, 286, 1967.

63. **Gaillot, J., Laroche, M. J., Rohrback, P., Rousselet, F., and Girard, M. L.**, Application of atomic absorption to the study of the intestinal assimilation of copper in the rat, *Ann. Biol. Clin. (Paris)*, 25, 1037, 1967 (in French).

64. **Gomez-Garcia, G. G. and Matrone, G.**, Copper metabolism in the early postnatal period of the piglet, *J. Nutr.*, 92, 237, 1967.

65. **Rawling, R. S. and Sullivan, J. V.**, Atomic absorption analysis for copper using a resonance monochromator, *Miner. Processing Extraction Metall.*, 76, C-238, 1967.

66. **Premi, P. R. and Cornfield, A. H.**, Determination of total copper, zinc, iron, manganese and chromium in plant materials and organic residues by extraction with HCl followed by atomic absorption spectroscopy, *Spectrovision*, 19, 15, 1968.

67. **Plyushch, G. V.**, Atomic absorption determination of copper, and cobalt in nickel berrites, *Zavod. Lab.*, 34, 1741, 1968.

68. **Obiols, J. and Rafols Rovira, J. M.**, Determination of copper and potassium in salt by atomic absorption spectrophotometry, *Afinidad*, 25, 23, 1968.

69. **Nagura, M. and Iida, C.**, Atomic absorption spectrophotometric determination of cobalt, nickel, lead and copper in silicates with the absorption tube technique, *Bunseki Kagaku*, 17, 1513, 1968.

70. **Harrison, W. W., Netsky, M. G., and Brown, M. D.**, Trace elements in human brain: copper, zinc, iron and magnesium, *Clin. Chim. Acta*, 21, 55, 1968.

71. **Girard, M. L.**, Determination of copper and zinc by atomic absorption, *Clin. Chim. Acta*, 20, 243, 1968 (in French).

72. **Endo, Y., Hata, T., and Nakahara, Y.**, Atomic absorption determination of calcium, magnesium, manganese, copper, zinc and aluminum in iron ores, *Jpn. Analyst*, 17, 679, 1968 (in Japanese).

73. **Dawson, J. B., Ellis, D. J., and Newton-John, H.**, Direct estimation of copper in serum and urine by atomic absorption spectroscopy, *Clin. Chem. Acta*, 21, 33, 1968.

74. **O'Leary, J. B. and Spellacy, W. N.**, Serum copper and zinc by atomic absorption spectrophotometry, *Science*, 162, 682, 1968.

75. **Olson, A. D. and Hamlin, W. B.**, Serum copper and zinc by atomic absorption spectrophotometry, *At. Absorpt. Newsl.*, 7, 69, 1968.

76. **Potter, S. L., Ducay, E. D., and McCready, R. M.**, Determination of sugar in plant materials. Measurement of unreduced copper by atomic absorption spectrometry, *J. Assoc. Off. Anal. Chem.*, 51, 748, 1968.

77. **Roach, A. G., Sanderson, P., and Williams, D. R.**, Determination of trace amounts of copper, zinc and magnesium in animal feeds by atomic absorption spectrophotometry, *Analyst (London)*, 93, 42, 1968.

78. **Simonian, J. V.**, Determination of copper in textiles by atomic absorption spectrophotometry, *At. Absorpt. Newsl.*, 7, 63, 1968.

79. **Berge, D. G. and Pflaum, R. T.**, Determination of copper in urine by atomic absorption spectrophotometry, *Am. J. Med. Technol.*, 34, 725, 1968.

80. **Devoto, G.**, New method for determination of copper in biological liquids by atomic absorption spectrophotometry, *Boll. Soc. Ital. Biol. Sper.*, 44, 1249, 1968 (in Italian).

81. **Ramirez-Munoz, J. and Roth, M. E.**, Metallurgical applications of atomic absorption flame photometry. I. Interference by iron on cobalt, chromium, copper, manganese and nickel, *Flame Notes*, 3, 2, 1968.

82. **Roussos, G. G. and Morrow, B. H.**, Direct method for determination of micro quantities of molybdenum, iron, and copper in milk xanthine oxidase fractions by atomic absorption spectroscopy, *Appl. Spectrosc.*, 22, 769, 1968.

83. **Amati, A., Rastelli, R., and Minguzzi, A.**, Determination of iron, copper, zinc and manganese in must and wine by atomic absorption spectrophotometry, *Ind. Agrar.*, 6, 630, 1968 (in Italian).

84. **Robinson, J. L., Barnekow, R. G., and Lott, P. F.**, Rapid determination of cadmium and copper in plating wastes and river water by atomic absorption spectroscopy, *At. Absorpt. Newsl.*, 8, 60, 1969.

85. **Schmidt, W.**, Routine determination of copper in blood serum by means of atomic absorption spectrophotometry, *Z. Anal. Chem.*, 243, 198, 1969.

86. **Spencer, D. W. and Brewer, P. G.**, The distribution of copper, zinc and nickel in sea water of the gulf of Maine and the Sargaso Sea, *Geochim. Cosmochim. Acta*, 33, 325, 1969.

87. **Blomfield, J. and MacMahon, R. A.,** Micro determination of plasma and erythrocytes copper by atomic absorption spectrophotometry, *J. Clin. Pathol.,* 22, 136, 1969.
88. **Dalton, E. F. and Malanoski, A. J.,** Atomic absorption analysis of copper and lead in meat and meat products, *J. Assoc. Off. Anal. Chem.,* 52, 21, 1969.
89. **Dewaele, M. and Harjadi, W.,** Effect of concentration on the optimal height of measure in flame atomic absorption. Experience on copper and manganese, *Anal. Chim. Acta,* 45, 21, 1969.
90. **Husler, J. W.,** The determination of copper and nickel in tungsten metal by atomic absorption spectrophotometry, *At. Absorpt. Newsl.,* 8, 1, 1969.
91. **Ichida, T. and Nobuoka, M.,** Determination of serum copper with atomic absorption spectrophotometry, *Clin. Chim. Acta,* 24, 299, 1969.
92. **Knight, D. M. and Pyzyna, M. K.,** Determination of Cu, Cr, Co, Mn, Mo, Ni, Si, W and vanadium in tool steel by atomic absorption spectrometry, *At. Absorpt. Newsl.,* 8, 129, 1969.
93. **Pemster, J. P. and Rapperport, E. J.,** Thermodynamic activity measurements using atomic absorption copper-zinc, *Trans. Metall. Soc. AIME,* 245, 1395, 1969.
94. **Lambert, M. J.,** The determination of copper, chromium and arsenic in preservation treated timber by the method of atomic absorption spectrophotometry, *J. Inst. Wood Sci.,* 27, 27, 1969.
95. **Yokoyama, U. and Ikeda, S.,** Atomic absorption spectrometry by a pulse technique and measurement of half life of copper and magnesium alloys, *Spectrochim. Acta,* 24b, 117, 1969.
96. **Backer, E. T.,** Chloric acid digestion in the determination of trace metals (iron, zinc, and copper) in brain and hair by atomic absorption spectrophotometry, *Clin. Chim. Acta,* 24, 233, 1969.
97. **Burke, K. E. and Albright, C. H.,** Atomic absorption spectrometric determination of copper and nickel in tea, *J. Assoc. Off. Anal. Chem.,* 53, 531, 1969.
98. **Cameron, A. G. and Hackett, D. R.,** Determination of copper in foods by atomic absorption spectrophotometry, *J. Sci. Food Agric.,* 21, 535, 1970.
99. **Gulcz, M. and Racznska, D.,** Determination of copper and manganese in raw rubber by atomic absorption spectrometry, *Chem. Anal. (Warsaw),* 15, 95, 1970.
100. **Semenova, O. P. and Sukhanova, G. B.,** Characteristics of the radiation of copper atoms and ions in a discharge from a hot hollow cathode, *Zh. Prikl. Spektrosk.,* 13, 956, 1970.
101. **Tusl, J. and Krska, M.,** Determination of magnesium, iron, manganese, copper and zinc in foods by atomic absorption spectrophotometry, *Prum. Potravin,* 21, 119, 1970 (in Czech).
102. **Nix, J. and Goodwin, T.,** The simultaneous extraction of iron, manganese, cobalt, nickel, chromium, lead and zinc from natural water for determination by atomic absorption spectroscopy, *At. Absorpt. Newsl.,* 9, 119, 1970.
103. **Yanagisawa, M., Kihara, H., Suzuki, M., and Takeuchi, T.,** Interferences in atomic absorption spectrometry and extraction of iron and copper, *Talanta,* 17, 888, 1970.
104. **Ure, A. M. and Berrow, M. L.,** Analysis of EDTA extracts of soil for copper, zinc and manganese by atomic absorption spectrophotometry with a mechanically separated flame, *Anal. Chim. Acta,* 52, 247, 1970.
105. **Lerner, L. A. and Nedler, V. V.,** Atomic absorption determination of available copper and cobalt in soils, *Pochvovedenie,* 11, 105, 1970.
106. **Mathieu, G. and Guiot, S.,** Determination of traces of copper and iron in silver, *Anal. Chim. Acta,* 52, 335, 1970 (in French).
107. **Miksovsky, M. and Moldan, B.,** Determination of copper, zinc, manganese and iron in silicate samples by atomic absorption spectroscopy, *Chem. Zvesti,* 24, 128, 1970.
108. **Heckman, M.,** Collaborative study of copper, sodium and potassium in feeds by atomic absorption spectrophotometry and sodium and potassium by flame emission spectrophotometry, *J. Assoc. Off. Anal. Chem.,* 53, 923, 1970.
109. **Jamro, G. H. and Frei, R. W.,** Determination of copper impurities in nickel and cobalt salts by atomic absorption spectroscopy, *Mikrochim. Acta,* 429, 1970.
110. **Kazimerczyk, S., Michalewska, M., and Jarosz, R.,** Determination of silver, copper and zinc by atomic absorption spectroscopy, *Chem. Anal. (Warsaw),* 15, 553, 1970.
111. **Knauer, G. A.,** The determination of magnesium, manganese, iron, copper and zinc in marine shrimp, *Analyst (London),* 95, 476, 1970.
112. **Lerner, L. A. and Ivanov, D. N.,** Determination of the total amount of zinc, copper, cobalt and manganese in soils by atomic absorption spectrophotometry, *Agrokhimiya,* 3, 133, 1970.
113. **Pirke, K. M. and Stamm, D.,** Measurement of urinary copper excretion by atomic absorption spectrophotometry, *Z. Klin. Chem. Klin. Biochem.,* 8, 449, 1970.
114. **Varju, M.,** Determination of soluble Cu-content in some salt affected soils with atomic absorption method, *Agrokem. Talajtan,* 19, 323, 1970 (in Hungarian).
115. **Zook, E., Greene, F., and Morris, E.,** Nutrient composition of selected wheats and wheat products. Distribution of Mn, Cu, Ni, Zn, Mg, Pb, Sn, Cd, Cr and Se as determined by atomic absorption spectroscopy and colorimetry, *Cereal Chem.,* 47, 720, 1970.

116. **Varju, M. E. and Elek, E.,** Determination of soluble copper in soils by atomic absorption spectroscopy, *At. Absorpt. Newsl.,* 10, 128, 1970.

117. **Willis, J. B.,** Atomization problems in atomic absorption spectroscopy. III. Absolute atomization efficiencies of sodium, copper, silver and gold in a meker type air-acetylene flame, *Spectrochim. Acta,* 26b, 177, 1971.

118. **Yamamoto, Y., Kumamaru, T., Hayashi, Y., and Kanke, M.,** Determination of a trace amount of cadmium, zinc, lead and copper in water by atomic absorption spectrophotometry combined with solvent extraction, *Bunseki Kagaku,* 20, 347, 1971.

119. **Svejda, H.,** Flame spectrophotometric determination of copper in glazes of the system silicon-potassium-lead, *Ber. Dtsch. Keram. Ges.,* 48, 116, 1971 (in German).

120. **Gallo, J. R., Bataglia, O. C., and Nayme Miguel, P. T.,** Determination of copper, iron, manganese and zinc in a bulk plant extract by atomic absorption spectrophotometry, *Bragantia,* 30, 155, 1971 (in Portuguese).

121. **Heckman, M.,** Collaborative study of copper in feeds by atomic absorption spectrophotometry, *J. Assoc. Off. Anal. Chem.,* 54, 666, 1971.

122. **Hoffman, G. L. and Duce, A. A.,** Copper contamination of atmospheric particulate samples collected with gelman hurricane air samples, *Environ. Sci. Technol.,* 5, 1134, 1971.

123. **Kudd, Y., Hasegawa, N., and Yamashita, T.,** Determination of aluminum, cobalt, chromium, copper, iron, manganese, molybdenum, nickel, tin and vanadium in titanium alloys by atomic absorption spectrometry, *Bunseki Kagaku,* 20, 1319, 1971.

124. **Kuwata, K., Hisatomi, K., and Hasegawa, T.,** The rapid determination of trace amounts of cadmium and copper in river and sea water by atomic absorption spectroscopy, *At. Absorpt. Newsl.,* 10, 111, 1971.

125. **Leonard, E. N.,** The determination of copper in fish tissues by atomic absorption spectrophotometry, *At. Absorpt. Newsl.,* 10, 84, 1971.

126. **Mikhailova, T. P., Gvozdenko, O. I., and Shafran, T. A.,** Atomic absorption determination of copper in plant liquors, *Izv. Sib. Otd. Akad. Nauk SSSR Ser. Khim. Nauk,* 5, 101, 1971.

127. **Baker, A. D.,** Determination of copper in alfalfa, *J. Assoc. Off. Anal. Chem.,* 54, 951, 1971.

128. **Burke, K. E. and Albright, C. E.,** Atomic absorption spectroscopy for determination of ppm quantities of nickel and copper in tea, *J. Assoc. Off. Anal. Chem.,* 54, 658, 1971.

129. **Coles, L. E.,** Determination of small amounts of copper in organic matter by atomic absorption spectroscopy, *Analyst (London),* 96, 741, 1971.

130. **Kulp, J. H., Windham, R. L., and Whealy, R. D.,** Atomic absorption spectrometry of copper with selected organic solvents after extraction from aqueous solution with 8-hydroxyquinoline, *Anal. Chem.,* 43, 1321, 1971.

131. **Follett, R. H. and Lindsay, W. L.,** Changes in DTPA-extractable zinc, iron, manganese and copper in soils following fertilization, *Soil Sci. Soc. Am. Proc.,* 35, 600, 1971.

132. **Ghanbari, H. A. and Mameesh, M. S.,** Iron, zinc, manganese and copper content of semidwarf wheat varieties grown under different agronomic conditions, *Am. Assoc. Cereal Chem.,* 48, 411, 1971.

133. **Kuwata, K., Hisatomi, K., and Hasegawa, T.,** The rapid determination of trace amounts of cadmium and copper in river and sea water by atomic absorption spectroscopy, *At. Absorpt. Newsl.,* 10, 111, 1971.

134. **Lerner, L. A., Orlova, L. P., and Ivanov, D. N.,** Use of rapid method of sample decompositions during the atomic absorption determination of the total copper, zinc, and manganese content in soils, *Agrokhimiya,* 5, 138, 1971.

135. **List, G. R., Evans, C. D., and Kwolek, W. F.,** Copper in edible oils. Trace amounts determined by atomic absorption spectroscopy, *J. Am. Oil Chem. Soc.,* 48, 438, 1971.

136. **Mahler, D. J., Walsh, J. R., and Haynie, G. D.,** Magnesium, zinc and copper in dialysis patients, *Am. J. Clin. Pathol.,* 56, 17, 1971.

137. **Matousek, J. P. and Stevens, B. J.,** Biological applications of the carbon rod atomizer in atomic absorption spectroscopy. Preliminary studies on magnesium, iron, copper, lead and zinc in blood and plasma, *Clin. Chem. (Winston-Salem, N.C.),* 17, 363, 1971.

138. **Meret, S. and Henkin, R. I.,** Simultaneous direct estimation by atomic absorption spectrophotometry of copper and zinc in serum, urine and cerebrospinal fluid, *Clin. Chem. (Winston-Salem, N.C.),* 17, 369, 1971.

139. **Orren, M. J.,** Determination of copper, zinc, iron, manganese, potassium, lithium and rubidium in seawater by atomic absorption spectrophotometry, *J. S. Afr. Chem. Inst.,* 24, 96, 1971.

140. **Schlewitz, J. H. and Shields, M. G.,** Atomic absorption analysis of zirconium alloys for chromium, copper, iron, nickel and tin, *At. Absorpt. Newsl.,* 10, 39, 1971.

141. **Spector, H., Glusman, S., Jatlow, P., and Seligson, D.,** Direct determination of copper in urine by atomic absorption spectrophotometry, *Clin. Chim. Acta,* 31, 5, 1971.

142. **Evans, C. D., List, G. R., and Black, L. T.,** Char ashing of glyceride oils preliminary to the atomic absorption determination of their copper and iron contents, *J. Am. Oil Chem. Soc.,* 48, 840, 1971.

143. **Glenn, M., Savory, J., Hart, L., Glenn, T., and Winefordner, J. D.,** Determination of copper in serum with a graphite rod atomizer for atomic absorption spectrophotometry, *Anal. Chim. Acta,* 52, 263, 1971.

144. **Kanke, M., Hayashi, Y., Kumamaru, T., and Yamamoto, Y.,** Determination of ppb levels of cadmium, zinc, lead and copper extracted as their dithizonates into nitrobenzene by atomic absorption spectrophotometry, *Nippon Kagaku Zasshi,* 92, 983, 1971.

145. **Salvesen, B.,** Atomic absorption spectroscopy in analytical pharmaceutical chemistry. Trace determination of copper in drugs, *Medd. Nor. Farm. Selsk.,* 33, 79, 1971.

146. **Toma, O. and Crisan, T.,** Determination of copper, tin, lead, zinc, nickel and iron in bronze by atomic absorption spectrophotometry, *Metallurgia,* 23, 715, 1971 (in Romanian).

147. **Bergner, K. C. and Lang, B.,** The determination of iron, copper, zinc, manganese and cadmium in grape juice and wine by atomic absorption spectrophotometry, *Dtsch. Lenensmittel-Edsch.,* 67, 121, 1971 (in German).

148. **Franklin, Z. G., Sokolov, A. A., and Bogdanova, N. N.,** Atomic absorption determination of zinc, copper, nickel, and cobalt in compounds of phosphorus (used as fertilizers), *Tr. Leningr. Nauchno-Issled. Proekt. Inst. Osnov. Khim. Prom.,* 4, 283, 1971 (in Russian).

149. **Rebmann, H. and Hoth, H. J.,** Determination of Na, K, Ca, Mg, Cu and Fe in milk with an atomic absorption spectrophotometer, *Milchwissenschaft,* 26, 411, 1971.

150. **Norvell, W. A. and Lindsay, W. L.,** Reactions of DTPA chelates of iron, zinc, copper and manganese with soils, *Soil Sci. Soc. Am. Proc.,* 36, 778, 1972.

151. **Reeves, R. D., Molnar, C. J., and Winefordner, J. D.,** Rapid atomic absorption determination of silver and copper by subsequential atomization from a graphite rod, *Anal. Chem.,* 44, 1913, 1972.

152. **Barnett, W. B. and Kahn, H. L.,** Determination of copper in fingernails by atomic absorption with the graphite furnace, *Clin. Chem. (Winston-Salem, N.C.),* 18, 923, 1972.

153. **Fell, G. S., Canning, E., Husain, S. L., and Scott, S.,** Copper and zinc in human health and disease, in *Trace Substances in Environmental Health,* Vol. 5, Hemphill, D. D., Ed., University of Missouri, Columbia, 1972.

154. **Hag, A. U. and Miller, M. H.,** Prediction of available soil Zn, Cu, and Mn using chemical extractants, *Agron. J.,* 64, 779, 1972.

155. **Ishizuka, T., Sunahara, H., and Tanaka, K.,** Determination of copper, iron, lead and zinc, in yttrium oxide and yttrium oxide sulfide by atomic absorption spectrometry, *Bunseki Kagaku,* 21, 847, 1972.

156. **Mills, C. F. and Dalgarno, A. C.,** Copper and zinc status of ewes and lambs receiving increased dietary concentrations of cadmium, *Z. Anal. Chem.,* 239, 171, 1972.

157. **Piscator, M. and Lind, B.,** Cadmium, zinc, copper and lead in human renal cortex, *Arch. Environ. Health,* 24, 426, 1972.

158. **Alder, J. F. and West, T. S.,** Atomic absorption and fluorescence spectrophotometry with a carbon filament atom reservoir. IX. The direct determination of silver and copper in lubricating oils, *Anal. Chim. Acta,* 58, 331, 1972.

159. **Kuboto, N. and Imai, Y.,** Silver platings obtained from cyanide baths. Determination of copper and iron in silver plating solutions by atomic absorption, *Kinzoku Hyomen Gijutsu,* 23, 95, 1972.

160. **Sapek, A. and Sapek, B.,** Determination of copper in hay and peat by atomic absorption spectrometry after preconcentration in the organic base, *Chem. Anal.,* 17, 339, 1972.

161. **Varju, M.,** The determination of copper, iron and magnesium in alcoholic products using atomic absorption, *At. Absorpt. Newsl.,* 11, 45, 1972.

162. **Williams, A. I.,** Use of atomic absorption spectrophotometry for the determination of copper, chromium and arsenic in preserved wood, *Analyst (London),* 97, 104, 1972.

163. **Roschnik, R. K.,** Determination of copper in butteroil by atomic absorption spectroscopy, *J. Dairy Sci.,* 55, 750, 1972.

164. **Stevens, B. J.,** Biological applications of the carbon rod atomizer in atomic absorption spectroscopy. II. Determination of copper in small samples of tissue, *Clin. Chem. (Winston-Salem, N.C.),* 18, 1379, 1972.

165. **Varju, M.,** Determination of iron, copper, calcium, magnesium and sodium in Hungarian alcoholic beverages by atomic absorption spectroscopy, *Z. Phys.,* 252, 268, 1972.

166. **Fuller, C. W.,** Determination of iron and copper in high purity silica by flameless atomic absorption spectrometry, *Anal. Chim. Acta,* 62, 261, 1972.

167. **Fuller, C. W.,** Loss of copper and nickel during pre-atomization heating periods in flameless atomic absorption determinations, *Anal. Chim. Acta,* 62, 442, 1972.

168. **Heinemann, G.,** Iron, copper and zinc analysis by atomic absorption spectrophotometry, *Z. Khim. Chem. Klin. Biochem.,* 10, 467, 1972.

169. **Varju, M.,** Determination of iron, copper, calcium, magnesium and sodium in spirits by atomic absorption, *Elemiszervizsgalati Kozl.,* 18, 207, 1972 (in Hungarian).

170. **Kidani, Y., Noji, M., and Koike, H.,** Quantitative determination of copper, nickel, cobalt and zinc in metal complexes by atomic absorption spectrometry, *Bunseki Kagaku,* 21, 1652, 1972.

171. **Kurz, D., Roach, J., and Eyring, E. J.,** Direct determination of serum zinc and copper by atomic absorption spectrophotometry, *Biochem. Med.,* 6, 274, 1972.

172. **Roschnik, M. R.,** Determination of traces of copper in liquid milk and milk powder by atomic absorption spectrophotometry, *Mitt. Geb. Lebensmittelunters. Hyg.,* 63, 206, 1972 (in French).

173. **Varju, M.,** Determination of the iron, copper, calcium, magnesium and sodium contents of Hungarian brandies by the atomic absorption method, *Z. Lebensm. Unters. Forsch.,* 148, 268, 1972.

174. **Halirova, D. and Musil, J.,** Atomic absorption spectrophotometry in metallurgical analysis. II. Determination of manganese, copper and nickel in ferrovanadium, *Hutn. Listy,* 27, 888, 1972 (in Czech).

175. **Zolotavin, V. I., Bukreev, U. F., and Barsukov, V. I.,** Effect of the composition of the test solution on the results of flame photometric analysis (for cesium, sodium, strontium and copper), *Tr. Tambovskogo Inst. Khim. Mashinostr.,* 8, 111, 1972 (in Russian).

176. **Ishizuka, T., Sunahara, H., and Tanaka, K.,** Determination of copper, iron, lead and zinc in yttrium oxide and yttrium oxide sulfide by atomic absorption spectrometry, *Nagoya Kogyo Gijutsu Shikensho Hokoku,* 21, 385, 1972.

177. **Zolotavin, V. L., Barsukov, V. I., and Bukreev, U. F.,** Application of heating of the (atomizer) chamber and the body of the burner in emission and atomic absorption flame photometry (of sodium, strontium and copper), *Tr. Tambovskogo Inst. Khim. Mashinostr.,* 8, 105, 1972.

178. **Kubo, Y., Nakazawa, N., and Sato, M.,** Determination of trace amounts of lead, chromium and copper in sea water by solvent extraction-atomic absorption spectrophotometry, *Sekiyu Gakkai Shi,* 16, 588, 1973.

179. **Lagerwerff, J. V., Brower, D. L., and Biersdorf, G. T.,** Accumulation of cadmium, copper, lead and zinc in soil and vegetation in the proximity of a smelter, in *Trace Substances in Environmental Health,* Vol. 6, Hemphill, D. D., Ed., University of Missouri, Columbia, 1973, 71.

180. **Martin, P. A.,** Institute of brewing analysis committee determination of iron and copper in beer by atomic absorption spectroscopy, *J. Inst. Brew. London,* 79, 289, 1973.

181. **Smeyers-Verbeke, J., Massart, D. L., Versieck, J., and Speecke, A.,** The determination of copper and zinc in biological materials. A comparison of atomic absorption with spectrophotometry and neutron activation, *Clin. Chim. Acta,* 44, 243, 1973.

182. **Sorensen, J. R. J., Levin, L. S., and Petering, H. G.,** Cadmium, copper, lead, mercury and zinc concentrations in the hair of individuals living in the United States, *Interface,* 2, 17, 1973.

183. **Stahlavska, A., Slansky, V., and Kocandova, E.,** Spectral analytical methods in drug analysis. Determination of copper in reagent solutions using atomic absorption spectrophotometry, *Pharmazie,* 28, 240, 1973.

184. **Baker, D. E.,** A new approach to soil testing. II. Ionic equilibria involving H, K, Ca, Mg, Mn, Fe, Cu, Zn, Na, P and S, *Soil Sci. Soc. Am. Proc.,* 37, 537, 1973.

185. **Briska, M. and Hoffmeister, W.,** Determination of copper in ammonium fluoride solutions by extraction and atomic absorption spectrophotometry, *Talanta,* 20, 805, 1973.

186. **Sinha, R. C. P. and Banerjee, B. K.,** Use of aqueous standards for estimation of trace amounts of copper in naphtha by atomic absorption spectrophotometry, *Technol. Sindri,* 10, 297, 1973.

187. **Chao, T. T. and Sanzolone, R. F.,** The atomic absorption spectrophotometric determination of microgram levels of cobalt, nickel, copper, lead and zinc, in soil and sediments extracts containing large amounts of manganese and iron, *J. Res. U.S. Geol. Surv.,* 1, 681, 1973.

188. **Osiname, O. A., Schulte, E. E., and Corey, R. B.,** Tests for available copper and zinc in soils of Western Nigeria, *J. Sci. Food Agric.,* 24, 1341, 1973.

189. **Peaston, R. T.,** Determination of copper and zinc in plasma and urine by atomic absorption spectrophotometry, *Med. Lab. Technol.,* 30, 249, 1973.

190. **Renshaw, G. D., Pounds, C. A., and Pearson, E. F.,** Determination of lead and copper in hair by non-flame atomic absorption spectrophotometry, *J. Forensic Sci.,* 18, 143, 1973.

191. **Buchauer, M. J.,** Contamination of soil and vegetation near a zinc smelter by zinc, cadmium, copper and lead, *Environ. Sci. Technol.,* 7, 131, 1973.

192. **Dybczynska, I., Fijalkowski, J., Chruscinska, T., and Myszka, E.,** Determination of copper, zinc and iron in platinum-rhodium and platinum-iridium alloys by atomic absorption spectroscopy, *Chem. Anal. (Warsaw),* 18, 169, 1973.

193. **Harms, U.,** Determination of transition metals, Mn, Fe, Co, Cu, and Zn in river fish with the help of X-ray fluorescence and flameless atomic absorption, *Mittelungsbl. GDCH (Ges. Dtsch. Chem.) Fachgruppe Lebensmittelchem. Gerichtl. Chem.,* 27, 271, 1973.

194. **Hoeschler, M. E., Kanabrocki, E. L., Moore, C. E., and Hattori, D. M.,** The determination of calcium, manganese, copper and iron in airborne particulates, *Appl. Spectrosc.,* 27, 185, 1973.

195. **Jackwerth, E., Hoehn, R., and Koos, K.,** Trace enrichment by partial dissolution of the matrix in presence of mercury. Determination of bismuth, copper, lead, nickel, silver, gold and palladium in high purity cadmium by atomic absorption spectrometry, *Z. Anal. Chem.,* 264, 1, 1973 (in German).

196. **Murthy, L., Menden, E. E., Eller, P. M., and Petering, H. G.,** Atomic absorption determination of zinc, copper, cadmium and lead in tissues solubilized by aqueous tetra methyl ammonium hydroxide, *Anal. Biochem.,* 53, 365, 1973.

197. **Nemets, A. M. and Nikolaev, G. I.,** Determination of the pressure of copper, titanium and vanadium saturated vapor by an atomic absorption method, *Zh. Prikl. Spektrosk.,* 18, 571, 1973.

198. **Okusu, H., Ueda, Y., Ota, K., and Kawano, K.,** Determination of cadmium, zinc, copper and lead in waste water by atomic absorption spectrometry, *Bunseki Kagaku,* 22, 84, 1973.

199. **Brivot, F., Cohort, I., Legrand, G., Louvrier, J., and Voinovitch, I.,** Comparative determination by atomic absorption spectrometry and by spectrophotocolorimetry of nickel, copper, chromium, manganese, aluminum and silicon in certain steels, *Analusis,* 2, 570, 1973.

200. **Fairless, C. and Bard, A. J.,** Hanging mercury drop electrodeposition technique for carbon filament flameless atomic absorption analysis. Application to the determination of copper in seawater, *Anal. Chem.,* 45, 2289, 1973.

201. **Helsby, C. A.,** Determination of copper and molybdenum in the hard dental tissues of rats by atomic absorption spectrophotometry, *Talanta,* 20, 779, 1973.

202. **Hohnadel, D. C., Sunderman, F. W., Jr., Nechay, M. W., and McNeely, M. D.,** Atomic absorption spectrometry of nickel, copper, zinc and lead in sweat collected from healthy subjects during sauna bathing, *Clin. Chem. (Winston-Salem, N.C.),* 19, 1288, 1973.

203. **Koester, H. M.,** The atomic absorption determination of copper (II), iron (III) and zinc in silicate rocks after separation on chloride-form Dowex I-X8 or Amberlite CG 400-I ion exchanger, *Neues Jahrb. Mineral. Abh.,* 119, 145, 1973 (in German).

204. **Mikhailova, T. P., Klimashova, B. P., and Babeuva, L. V.,** Determination of zinc, cadmium, copper and iron in aluminum and gallium nitrites using an atomic absorption method, *Izv. Sib. Otd. Akad. Nauk SSSR Ser. Khim. Nauk,* 135, 1973.

205. **Mizuno, T.,** Determination of traces of iron, copper, zinc, nickel and cobalt in water by atomic absorption spectrometry using PAN-MIBK (1-[2-pyridyl azo] -2-naphthol)-(methyl isobutyl ketone) extraction, *Nippon Kagaku Kaishi,* 10, 1904, 1973.

206. **Murthy, G. K., Rhea, U. S., and Peeler, J. T.,** Levels of copper, nickel, rubidium and strontium in institutional total diets, *Environ. Sci. Technol.,* 7, 1042, 1973.

207. **Postel, W., Drawert, F., and Guveric, U.,** Determination of trace elements in foods by atomic absorption spectrophotometry. II. Copper in beer, *Brauwissenschaft,* 25, 391, 1973 (in German).

208. **Sorensen, J. R. J., Melby, E. G., Nord, P. J., and Petering, H. G.,** Interferences in the determination of metallic elements in human hair. Evaluation of zinc, copper, lead and cadmium using atomic absorption spectrophotometry, *Arch. Environ. Health,* 27, 36, 1973.

209. **Chae, Y. S., Vacik, J. P., and Shelver, W. H.,** Determination of copper and manganese in vitamin-mineral tablets by atomic absorption spectrophotometry, *J. Pharm. Sci.,* 62, 1838, 1973.

210. **Facchinetti, M., Grassi, R. L., and Diez, A. L.,** Atomic absorption spectrophotometry of extractable copper, zinc, manganese and iron in soils, *Agrochimica,* 17, 413, 1973.

211. **Tessmer, C. F., Krohn, W., Johnston, D., Thomas, F. B., Hrgovic, M., and Brown, B.,** Serum copper in children (6-12 years old). An age correction factor, *Am. J. Clin. Pathol.,* 60, 870, 1973.

212. **Muzzarelli, R. A. and Rochetti, A.,** Determination of copper in sea water by atomic absorption spectrometry with a graphite atomizer after elution from chitosan, *Anal. Chim. Acta,* 69, 35, 1974.

213. **Rifkind, J. M.,** Copper and the autoxidation of hemoglobin, *Biochemistry,* 13, 2475, 1974.

214. **Smeyers-Verbeke, J., Defrise-Gussenhoven, E., Ebinger, G., Loewenthal, A., and Massart, D. L.,** Distribution of Cu and Zn in human brain tissue, *Clin. Chim. Acta,* 51, 309, 1974.

215. **Sinha, R. C. P. and Banerjee, B. K.,** Interferences in estimation of trace amounts of cobalt, copper and zinc in soils by atomic absorption spectrophotometry, *Technol. Sindri,* 11, 263, 1974.

216. **Baker, A. S. and Smith, R. L.,** Preparation of solutions for atomic absorption analysis of iron, manganese and copper in plant tissue, *J. Agric. Food Chem.,* 22, 103, 1974.

217. **Deck, R. and Kaiser, K.,** Analytical method for determining copper in edible shortening and oils, *J. Am. Oil Chem. Soc.,* 47, 126, 1974.

218. **DeJong, G. J. and Piepmeier, E. H.,** Self reversal in a copper pulsed hollow cathode lamp, *Anal. Chem.,* 46, 318, 1974.

219. **Hartley, T. F., Dawson, J. B., and Hodgkinson, A.,** Simultaneous measurement of Na, K, Ca, Mg, Cu and Zn balances in man, *Clin. Chim. Acta,* 52, 321, 1974.

220. **Ichinose, N.,** Extraction and atomic absorption sepctrometric determination of trace copper with zinc dibenzyl dithiocarbamate, *Anal. Chim. Acta,* 70, 222, 1974.

221. **Kremling, K. and Petersen, H.,** Ammonium pyrrolidine dithiocarbamate-methyl isobutyl ketone extraction system for the determination of copper and iron in 1 cm^3 of sea water by flameless atomic absorption spectrometry, *Anal. Chim. Acta,* 70, 35, 1974.

222. **Britske, M. E. and Slabodenyuk, I. V.,** Determination of copper in copper concentrates and matter by atomic absorption, *Nauchn. Tr. Nauchno-Issled. Inst. Tsvet. Metall.,* 35, 38, 1974 (in Russian).

223. **Magyar, B. and Wechsler, P.,** Application of flameless atomic absorption in determining the chemistry of complexes. Distribution of copper (II) between chloroform and water in the presence of 8-hydroxy quinoline, *Talanta,* 21, 539, 1974.

224. **McIntyre, N. S., Cook, M. G., and Boase, D. G.,** Flameless atomic absorption determination of cobalt, nickel, and copper. Comparison of tantalum and molybdenum evaporation surfaces, *Anal. Chem.,* 46, 1983, 1974.

225. **Popova, S. A., Bezur, L., and Pungor, E.,** Determination of cobalt and copper in animal feeds by extraction and atomic absorption spectroscopy, *Z. Anal. Chem.,* 271, 269, 1974.

226. **Rosenthal, R. W. and Blackburn, A.,** Higher copper concentrations in serum than in plasma, *Clin. Chem. (Winston-Salem, N.C.),* 20, 1233, 1974.

227. **Tinsley, D. A. and Iddon, A.,** Use of a liquid ion exchanger in the solvent extraction and atomic absorption determination of trace copper in waters, *Talanta,* 21, 633, 1974.

228. **Tomkins, D. F. and Frank, C. W.,** Investigation of metallic copper-chloride interaction in a hydrogen-air flame, *Anal. Chem.,* 46, 1187, 1974.

229. **Begak, O. U., Nikolaev, G. I., and Pokrovskaya, K. A.,** Atomic absorption determination of copper in steels, *Zh. Prikl. Khim. (Leningrad),* 47, 1171, 1974.

230. **Diez, A. L., Grassi, R. L., and Facchinetti, M.,** Determination of copper, zinc, manganese and iron in garlic plants by atomic absorption spectrophotometry, *Agrochimica,* 18, 128, 1974.

231. **Falchuk, K. H., Evenson, M., and Vallee, B. L.,** Multichannel atomic absorption instrument. Simultaneous analysis of zinc, copper and cadmium in biologic materials, *Anal. Biochem.,* 62, 255, 1974.

232. **Ichinose, N.,** Extraction and determination of metals. Extraction and atomic absorption spectrophotometric determination of micro amounts of copper in iron and steel with zinc dibenzyl dithiocarbamate, *Bunseki Kagaku,* 23, 348, 1974.

233. **Kundu, M. K. and Prevot, A.,** Oxygen rich atmosphere for direct determination of copper in oils by nonflame atomic absorption spectrometry, *Anal. Chem.,* 46, 1591, 1974.

234. **Kubota, J., Mills, E. L., and Oglesby, R. T.,** Lead, cadmium, zinc, copper and cobalt in streams and lake waters of Cayuga Lake Basin, New York, *Environ. Sci. Technol.,* 8, 243, 1974.

235. **Langmyhr, F. J., Thomassen, Y., and Massoumi, A.,** Atomic absorption spectrometric determination of copper, lead, cadmium and manganese in pulp and paper by the direct atomization technique, *Anal. Chim. Acta,* 68, 305, 1974.

236. **Taylor, J. D., Krahn, P. M., and Higgins, T. N.,** Serum copper levels and diphenylhydantoin, *Am. J. Clin. Pathol.,* 61, 577, 1974.

237. **Venable, R. L. and Ballad, R. V.,** Effects of surfactants on atomic absorption analysis of dilute aqueous copper and nickel solutions, *Anal. Chem.,* 46, 131, 1974.

238. **Brovko, I. A., Nazarov, S. N., and Rish, M. A.,** Atomic absorption determination of zinc, cadmium, cobalt, copper and nickel after their extraction concentration in the diphenyl carbazone-pyridine-toluene system, *Zh. Anal. Khim.,* 29, 2387, 1974.

239. **Fletcher, G. E. and Collins, A. G.,** Atomic absorption methods of analysis of oilfield brines: Ba, Ca, Cu, Pb, Li, Mg, Mn, K, Na, Sr and Zn, *Rep. Invest. U.S. Bur. Mines,* RI 7861, 1974.

240. **Fuller, C. W.,** Kinetic theory of atomization for nonflame atomic absorption spectrometry with a graphite furnace kinetics mechanism of atomization for copper, *Analyst (London),* 99, 739, 1974.

241. **Fazakas, J., German, A., Baiulescu, G., and Mullins, C.,** Determination of microelements in carbon dioxide-containing mineral water by atomic absorption and emission spectrophotometry. II. Determination of copper and chromium by flameless atomic absorption spectrophotometry, *Rev. Chim.,* 25, 917, 1974.

242. **Lofberg, R. T. and Levri, E. A.,** Analysis of copper and zinc in hemolyzed serum samples, *Anal. Lett.,* 7, 775, 1974.

243. **Stanton, R. E.,** Determination of calcium, cobalt, copper, iron, molybdenum, phosphorus, selenium and sulfur in plant material, *Lab. Pract.,* 23, 233, 1974.

244. **Maurer, L.,** Rapid, simple procedures for the determinations of copper in cheese with a graphite furnace, *Z. Lebensm. Unters. Forsch.,* 156, 284, 1974.

245. **Evenson, M. A. and Warren, B. L.,** Determination of serum copper by atomic absorption with use of the graphite cuvette, *Clin. Chem. (Winston-Salem, N.C.),* 21, 619, 1975.

246. **Fuller, C. W.,** Kinetic theory of atomization for nonflame atomic absorption spectrometry with a graphite furnace. II. Analytical applications of kinetic information for copper, *Analyst (London),* 100, 229, 1975.

247. **Jacob, R. A. and Klevay, L. M.,** Determination of trace amounts of copper and zinc in edible fats and oils by acid extraction and atomic absorption spectrophotometry, *Anal. Chem.,* 47, 741, 1975.

248. **Korkisch, J., Goedl, J., and Gross, H.,** Use of ion exchange for the determination of trace amounts in natural waters. VII. Copper, *Talanta,* 22, 289, 1975 (in German).

249. **Lott, I. T., DiPaolo, R., Schwartz, D., Janowska, S., and Kanfer, J. N.,** Copper metabolism in the steely-hair syndrome, *Med. Intelligence,* 292, 197, 1975.

250. **Smeyers-Verbeke, J., Segebarth, G., and Massart, D. L.,** The determination of copper and manganese in small biological samples in the graphite furnace atomic absorption spectrometry, *At. Absorpt. Newsl.,* 14, 153, 1975.

251. **Luzar, O. and Sliva, V.**, Determination of calcium oxide magnesium oxide, alumina, silica, iron, copper, zinc, lead, cadmium, sodium oxide and potassium oxide in iron ores and agglomerates, *Hutn. Listy*, 30, 55, 1975 (in Czechoslovakian).

252. **Muzzarelli, R. A. and Rocchetti, R.**, Atomic absorption determination of manganese, cobalt and copper in whole blood and serum with a graphite atomizer, *Talanta*, 22, 683, 1975.

253. **Piepmeier, F. H. and DeGalan, L.**, Line profile emitted by copper and calcium hollow cathode lamps pulsed to one ampere, *Spectrochim. Acta*, 30b, 211, 1975.

254. **Schweizer, V. B.**, Determination of Co, Cr, Cu, Mo, Ni and V in carbonate rocks with the HGA-70 graphite furnace, *At. Absorpt. Newsl.*, 14, 137, 1975.

255. **Shigematsu, T., Matsui, M., Fujimo, O., Mitsuno, S., and Nagahiro, T.**, Determination of copper in sea water and shell fishes by atomic absorption spectrometry with a carbon tube atomizer, *Nippon Kagaku Kaishi*, 8, 1328, 1975.

256. **Stephens, B. G. and Felkel, H. L.**, Extraction of copper (II) from aqueous thiocyanate solutions into propylene carbonate and subsequent atomic absorption spectrophotometric determination, *Anal. Chem.*, 47, 1676, 1975.

257. **Sychra, V., Janouskova, J., Kolihova, D., Dudova, N., and Marek, S.**, Analysis of ilmenite and inorganic pigments on titaniium (IV) oxide by atomic absorption spectrometry. II. Determination of copper, *Chem. Listy*, 69, 623, 1975.

258. **Wagenaar, H. C. and DeGalan, L.**, Influence of line profile upon analytical curves for copper and silver in atomic absorption spectroscopy, *Spectrochim. Acta*, 30b, 361, 1975.

259. **Glaeser, E.**, Determination of copper in the ppb range at water stations using an atomic absorption spectrometer, *Jenaer Rundsch.*, 20, 244, 1975 (in German).

260. **Iosof, V., Mihalka, S., and Colios, E.**, Determination of copper, lead, zinc in mineral products by atomic absorption spectrophotometry, *Rev. Chim.*, 26, 680, 1975.

261. **Mehlich, C. and Bowling, S. S.**, Advances in soil test methods for copper by atomic absorption spectrophotometry, *Commun. Soil Sci. Plant Anal.*, 6, 113, 1975.

262. **Moll, M., Flayeux, R., Bazard, D., and Lehuede, J. M.**, Determination of iron and copper in beer by flame and flameless atomic absorption spectrophotometry *Bios. Franc*, 6, 245, 1975 (in French).

263. **Robbins, W. B., Dekoven, B. M., and Caruso, J. A.**, Copper in erythrocytes by flameless atomic absorption spectroscopy, *Biochem. Med.*, 14, 184, 1975.

264. **Van Campenhausen, H. and Mueller-Plathe, O.**, Determination of serum copper by atomic absorption spectroscopy, *Z. Klin. Chem. Klin. Biochem.*, 13, 489, 1975.

265. **Yamamoto, Y., Kumamaru, T., Kamada, T., and Tanaka, T.**, Determination of parts per 10^9 levels of cadmium, lead and copper in water by carbon tube flameless atomic absorption spectrophotometry combined with ammonium pyrrolidine-1-carbodithioate-isobutyl methyl ketone extraction, *Eisei Kagaku*, 21, 71, 1975 (in Japanese).

266. **Acatini, C., de Berman, S. N., Colombo, O., and Fondo, O.**, Determination of silver, copper, lead, tin, antimony, iron, calcium, zinc, magnesium, potassium and manganese in canned tomatoes by atomic absorption spectrophotometry, *Revta. Asoc. Bioquim. Argent.*, 40, 175, 1975 (in Spanish).

267. **Downes, T. E. H. and Labuschagne, J. H.**, Determination of copper in milk and butter with the graphite furnace atomic absorption apparatus, *S. Afr. J. Dairy Technol.*, 7, 167, 1975.

268. **Beyer, M. E. and Bond, A. M.**, Simultaneous determination of cadmium, copper, lead and zinc in lead and zinc concentrates by a.c. polarographic methods. Comparison with atomic absorption spectrometry, *Anal. Chim. Acta*, 75, 409, 1975.

269. **Boyle, E. and Edmond, J. M.**, Copper in surface waters of New Zealand, *Nature (London)*, 253, 109, 1975.

270. **Briska, M.**, Rapid determination of copper, silver and palladium in lead-tin tinning baths by atomic absorption spectroscopy, *Z. Anal. Chem.*, 273, 283, 1975.

271. **Burch, R. E., Hahn, H. K. J., and Sullivan, J. F.**, Newer aspects of the roles of zinc, manganese and copper in human nutrition, *Clin. Chem. (Winston-Salem, N.C.)*, 21, 501, 1975.

272. **Eroshevich, T. A. and Makarov, D. F.**, Analysis of alloys based on nickel, copper, iron and cobalt by atomic absorption, *Zavod. Lab.*, 41, 186, 1975.

273. **Evenson, M. A. and Anderson, C. T., Jr.**, Ultramicro analysis for copper, cadmium and zinc in human liver tissue by use of atomic absorption spectrophotometry and the heated graphite tube atomizer, *Clin. Chem. (Winston-Salem, N.C.)*, 21, 537, 1975.

274. **Noguchi, C., Hirayama, H., Shige, T., Jinbo, M., Takahashi, T., Tsuji, K., Nakasato, S., Matsubara, S., Murase, Y., Murui, T., Yamashita, Y., and Yoshida, J.**, Determination of microamounts of cadmium, copper, nickel and manganese in fats and oils by atomic absorption spectrophotometry, *Yukagaku*, 24, 100, 1975.

275. **Pekarek, P. S., Kluge, R. M., DuPont, H. L., Wannamacher, R. W., Jr., Hornick, R. B., Bostian, K. A., and Beisel, W. R.**, Serum zinc, iron and copper concentrations during typhoid fever in man: effect of chloramphenicol therapy, *Clin. Chem. (Winston-Salem, N.C.)*, 21, 528, 1975.

276. **Simmons, W. and Loneragan, J. F.,** Determination of copper in small amounts of plant material by atomic absorption spectrophotometry using a heated graphite atomizer, *Anal. Chem.,* 47, 566, 1975.

277. **Titova, I. N., Novikov, Y. V., and Yudina, T. V.,** Determination of copper in air and biological materials using atomic absorption spectral analysis, *Gig. Sanit.,* 1, 57, 1975.

278. **Van Stekelenburg, G. J., Van De Laar, A. J. B., and Van Der Laag, J.,** Copper analysis of nail clippings. An attempt to differentiate between normal children and patients suffering from cystic fibrosis, *Clin. Chim. Acta,* 59, 233, 1975.

279. **Vasiliades, J. and Sahawneh, T.,** Effect of diphenylhydantoin on serum copper, zinc and magnesium, *Clin. Chem. (Winston-Salem, N.C.),* 21, 637, 1975.

280. **Warren, J. and Carter, D.,** The determination of trace amounts of copper. vanadium. chromium. nickel. cobalt and barium in silicate rock using atomic absorption spectrometry, *Can. J. Spectrosc.,* 20, 1, 1975.

281. **Aihara, M. and Kiboku, M.,** Atomic absorption spectrophotometry of cadmium and copper using solvent extraction with potassium ethyl xanthate-methyl isobutyl ketone, *Bunseki Kagaku,* 24, 447, 1975.

282. **Carr, G. and Wilkinson, A. W.,** Zinc and copper urinary excretions in children with burns and scalds, *Clin. Chim. Acta,* 61, 199, 1975.

283. **Fujiwara, K., Haraguchi, H., and Fuwa, K.,** Response surface and atomization mechanism in air-acetylene flames. Acid interference in atomic absorptions of copper and indium, *Anal. Chem.,* 47, 1670, 1975.

284. **Greiner, A. C., Chan, S. C., and Nicolson, G. A.,** Human brain contents of calcium, copper, magnesium and zinc in some neurological pathologies, *Clin. Chim. Acta,* 64, 211, 1975.

285. **Greiner, A. C., Chan, S. C., and Nicolson, G. A.,** Determination of calcium, copper, magnesium and zinc content of identical areas in human cerebral hemispheres of normals, *Clin. Chim. Acta,* 64, 335, 1975.

286. **Hiro, K., Kawahara, A., Tanaka, T., and Hirai, A.,** Interference of alkali salts in the atomic absorption spectrometric determination of copper and iron, *Bunseki Kagaku,* 24, 275, 1975.

287. **Kono, T. and Nemori, A.,** Extraction of copper, cadmium, lead, silver and bismuth with iodide methyl isobutyl ketone in atomic absorption spectrophotometric analysis, *Bunseki Kagaku,* 24, 419, 1975.

288. **Korkisch, J. and Sorio, A.,** Determination of cadmium. copper and lead in natural waters after anion exchange separation, *Anal. Chim. Acta,* 76, 393, 1975.

289. **Korkisch, J., Steffan, I., and Gross, H.,** Analysis of nuclear raw materials. IX. Atomic absorptio spectrophotometric determination of copper in tri uranium octoxide and yellow cake samples, *Mikrochim. Acta,* 2, 569, 1975.

290. **Machata, G.,** The normal level of cadmium, copper and zinc in blood of the Viennese population, *Wien. Klin. Wochenschr.,* 87, 494, 1975 (in German).

291. **Vondenhoff, T.,** Determination of lead, cadmium, copper and zinc in plant and animal material by atomic absorption in a flame and in a graphite tube after sample decomposition by the Schoniger technique, *Mitteilungsbl. GDch (Ges. Dtsch. Chem.) Fachgruppe Lebensmittelchem. Gerichtl. Chem.,* 29, 341, 1975.

292. **Rudnevskii, N. K., Demarin, V. T., Molyanov, A. I., and Sklemina, L. V.,** Effect of heating the aerosol on sensitivity of atomic absorption determination of certain elements (copper, iron, magnesium and zinc), *Tr. Khim. Khim. Tekhnol.,* 1, 103, 1975.

293. **Rudnevskii, N. K., Demarin, V. T., Sklemina, L. V., and Tumanova, A. N.,** Influence of some organic solvents on sensitivity of atomic absorption determination of iron, cadmium, copper, magnesium, sodium, vanadium, selenium and zinc, *Tr. Khim. Khim. Tekhnol.,* 1, 106, 1975.

294. **Korkisch, J. and Huebner, H.,** Atomic absorption spectrophotometric determination of cobalt, copper. manganese. and zinc. in multivitamin preparations after separation by means of anion exchange. *Mikrochim. Acta,* 2, 311, 1976.

295. **Lundberg, E. and Johansson, G.,** Simultaneous determination of manganese, cobalt and copper with a computer controlled flameless atomic absorption spectrophotometer, *Anal. Chem.,* 48, 1922, 1976.

296. **Mantel, M., Aladjen, A., and Nothmann, R.,** Determination of copper and iron in niobium by electrolytic dissolution and atomic absorption spectroscopy, *Anal. Lett.,* 9, 671, 1976.

297. **Olejko, J. T.,** Determination of iron, copper, nickel and manganese in fats and oils by flameless atomic absorption spectroscopy, *J. Am. Oil Chem. Soc.,* 53, 480, 1976.

298. **Pakalns, P. and Farrar, Y. J.,** Effects of fats, mineral oils and creosote on the extraction-atomic absorption determination of copper, iron, lead and manganese in water, *Water Res.,* 10, 1027, 1976.

299. **Sanzolone, R. F. and Chao, T. T.,** Atomic absorption spectrometric determination of copper, zinc and lead in geological materials, *Anal. Chim. Acta,* 86, 163, 1976.

300. **Berger, S. A.,** Solvent extraction of copper (II) with chlorendic acid, *Talanta,* 23, 475, 1976.

301. **Berndt, H. and Jackwerth, E.,** Automated injection method for dispensing small volume samples in flame atomic absorption. Multielement analysis of high purity aluminum and determination of copper, iron and zinc, in serum, *At. Absorpt. Newsl.,* 15, 109, 1976.

302. **Bezur, L., Sivosh, K., Popova, S., and Pungor, E.,** Atomic absorption determination of copper, manganese, zinc, iron, potassium, sodium, calcium and magnesium in some animal foods, *Khim. Ind.,* 48, 204, 1976.

303. **Chambers, J. C. and McClellan, B. E.,** Enhancement of atomic absorption sensitivity for copper, cadmium, antimony, arsenic and selenium by means of solvent extraction, *Anal. Chem.,* 48, 2061, 1976.

304. **Delves, H. T.,** The microdetermination of copper in plasma protein fractions, *Clin. Chim. Acta,* 71, 495, 1976.

305. **Benjamin, M. M. and Jenne, E. A.,** Trace element contamination. I. Copper from plastic microliter pipet tips, *At. Absorpt. Newsl.,* 15, 53, 1976.

306. **Bower, N. W. and Ingle, J. D.,** Precision of flame atomic absorption measurements of copper, *Anal. Chem.,* 48, 686, 1976.

307. **Brenner, I. B., Gleit, L., and Harel, A.,** Interlaboratory and interinstrumental spectrochemical precision. Comparison of dc carbon arc optical emission spectrographic, atomic absorption and X-ray fluorescence spectrometric procedures, *Appl. Spectrosc.,* 30, 335, 1976.

308. **Korkisch, J., Steffan, I., and Gross, H.,** Contributions to the analysis of nuclear raw materials. XI. Atomic absorption spectrophotometric determination of copper in radioactive ores and geological standard samples, *Mikrochim. Acta,* 1, 263, 1976.

309. **Lafargue, P., Couture, J. C., Monteil, R., Guilbaud, J., and Saliou, L.,** Evolution of serum copper and zinc levels in burn patients, *Clin. Chim. Acta,* 66, 181, 1976 (in French).

310. **Mzhel'skaya, T. I.,** Determination of copper, iron and zinc content of blood serum using the atomic absorption spectrophotometer "Spektr-I", *Lab. Delo,* 4, 229, 1976.

311. **Shaburova, V. P., Yudelevich, I. G., Seryokavo, I. V., and Zolotov, Y. A.,** Extraction-atomic absorption determination of copper, silver and thallium in some metal halides, *Zh. Anal. Khim.,* 31, 255, 1976.

312. **Smeyers-Verbeke, J., Michotte, Y., Vanden Winkel, P., and Massart, D. L.,** Matrix effects in the determination of copper and manganese in biological materials using furnace atomic absorption spectrometry, *Anal. Chem.,* 48, 125, 1976.

313. **Tello Rosales, A.,** The determination of copper from copper flotation products using low sensitivity resonant lines, *At. Absorpt. Newsl.,* 15, 51, 1976.

314. **Tsutsumi, C., Koizmui, H., and Yoshikawa, S.,** Atomic absorption spectrophotometric determination of lead, cadmium and copper in foods by simultaneous extraction of the iodides with methyl isobutyl ketone, *Bunseki Kagaku,* 25, 150, 1976.

315. **Satake, M., Asano, T., Takagi, Y., and Yonekobu, T.,** Determination of copper, zinc, lead, cadmium and manganese in brackish and coastal waters by the combination of chelating ion exchange separation and atomic absorption spectrophotometry, *Nippon Kagaku Kaishi,* 5, 762, 1976.

316. **Sychra, V., Kolihova, D., and Dudova, N.,** A rapid determination of copper and iron in a capacitor paper by flameless atomic absorption spectrometry using a solid sampling technique, *Chem. Listy,* 70, 737, 1976.

317. **Wuyts, L., Smeyers-Verbeke, J., and Massart, D. L.,** Atomic absorption spectrophotometry of copper and zinc in human brain tissue. A critical investigation of two digestion techniques, *Clin. Chim. Acta,* 72, 405, 1976.

318. **Jonsson, H.,** Determination of copper, iron and manganese in milk with flameless AAS and a survey of the contents of these metals in Swedish market milk, *Milchwissenschaft,* 31, 210, 1976.

319. **Orlschlager, W., Schmidt, S., and Bestenlehner, L.,** Determination of copper in vegetable and animal materials and in mineral fodders by atomic absorption spectrophotometry, *Landwirtsch. Forsch.,* 29, 70, 1976 (in German).

320. **Wall, G.,** Determination by atomic absorption spectrophotometry of copper, lead and zinc in sulfide concentration, *Natl. Inst. Metall. Repub. S. Afr. Rep.,* No. 1978, 1976.

321. **Haenni, H., Hulstkamp, J., and Rothenbuehler, A.,** Determination of copper and iron in milk and milk products by the method of flameless atomic absorption, *Mitt. Geb. Lebensmittelunters. Hyg.,* 67, 448, 1976 (in German).

322. **Pakalns, P. and Farrar, Y. J.,** The effect of surfactants on the extraction-atomic absorption spectrophotometric determination of copper, iron, manganese, lead, nickel, zinc, cadmium and cobalt, *Water Res.,* 11, 145, 1977.

323. **Pandey, L. P., Ghose, A., and Dasgupta, P.,** Determination of zinc, silver, copper, iron and antimony in lead metal by atomic absorption spectrophotometry, *J. Inst. Chem.,* 49, 35, 1977.

324. **Putov, I., Popova, S., and Brashnarova, A.,** Determination of copper, nickel, cobalt, iron, zinc, and manganese in rubber stocks and adhesives by an atomic absorption spectrophotometric method, *Khim. Ind.,* 49, 16, 1977.

325. **Razumov, V. A. and Chuikov, V. A.,** Atomic absorption method for determining manganese and copper without previous ashing of plant material, *Agrokhimiya,* 1, 142, 1977.

326. **Schock, M. R. and Mercer, R. B.,** Direct determination of copper, manganese and iron in soils and sediments using secondary absorption lines, *At. Absorpt. Newsl.,* 16, 30, 1977.

327. **Florence, T. M. and Batley, G. E.,** Determination of the chemical forms of trace metals in natural waters with special reference to copper, lead, cadmium and zinc, *Talanta,* 24, 151, 1977.

328. **Janssen, A., Melchior, H., and Scholz, D.,** Application of flameless atomic absorption (heated graphite atomizer) to the determination of traces of copper in steel and other metal alloys after extraction, *Z. Anal. Chem.,* 283, 1, 1977.

329. **Kono, T.,** Determination of copper by atomic absorption spectrophotometry using an oxygen-sandwiched air-acetylene flame, *Bunseki Kagaku,* 26, 162, 1977.

330. **McCullars, G. M., O'Reilly, S., and Brennan, M.,** Pigment binding of copper in human bile, *Clin. Chim. Acta,* 74, 33, 1977.

331. **Mitchell, D. G., Mills, W. N., Ward, D. F., and Aldous, K. M.,** Determination of cadmium, copper and lead in sludges by microsampling cup atomic absorption spectrometry in a nitrous oxide-acetylene flame, *Anal. Chim. Acta,* 90, 275, 1977.

332. **Muzzarelli, R. A. and Rocchetti, R.,** Atomic absorption determination of manganese, cobalt and copper in serum and whole blood, *Talanta,* 24, 77, 1977.

333. **Alder, J. F., Baker, A. E., and West, T. S.,** Determination of copper in alloys by electrography and atomic absorption spectrometry, *Anal. Chim. Acta,* 90, 267, 1977.

334. **Armannsson, H.,** The use of dithizone extraction and atomic absorption spectrometry for the determination of cadmium, zinc, copper, nickel and cobalt in rocks and sediments, *Anal. Chim. Acta,* 88, 89, 1977.

335. **Bogden, J. D., Troiano, R. A., and Joselow, M. M.,** Copper, zinc, magnesium and calcium in plasma and cerebrospinal fluid of patients with neurological diseases, *Clin. Chem. (Winston-Salem, N.C.),* 23, 485, 1977.

336. **Bower, N. W. and Ingle, J. D.,** Precision of flame atomic absorption measurements of arsenic, cadmium, calcium, copper, iron, magnesium, molybdenum, sodium and zinc, *Anal. Chem.,* 49, 524, 1977.

337. **Chowdhury, A. N., De, D. K., and Das, A. K.,** Determination of traces of copper, nickel and cobalt by solvent extraction and atomic absorption spectrophotometry, *Z. Anal. Chem.,* 283, 41, 1977.

338. **Aznarez Alduan, J. L. and Castillo Suarez, J. R.,** Atomic absorption spectrophotometric determination of copper following extraction with salicylaldoxime in methyl isobutyl ketone, *An. Quim.* 73, 699, 1977 (in Spanish).

339. **Boline, D. R. and Schrenk, W. G.,** Atomic absorption spectrometry of copper and iron in plant material, *J. Assoc. Off. Anal. Chem.,* 60, 1170, 1977.

340. **Boyle, E. A. and Edmond, J. M.,** Determination of copper, nickel and cadmium in sea water by APDC chelate coprecipitation and flameless atomic absorption spectrometry, *Anal. Chim. Acta,* 91, 189, 1977.

341. **Capar, S. G.,** Atomic absorption spectrophotometric determination of lead, cadmium, zinc and copper in clams and oysters: collaborative study, *J. Assoc. Off. Anal. Chem.,* 60, 1400, 1977.

342. **Deguchi, M., Kinoshita, T., Yamaguchi, K., and Okumura, I.,** Atomic absorption spectrophotometric determination of copper using solvent extraction with thiothenoyl trifluoroacetone, *Bunseki Kagaku,* 26, 507, 1977.

343. **Inoue, S., Yotsuyanagi, T., Sasaki, M., and Aomura, K.,** Atomic absorption method for the determination of trace amounts of cadmium, copper, lead and zinc by ion-pair extraction of dimercaptomaleonitrile complexes, *Bunseki Kagaku,* 26, 550, 1977.

344. **Kang, H. K., Harvey, P. W., Valentine, J. L., and Swendsied, M. E.,** Zinc, iron, copper and magnesium concentrations in tissues of rats fed various amounts of zinc, *Clin. Chem. (Winston-Salem, N. C.),* 23, 1834, 1977.

345. **Kirchner, S. J. and Fernando, Q.,** Rate of extraction of copper from aqueous solutions, *Anal. Chem.,* 49, 1636, 1977.

346. **Lester, J. N., Harrison, R. M., and Perry, R.,** Rapid flameless atomic absorption analysis of the metallic content of sewage sludges. I. Lead, cadmium and copper, *Sci. Total Environ.,* 8, 153, 1977.

347. **Lorber, K. and Mueller, K.,** Atomic absorption spectrometry determination of copper and zinc in high purity tungsten and molybdenum compounds after column extraction chromatographic enrichment with dithizone in *o*-dichlorobenzene, *Mikrochim. Acta,* 2, 5, 1977.

348. **Manoliu, C. and Tomi, B.,** The determination of lithium, cadmium and copper in aluminum alloys by atomic absorption spectrophotometry, *Rev. Chim.,* 28, 370, 1977.

349. **Maurer, J.,** Extraction method for the simultaneous determination of sodium, potassium, calcium, magnesium, iron, copper, zinc and manganese in organic material using atomic absorption spectrophotometry, *Z. Lebensm. Unters. Forsch.,* 165, 1, 1977.

350. **Menden, E. E., Brockman, D., Choudhury, H., and Petering, H. G.,** Dry ashing of animal tissues for atomic absorption spectrometric determination of zinc, copper, cadmium, lead, iron, manganese, magnesium and calcium, *Anal. Chem.,* 49, 1644, 1977.

351. **Petering, H. G., Murthy, L., and O'Flaherty, F.,** Influence of dietary copper and zinc on rat lipid metabolism, *J. Agric. Food Chem.,* 25, 1105, 1977.

352. **Proshova, M. and Kovac, G.,** Atomic absorption spectrophotometry in veterinary medicine. Comparison of several methods during the determination of copper and iron levels in blood serum of sheep, *Chem. Listy,* 71, 978, 1977.

353. **Stoffers, P., Summerhayes, C., Forstner, U., and Patchineelam, S. R.,** Copper and other heavy metal contamination in sediments from New Bedford Harbor, Massachusetts: a preliminary note, *Environ. Sci. Technol.,* 11, 819, 1977.

354. **Tatro, M. E., Raynolds, W. L., and Costa, F. M.,** Determination of copper and iron in biological specimens by flameless atomic absorption, *At. Absorpt. Newsl.,* 16, 143, 1977.

355. **Zawadzka, H., Baralkieweiz, D., and Elbanowska, H.,** Flame atomic absorption spectrometric determination of cobalt, cadmium, lead, nickel, copper and zinc in natural waters, *Chem. Anal.,* 22, 913, 1977.

356. **Berenguer-Navarro, V., Espla-Moncho, M., and Hernandez-Mendez, J.,** Atomic absorption analysis with emulsion formation. II. Determination of iron and copper, previous extraction as oxinates, *Quim. Anal.,* 31, 329, 1977.

357. **Moll, M.,** Determination of copper, iron and calcium in beer, *Brauwissenschaft,* 30, 347, 1977.

358. **Pellerin, F. and Goulle, J. P.,** Dissolution and rapid determination of cadmium, copper, lead and zinc in dyes and antioxidants authorized for use in drugs and foodstuffs, *Ann. Pharm. Fr.,* 35, 189, 1977 (in French).

359. **Lanning, F. C. and Schrenk, W. G.,** Comparison of atomic absorption flame spectroscopy and official AOAC method for determining copper in plant tissues, *Trans. Kans. Acad. Sci.,* 80, 41, 1977.

360. **Churella, D. J. and Copeland, T. R.,** Interference of salt matrices in the determination of copper by atomic absorption spectrometry with electrothermal atomization, *Anal. Chem.,* 50, 309, 1978.

361. **Geladi, P. and Adams, F.,** The determination of cadmium, copper, iron, lead and zinc in aerosols by atomic absorption spectrometry, *Anal. Chim. Acta,* 96, 229, 1978.

362. **Glaeser, E.,** Atomic absorption spectroscopic determination of iron, copper, zinc, cadmium, manganese and cobalt in waters by the ''injection method'' after an extraction concentration, *Acta Hydrochim. Hydrobiol.,* 6, 83, 1978.

363. **Hannaford, P. and McDonald, D. C.,** Determination of relative oscillator strengths of the copper resonance lines by atomic absorption spectroscopy, *J. Phys.,* B11, 1177, 1978.

364. **Hudnik, V., Gomiscek, S., and Gorenc, B.,** The determination of trace metals in mineral waters. I. Atomic absorption spectrometric determination of cadmium, cobalt, chromium, copper, nickel and lead by electrothermal atomization after concentration by coprecipitation, *Anal. Chim. Acta,* 98, 39, 1978.

365. **Ishino, F., Matsumae, H., Shibata, K., Ariga, N., and Goshima, F.,** Determination of copper, cadmium and zinc in bone by atomic absorption spectrophotometry, *Bunseki Kagaku,* 27, 232, 1978.

366. **Iwata, Y., Fujiwara, K., and Fuwa, K.,** Noise distribution in atomic absorption flame of chromium, copper and cobalt, *Bunseki Kagaku,* 27, 62, 1978.

367. **Norval, E.,** A tungsten carbide-coated crucible for electrothermal atomization. Determination of copper in some biological standards, *Anal. Chim. Acta,* 97, 399, 1978.

368. **O'Brien, J. F. and Emmerling, M. E.,** Estimation of glycosaminoglycans by atomic absorption determination of copper in their alcian blue complexes, *Anal. Biochem.,* 85, 377, 1978.

369. **Ryabinin, A. I. and Lazareva, E. A.,** Solvent extraction-atomic absorption determination of copper, silver and cadmium in Black Sea water, *Zh. Anal. Khim.,* 33, 298, 1978.

370. **Simmons, W. J.,** Background absorption error in determination of copper in plants by flame atomic absorption spectrometry, *Anal. Chem.,* 50, 870, 1978.

371. **Sire, J., Collin, J., and Voinovitch, I. A.,** Physical and chemical conditions for determination of vanadium in steel by atomic absorption spectrometry and extension to copper, chromium, nickel and manganese, *Spectrochim. Acta,* 33b, 31, 1978.

372. **Smeyers-Verbeke, J., Michotte, Y., and Massart, D. L.,** Influence of some matrix elements on the determination of copper and manganese by furnace atomic absorption spectrometry, *Anal. Chem.,* 50, 10, 1978.

373. **Bogden, J. D. and Troiano, R. A.,** Plasma calcium, copper, magnesium and zinc concentrations in patients with the alcohol withdrawl syndrome, *Clin. Chem. (Winston-Salem, N.C.),* 24, 1553, 1978.

374. **Davidoff, G. N., Votaw, M. L., Coon, W. V., Hultquist, D. E., Filter, B. J., and Wexler, S. A.,** Elevations in serum copper, erythrocytic copper and ceruloplasmin concentrations in smokers, *Am. J. Clin. Pathol.,* 70, 790, 1978.

375. **Fudagawa, N. and Kawase, A.,** Determination of copper and zinc in tea by atomic absorption spectrometry, *Bunseki Kagaku,* 27, 353, 1978.

376. **Healy, P. J., Turvey, W. S., and Willats, H. G.,** Interference in estimation of serum copper concentration resulting from use of silicone-coated tubes for collection of blood, *Clin. Chim. Acta,* 88, 573, 1978.

377. **Hoenig, M. and Vanderstappen, R.,** Determination of cadmium, copper, lead, zinc and mangenese in plants by flame atomic absorption spectroscopy: mineralization effects, *Analusis,* 6, 312, 1978.

378. **Damel, H., Teape, J., Brown, D. H., Ottaway, J. M., and Smith, W. E.,** Determination of copper in plasma ultrafiltrate by atomic absorption spectrometry using carbon furnace atomization, *Analyst (London),* 103, 921, 1978.

379. **Kolihova, D., Sychra, V., and Dudova, N.,** Atomic absorption spectrometric analysis of ilmenite and inorganic pigments based on titanium dioxide. III. Determination of copper, manganese, chromium and iron by atomic absorption spectrometry with electrothermal atomization, *Chem. Listy,* 72, 108, 1978.

380. **Kosonen, P. O., Salonen, A. M., and Nieminen, A.,** Determination of manganese, copper, cobalt, iron and molybdenum in multivitamin mineral tablets by flameless atomic absorption spectrophotometry, *Finn. Chem. Lett.,* 4, 136, 1978.

381. **Maruta, T., Minegishi, K., and Sudoh, G.,** Atomic absorption spectrometric determination of trace amounts of copper, manganese, lead and chromium in cements by direct atomization in a carbon furnace, *Yogyokyokai Shi,* 86, 532, 1978.

382. **Sukhanova, G. B.,** Mechanism of the population of highly excited states of copper and silver in a hollow cathode discharge, *Izv. Vyssh. Uchebn. Zaved. Fiz.,* 21, 105, 1978.

383. **Tanaka, T., Hayashi, Y., and Ishizawa, M.,** Simultaneous determination of cadmium and copper in water by a graphite furnace dual channel atomic absorption spectrophotometry, *Bunseki Kagaku,* 27, 499, 1978.

384. **Viets, J. G.,** Determination of silver, bismuth, cadmium copper, lead and zinc in geologic materials by atomic absorption spectrometry with tricaprylylmethyl ammonium chloride, *Anal. Chem.,* 50, 1097, 1978.

385. **Zaguzin, V. P., Karmanova, N. G., and Pograbnyak, Y. F.,** Use of the graphite capsule-flame atomizer for determining copper and lead in powdered rock samples, *Zh. Prikl. Spektrosk.,* 28, 963, 1978.

386. **Emara, M. M., Farid, N. A., Ali, M. M., and Gharib, A. E.,** Investigation on possible acid and cation interference in the determination of copper by atomic absorption spectroscopy in an air-acetylene flame, *J. Indian Chem. Soc.,* 55, 992, 1978.

387. **Erkovich, G. E., Yakovuk, V. I., and Malykh, V. D.,** Atomic absorption determination of impurities in silver-copper alloys, *Zavod. Lab.,* 44, 1480, 1978.

388. **Hioki, T., Dokiya, Y., Notsu, K., and Fuwa, K.,** Analytical condition for the determination of copper, zinc and nickel in silicates by atomic absorption spectrometry, *Bunseki Kagaku,* 27, 487, 1978.

389. **Korajewski, J. and Matczak, W.,** Determination of iron, manganese, nickel, copper and chromium oxides in air by atomic absorption spectrometry, *Chem. Anal.,* 23, 1019, 1978.

390. **Potter, N. M.,** Determination of copper in gasoline by atomic absorption spectrometry with electrothermal atomization, *Anal. Chim. Acta,* 102, 201, 1978.

391. **Schutze, I. and Mueller, W.,** Determination of bound heavy metals, copper, iron, nickel, zinc, lead and chromium in dietary fats, *Nahrung,* 22, 277, 1978.

392. **Zeluykova, Y. V., Kravchenko, T. B., and Kucher, A. A.,** Atomic absorption determination of iron and copper impurities in rare earth metal products, *Zavod. Lab.,* 44, 677, 1978 (in Russian).

393. **Vackova, M. and Zemberyova, M.,** Determination of copper lead, cobalt, nickel and zinc in ferromanganese by atomic absorption spectrometry, *Hutn. Listy,* 33, 890, 1978 (in Slovak).

394. **Terashima, S.,** Atomic absorption determination of Mn, Fe, Cu, Ni, Co, Pb, Zn, Si, Al, Ca, Mg, Na, K, Ti and Sr in manganese nodules, *Chishitsu Chosasho Geppo,* 29, 401, 1978 (in Japanese).

395. **Kumina, D. M. and Karyakin, A. V.,** Reduction of limits of detection for trace amounts of magnesium, zinc and copper in atomic absorption analysis using mixtures of organic solvents, *Zh. Anal. Khim.,* 34, 1411, 1979.

396. **Langmyhr, F. J., Eyde, B., and Jonsen, J.,** Determination of the total content and distribution of cadmium, copper and zinc in human parotid saliva, *Anal. Chim. Acta,* 107, 211, 1979.

397. **Page, A. G., Godbole, S. V., Kulkarni, M. J., Shelar, S. S., and Joshi, B. D.,** Direct AAS determination of cobalt, chromium, copper, manganese and nickel in uranium oxide (U_3O_8) by electrothermal atomization, *Z. Anal. Chem.,* 296, 40, 1979.

398. **Teape, J., Kamel, H., Brown, D. H., Ottaway, J. M., and Smith, W. E.,** An evaluation of the use of electrophoresis and carbon furnace atomic absorption spectrometry to determine the copper level in separated serum protein fractions, *Clin. Chim. Acta,* 94, 1, 1979.

399. **Tsushida, T. and Takeo, T.,** Direct determination of copper, lead and cadmium in tea infusions by flameless atomic absorption spectrometry, *Agric. Biol. Chem.,* 43, 1347, 1979.

400. **Wawschinek, O.,** Determination of copper and gold in serum by flameless atomic absorption, *Mikrochim. Acta,* 2, 111, 1979.

401. **Ichinose, N. and Inui, T.,** Effect of solvent extraction on the atomic absorption spectrophotometric determination of traces of copper with zinc dibenzyl dithiocarbamate, *Z. Anal. Chem.,* 295, 352, 1979.

402. **Jakutowicz, K. and Korpaczewka, W.,** Determination of the copper concentration in seven parasite species by the atomic absorption spectrometry, *Bull. Acad. Pol. Sci. Ser. Sci. Biol.,* 27, 69, 1979.

403. **Katskov, D. A. and Grinshtein, I. L.,** Study of the chemical interaction of copper, gold and silver with carbon and atomic absorption method using an electrothermal atomizer, *Zh. Prikl. Spektrosk.,* 30, 787, 1979.

404. **Khalighie, J., Ure, A. M., and West, T. S.,** An investigation of atom collection phenomena in the atomic absorption spectrometry of copper, *Anal. Chim. Acta,* 107, 191, 1979.

405. **Kumina, D. M. and Karyakin, A. V.**, Reduction of limits of detection for trace amounts of magnesium, zinc, and copper in atomic absorption analysis using mixtures of organic solvents, *Zh. Anal. Khim.*, 34, 1411, 1979.

406. **Armannsson, H.**, Dithiozone extraction and flame atomic absorption spectrometry for the determination of cadmium zinc, lead, copper, nickel, cobalt and silver in sea water and biological tissues, *Anal. Chim. Acta*, 110, 21, 1979.

407. **Arpadjan, S. and Alexandrova, I.**, Determination of trace elements (lead, bismuth, copper, zinc and silver) in high-purity tin by flame atomic absorption spectrophotometry, *Z. Anal. Chem.*, 298, 159, 1979.

408. **Berndt, H. and Jackwerth, E.**, Determination of iron, copper and zinc by a mechanized flame photometric micromethod ("injection method") of flame photometry (atomic absorption-atomic emission) for the determination of serum electrolytes and trace elements. II., *J. Clin. Chem. Clin. Biochem.*, 17, 489, 1979.

409. **Bozsai, G. and Csanady, M.**, Systematic investigations on the heavy metal pollution (cadmium, lead, copper, zinc, chromium, and barium) of drinking water using atomic absorption spectrometric methods, *Z. Anal. Chem.*, 297, 370, 1979.

410. **Gregorczyk, S. and Wycislik, A.**, Determination of manganese, chromium, nickel, molybdenum, cobalt and copper in low alloy and medium-alloy steel by atomic absorption spectroscopy with acetylene-air flame, *Chem. Anal.*, 24, 529, 1979.

411. **Vratkovskaya, S. V. and Pogrebnyak, Y. F.**, Determination of copper, lead and zinc in slightly mineralized water by flame atomic absorption spectrophotometry, *Zh. Anal. Khim.*, 34, 759, 1979.

412. **Weisel, C. P., Fasching, J. L., Piotrowicz, S. R., and Duce, R. A.**, A modified standard addition method for determining cadmium, lead, copper and iron in sea water derived samples by atomic absorption spectroscopy, *Adv. Chem. Ser.*, No. 172, American Chemical Society, Washington, D.C., 1979, 134.

413. **Weiss, H. V., Kenis, P. R., Korkisch, J., and Steffan, I.**, Determination of copper and manganese in sea water by neutron activation analysis and atomic absorption spectrometry, *Anal. Chim. Acta*, 104, 337, 1979.

414. **Alder, J. F. and Bucklow, P. L.**, Determination of copper, zinc and chromium in carbon cloth by atomic absorption spectrometry without sample ashing, *At. Absorpt. Newsl.*, 18, 123, 1979.

415. **Alevato, S. J. and Curtius, A. J.**, Determination of copper and antimony in lead alloy by atomic absorption spectrophotometry, *Mikrochim. Acta*, 1, 361, 1979.

416. **Lu, K. L., Pulford, I. D., and Duncan, H. J.**, Determination of cadmium, cobalt, copper, nickel and lead in soil extracts by dithizone extraction and atomic absorption spectrometry with electrothermal atomization, *Anal. Chim. Acta*, 106, 319, 1979.

417. **Kuga, K., Sugaya, I., and Tsujii, K.**, Determination of copper, iron, potassium and sodium in polyimide resins by atomic absorption spectrometry, *Bunseki Kagaku*, 28, 201, 1979.

418. **Manoliu, M. G. and Manoliu, C.**, Direct determination of copper traces in petroleum products by atomic absorption spectrophotometry, *Rev. Roum. Chim.*, 24, 95, 1979.

419. **Meranger, J. C., Subramanian, K. S., and Chalifoux, C.**, A national survey for cadmium, chromium, copper, lead, zinc, calcium and magnesium in Canadian drinking water supplies, *Environ. Sci. Technol.*, 13, 707, 1979.

420. **Sanzolone, R. F., Chao, T. T., and Crenshaw, G. L.**, Atomic absorption spectrometric determination of cobalt, nickel and copper in geological materials with matrix masking and chelation-extraction, *Anal. Chim. Acta*, 105, 247, 1979.

421. **Arnac, M. and Chanut, J. P.**, Statistical comparison of three methods for the determination of copper, *Talanta*, 26, 181, 1971 (in French).

422. **Berndt, H. and Jackwerth, E.**, Mechanized micromethod ("injection method") of flame photometry (atomic absorption-atomic emission) for determination of serum electrolytes and trace elements (iron, copper, zinc). Determination of lithium, sodium, potassium, magnesium, calcium with a mechanized flame spectrometric method, *J. Clin. Chem. Clin. Biochem.*, 17, 71, 1979.

423. **Bruland, K. W., Franks, R. P., Knauer, G. A., and Martin, J. H.**, Sampling and analytical methods for the determination of copper, cadmium, zinc and nickel at the nanogram per liter level in sea water, *Anal. Chim. Acta*, 105, 233, 1979.

424. **Isozaki, A., Soeda, N., Okutani, T., and Utsumi, S.**, Flameless atomic absorption spectrophotometry of copper by the introduction of chelating resin into a carbon tube atomizer, *Nippon Kagaku Kaishi*, 4, 549, 1979.

425. **Gregorczyk, S. and Wycislik, A.**, Chemical analysis of high-alloy manganese-aluminum steels by atomic absorption spectrophotometry, *Wiad. Hutn.*, 35, 261, 1979.

426. **Rockland, L. B., Wolf, W. R., Hahn, D. M., and Young, R. S.**, Estimation of zinc and copper in raw and cooked legumes: interlaboratory study of atomic absorption and X-ray fluorescence spectroscopy. *J. Food Sci.*, 44, 1711, 1979.

427. **Ihida, M., Ishii, T., and Ohnishi, R.**, Precision on the determination of trace elements in coal, *Am. Chem. Soc. Div. Fuel Chem.*, 247, 262, 1979.

428. **Anonymous,** Determination of Trace Metals in Liquid Coke-oven Effluents by Atomic Absorption Spectrophotometry, Carbonization Res. Rep., British Carbonization Research Association, Chesterfield, Derbyshire, England, 76, 1979, 17.

429. **Fonds, A. W., Kempf, T., Minderhoud, A., and Sonneborn, M.,** Heavy metal content of various kinds of water, *Reinhalt Wassers,* 78—88, 1979.

430. **Boyer, K. W., Capar, S. G., Jones, J. W., Suddendorf, R. S., and Forwalter, J.,** Multielement/trace analysis identifies 48 elements in foods. Simultaneous element analysis and data reduction reaches parts per billion, *Food Process.,* 40, 72, 1979.

431. **Capar, S. G. and Gould, J. H.,** Lead, fluoride and other elements in bonemeal supplements, *J. Assoc. Off. Anal. Chem.,* 62, 1054, 1979.

432. **Carmichael, N. G., Squibb, K. S., and Fowler, B. A.,** Metals in the molluscan kidney: a comparison of two closely related bivalve species (Argopecten) using X-ray microanalysis and atomic absorption spectroscopy, *J. Fish. Res. Board. Can.,* 36, 1149, 1979.

433. **Flanjak, J. and Lee, H. Y.,** Trace metal content of livers and kidneys of cattle, *J. Sci. Food Agric.,* 30, 503, 1979.

434. **Kitagawa, K., Shigeyasu, T., and Takeuchi, T.,** Application of the atomic faraday effect to the trace determination of elements (cadmium, silver and copper): effect of the hyperfine structure on the Zeeman splitting and line crossing, *Spectrochim. Acta,* 34b, 389, 1979.

435. **Komarek, J., Havel, J., and Sommer, L.,** The use of chelates of copper, nickel, cobalt, cadmium and zinc with heterocyclic azo dyes in the AAS determination of these elements, *Coll. Czech. Chem. Commun.,* 44, 3241, 1979.

436. **Lambert, J. B., Szupnar, C. B., and Buikstra, J. E.,** Chemical analysis of excavated human bone from middle and late Woodland sites, *Archaeometry,* 21, 115, 1979.

437. **Okubo, N., Kojima, H., Inoue, T., Sugiyama, H., and Miyazaki, M.,** Studies on metals in the environment. II. Preconcentration-atomic absorption spectrophotometry analysis of micro amounts of copper and its application for measurement of copper in dust, *Eisei Kagaku,* 25, 136, 1979.

438. **Rezchikov, V. G. and Usvatov, V. A.,** Atomic absorption determination of copper, lead and bismuth in silver nitrate, *Zavod. Lab.,* 45, 1112, 1979.

439. **Schuller, P. L. and Coles, L. E.,** The determination of copper in foodstuffs, *Pure Appl. Chem.,* 51, 385, 1979.

440. **Taddia, M.,** Atomic absorption determination of copper impurities in zirconium salts using an air-acetylene flame, *Z. Anal. Chem.,* 299, 261, 1979.

441. **Fricke, F. L., Robbins, W. B., and Caruso, J. A.,** Trace element analysis of food and beverages by atomic absorption spectrometry, *Prog. Anal. At. Spectrosc.,* 2, 85, 1979.

442. **Abo-Rady, M. D. K.,** Aquatic macrophytes as indicator for heavy metal pollution in the river Seine, *Arch. Hydrobiol.,* 89, 387, 1980 (in German).

443. **Balaes, G. and Robert, R. V. D.,** Analysis by atomic absorption spectrophotometry of activated charcoal, *Natl. Inst. Metall. Repub. S. Afr. Rep.,* No. 2060, 1980.

444. **Belyaev, Y. I. and Issers, V. V.,** Effect of sample mass on the precision of determining some trace elements in rocks, *Zh. Anal. Khim.,* 35, 2374, 1980 (in Russian).

445. **Brovko, I. A.,** Diphenyl carbazone as a reagent for extraction-atomic absorption determination of cadmium, cobalt, copper, manganese, nickel and zinc, *Zh. Anal. Khim.,* 35, 2095, 1980 (in Russian).

446. **Budniok, A.,** Determination of cadmium and copper in galvanic coatings by means of atomic absorption, *Microchem. J.,* 25, 531, 1980.

447. **Cool, M., Marcoux, F., Paulin, A., and Mehra, M. C.,** Metallic contaminants in street soils of Moncton, New Brunswick, Canada, *Bull. Environ. Contam. Toxicol.,* 25, 409, 1980.

448. **Langmyhr, F. J. and Aadalen, U.,** Direct atomic absorption spectrometric determination of copper, nickel and vanadium in coal and petroleum coke, *Anal. Chim. Acta,* 115, 365, 1980.

449. **Lerner, L. A. and Igoshina, E. V.,** Atomic absorption determination with a graphite furnace of copper, cobalt, and nickel extracted from soil with ammonium acetate buffer solutions, *Pochvovedenie,* 3, 106, 1980.

450. **Pedersen, B., Williams, M., and Storgaard-Joergensen, S.,** Determination of copper, lead, cadmium, nickel and cobalt in EDTA extracts of soil by solvent extraction and graphite furnace atomic absorption spectrophotometry, *Analyst (London),* 105, 119, 1980.

451. **Petrov, I. I., Tsalev, D. L., and Barsev, A. I.,** Atomic absorption spectrometric determination of cadmium, cobalt, copper, manganese, nickel, lead and zinc in acetate soil extracts, *At. Spectrosc.,* 1, 47, 1980.

452. **Smith, R. G. and Windom, H. L.,** A solvent extraction technique for determining nanogram per liter concentrations of cadmium, copper, nickel and zinc in sea water, *Anal. Chim. Acta,* 113, 39, 1980.

453. **Chakrabarti, C. L., Wan, C. C., and Li, W. C.,** Direct determination of traces of copper, zinc, lead, cobalt, iron and cadmium in bovine liver by graphite furnace atomic absorption spectrometry using the solid sampling and the platform technique, *Spectrochim. Acta,* 35b, 93, 1980.

454. **Feinberg, M. and Ducauze, C.,** High temperature dry ashing of foods for atomic absorption spectrometric determination of lead, cadmium and copper, *Anal. Chem.,* 52, 207, 1980.

455. **Guerra, R.,** Rapid determination of gold and copper in lead tin based solder, *At. Spectrosc.,* 1, 58, 1980.

456. **Hydes, D. J.,** Reduction of matrix effects with a soluble organic acid in the carbon furance atomic absorption spectrometric determination of cobalt, copper and manganese in seawater, *Anal. Chem.,* 52, 959, 1980.

457. **Iwasa, A., Nakagawa, K., and Yonemoto, T.,** Effect of sample flow rate on the atomic absorption spectrophotometry of copper, *Bunseki Kagaku,* 29, 86, 1980.

458. **Kuga, K.,** Rapid determination of copper, iron and manganese in polyimide resins by atomic absorption spectrometry using a graphite furnace atomizer, *Bunseki Kagaku,* 29, 342, 1980.

459. **Danchev, M. D., Ekivina, N. I., and Belyaev, V. P.,** Atomic absorption determination of copper, nickel and iron in concentrates of the noble metals, *Zavod. Lab.,* 46, 1110, 1980.

460. **Davison, W.,** Ultratrace analysis of soluble zinc, cadmium, copper and lead in Windermere Lake water using anodic stripping voltammetry and atomic absorption spectroscopy, *Freshwater Biol.,* 10, 223, 1980.

461. **Emara, M. M., Ashy, M. A., Farid, N. A., and Gharib, A.,** Effect of mixed acids on the determination of copper by atomic absorption spectroscopy. *Ann. Chim. (Rome),* 70, 611, 1980 (in English).

462. **Fang, Q. Y., Hung, Z. H., and Zhou, H. D.,** Analysis of trace elements in Zhumulangmafeng High Mountain water, *Tzu Jan Tsa Chih,* 3, 797, 1980 (in Chinese).

463. **Freeland-Graves, J. H., Ebangit, L. M., and Bodzy, P. W.,** Zinc and copper content of foods used in vegetarian diets, *J. Am. Diet. Assoc.,* 77, 648, 1980.

464. **Hrabovecka, G. and Matherny, M.,** Determination of copper, molybdenum and manganese by atomic absorption spectrometry in mineral waters, *Chem. Zvesti,* 34, 465, 1980 (in German).

465. **Hughes, M. J.,** Analysis of Roman tin and pewter ingots, *Occas. Pap., British Museum,* 17, 41, 1980.

466. **Ishii, T., Hirano, S., Matsuba, M., and Koyangai, T.,** Determination of trace elements in shell fish, *Nippon Suisan Gakkaishi,* 46, 1375, 1980 (in English).

467. **Itokawa, H., Watanabe, K., Tazaki, T., Hayashi, T., and Hayashi, Y.,** Quantitative analysis of metals in crude drugs, *Shoyakugaku Zasshi,* 34, 155, 1980.

468. **Ivanova, E., Mareva, S., and Iordanov, N.,** Extraction-flame atomic absorption determination of microtraces of copper, zinc, nickel, cobalt, manganese and iron in some alkali salts, *Z. Anal. Chem.,* 303, 378, 1980.

469. **Kosa, F., Foldes, V., Viragos-Kis, E., Rengei, B., and Ferke, A.,** Atomic absorption spectrophotometric study on content of inorganic substances in fetal bone for the determination of age, *Arch. Kriminol.,* 166, 44, 1980 (in German).

470. **Lindh, U., Brune, D., Nordberg, G., and Wester, P. O.,** Levels of antimony, arsenic, cadmium, copper, lead, mercury, selenium, silver, tin and zinc in bone tissue of industrially exposed workers, *Sci. Total Environ.,* 16, 109, 1980.

471. **Ludany, A., Kellermayer, M., and Jobst, K.,** Metal content of rat liver cell organelles, *Acta Biochim. Biophys. Acad. Sci. Hung.,* 15, 229, 1980.

472. **Maeda, T., Nakagawa, M., Kawakatsu, M., and Tanimoto, Y.,** Determination of metallic elements in serum and urine by flame and graphite furnace atomic absorption spectrophotometry, *Shimadzu, Hyoron,* 37, 1980 (in Japanese).

473. **McCamey, D. A. and Niemczyk, T. M.,** Interference in sputter atomization atomic absorption spectroscopy, *Appl. Spectrosc.,* 34, 692, 1980.

474. **Mendelson, R. A. and Huber, A. M.,** Effect of duration of alcohol administration of deposition of trace elements in the fetal rat, *Adv. Exp. Med. Biol.,* 132, 295, 1980.

475. **Ogihara, K., Seki, H., and Nagase, K.,** Comparison of the method of bottom sediment analysis and the method of the agricultural soil pollution prevention law, *Nagano-Ken Eisei Kogai Kenkyusho Kenkyu Hokoku,* 2, 188, 1980 (in Japanese).

476. **Ospanov, K. K., Siromakha, L. N., and Sutanbaeva, R. S.,** Rapid methods for the selective dissolution and determination of copper from chalcocite in ores and beneficiation products, *Zavod. Lab.,* 46, 902, 1980.

477. **Posta, J.,** Determination of the optimum parameters of a combined atomic absorption spectrophotometric plus arc flame method for the analysis of floating dust samples, *Hung. Sci. Instrum.,* 47, 33, 1980.

478. **Roguljic, A., Mikac-Devic, D., and Krusic, J.,** Copper, zinc, and magnesium levels in healthy tissues and benign and malignant tumors of the uterus, *Period. Biol.,* 82, 213, 1980.

479. **Sourova, J. and Capkova, A.,** Determination of trace elements in water with a high iron content by AAS, *Vodni Hospod.,* 303, 133, 1980.

480. **Das, A. K.,** AAS Determination of trace metals in raw sewage, *Indian J. Environ. Health,* 22, 130, 1980.

481. **Akama, Y., Nakai, T., and Kawamura, F.,** Determination of traces of cobalt, copper, manganese, nickel and lead in solar salt by AA spectrometry combined with extraction, *Nippon Kaishi Gakkaishi,* 34, 196, 1980.

482. **Acebal, G. S., Grassi, R. L., and Gutierrez, D. O.,** Behavior of oxalic, citric and nitrilotriacetic acids as extractants of trace elements in the soil, *Agrochimica,* 24, 462, 1980.

483. **Jarkovsky, J., Plsko, E., and Stresko, V.,** Contribution to the analytical geochemistry of antimonites, *Acta Geol. Geogr. Univ. Comeniance Geol.,* 34, 71, 1980 (in English).

484. **Chakrabarti, C. L., Wan, C. C., and Li, W. C.,** Atomic absorption spectrometric determination of cadmium, lead, zinc, copper, cobalt and iron in oyster tissue by direct atomization from the solid state using the grahite furnace platform technique, *Spectrochim. Acta,* 35b, 547, 1980.

485. **Stendal, H.,** Leaching studies for the determination of copper, zinc, lead, nickel and cobalt in geological materials by atomic absorption spectrophotometry, *Chem. Erde,* 39, 276, 1980 (in English).

486. **Tsushida, T. and Takeo, T.,** Determination of copper, lead and cadmium in tea by graphite furnace atomic absorption spectrophotometry, *Nippon Shokuhin Kogyo Gakkaishi,* 27, 585, 1980.

487. **Tursunov, A. T., Brovko, I. A., and Nazarov, S. N.,** Use of a furnace-flame system for the atomic absorption determination of some trace elements in soil extracts, *Uzb. Khim. Zh.,* 5, 5, 1980 (in Russian).

488. **Uchida, T., Kojima, I., and Iida, C.,** Application of an automatically triggered digital integrator to flame atomic absorption spectrometry of copper using a discrete nebulization technique, *Analyst (London),* 106, 206, 1980.

489. **Van Willis, W., El-Ahraf, A., Vinjamoori, D. V., and Aref, K.,** Analysis of animal feed ingredients and soil amendment products produced from beef cattle manure for selected trace metals using atomic absorption spectrophotometry, *J. Food Prot.,* 43, 834, 1980.

490. **Freeland-Graves, J. H., Ebangit, M. I., and Bodzy, P. W.,** Zinc and copper content of foods used in vegetarian diets, *J. Am. Diet. Assoc.,* 77, 648, 1980.

491. **Saito, S. and Kamoda, M.,** Sample preparation by dry ashing prior to the determination of metals in sugars, *Seito Gijutsu Kenkyu Kaishi,* 29, 36, 1980.

492. **Levi, S. and Purdy, W. C.,** The AAS determination of copper and zinc levels in the serum of hemodialysis patients, *Clin. Biochem.,* 3, 253, 1980.

493. **Fink, L. K., Jr., Harris, A. B., and Schick, L. L.,** Trace Metals in Suspended Particulates, Biota and Sediments of the St. Croix, Narraguagus and Union Estuaries and the Goose Cove Region of Penobscot Bay, Rep. W80-4809, OWRT-A-041-ME (1), 1980; available from the NTIS, Springfield, Va.; *Govt. Rep. Announce. Index, (U.S.),* 80, 3372, 1980.

494. **Huang, H. M., Wu, C. C., Chou, M. C., and Liu, C. C.,** Flameless atomic absorption determination of sodium, iron, and copper in ultrapure water and eight other reagents, *Fen Hsi Hua Hsueh,* 8, 233, 1980 (in Chinese).

495. **Raj, K. P. S., Agrawal, Y. K., and Patel, M. R.,** Analysis of garlic for its metal contents, *J. Indian Chem. Soc.,* 57, 1121, 1980.

496. **Hoshino, Y., Utsunomiya, T., and Fukui, K.,** Graphite furnace atomic absorption spectrometry utilizing selective concentration onto tungsten wire, *Eng. Mater. Tokyo Inst. Technol.,* 5, 109, 1980 (in English).

497. **Caristi, C., Cimino, G., and Ziino, M.,** Heavy metal pollution. II. Determination of trace heavy elements in lemon, orange and mandarin juices by flameless atomic absorption spectrophotometry, *Essenz Deriv. Agrum.,* 50, 165, 1980.

498. **Chebotareva, N. A. and Samokhvalov, S. G.,** Atomic absorption determination of copper, zinc, manganese, iron and cobalt in plant materials, *Byul. Pochv. In-ta Vaskhnil.,* 24, 49, 1980 (in Russian).

499. **Inhat, M., Gordon, A. D., Gaynor, J. D., Berman, S. S., Desauliners, A., Stoeppler, M., and Valenta, P.,** Interlaboratory analysis of natural fresh waters for copper, zinc, cadmium and lead, *Int. J. Environ. Anal. Chem.,* 8, 259, 1980.

500. **Motuzova, G. V. and Obukhov, A. I.,** Effect of conditions for soil decomposition on the results of the determination of trace nutrient content on it, *Biol. Nauk (Moscow),* 11, 87, 1980.

501. **Jonsen, J., Helgeland, K., and Steinnes, E.,** Trace element in human serum. Regional distribution in Norway, Geomed. Aspects Present Future Res. 189—95, 1978, published 1980.

502. **Knight, M. J.,** Comparison of Four Digestion Procedures Not Requiring Perchloric Acid for the Trace Element Analysis of Plant Material, Rep. ANL/LRP-TM-18, Argonne National Laboratory, Argonne, Ill., 1980; availble from the NTIS, Springfield, Va.

503. **Orpwood, B.,** Use of Chelating Ion-Exchange Resins for the Determination of Trace Metals in Drinking Waters, Tech. Rep. TR-Water Res. Cent., Tech. Rep. 153, Medmenham Laboratory, Medmenham, Marlow, Bucks, U.K., 1980.

504. **Calixto, F. S. and Bauza, M.,** Determination of inorganic elements in the almond (*Prunus amygdalus*), *An. Bromatol.,* 32, 119, 1980.

505. **Iyengar, S. S., Martens, D. C., and Miller, W. P.,** Determination of copper and zinc in soil extracts by atomic absorption spectrophotometry using APDC-MIBK solvent extraction, *Soil Sci.,* 131, 95, 1981.

506. **Laxen, D. P. and Harrison, R. M.,** Cleaning methods for polythene containers prior to the determination of trace metals in fresh water samples, *Anal. Chem.,* 53, 345, 1981.

507. **Meranger, J. C., Hollebone, B. R., and Blanchette, G. A.,** Effect of storage times, temperatures and container types on the accuracy of atomic absorption determination of Cs, Cu, Hg, Pb, and Zn in whole heparized blood, *J. Anal. Toxicol.,* 5, 33, 1981.

508. **Meranger, J. C., Subramanian, K. S., and Chalifoux, C.,** Survey for cadmium, cobalt, chromium, copper, nickel, lead, zinc, calcium and magnesium in Canadian drinking water supplies, *J. Assoc. Off. Anal. Chem.,* 64, 44, 1981.

509. **Mohamed, N. and Fry, R. C.,** Slurry atomization direct atomic spectrochemical analysis of animal tissue, *Anal. Chem.,* 53, 450, 1981.

510. **Murakami, K., Ito, Y., Taguchi, K., Ogata, K., and Imanari, T.,** Determination of copper in rabbit plasma and red cells by flameless atomic absorption spectrometry, *Bunseki Kagaku,* 30, 200, 1981.

511. **Carpenter, R. C.,** Determination of cadmium, copper, lead and thallium in human liver and kidney tissue by flame atomic absorption spectrometry after enzymic digestion, *Anal. Chim. Acta,* 125, 209, 1981.

512. **Clark, J. R. and Viets, J. G.,** Multielement extraction system for the determination of 18 trace elements in geochemical samples, *Anal. Chem.,* 53, 61, 1981.

513. **Clark, J. R. and Viets, J. G.,** Back-extraction of trace elements from organometallic-halide extracts for determination by flameless atomic absorption spectrometry, *Anal. Chem.,* 53, 65, 1981.

514. **Douglas, D. J. and French, J. B.,** Elemental analysis with a microwave-induced plasma quadrupole mass spectrometer system, *Anal. Chem.,* 53, 37, 1981.

515. **Gardiner, P. E., Ottaway, J. M., Fell, G. S., and Burns, R. R.,** Application of gel filtration and electrothermal atomic absorption spectrometry to the speciation of protein-bound zinc and copper in human blood serum, *Anal. Chim. Acta,* 124, 281, 1981.

516. **Goldberg, W. J. and Allen, N.,** Determination of copper, manganese, iron and calcium in six regions of human brain by atomic absorption spectroscopy, *Clin. Chem. (Winston-Salem, N.C.),* 27, 562, 1981.

517. **Pabalkar, M. A., Naik, S. V., and Sanjana, N. R.,** Determination of copper, nickel and iron in heavy alloy using atomic absorption spectrophotometry, *Analyst (London),* 106, 47, 1981.

518. **Pakalns, P.,** Effect of surfactants on mixed chelate extraction-atomic absorption spectrophotometric determination of copper, nickel, iron, cobalt, cadmium, zinc and lead, *Water Res.,* 15, 7, 1981.

519. **Pleban, P. A., Kerkay, J., and Pearson, K. H.,** Polarized Zeeman effect flameless atomic absorption spectrometry of cadmium, copper, lead and manganese in human kidney cortex, *Clin. Chem. (Winston-Salem, N.C.),* 27, 68, 1981.

520. **Sandiez-Rasero, F.,** Atomic absorption spectrophotometric method for determination of water-soluble copper in water-insoluble copper fungicides: CIPAC collaborative study. Comparison with bathocuproine method, *J. Assoc. Off. Anal. Chem.,* 64, 75, 1981.

521. **Suzuki, M., Ohta, K., and Yamakita, T.,** Elimination of alkali chloride interference with thiourea in electrothermal atomic absorption spectrometry of copper and manganese, *Anal. Chem.,* 53, 9, 1981.

522. **Takada, T., Okano, H., Koide, T., Fujita, K., and Nakano, K.,** Determination of trace copper by electrothermal atomic absorption spectrometry with direct heating of metal-absorbed ion exchange resins, *Nippon Kagaku Kaishi,* 1, 13, 1981.

523. **Taylor, A. and Bryant, T. N.,** Comparison of procedures for determination of copper and zinc in serum by atomic absorption spectroscopy, *Clin. Chim. Acta,* 110, 83, 1981.

524. **Yamada, T., Kashima, J., and Naganuma, K.,** Sputtering and emission intensity of copper alloys in a grim glow lamp, *Anal. Chim. Acta,* 124, 275, 1981.

525. **Arpadjan, S. and Nakova, D.,** Direct determination of iron, zinc, copper and manganese in milk pwoder by atomic absorption, *Nahrung,* 25(4), 359, 1981.

526. **Engerbretson, J. A. and Mason, W. H.,** Depletion of trace elements in mated male *Heliothis virescens* and *Drosophila melanogaster, Comp. Biochem. Physiol. A,* 68a(3), 523, 1981 (in English).

527. **DeAntonio, S. M., Katz, S. A., Scheiner, D. M., and Wood, J. D.,** Anatomical variations of trace metal levels in hair, *Anal. Proc. (London),* 18(4), 162, 1981.

528. **Rasmussen, L.,** Determination of trace metals in sea water by Chelax-100 or solvent extraction technique and atomic absorption spectrometry, *Anal. Chim. Acta,* 125, 117, 1981.

529. **Spachidis, C., Weitz, A., and Baechmann, K.,** Determination of trace elements in photoconductor materials by matrix volatilization and flameless atomic absorption spectrometry, *Fresenius' Z. Anal. Chem.,* 306, 268, 1981.

530. **Takada, T., Okano, H., Koide, T., and Fuji, K.,** Determination of trace copper by electrothermal atomic absorption spectrometry with direct heating of metal-adsorbed ion exchange resin, *Nippon Kagaku Kaishi,* 1, 31, 1981 (in Japanese).

531. **Slovak, Z., Docekal, B., and Bohumil, N.,** Determination of trace metals in aluminum oxide by electrothermal atomic absorption spectrometry with direct injection of aqueous suspensions, *Anal. Chim. Acta,* 129, 263, 1981.

532. **Watson, M. E.,** Interlaboratory comparison in the determination of nutrient concentrations of plant tissue (9 plants studied), *Commun. Soil Sci. Plant Anal.,* 12(6), 601, 1981.

533. **Capel, I. D., Pinnock, M. H., Dorrell, H. M., Williams, D. C., and Grant, E. C. G.,** Comparison of concentrations of some trace, bulk and toxic metals in the hair of normal and dyslexic children, *Clin. Chem. (Winston-Salem, N.C.),* 27, 879, 1981.

534. **Kimura, M. and Kawanami, K.,** Separation and preconcentration of trace amounts of several metals in sodium perchlorate using activated carbon as a collector, *Nippon Kagaku Kaishi,* 1, 1, 1981 (in Japanese).

535. **Hoffman, M. R., Yost, E. C., Eisenreich, S. J., and Maier, W. J.,** Characterization of soluble and colloidal phase metal complexes in river water by ultrafiltration. A mass balance approach, *Environ. Sci. Technol.,* 15(6), 655, 1981.

536. **Bewers, J. M., Dalziel, J., Yeats, P. A., and Barren, J. L.,** An intracalibration for trace metals in sea water, *Mar. Chem.,* 10, 173, 1981.

537. **Carleer, R., Francois, J. P., and Van Poucke, L. C.,** Determination of the main, minor and trace elements in lead/tin based solder by atomic absorption spectrophotometry, *Bull. Soc. Chim. Belg.,* 90, 357, 1981.

538. **Petrov, I., Tsalev, D., and Vasileva, E.,** Pulse nebulization atomic absorption spectrometry after pre-concentration from acidic media, *Dokl. Bolg. Akad. Nauk,* 34, 679, 1981; CA 95, 146378n, 1981.

539. **Amini, M. K., Defreese, J. D., and Hathaway, L. R.,** Comparison of three instrumental spectroscopic techniques for elemental analysis of Kansas shales, *Appl. Spectrosc.,* 35, 497, 1981.

540. **Takiyama, K., Ishii, Y., and Yoshimura, I.,** Determination of metals in vegetables by ashing in a teflon crucible, *Mukogawa, Joshi Daigaku Kiyo, Shokumotsu-hen,* 28, F15, 1980; CA 95, 148770h, 1981.

541. **Wandiga, S. O.,** The concentrations of Zn, Cu, Pb, Mn, Ni and F in rivers and lakes of Kenya, *Sinet,* 3, 67, 1981.

542. **Henrion, G., Gelbrecht, J., Hoffmann, T., and Marquardt, D.,** AAS trace determination tungsten after preconcentration on chelate resin WOFATIT MC 50, *Z. Chem.,* 21, 1981.

543. **Pihlaja, H.,** Determination of traces of metals in Finnish margarines by the flameless atomic absorption spectrophotometry method, *Fette Seifen Anstrichm.,* 83, 294, 1981.

544. **Zolotov, Y. A., Larikova, G. A., Bodnya, V. A., Efremova, O. A., Davydova, S. L., Yatsimirskii, K. B., and Kol'chinskii, A. G.,** Nitrogen containing macrocyclic compounds as extractants for the selective separation of copper, *Dokl. Akad. Nauk SSSR,* 258, 889, 1981 (in Russian).

545. **Sturgeon, R. E., Berman, S. S., Willie, S. N., and Desauliners, J. A. H.,** Preconcentration of trace elements from seawater with silica-immobilized 8-hydroxy-quinoline, *Anal. Chem.,* 53, 2337, 1981.

546. **Farmer, J. G.,** The analytical chemist in studies of metal pollution in sediment cares, *Anal. Proc. (London),* 18, 249, 1981.

547. **Normakhmatov, R. N. and Muradova, S. B.,** Mineral composition of pecan kernels, *Khlebopek. Konditer. Prom.,* 7, 37, 1981.

548. **Berndt, H. and Messerschmidt, J.,** *o,o*-Diethyldithiophosphate for trace enrichment as activated carbon. I. Analysis of high purity gallium and aluminum determination of element traces by flame AAS (injection method and loop AAS), *Fresenius' Z. Anal. Chem.,* 308, 104, 1981.

549. **Julshamn, K.,** Studies on major and minor trace elements in molluscs in Western Norway. VII. The contents of 12 elements including copper, zinc, cadmium and lead in common mussel (*Mytilus edulis*) and brown seaweed (*Ascophyllum nodosum*) relative to the distance from the industrial sites in Sorfjorden, inner Hardangerfjord, *Fiskeridir. Skr. Ser. Ernaer.,* 1, 267, 1981.

550. **Farmer, J. G. and Gibson, M. J.,** Direct determination of cadmium, chromium, copper and lead in siliceous standard reference materials from a fluoboric acid matrix by graphite furnace atomic absorption spectrometry, *At. Spectrosc.,* 2, 176, 1981.

551. **Klein, A. A.,** Analysis of low-alloy steel using a sequential atomic absorption spectrophotometer equipped with an autosampler, *ASTM STP,* 747, 29, 1981.

552. **Chang, J. G. and Graff, R. L.,** Spectrochemical determination of impurities in uranium following solvent extraction using tributyl phosphate in methyl isobutyl ketone, *ASTM STP,* 747, 106, 1981.

553. **Lin, I. S., Yang, Y. F., and Hsu, Y. E.,** Determination of trace copper in human tissues and its preliminary clinical application, *Shanghai I Hseuh,* 4, 38, 1981.

554. **Mitchell, G. E.,** Trace metal levels in Queensland dairy products, *Aust. J. Dairy Technol.,* 36, 70, 1981.

555. **Takahashi, K., Minami, S., and Ohyagai, Y.,** Determination of copper and tin in antifouling coatings for ship bottom by atomic absorption spectrophotometry, *Shikizai Kyokaishi,* 54, 606, 1981 (in Japanese).

556. **Britske, M. E., Samakhvalova, L. G., and Slabodenyuk, I. V.,** Atomic absorption determination of copper in solutions obtained in the phase analysis of ores and their treatment products, *Nauch. Tr. NII Tsvet. Met.,* 48, 56, 1981 (in Russian); from *Ref. Zh. Metall.,* Abstr. No. 10k69, 1981.

557. **Rawat, N. S., Sahoo, B., and Sinha, J. K.,** Trace metals in respirable coal dust. An atomic absorption spectrophotometric study, *Chem. Ind. (London),* 13, 470, 1981.

558. **Ranchet, J., Fenissier, F., Lamathe, J., and Voinovitch, I.,** Interlaboratory comparison of the determination of cadmium, chromium, copper and lead by flameless atomic absorption spectrometry, *Bull. Liaison Lab. Ponts Chaussees,* 114, 81, 1981 (in French).

559. **Magnusson, B. and Westerlund, S.,** Solvent extraction procedures combined with back-extraction for trace metal determinations by atomic absorption spectrometry, *Anal. Chim. Acta,* 131, 63, 1981.

560. **Jaros, J. and Radil, J.,** Use of Dithiocarbamate for Determining Selected Zirconium Impurities by Atomic Absorption Spectrometry, Report UJP-496, 1980 (in Czech); available from Ustav. Jad. Paliv. Prague-Zbraslav. Czechoslovakia; INIS Atomindex 12(8), Abstr. No. 594651, 1981.

561. **Wesenberg, G. B. R., Fosse, G., and Rasmussen, P.,** The effect of graded doses of cadmium on lead, zinc and copper content of target and indicator organs in rats, *Int. J. Environ. Stud.,* 17, 191, 1981.

562. **Mueller, W. and Iffland, R.,** Studies on metals in meningiomas by atomic absorption spectrometry, *Acta Neuropathol.,* 55, 53, 1981.

563. **Haluska, M., Smrhova, A., and Kousal, M.,** Study of the mechanical state of a device by means of lubricating oils, *Ropa Uhlie,* 23, 549, 1981 (in Czech).

564. Copper in Potable Waters by Atomic Absorption Spectrophotometry 198D, U.K. Department of the Environment, Methods Exam. Waters Assoc. Mater., London, 1981.

565. **Kirleis, A. W., Sommers, L. E., and Nelson, D. W.,** Heavy metal content of groats and hulls of oats grown on soil treated with sewage sludge, *Cereal Chem.,* 58, 530, 1981.

566. **Adelman, H., Jenniss, S. W., and Katz, S. A.,** Interlaboratory analysis of sewage sludge, *Am. Lab.,* 31, 1981.

567. **Blakemore, W. M. and Billedeau, S. M.,** Analysis of laboratory animal feed for toxic and essential elements by atomic absorption and inductively coupled argon plasma emission spectrometry, *J. Assoc. Off. Anal. Chem.,* 64, 1284, 1981.

568. **Bank, H. L., Robson, J., Bigelow, J. B., Morrison, J., Spell, L. H., and Kantor, R.,** Preparation of fingernails for trace element analysis, *Clin. Chim. Acta,* 116, 179, 1981.

569. **Mattsson, P., Albanus, L., and Frank, A.,** Cadmium and some other elements in liver and kidney from moose *(Alces alces). Var Foeda,* 33, 335, 1981 (in Swedish).

570. **Brovko, I. A. and Tursunov, A. T.,** Extraction-atomic absorption determination of cadmium, cobalt, copper, manganese, nickel and zinc in natural waters using diphenyl carbazole as the extraction reagent, *Uzb. Khim. Zh.,* 3, 18, 1981.

571. **Ejaz, M., Dil, W., Akhtar, A., and Chaudhuri, S. A.,** Extraction and preconcentration of copper from water, soils, lubricating oils and plant materials and its subsequent determination by atomic absorption spectrophotometry, *Talanta,* 28, 441, 1981.

572. **Halls, D. J., Fell, G. S., and Dunbar, P. M.,** Determination of copper in urine by graphite furnace atomic absorption spectrometry, *Clin. Chim. Acta,* 114, 21, 1981.

573. **Isozaki, A., Soeda, N., and Utsumi, S.,** Sensitive atomic absorption spectrometric method for copper employing the direct introduction of chelating resin into a carbon tube atomizer, *Bull. Chem. Soc. Jpn.,* 54, 1364, 1981.

574. **Iwata, Y., Matsumoto, K., Haraguchi, H., Fuwa, K., and Okamoto, K.,** Proposed certified reference material for pond sediment, *Anal. Chem.,* 53, 1136, 1981.

575. **Jastrow, J. D., Zimmerman, C. A., Dvorak, A. J., and Hinchman, R. R.,** Plant growth and trace element uptake on acidic coal refuse amended with lime or fly ash, *J. Environ. Qual.,* 10, 154, 1981.

576. **Katz, S. A., Jenniss, S. W., Mount, T., Tout, R. E., and Chatt, A.,** Comparison of sample preparation methods for the determination of metals in sewage sludges by atomic absorption spectrometry, *Int. J. Environ. Anal. Chem.,* 9, 209, 1981.

577. **Kozma, M., Szerdahelyi, P., and Kasa, P.,** Histochemical detection of zinc and copper in various neurons of the central nervous system, *Acta Histochem.,* 69, 12, 1981.

578. **Krivan, V., Geiger, H., and Franz, H. E.,** Determination of iron, cobalt, copper, zinc, selenium, rubidium and cesium, in NBS bovine liver, blood plasma and erythrocytes by INAA and AAS, *Z. Anal. Chem.,* 305, 399, 1981.

579. **Levi, S., Fortin, R. C., and Purdy, W. C.,** Electrothermal atomic absorption spectrometric techniques for the determination of zinc and copper in microliter and submicroliter volumes of aqueous and serum matrixes, *Anal. Chim. Acta,* 127, 103, 1981.

580. **Makino, T. and Takahara, K.,** Direct determination of plasma copper and zinc in infants by atomic absorption with discrete nebulization, *Clin. Chem. (Winston-Salem, N.C.),* 27, 1445, 1981.

581. **Sasaki, Y. and Kawae, M.,** Atomic absorption spectrophotometric determination of iron and copper in water by extraction with aluminum cupferrate-methyl salicylate, *Bunseki Kagaku,* 30, 577, 1981.

582. **Weinstock, N. and Ihlemann, M.,** Automated determination of copper in undiluted serum by atomic absorption spectroscopy, *Clin. Chem. (Winston-Salem, N.C.),* 27, 1438, 1981.

583. **Wolff, E. W., Landy, M. P., and Peel, D. A.,** Preconcentration of cadmium, copper, lead and zinc in water at the 10^{-12} g/g level by absorption onto tungsten followed by flameless atomic absorption spectrometry, *Anal. Chem.,* 53, 1030, 1981.

584. **Young, R. S.,** Analysis of nickel refinery slimes and residues, *Talanta,* 28, 25, 1981.

585. **Burba, P. and Schaefer, W.,** Atomic absorption spectrometric determination (flame-AAS) of heavy metals in bacterial leach liquors (Jarosite) after their analytical separation on cellulose, *Erzmetall,* 34, 582, 1981.

586. **Miyagawa, H.,** Evaluation (chemical evaluation) of the performance of eyeglass frame materials. I. Analytical methods for the materials, *Nenpo-Fukui-Ken Kogyo Shikenjo,* 1981.

587. **Clegg, M. S., Keen, C. L., Leonnerdal, B., and Hurley, L. S.,** Influence of ashing techniques on the analysis of trace elements in biological samples. II. Dry ashing, *Biol. Trace Elem. Res.*, 3, 237, 1981.

588. **Dyakova, N. P., Kulachenko, S. P., Kulachenko, V. P., and Schastlivenko, V. A.,** Determination of trace elements in hair, *Veterinariya (Moscow)*, 10, 62, 1981.

589. **Kayakirilmaz, K. and Kosal, O.,** Determination of copper, iron and zinc in breast milk by atomic absorption spectrophotometry and effect of lactation stages and cultural, social and economical conditions of mother on the amount of these minerals, *Doga Ser. C.*, 5, 151, 1981 (in Turkish).

590. **Jin, K., Matsuda, K., and Chiba, Y.,** Concentrations of some metals in whole blood and plasma of normal adult subjects, *Hokkaidoritsu Eisei Kenkyushoho*, 31, 16, 1981.

591. **Okamoto, A., Ohmori, M., and Ishibashi, T.,** Analysis of the trace elements in natural food dyes, *Annu. Rep. Osaka City Inst. Public Health Environ. Sci.*, 43, 98, 1981.

592. **Karring, M., Pohjanvirta, R., Rahko, T., and Korpela, H.,** The influence of dietary molybdenum and copper supplementation on the contents of serum uric acid and some trace elements in cocks, *Acta Vet. Scand.*, 22, 289, 1981.

593. **Takahashi, Y., Shiozawa, Y., Kawai, T., Shimizu, K., Kato, H., and Hasegawa, J.,** Elements dissolved from three types of amalgams in synthetic saliva, *Aichi Gakuin Daigaku Shigakkaishi*, 19, 107, 1981.

594. **Tateuchi, S.,** Determination of copper, zinc, iron and manganese in commercial premix by atomic absorption photometry, *Shiryo Kenkyu Hokoku*, 7, 183, 1981.

595. **Peters, H. J. and Koehler, H.,** Direct determination by flame atomic absorption spectrometry of the trace elements manganese, copper and zinc in biological tissues, *Dtsch. Gesundheitswes.*, 36, 1919, 1981.

596. **Fiala, K. and Studeny, M.,** Use of a dry method for mineralization of plant material to improve chemical analysis for determination of major and trace elements from a common weighed sample, *Ved. Pr. Vysk. Ustavu Podoznalectva Vyz. Rastl. Bratislave*, 10, 167, 1981.

597. **Rao, N. C. and Rao, B. S. N.,** Trace element content of Indian foods and the dietaries, *Ind. J. Med. Res.*, 73, 904, 1981.

598. **West, T. S.,** The atomic spectroscopy of biosignificant trace element in soils in relation to plant and animal nutrition, *Bunseki Kagaku*, 30, S103, 1981.

599. **Gregorczyk, S., Matysik, S., and Wycislik, A.,** Chemical analysis of cobalt- and iron-based alloys by atomic absorption, *Wiad. Hutn.*, 37, 248, 1981.

600. **Suzuki, M., Ohta, K., Yamakita, T., and Katsuno, T.,** Electrothermal atomization with a metal microtube in atomic absorption spectrometry, *Spectrochim. Acta*, 36B, 679, 1981.

601. **Crespo, S., Soriano, E., Sampera, C., and Balasch, J.,** Zinc and copper distribution in excretory organs of the dogfish *Scyliorhinus canicula* and chloride cell response following treatment with zinc sulfate, *Mar. Biol.*, 65, 117, 1981.

602. **Arafat, N. M. and Glooschenko, W. A.,** Method for the simultaneous determination of arsenic, aluminum, iron, zinc, chromium and copper in plant tissue without the use of perchloric acid, *Analyst (London)*, 106, 1174, 1981.

603. **Sekiya, T., Tanimura, H., and Hikas, Y.,** Simplified determination of copper, zinc and manganese in plasma and bile by flameless atomic absorption spectrometry, *Arch. Jpn. Chir.*, 50, 729, 1981.

604. **Satsmadjis, J. and Voutsinou-Taliadouri, F.,** Determination of trace metals at concentrations above the linear range by electrothermal atomic absorption spectrometry, *Anal. Chim. Acta*, 131, 83, 1981.

605. **Elson, C. M., Bem, E. M., and Ackman, R. G.,** Determination of heavy metals in a menhaden oil after refining and hydrogenation using several analytical methods, *J. Am. Oil Chem. Soc.*, 58, 1024, 1981.

606. **Pilipenko, A. T. and Samchuk, A. I.,** Extraction-atomic absorption determination of trace elements in natural waters, *Khim. Tekhnol. Vody*, 3, 343, 1981.

607. **Kovarskii, N., Kovekovdova, L. T., Pryazhevskaya, I. S., Belen'kii, V. S., Shapovlov, E. N., and Popkova, S. M.,** Preconcentration of trace elements from sea water by electrodeposited magnesium hydroxide, *Zh. Anal. Khim.*, 36, 2264, 1981.

608. **Ohta, K., Smith, B., and Winefordner, J. D.,** High temperature-gas chromatography with an atomic absorption spectrometric detector, *Anal. Chem.*, 54, 320, 1981.

609. **Danielson, L. G., Magnusson, B., and Zhang, K.,** Matrix interference in the determination of trace metals by graphite furnace AAS after Chelax-100 preconcentration, *At. Spectrosc.*, 2, 39, 1981.

610. **Borg, H.,** Trace Metals in Natural Waters. An Analytical Intercomparison, Rapp.-Naturvardsverket (Swed.) SNV PM 1463, Forskningssekretariatet, Statens Naturvaardsverk, Solna, Sweden, 1981.

611. **Pape, H.,** Development of a geochemical mapping method for prospecting deposits, environmental research and regional planning on the basis of multielement studies of plant ashes, *Monogr. Ser. Miner. Deposits*, 19, 1, 1981.

612. **Masuda, N.,** Determination of total dissolved copper in seawater by carbon furnace atomic absorption spectrometry, *Hokkaido Daigaku Suisangakubu, Kenkyu Iho*, 32, 425, 1981.

613. **Greenwood, R.,** Distribution of heavy metals (copper, lead and zinc) in rural areas of Rio De Janiero (Brazil), *Rev. Bras. Geocience*, 11, 98, 1981.

614. **Bagdach, S.,** Method for measuring the deposition rate during chemical copper planting, *Tagungscand-Kammer Tech. Suhl.,* 64, 79, 1981 (in German).

615. **Senesi, N. and Polemio, M.,** Trace elements addition to soil by application of NPK fertilizers, *Fert. Res.,* 2, 289, 1981.

616. **Costantini, S., Macri, A., and Vernillo, I.,** Atomic absorption spectrophotometric analysis of mineral elements in milk-base feeds for zootechnical use, *Riv. Soc. Ital. Sci. Aliment.,* 10, 231, 1981 (in Italian).

617. **Ogawa, T. and Goto, T.,** Atomic absorption spectrophotometry of copper after thiosulfato complex solvent extraction with capriquat, *Nihon Daigaku Kogakubu Kiyo Bunrui A,* 22, 227, 1981.

618. **Fosse, G., Justesen, N. P. B., and Wesenberg, G. B. R.,** Microstructure and chemical composition of fossil mammalian teeth, *Calif. Tissue Int.,* 33, 521, 1981.

619. **Yuffa, A. Y., Ryazanova, L. M., Gvozdeva, G. M., and Turova, M. Y.,** Heterogenised transition metal halides in the synthesis of highly dispersed metallic and metal-complex catalysis. I. Cyclohexene conversion catalysts made of fixed transition metal chlorides, *Kinet. Katal.,* 22, 1465, 1981.

620. **Florio, J. V., Matysik, K. J., and Ramos, Z. Q.,** Atomic absorption analysis and auger depth profiles of heat-treated polycrystalline copper sulfide (Cu_xS)/cadmium sulfide solar cells, *Conf. Rec. IEEE Photovoltaic Spec. Conf.,* 15th, IEEE, Piscataway, N.J., 1981, 793.

621. **Kawahara, H., Yamada, T., Nakamura, M., Tomoda, T., Kobayashi, H., Saijo, A., Kawata, Y., and Hikari, S.,** Solubility of metal components into tissue culture medium from dental amalgams, *Shika Rikogaku Zasshi,* 22, 285, 1981.

622. **Ohtake, M., Chiba, R., Mochizuki, K., and Tada, K.,** Zinc and copper concentrations in human milk and in serum exclusively breast-fed infants during the first 3 months of life, *Tohoku J. Exp. Med.,* 135, 335, 1981.

623. **Rohbock, E.,** The effect of airborne heavy metals on automobile passengers in Germany, *Environ. Int.,* 5, 133, 1981.

624. **Xia, L.,** Flame atomic absorption spectrophotometric determination of copper, cadmium, lead and zinc in tea leaves, *Fenxi, Huaxue,* 9, 498, 1981 (in Chinese).

625. **Jones, E. A. and Dixon, K.,** The separation of trace elements in manganese dioxide, *Natl. Inst. Metall. Repub. So. Afr. Rep.,* No. 2131, 1981.

626. **Henrion, G. and Gelbrecht, J.,** Determination of trace metals in molybdenum (VI) compounds using extraction/flame. AAS, *Z. Chem.,* 21, 453, 1981.

627. **Jarabak, R.,** 3-Mercaptopyruvate sulfurtransferase, *Methods Enzymol.,* 77, 291, 1981.

628. **Borg, H., Edin, A., Holm, K., and Skoeld, E.,** Determination of metals in fish livers by flameless atomic absorption spectroscopy, *Water Res.,* 15, 1291, 1981.

629. **Garcia-Baez, M., Gonzalez-Espinosa, C., Moya, M., and Domenech, E.,** Study of copper and ceruloplasmin at parturition and in the newborn, *An. Esp. Pediatr.,* 15, 544, 1981 (in Spanish).

630. **Wei, J. and Geng, T.,** Determination of available cadmium, lead, copper and zinc in calcereous soils, *Fenxi Huaxue,* 9, 565, 1981.

631. **Heider, P. J.,** Uranium Occurrence in (the Mining District of) California near Bucaramanga (Colombia), Rep. INIS-mf-6712, 1980 (in German); available from INIS Atomindex 12, Abstr. No. 641045, 1981.

632. **Koroschetz, F., Hoke, E., and Grasserbauer, M.,** Diffusion processes in electrodeposited lead-tin-copper bearing overlays, *Mikrochim. Acta Suppl.,* 9, 139, 1981.

633. **Fugas, M.,** Metals in airborne atmospheric particles, *Zast. Atmos.,* 9, 13, 1981 (in Serbo-Croatian).

634. **Marleer-Geets, O., Heck, J. P., Barideau, L., and Rocher, M.,** Method for Determination of Heavy Metals in sewage Sludge, their Distribution according to their Origin and their Concentration Variation with Time, Comm. Eur. Communities, (Rep.) EUR 7076, Charact., Treat. Use Sewage Sludge, 289—90, Serv. Sci. Sol. Fac. Sci., Agrono. Etat. Gembloux, Belgium, 1981 (in French).

635. **Hirano, K., Iida, K., Shimada, T., Iguchi, K., and Nagasaki, Y.,** Study on practical application of monitoring method of total amounts of heavy metals in waste water on adsorption of chelating resins and ion-exchange resins, *Zenkoku Kogaiken Kaishi,* 6, 9, 1981.

636. **Inhat, M.,** Analytical approach to the determination of copper, zinc, cadmium and lead in natural fresh waters, *Int. J. Environ. Anal. Chem.,* 10, 217, 1981.

637. **Foerster, M. and Lieser, K. H.,** Determination of traces of heavy metals in inorganic salts and organic solvents by energy-dispersive x-ray fluorescence analysis or flameless atomic absorption spectrometry, after enrichment on a cellulose exchanger, *Fresenius' Z. Anal. Chem.,* 309, 355, 1981.

638. **Monteil, A. and Welte, B.,** Comparison of the different methods of attack during the determination of micropollutants in the sediments, *Congr. Mediterr. Ing. Quim.* (Actas), 2nd, C20-1/C20-13, F.O.I.M., Barcelona, Spain, 1981 (in French).

639. **Hayashi, T., Nishizawa, M., and Yamagishi, T.,** Chemical studies on Paeoniae radix. IV. Relation between gallotannin contents and discoloration of Paeoniae radix, *Hokkaidoritsu Eisei Kenkyusho,* 31, 23, 1981.

640. **Hwang, H. L., Ho, J. S., Ou, H. J., Lee, Y. K., Sun, C. Y., Chen, C. J., and Loferski, J. J.,** Copper (I) sulfide/cadmium sulfide solar cells prepared by organometallic chemical vapor deposition: preliminary stage, *Conf. Rec. IEEE Photovoltaic Spec. Conf.*, 15th, IEEE, Piscataway, N.J., 1035, 1981.

641. **Balaes, G. E. E. and Robert, R. V. D.,** Determination by atomic absorption spectrophotometry of impurities in manganese dioxide, *Natl. Inst. Metall. Repub. S. Afri. Rep.*, No. 2094, 1981.

642. **Boyer, K. W., Jones, J. W., Linscott, D., Wright, S. K., Stroube, W., and Cunningham, W.,** Trace element levels in tissues from cattle fed a sewage sludge-amended diet, *J. Toxicol. Environ. Health*, 8, 281, 1981.

643. **Brzozowska, B. and Zawadzka, T.,** Determination of lead cadmium, zinc, and copper in vegetable products by atomic absorption spectrophotometry, *Rocz. Panstw. Zakl. Hig.*, 32, 9, 1981.

644. **Mattera, V. D., Jr., Arbige, V. A., Jr., Tomellini, S. A., Erbe, D. A., Doxtader, M. M., and Force, R. K.,** Evaluation of wash solutions as a preliminary step for copper and zinc determinations in hair, *Anal. Chim. Acta*, 124, 409, 1981.

645. **Mitchell, G. E.,** Trace metal levels in Queensland dairy products, *Aust. J. Dairy Technol.*, 36, 70, 1981.

646. **Mueller, W. and Iffland, R.,** Studies on metals in meningiomas by atomic absorption spectrometry, *Acta Neuropathol.*, 55, 53, 1981.

647. **Salmela, S., Vouri, E., and Kilpio, J. O.,** Effect of washing procedures on trace element content of human hair, *Anal. Chim. Acta*, 125, 131, 1981.

648. **Scheuhammer, A. M. and Cherian, M. G.,** Influence of manganese on the distribution of essential trace elements. I. Regional distribution of manganese, sodium, potassium, magnesium, zinc, iron, and copper in rat brain after chronic manganese exposure, *Toxicol. Appl. Pharmacol.*, 61, 227, 1981.

649. **Takahashi, K., Minami, S., and Ohyagi, Y.,** Determination of copper and tin in antifouling coatings for shipbottom by atomic absorption spectrophotometry, *Shikizai Kyokaishi*, 54, 606, 1981.

650. **Varma, A.,** Analytical techniques for corrosion studies on noble metals, *Talanta*, 28, 701, 1981.

651. **Wesenberg, G. B. R., Fosse, G., and Rasmussen, P.,** Effect of graded doses of cadmium on lead, zinc and copper content of target and indicator organs in rats, *Int. J. Environ. Stud.*, 17, 191, 1981.

652. **Bourcier, D. R., Sharma, R. P., and Brinkerhoff, C. R.,** Cadmium-copper interaction: tissue accumulation and subcellular distribution of cadmium in mice after simultaneous administration of cadmium and copper, *Trace Subst. Environ. Health*, 15, 190, 1981.

653. **Li, Q.,** Atomic absorption spectrometric determination of microamounts of copper, zinc, nickel, cobalt, cadmium and lead, *Fenxi Huaxue*, 9, 718, 1981 (in Chinese).

654. **Taliadouri-Voutsinou, F.,** Trace metals in marine organisms from the Saronikos Gulf (Greece), *J. Etud. Pollut. Mar. Mediterr.*, 5, 275, 1981.

655. **Uysal, H.,** Levels of trace elements in some food-chain organisms from the Aegean coasts, *J. Etud. Pollut. Mar. Mediterra.*, 5, 503, 1981.

656. **Djujic, I., Djordjevic, V., and Rada, N.,** Lead, cadmium, copper, arsenic and mercury in some additives (for meat), *Technol. Mesa*, 22, 355, 1981.

657. **Brovko, I. A.,** Diphenylcarbazone as an extractant reagent in atomic absorption, *Deposited Doc. VINITI*, 2842-81, VINITI, Moscow, U.S.S.R., 1981.

658. **Baranowska-Dutkiewicz, B., Choinska, J., and Rozanska, R.,** Choice of conditions for collecting urine samples for evaluating the occupational exposure of workers of an electrosteel plant to nickel, chromium, manganese and copper, *Med. Pr.*, 32, 451, 1981.

659. **Sutton, D. C., Rosa, W. C., and Legotte, P. A.,** Analytical measurements of selected metals in samples from a human metabolic study, *Trace Subst. Environ. Health*, 15, 270, 1981.

660. **Roberdo, F., Carrondo, M. J. T., Ganho, R. M. B., and Oliveira, J. F. S.,** Use of a Rapid Flameless Atomic Absorption Method for the Determination of the Metallic Content of Sediments in the Tejo Estuary, Portugal, *Heavy Met. Environ.*, Int. Conf., 3rd CEP Consultants, Ltd., Edinburgh, U.K., 1981, 587.

661. **Popov, G. K. and Shamgunov, A. N.,** Determination of iron, copper and zinc in a small amount of a biological material by atomic absorption spectrophotometric analysis, *Deposited Doc. VINITI*, 4618-80, VINITI, Moscow, U.S.S.R., 1980 (in Russian); CA 96,65145c, 1982.

662. **Kasperek, K., Iyengar, G. V., Feinendgen, L. E., Hashish, S., and Mahfouz, M.,** Multielement analysis of fingernail, scalp hair and water samples from Egypt (a preliminary study), *Sci. Total Environ.*, 22, 149, 1982.

663. **Sekiya, T., Tanimura, H., and Hikasa, Y.,** Study on the dosage of trace elements in total parenteral nutrition, *Jutsugo Taisha Kenkyu Kaishi*, 14, 114, 1980 (in Japanese); CA 96,18923v, 1982.

664. **Ohta, K., Smith, B. W., and Winefordner, J. D.,** High temperature gas chromatography with an atomic absorption spectrometric detector, *Anal. Chem.*, 54, 320, 1982.

665. **Suwirma, S., Surtipanti, S., and Thamzil, L.,** Distribution of heavy metals mercury, lead, cadmium, chromium, copper and zinc in fish, *Majalah BATAN*, 13, 9, 1980 (in Indonesian); CA 96,15797w, 1982.

666. **Sturgeon, R. E., Desauliners, J. A. H., Berman, S. S., and Russell, D. S.,** Determination of trace metals in estuarine sediments by graphite furnace-atomic absorption spectrometry, *Anal. Chim. Acta*, 134, 283, 1982.

667. **Olney, C. E., Schauer, P. S., McLean, S., Lu, Y., and Simpson, K. L.,** International study on Artemia. VIII. Comparison of the chlorinated hydrocarbons and heavy metals in five different strains of newly hatched artemia and laboratory-reared marine fish, *Brine Shrimp Artemia, Proc. Int. Symp.,* 3, 343, Universa Press, Wetteren, Belgium, 1980; CA 96,81005n, 1982.

668. **Rovid, K., Graf Harsanyi, E., Polos, L., Fodor, P., and Pungor, E.,** Determinatin of metallic 14 components in lubricating oils by absorption spectrometry, *Magy. Kem. Foly.,* 88, 39, 1982 (in Hungarian).

669. **Jacyszyn, K., Walas, J., Malinowski, A., Latkowski, T., and Cwynar, L.,** Concentration of heavy metals in pregnant women, *Zentralbl. Gynaekol.,* 104, 117, 1982.

670. **Briggs, R. W. and Armitage, I. M.,** Evidence of site-selective metal binding in calf liver metallothionein, *J. Biol. Chem.,* 257, 1259, 1982.

671. **Kumina, D. M. and Karyakin, A. V.,** Effect of the nature of organic compounds on the atomic absorption of certain elements, *Zh. Prikl. Spektrosk.,* 36, 143, 1982.

672. **Velghe, N., Campe, A., and Claeys, A.,** Determination of copper in undiluted serum and whole blood by atomic absorption spectrophotometry with graphite furnace, *At. Spectrosc.,* 3, 48, 1982.

673. **Saura-Calixto, F. and Canellas, J.,** Mineral composition of almond varieties (*prunus amygdalus*), *Z. Lebesm. Unters. Forsch.,* 174, 129, 1982.

674. **Khalighie, J., Ure, A. M., and West, T. S.,** Atom-trapping absorption spectrometry with water-cooled metal collector tubes, *Anal. Chim. Acta,* 134, 271, 1982.

675. **DeCarlo, E. H., Zeitlin, H., and Fernando, Q.,** Separation of copper, nickel and manganese from deep-sea ferromanganese nodules by absorbing colloid flotation, *Anal. Chem.,* 54, 898, 1982.

676. **Kauffman, R. E., Saba, C. S., Rhine, W. E., and Eisentraut, K. J.,** Quantitative multielement determination of metallic wear species in lubricating oils and hydraulic fluids, *Anal. Chem.,* 54, 975, 1982.

677. **Sakla, A. B., Bradran, A. H., and Shalaby, A. M.,** Determination of elements by atomic absorption spectrometry after destruction of blood in the oxygen flask, *Mikrochim. Acta,* 1, 483, 1982.

678. **Szerdahelyi, P., Kozma, M., and Ferke, A.,** Zinc deficiency-induced trace element concentration and localization changes in the central nervous system of albino rat during postnatal development. II. Atomic absorption spectrophotometric examinations, *Acta Histochem.,* 70, 173, 1982.

679. **Rawat, N. S., Sinha, J. K., and Sahoo, B.,** Atomic absorption spectrophotomeric and X-ray studies of respirable dusts in Indian coal mines, *Arch. Environ. Health,* 37, 32, 1982.

680. **Ikebe, K. and Tanaka, R.,** Determination of heavy metals in blood by atomic absorption spectrometry. II. Osaka-Furitsu Koshu Eisei Kenkyusho Kenkyu Hokoku, *Shokuhin Eisei Hen,* 11, 43, 1980; CA 96, 137220y, 1982.

681. **Ranchet, J., Menissier, F., Lamathe, J., and Voinovitch, I.,** Interlaboratory comparison: the determination of cadmium, chromium, copper and lead in standard solutions by flameless atomic absorption spectrometry, *Analusis,* 10, 71, 1982.

682. **Harada, Y.,** Determination of trace impurities in high-purity aluminum by flame atomic absorption spectrometry after coprecipitation with nickel hydroxide, *Bunseki Kagaku,* 31, 130, 1982.

683. **Lappalainen, R., Knuuttila, M., Lammi, S., Alhava, E. M., and Olkkonen, H.,** Zinc and copper content in human cancellous bone, *Acta Orthop. Scand.,* 53, 51, 1982.

684. **Bangia, T. R., Kartha, K. N. K., Varghese, M., Dhawale, B. A., and Joshi, B. D.,** Chemical separation and electrothermal atomic absorption spectrophotometric determination of cadmium, cobalt, copper and nickel in high-purity uranium, *Fresenius' Z. Anal. Chem.,* 310, 410, 1982.

685. **Fischer, W. R. and Fechter, H.,** Analytical determination and fractionation of copper, zinc, lead, cadmium, nickel and cobalt in soils and underwater soils, *Z. Pflanzenernaehr. Bodenkd.,* 145, 151, 1982.

686. **Katskov, D. A. and Grinshtein, I. L.,** Formation of copper, silver and calcium acetylides in graphite furnaces for atomic-absorption analysis, *Zh. Prikl. Spektrosk.,* 36, 181, 1982.

687. **Kennedy, J. T. and Svehla, G.,** Determination of copper traces in grass using a catalytic method, *Fresenius' Z. Anal. Chem.,* 311, 218, 1982.

688. **Velghe, N., Campe, A., and Claeys, A.,** Determination of copper in undiluted serum and whole blood by atomic absorption spectrophotometry with graphite furnace, *At. Spectrosc.,* 3, 48, 1982.

689. **Jelinke, P., Illek, J., and Jagos, P.,** Levels of zinc, manganese and copper in blood plasma, liver, hair, gonads and accessory sexual glands of coypus males, *Zivocisna Vyroba,* 27, 223, 1982.

690. **Sheedlo, H. J. and Beck, M. L.,** Altered copper metabolism in heterozygous tortoise shell (Mo^{to}/ +) female mice. *Mus musculus, Comp. Biochem. Physiol. A,* 71a, 341, 1982.

691. **Abdulla, M., Norden, A., Schersten, B., Svensson, S., Thulin, T., and Geckerman, P. A.,** The intake and urinary excretion of electrolytes and trace elements, in *Proc. 4th Int. Symp. Trace Element Metabolism of Man and Animals,* Gawthorne, J. M., Howell, J. M., and White, C. L., Eds., Springer Verlag, Berlin, 1982, 81.

692. **Gonzalez, O., Baez, B. C., Maria, R. A., Florencia, M. L., and Avelina, Z. Q.,** Method for the determination of copper in plant fibers by flameless atomic absorption spectrophotometry, *Bol. Soc. Chil. Quim.,* 27, 259, 1982.

693. **Kojima, I., Iida, C., and Yamasaki, K.,** Determination of five elements in a milligram amount of lead glasses by flame atomic absorption and emission spectrometry using a one-drop method, *Bunseki Kagaku,* 31, E167, 1982.

694. **Sasaki, Y., Kawae, M., and Koizumi, S.,** The analysis of dissolved metal ions (iron and copper) in solution and adsorbed metal ions in suspended particles in the river water of Tannan region, Kenkyu Kiyo-Fukui Kogyo Koto Senmon Gakko, *Shizen Kagaku Kagaku,* 15, 71, 1982.

695. **Baldini, M., Grossi, M., Micco, C., and Stacchini, A.,** Presence of metal in cereal. II. Contamination of 1978 Italian rice, *Riv. Soc. Ital. Sci. Aliment.,* 11, 23, 1982.

696. **Helgeland, K., Haider, T., and Jonsen, J.,** Copper and zinc in human serum in Norway. Relationship to geography, sex and age, *Scand. J. Clin. Lab. Invest.,* 42, 35, 1982.

697. **Ough, C. S., Crowell, E. A., and Benz, J.,** Metal content of California wines, *J. Food Sci.,* 47, 825, 1982.

698. **Yamagata, N.,** Interlaboratory comparison study on the reliability of environmental analyses. Soil and sediment 1978-80, *Bunseki Kagaku,* 31, T1, 1982 (in Japanese).

699. **Johnson, A. C., Wibetoe, G., Langmyhr, F. J., and Aaseth, J.,** Atomic absorption spectrometric determination of the total content and distribution of copper and gold in synovial fluid from patients with rheumatoid arthritis, *Anal. Chim. Acta,* 135, 243, 1982.

700. **Tarui, T., Nakatani, S., and Tokairin, H.,** Determination of copper in petroleum by graphite furnace atomic absorption spectrometry, *Bunseki Kagaku,* 31, T29, 1982.

701. **Stupar, J. and Ajlec, R.,** Study of the use of soil suspensions in the determination of iron, manganese, magnesium and copper in soils by flame atomic absorption spectrometry, *Analyst (London),* 107, 144, 1982.

702. **Klein, A. A.,** Analysis of low-alloy steel using a sequential atomic absorption spectrophotometer equipped with an autosampler, *At. Spectrosc.,* 3, 133, 1982.

703. **Eaton, A., Oelker, G., and Leong, L.,** A comparison of AAS and ICP for analysis of natural waters, *At. Spectrosc.,* 3, 152, 1982.

704. **Alder, J. F. and Batoreu, M. C.,** Ion exchange resin beads as solid standards for electrothermal atomic absorption spectrometric determination of metals in hair, *Anal. Chim. Acta,* 135, 229, 1982.

705. **Bag, S. O. and Freiser, H.,** Kinetics and mechanism of solvent extraction of copper with Chelex-100 in presence of nitrilotriacetic acid, *Anal. Chim. Acta,* 134, 333, 1982.

706. **Balaes, G. E., Dixon, K., Russell, G. M., and Wall, G. J.,** Analysis of activated carbon as used in the carbon-in-pulp process, for gold and eight other constituents, *S. Afr. J. Chem.,* 35, 4, 1982.

707. **Danielsson, L. G., Magnusson, B., and Zhang, K.,** Matrix interference in the determination of trace metals by graphite furnace AAS after Chelex-100 preconcentration, *At. Spectrosc.,* 3, 39, 1982.

708. **Fagioli, F., Landi, S., and Lucci, G.,** Determination of manganese and copper in small amounts of maize roots by graphite tube furnace atomic absorption spectroscopy with liquid and solid sampling techniques, *Ann. Chim.,* 72, 63, 1982.

709. **Gardiner, P. E., Roesick, Roesick, U., Braetter, P., and Kynast, G.,** Application of GEL filtration, immunonephelometry and electrothermal atomic absorption spectrometry to the study of the distribution of copper, iron, and zinc bound constituents in human amniotic fluid, *Clin. Chim. Acta,* 120, 103, 1982.

710. **Hernandez Mendez, J., Alonso Mateos, A., and Martin Mateos, E. J.,** Extraction of the ion-pair $[(C_4H_9)_4N^+]_2$ $[Cu (SCN)_4{}^{2-}]$ by MIBK: spectrophotometric determination of copper, *Anal. Lett.,* 15, 67, 1982.

711. **Jackwerth, E. and Salewski, S.,** Contribution to the multielement preconcentration from pure cadmium, *Z. Anal. Chem.,* 310, 108, 1982 (in German).

712. **Lendermann, B. and Hundeshagen, D.,** Use of multielement standards for calibration in water analysis by AAS *Z. Anal. Chem.,* 310, 415, 1982.

713. **Olsen, J. B., Wilkerson, C. L., Toste, A. P., and Hays, D. J.,** Isolation of metallic complexes in shale oil and shale oil retort waters, *NBS Spec. Publ.,* 618, 105, 1982.

714. **Salmon, S. G. and Holcombe, J. A.,** Alteration of metal release mechanisms in graphite furnace atomizers by chemisorbed oxygen, *Anal. Chem.,* 54, 630, 1982.

715. **Krivan, V. and Lang, M.,** Radiotracer studies on the direct determination of copper in biological matrices by flameless AAS, *Fresenius' Z. Anal. Chem.,* 312, 324, 1982.

716. **Ishii, M.,** A new approach to a clinical diagnosis of diseases by atomic absorption/high performance aqueous gel permeation chromatography, *Kyorin Igakkai Zasshi,* 13, 47, 1982.

717. **Hoenig, M. and Wollast, R.,** The possibilities and limitations of electrothermal atomization in atomic absorption spectrometry for the direct determination of trace metals in seawater, *Spectrochim. Acta,* 37b, 399, 1982.

718. **Moriyama, K., Watanabe, A., Sugiura, S., Arayashiki, H., and Mori, Y.,** Atomic absorption spectrophotometric determination of cadmium, lead, nickel and copper in sewage sludge, *Gesuido Kyokaishi,* 19, 68, 1982.

719. **Witkowski, S. A. and Frazier, J. G.,** Heavy metals in sea turtles, *Mar. Pollut. Bull.,* 13, 254, 1982.

720. **Egorova, K. A., Ermakova, T. I., Shkol'nikov, V. M., and Shirokova, G. B.,** Rapid micromethod for testing of hydraulic oils, *Khim. Tekhnol. Topl. Masel,* 7, 38, 1982 (in Russian).

721. **Adam, J.,** Determination of copper, zinc and lead in mine waters, *Hutn. Listy,* 37, 276, 1982.

722. **Prokof'ev, A. K., Oradovskii, S. G., and Georgievskii, V. V.,** Flameless atomic absorption method for the determination of copper, lead and cadmium in marine bottom sediments, *Tr. Gos. Okeanogr. Inst.,* 162, 51, 1981 (in Russian); from *Ref. Zh. Khim. Abstr.,* 11G258, 1982.

723. **Takada, K. and Hirokawa, K.,** Atomization and determination of traces of copper, manganese, silver and lead in microamounts of steel. Atomic absorption spectrometry using direct atomization of solid sample in a graphite-cup cuvette, *Fresenius' Z. Anal. Chem.,* 312, 109, 1982.

724. **Matsuzaki, A., Kondo, O., and Saito, N.,** Determination of trace copper in electrical insulating oil by atomic absorption spectrometry with a graphite furnace atomizer, *Sekiyu Gakkaishi,* 25, 255, 1982.

725. **Bogden, J. D., Zadzielski, E., Weiner, B., Oleska, J. M., and Aviv, A.,** Release of some trace metals from disposable coils during hemodialysis, *Am. J. Clin. Nutr.,* 36, 403, 1982.

726. **Zhou, Z. and Mao, X.,** A new chelate-forming resin bearing xylenol orange group and its application, *Gaodeng Xuexiao Huaxue Xuebo,* 3, 181, 1982 (in Chinese).

727. **Gregorio, P., Siracusano, C., and Toscano, G.,** Contents of iron, copper, zinc, magnesium and manganese in infant foods. I. Homogenized baby food, *Ig. Mod.,* 77, 348, 1982 (in Italian).

728. **Travkina, V. I., Mozherina, L. V., and Kalashnikova, N. G.,** Experience using the AAS-1 atomic absorption spectrometer, *Zavod. Lab.,* 48, 83, 1982.

729. **Kuzenko, S. V. and Kucher, V. N.,** Extraction-atomic absorption determination of nickel, copper and cobalt during phase analysis of rocks, *Ukr. Khim. Zh.,* 48, 510, 1982.

730. **Iyengar, G. V., Kasperek, K., Feinendegen, L. E., Wang, Y. X., and Weese, H.,** Determination of cobalt, copper, iron, mercury, manganese, antimony, selenium and zinc in milk samples, *Sci. Total Environ.,* 24, 267, 1982.

731. **Brzezinska, A.,** Some remarks on the determination of trace metals in marine samples by electrothermal atomization, *Eur. Spectrosc. News,* 40, 19, 1982 (in English).

732. **Silva, M. and Valcarcel, M.,** Liquid-liquid extraction combined with atomic absorption spectrometry for determination of copper in waters, foods and analytical reagents using 1,2-naphthoquinone thiosemicarbazone, *Analyst (London),* 107, 511, 1982.

733. **Ines, G., Balabanoff, K., Leonardo, V. S., Rita, M., and Luz, V. Q.,** Technique for the preparation of mixed human saliva samples for the determination of copper, zinc and manganese by flameless atomic absorption spectrometry, *Bol. Soc. Chil. Quim.,* 27, 340, 1982.

734. **Schmidt, L. H., Meissner, D., and Lenski, K. H.,** Results and perspectives for standardization of trace element analysis in biological material, *Zentralbl. Pharm., Pharmakother. Laboratoriumsdiagn.,* 121, 444, 1982.

735. **Jones, E. A.,** The Separation and Determination of Trace Elements in Chromic Oxide, Rep.-MINTEK M15, Anal. Chem. Div., Mintek, Randburg, South Africa, 1982.

736. **Ueno, G.,** Atomic absorption spectrophotometric determination of cadmium, lead and copper in sugars using iodide-methyl isobutyl keytone extraction, *Seito Gijutsu Kenkyu Kaishi,* 30, 11, 1982.

737. **Koval'skii, V. V., Letunova, S. V., and Alekseeva, S. A.,** Accumulation of nickel and other elements in soil microflora biomass in the South Urals subregion biosphere, *Superenelem.-Symp.: Nickel,* 3rd, Anke, M., Schneider, H., and Brueckner, C., Eds., Friedrich-Schiller University, Jena, East Germany, 1980, 163, (in Russian); CA 97,95913q, 1982.

738. **Fudagawa, N., Nakamura, S., and Kawase, A.,** Comparison of tungsten, molybdenum and tantalum ribbon atomizers in atomic absorption spectrometry, *Bunseki Kagaku,* 31, 324, 1982.

739. **Teramoto, K., Ninomiya, K., Kurono, T., Horiguchi, S., and Kim, D. K.,** Contents of copper, iron, manganese, zinc and lead in several foods from Korea and Japan, *Seikatsu Eisei,* 26, 221, 1982.

740. **Yasuda, H., Okuno, R., and Iuchi, I.,** Physical and chemical analysis in archaeology. X. Atomic absorption spectrophotimetric analysis of green glaze of tile excavated in Keong-ju, Korea, *Mukogawa Joshi Daigaku Keiyo Yakugaku Hen,* 29, 25, 1981; CA 97;161780x, 1982.

741. **Suzuki, M., Ohta, K., and Katsuno, T.,** Determinations of traces of lead and copper in foods by electrothermal atomic absorption spectrometry with metal atomizer, *Mikrochim. Acta,* 32, 225, 1982.

AUTHOR INDEX

Hattori, D. M., 195
Havel, J., 435
Hayashi, T., 467
Hayashi, Y., 118, 144, 383, 467
Haynie, G. D., 136
Hays, D. J., 713
Healy, P. J., 376
Heck, J. P., 634
Heckman, M., 108, 121
Heffman, G. L., 122
Heider, J., 631
Heinemann, G., 168
Helgeland, K., 501, 696
Heller, R. H., 50
Helsby, C. A., 201
Heneage, P., 43
Henkin, R. I., 138
Henrion, G., 542, 626
Hernandez-Mendez, J., 356, 710
Herrmann, R., 13
Higgins, T. N., 236
Hikari, S., 621, 663
Hikasa, Y., 570
Hinchman, R. R., 575
Hioki, T., 388
Hirai, A., 286
Hirano, K., 635
Hirano, S., 466
Hirayama, H., 274
Hiro, K., 286
Hirokawa, K., 723
Hisatomi, K., 124, 133
Ho, J. S., 640
Hodgkinson, A., 219
Hoehn, R., 195
Hoenig, M., 377, 717
Hoffman, M. R., 535
Hoffman, T., 542
Hoffmeister, W., 185
Hohnadel, D. C., 202
Hoke, E., 632
Holcombe, J. A., 714
Hollebone, B. R., 507
Holm, K., 628
Holtzman, N. A., 50
Horiguchi, S., 739
Hornick, R. B., 275
Hoschler, M. E., 194
Hoshino, Y., 497
Hoth, H. J., 149
Hrabovecka, G., 464
Hrgovic, M., 211
Hsu, Y. E., 553
Huang, H. M., 494
Huang, Z. H., 462
Huber, A. M., 474
Hudnik, V., 364
Huebner, H., 294
Hughes, M. J., 465
Hulstkamp, J., 321
Hultquist, D. E., 374

Hundeshagen, D., 712
Hurley, L. S., 587
Husain, S. L., 153
Hydes, D. J., 456
Hwang, H. L., 640
Ichida, T., 91
Ichinose, N., 220, 232, 401
Iddon, A., 227
Iffland, R., 562, 646
Igoshina, E. V., 449
Iguchi, K., 635
Ihida, M., 427
Ihlemann, M., 582
Iida, C., 69, 488, 693
Iida, K., 635
Ikebe, K., 680
Ikeda, S., 95
Illek, J., 689
Imai, Y., 159
Imanari, T., 510
Ines, G., 733
Ingle, J. D., 306, 336
Inhat, M., 499, 636
Inoue, S., 343
Inoue, T., 401, 437
Iordanov, N., 468
Ishibashi, T., 591
Ishii, M., 716
Ishii, T., 427, 466, 540
Ishino, F., 365
Ishizawa, M., 383
Ishizuka, T., 155, 176
Isozaki, A., 424, 573
Issers, V. V., 444
Ito, Y., 510
Itokawa, H., 467
Iuchi, I., 740
Ivanov, D. N., 112, 134
Ivanova, E., 468
Iwasa, A., 457
Iwata, Y., 366, 574
Iyengar, G. V., 662, 730
Iyengar, S. S., 505
Jacob, R. A., 247
Jackwerth, E., 195, 301, 408, 422, 711
Jacyszyn, K., 669
Jagos, P., 689
Jakutowicz, K., 402
James, G. S., 7
Jamro, G. H., 109
Janouskova, J., 257
Janowska, S., 249
Jansen, A. V., 7
Janssen, A., 328
Jarabak, R., 627
Jarkovsky, J., 483
Jaros, J., 560
Jarosz, R., 110
Jastrow, J. D., 575
Jelinek, P., 689
Jenne, E. A., 305

SUBJECT INDEX (PHYSICS)

SUBJECT INDEX (BIOLOGY)

SILVER Ag (ATOMIC WEIGHT 107.868)

Silver has been known since ancient times. It occurs natively as metallic silver as well as in combined form in copper, gold, lead, nickel, and zinc ores. It is also found associated with antimony, bismuth, mercury, and platinum. Silver is mainly used in the form of metal, a considerable amount of which is involved in the monetary system. It has been used as a standard value in coins for centuries. This metal is used in silver-plating, table silverware, jewelry and ornaments, dental alloys, solder and brazing alloys, and the manufacture of mirrors. The most important use is in the photographic industry. Silver has the highest electrical and thermal conductivity of all metals, therefore, it is used in electrical contacts, printed circuits, and high-capacity silver-zinc and silver-cadmium batteries.

Determination of copper is required in ores, alloys, copper and lead furnace by-products, copper-gold alloy assays, slimes, cyanide mill solutions, and plating baths, etc. Sample dissolution is usually achieved by treatment with nitric acid. Silver forms insoluble chlorides with hydrochloric acid. Sulfuric acid and alkalis also do not have much effect. Hard-to-dissolve samples can be treated with nitric acid followed by fusion of the residue.

No significant interferences have been reported in air-acetylene flame. However, 5% solutions of sulfuric and phosphoric acids depress the sensitivity. Iodate, permanganate, and tungstate interfere by precipitating the silver. Organic solvent effect has been observed with 5% acetic acid and it increases the sensitivity.

Standard Solution

To prepare 1000 mg/ℓ solution, dissolve 1.000 g of silver metal in 20 mℓ 1:1 nitric acid or dissolve 0.787 g of silver nitrate ($AgNO_3$) in 50 mℓ water. Dilute to 1 ℓ with 1% (v/v) of nitric acid. Store the solution in dark or amber glass bottle.

Instrumental Parameters

Wavelength	328.1 nm
Slit width	0.7 nm
Light source	Hollow cathode lamp
Flame	Air-acetylene, oxidizing (lean, blue)
Sensitivity	2.5 mg/ℓ
Detection limit	0.002 mg/ℓ
Optimum range	0.05—4 mg/ℓ
Secondary wavelength and its sensitivity	
338.3 nm	5—6 mg/ℓ

REFERENCES

1. **Rawling, B. S., Greaves, M. C., and Amos, M. D.,** The Determination of silver in lead concentrates by atomic absorption spectroscopy, *Nature (London)*, 188, 137, 1960.
2. **Rawling, B. S., Amos, M. D., and Greaves, M. C.,** The determination of silver in lead sulfide concentrates by atomic absorption spectroscopy, *Aust. Inst. Min. and Metall. Proc.*, Australian Institute of Mining and Metallurgy, Canberra, 1961, 199.
3. **Schuler, V. C. O., Jansen, A. V., and James, G. S.,** The development of atomic absorption methods for the determination of silver, copper, iron, lead and zinc in high purity gold and the role of organic additives, *J. S. Afr. Inst. Min. Metal.*, 62, 806, 1962.
4. **Sprague, S. and Slavin, W.,** The application of atomic absorption spectroscopy to the analysis of petroleum products, *At. Absorpt. Newsl.*, No. 12, April 1963.
5. **Ginzburg, V. L., Livshits, D. M., and Satarina, G. I.,** Atomic absorption flame photometric determination of silver, gold, palladium, platinum and rhodium, *Zh. Anal. Khim.*, 19, 1089, 1964.
6. **Belcher, R., Dagnall, R. M., and West, T. S.,** An examination of the atomic absorption spectroscopy of silver, *Talanta*, 11, 1257, 1964.
7. **Wilson, L.,** The determination of silver in aluminum alloys by atomic absorption spectroscopy, *Anal. Chim. Acta*, 30, 377, 1964.
8. **Khalifa, H., Erdey, L., and Svehla, G.,** Accuracy of silver determination by atomic absorption methods, *Acta Chim. Acad. Sci. Hung.*, 41, 187, 1964.
9. **Means, E. A. and Ratcliffe, D.,** Determination of wear metals in lubricating oils by atomic absorption spectroscopy, *At. Absorpt. Newsl.*, 4, 174, 1965.
10. **Gomez Coedo, A. and Jiminez Seco, J. L.,** Atomic absorption technique for silver in ores, *Rev. Met.*, 1, 158, 1965.
11. **Tindall, F. M.,** Silver and gold assay by atomic absorption spectrophotometry, *At. Absorpt. Newsl.*, 4, 339, 1965.
12. **Fixman, M. and Boughton, L.,** Mineral assay of silver, zinc and cadmium, *At. Absorpt. Newsl.*, 5, 33, 1966.
13. **Huffman, C., Mensik, J. D., and Rader, L. F.,** Determination of silver in mineralised rocks by atomic absorption spectrophotometry, *U.S. Geol. Surv. Prof. Pap.*, 550b, 189, 1966.
14. **Sattur, T. W.,** Routine atomic absorption analysis of nonferrous alloys and plant intermediates, *At. Absorpt. Newsl.*, 5, 37, 1966.
15. **Slavin, S. and Slavin, W.,** Full automatic analysis of used aircraft oils, *At. Absorpt. Newsl.*, 5, 106, 1966.
16. **Tindall, F. M.,** Notes on silver and gold assay by atomic absorption, *At. Absorpt. Newsl.*, 5, 140, 1966.
17. **Takeuchi, T., Sazaki, M., and Yanagisawa, M.,** Some observations on the determinations of metals by atomic absorption spectroscopy combined with extraction, *Anal. Chim. Acta*, 36, 258, 1966.
18. **Venghiattis, A.,** A technique for the direct sampling of solids without prior dissolution, *At. Absorpt. Newsl.*, 6, 19, 1967.
19. **Rubeska, I., Suleck, Z., and Moldan, B.,** The determination of silver in sulfide minerals by atomic absorption spectrophotometry, *Anal. Chim. Acta*, 37, 27, 1967.
20. **West, F. K., West, P. W., and Ramakrishna, T. V.,** Stabilization and determination of traces of silver in waters, *Environ. Sci. Technol.*, 1, 717, 1967.
21. **Hickey, L. J.,** The determination of silver in fine silver bullion by atomic absorption spectroscopy, *Anal. Chim. Acta*, 41, 546, 1968.
22. **Biancifiori, M. A., Bordonali, C., and Besazza, G.,** Determination by atomic absorption of traces of metals, silver, chromium and cobalt, *Chim. Ind. (Milan)*, 50, 423, 1968 (in Italian).
23. **Carr, R. A.,** Determination of silver in sea water matrix by atomic absorption spectrophotometry, *At. Absorpt. Newsl.*, 8, 69, 1969.
24. **Aldous, K. M., Dagnall, R. M., and West, T. S.,** Preparation and spectral characteristics of microwave excited electrodeless discharge tubes for palladium, silver, platinum and gold, *Analyst (London)*, 94, 347, 1969.
25. **Chao, T. T., Fishman, M. J., and Ball, J. W.,** Determination of traces of silver in water by anion exchange and atomic absorption spectrophotometry, *Anal. Chim. Acta*, 47, 189, 1969.
26. **Edwards, H. W.,** Direct determination of silver in air by atomic absorption spectrometry, *Anal. Chem.*, 41, 1172, 1969.
27. **Emmermann, R. and Luecke, W.,** Determination of trace amounts of lead, zinc and silver in soil samples by atomic absorption spectrometry using a tantalum sampling boat, *Z. Anal. Chem.*, 248, 325, 1969 (in German).
28. **Baranova, S. V., Ivanov, N. P., Pofralidi, L. G., Knyazev, V. V., Talalaev, B. M., and Vasil'ev, E. N.,** Electrodeless lamps with high frequency excitation of the spectrum as a radiation source in atomic absorption analysis. Iodide lamps for the atomic absorption determination of silver, lead and iron, *Zh. Anal. Khim.*, 24, 1649, 1969.

29. **Warburton, J. A.,** Trace silver detection in precipitation by atomic absorption spectrophotometry, *J. Appl. Meteorol.,* 8, 464, 1969.

30. **Eroshevich, T. A., Kukushkin, U. N., and Makarov, D. F.,** Atomic absorption determination of platinum metals, gold and silver in copper-nickel slurries and products of their processing, *Zh. Anal. Khim.,* 42, 2174, 1969.

31. **Garcia, D. J.,** Atomic absorption spectrophotometric analysis of silver in ores and other substances, *Inform. Quim. Anal.,* 23, 132, 1969.

32. **Roth, E. and Gilbert, E.,** Determination of trace metals in foodstuffs and other commodities by atomic absorption. I. Silver in wine, *Mitt. Rebewein Obstb. Fruechtverwert,* 19, 11, 1969 (in German).

33. **Goodwin, E.,** The determination of silver in copper concentrates and copper anodes by atomic absorption spectrometry, *At. Absorpt. Newsl.,* 9, 95, 1970.

34. **Andersen, J.,** Silver loss from contact switches, *At. Absorpt. Newsl.,* 9, 96, 1970.

35. **Kallmann, S. and Hobart, E. W.,** Determination of silver gold and palladium by a combined fire assay atomic absorption procedure, *Talanta,* 17, 845, 1970.

36. **Kazimerczyk, S., Michalewska, M., and Jarosz, R.,** Determination of silver, copper and zinc by atomic absorption spectroscopy, *Chem. Anal. (Warsaw),* 15, 553, 1970.

37. **Michailova, T. P. and Rezpina, V. A.,** Determination of gold and silver in plant liquors and electrolytes by atomic absorption spectrophotometry, *Analyst (London),* 95, 769, 1970.

38. **Michailova, T. P., Rezpina, V. A., and Dolsenko, L. G.,** Atomic absorption method for determining gold and silver in electrolytes and technological products, *Zh. Anal. Khim.,* 25, 1477, 1970.

39. **Talalaev, B. M. and Mironova, O. N.,** Atomic absorption determination of silver in industrial waste waters, *Zh. Anal. Khim.,* 25, 1317, 1970.

40. **Chao, T. T. and Ball, J. W.,** Determination of nanogram levels of silver in suspended materials of streams retained by a membrane filter with the "sampling boat" technique, *Anal. Chim. Acta,* 54, 168, 1971.

41. **Lee, R. F. and Pickering, W.,** Effect of precipitate and complex formation on the determination of silver by atomic absorption spectroscopy, *Talanta,* 18, 1083, 1971.

42. **Luecke, W. and Emmermann, R.,** The application of boat technique for lead, zinc, silver and cadmium in soil samples, *At. Absorpt. Newsl.,* 10, 45, 1971.

43. **Molughney, P. E. and Graham, J. A.,** Determination of silver in ores and metallurgical concentrates by a combination of fire assay preconcentration (using tin as collector) and atomic absorption spectrophotometry, *Talanta,* 18, 475, 1971.

44. **Reynolds, R. J. and Lagden, D. S.,** Potential hazards by formation of silver acetylide upon aspirating solutions containing high concentrations of silver to an atomic absorption spectrophotometer when acetylene is used as fuel, *Analyst (London),* 96, 319, 1971.

45. **Takahashi, M., Tsukahara, I., and Shibyua, S.,** Determination of microamounts of silver in copper, aluminum and their alloys by solvent extraction - atomic absorption spectrophotometry, *Bunseki Kagaku,* 20, 188, 1971.

46. **Thomas, B. G.,** Determination of silver, lead and zinc in high grade ores, *At. Absorpt. Newsl.,* 10, 73, 1971.

47. **Willis, J. B.,** Atomization problems in atomic absorption spectroscopy. III. Absolute efficiencies of sodium, copper, silver and gold in a meker type air-acetylene flame, *Spectrochim. Acta,* 26b, 177, 1971.

48. **Fishkova, N. L. and Kazarina, T. M.,** Atomic absorption determination of silver in ores, *Zavod. Lab.,* 37, 1447, 1971.

49. **Malykh, V. D., Erkovich, G. E., Goncharova, N. N., and Shipitsin, S. A.,** Determination of gold and silver in cyanide solutions by atomic absorption, *Nauk Tr. Irkutzk. Gos. Nauchno-Issled Inst. Redk. Tsvet. Metall.,* 22, 111, 1971.

50. **Hermon, S. E. and Rennie, R. J.,** Determination of silver in wrought aluminum alloys with particular reference to the use of atomic absorption spectrophotometry, *Metall. Met. Form.,* 38, 258, 1971.

51. **Dancheva, R. and Beleva-Naumova, S.,** Atomic absorption determination of silver in (copper or gold concentrates), *Rudodobiv Metal.,* 26, 18, 1971 (in Bulgarian).

52. **Adler, J. F. and West, T. S.,** Atomic absorption and fluorescence spectrophotometry with a carbon filament atom reservoir. IX. The direct determination of silver and copper in lubricating oils, *Anal. Chim. Acta,* 58, 331, 1972.

53. **Bratzel, M. P., Jr., Chakrabarti, C. L., Sturgeon, R. E., McIntyre, M. W., and Agemian, H.,** Determination of gold and silver in parts per billion or lower levels in geological and metallurgical samples by atomic absorption spectrometry with a carbon rod atomizer, *Anal. Chem.,* 44, 372, 1972.

54. **Purushottam, A., Lal, S. S., and Naidu, P. P.,** Rapid determination of traces of silver in sulfide ores by atomic absorption, *Talanta,* 19, 208, 1972.

55. **Reeves, R. D., Molnar, C. J., and Winefordner, J. D.,** Rapid atomizer absorption determination of silver and copper by sequential atomization from a graphite rod, *Anal. Chem.,* 44, 1913, 1972.

56. **Britske, M. E., Ioffe, V. P., Saveleva, A. N., Slabodenyuk, I. V., and Khairulina, N. P.,** Determination of silver and gold in copper industry products by atomic absorption, *Zavod. Lab.,* 38, 1458, 1972.

57. **Steele, T. W., Mallet, R. C., Pearton, D. C. G., and Ring, E. J.**, Determination by atomic absorption spectrophotometry of platinum, rhodium, ruthenium, gold, silver, and lead in drills, *Lab. Meth.*, No. 78/4, 1972.

58. **Fishkova, N. L.**, Atomic absorption determination of silver in ores, *Tr. Tsvet. Nauchno-Issled Gorn. Inst. Tsvet. Blagorod. Metall.*, 102, 178, 1972 (in Russian).

59. **Adriaenssens, E. and Verbeek, F. V.**, Atomic absorption spectrophotometry of silver, gold, palladium and platinum in a potassium cyanide medium, *At. Absorpt. Newsl.*, 12, 57, 1973.

60. **Ng, W. K.**, Determination of silver in some sulfide minerals by atomic absorption spectrometry, *Anal. Chim. Acta*, 63, 469, 1973.

61. **Woodriff, R., Culver, B. R., Shrader, D., and Super, A. B.**, Determination of subnanogram quantities of silver in snow by furnace atomic absorption spectrometry, *Anal. Chem.*, 45, 230, 1973.

62. **Walton, G.**, Determination of silver in ores and mineral products by atomic absorption spectroscopy, *Analyst (London)*, 98, 335, 1973.

63. **Jackwerth, E., Hoehn, R., and Koos, K.**, Trace enrichment by partial dissolution of the matrix in presence of mercury. Determination of bismuth, copper, lead, nickel, silver, gold and palladium in high purity cadmium by atomic absorption spectrometry, *Z. Anal. Chem.*, 264, 1, 1973 (in German).

64. **Galanova, A. P., Kudryavina, A. K., Pronin, V. A., Yudelevich, I. G., Vall, G. A., and Gilbert, E. W.**, Extraction-atomic absorption determination of gold and silver in antimony containing products, *Izv. Sib. Otd. Akad. Nauk SSSR Ser. Khim. Nauk*, 89, 1973.

65. **Sen Gupta, J. G.**, A review of the methods for the determination of platinum group metals, silver and gold by atomic absorption spectroscopy, *Miner. Sci. Eng.*, 5, 207, 1973.

66. **Makarov, D. F., Kukushkin, U. N., and Eroshevich, T. A.**, Atomic absorption determination of gold and silver in black and cathodic copper, *Zh. Prikl. Khim. Leningr.*, 46, 656, 1973 (in Russian).

67. **Pronin, V. A., Galanova, A. P., Kudryavina, A. K., Shastina, Z. N., Apolitskii, V. N., and Usol'teva, M. V.**, Stability of standard solutions of gold, silver and platinum used in atomic absorption analysis, *Zh. Anal. Khim.*, 28, 2328, 1973.

68. **Mizuike, A., Fukuda, K., and Sakamoto, T.**, Rapid collection of traces of silver (or gold or palladium) with mercury in an ultrasonic field, *Bull. Chem. Soc. Jpn.*, 46, 3596, 1973.

69. **Berezkin, O. P., Lukicheva, M. P., Karatsuba, T. I., and Dementeva, A. K.**, Determination of silver in products containing large amounts of iron, magnesium and aluminum, *Nauch. Tr. Sib. Nauchno-Issled Proekt. Inst. Tsvet. Metal.*, 6, 21, 1973.

70. **Chowdhury, A. N., Das, A. K., and Das, T. N.**, Determination of silver in rocks and minerals by atomic absorption spectrophotometry, *Z. Anal. Chem.*, 269, 284, 1974.

71. **Fishkova, N. L.**, Determination of the platinum metals, gold and silver by atomic absorption spectrophotometry, *Zh. Anal. Khim.*, 29, 2121, 1974.

72. **Adriaenssens, E. and Verbeek, F.**, The determination of silver, gold, palladium and platinum by a combined fire assay atomic absorption source, *At. Absorpt. Newsl.*, 13, 141, 1974.

73. **Galanova, A. P., Kudryavina, A. K., Pronin, V. A., Yudelevich, I. G., and Vall, G. A.**, Atomic absorption determination of silver in ores and products of their processing, *Izv. Sib. Otd. Akad. Nauk SSSR Ser. Khim. Nauk.*, 1, 1974.

74. **Lacaux, J. P., Van Dinh, P., and Beguin, J.**, The determination of silver iodide in air by flameless atomic absorption spectrophotometry, *At. Absorpt. Newsl.*, 13, 49, 1974.

75. **Langmyhr, F. J., Solberg, R., and Wold, L. T.**, Atomic absorption spectrometric determination of silver, bismuth and cadmium in sulfide ores by direct atomization from the solid state, *Anal. Chim. Acta*, 69, 267, 1974.

76. **Burke, K. E.**, Determination of microgram quantities of silver in aluminum, iron and nickel base alloys, *Talanta*, 21, 417, 1974.

77. **Rattonetti, A.**, Determination of soluble cadmium, lead, silver and indium in rain water and stream water with the use of flameless atomic absorption, *Anal. Chem.*, 46, 739, 1974.

78. **Langmyhr, F. J., Stuberg, J. R., Thomassen, Y., Hanssen, J. E., and Dolezal, J.**, Atomic absorption spectrometric determination of cadmium, lead, silver, thallium and zinc in silicate rocks by direct atomization from the solid state, *Anal. Chim. Acta*, 71, 35, 1974.

79. **Hofton, M. E.**, Determination of silver in iron and steels by atomic absorption, *Br. Steel Corp. Open Rep.*, GS/TE-CH/558/1/746, 1974.

80. **Mallet, R. C. and Kellermann, S.**, Assessment of the carbon rod atomizer for the determination of silver, *Natl. Inst. Metall. Repub. S. Afr. Rep.*, No. 1669, 1974.

81. **Belyaev, Y. I., Oreshkin, V. N., and Vnukovskaya, G. L.**, Atomic absorption determination of element traces in rocks by using pulse thermal atomization of solid samples. III. Suppression of the non-resonance absorption in determination of cadmium, silver and thallium, *Zh. Anal. Khim.*, 30, 503, 1975.

82. **Fishkova, N. L., Zdorova, E. P., and Popova, N. N.**, Determination of low gold and silver contents in mineral raw materials by the assay-atomic absorption method, *Zh. Anal. Khim.*, 30, 806, 1975.

83. **Keliher, P. N. and Wohlers, C. C.,** Spectral line profile measurements from calcium, silver and aluminum hollow cathode lamps, *Appl. Spectrosc.,* 29, 198, 1975.

84. **Minkkiner, P.,** A method for the correction of background absorption in silver analysis of calcareous samples, *At. Absorpt. Newsl.,* 14, 71, 1975.

85. **Greig, R. A.,** Comparison of atomic absorption and neutron activation analyses for the determination of silver, chromium and zinc in various marine organisms, *Anal. Chem.,* 47, 1682, 1975.

86. **Kirk, M., Perry, E. G., and Arritt, J. M.,** Separation and atomic absorption measurement of trace amounts of lead, silver, zinc, bismuth and cadmium in high nickel alloys, *Anal. Chim. Acta,* 80, 163, 1975.

87. **Kono, T. and Nemori, A.,** Extraction of copper, cadmium, lead, silver and bismuth with iodide methyl isobutyl ketone in an atomic absorption spectrophotometric analysis, *Bunseki Kagaku,* 24, 419, 1975.

88. **Rooney, R. C.,** Determination of silver in animal tissues by wet oxidation followed by atomic absorption spectrophotometry, *Analyst (London),* 100, 471, 1975.

89. **Wagenaar, H. C. and De Galan, L.,** Influence of line profiles upon analytical curves for copper and silver in atomic absorption spectroscopy, *Spectrochim. Acta,* 30b, 361, 1975.

90. **Iida, C. and Uchida, T.,** Atomic absorption spectrometry of silver with absorption tube having nebulization chamber, *Spectrosc. Lett.,* 8, 751, 1975.

91. **Rakhlina, M. L., Mikhailov, P. M., and Lomekov, A. S.,** Atomic absorption analysis of silver in ammonia solutions, *Zavod. Lab.,* 41, 1340, 1975.

92. **Shelton, B. J.,** Determination of silver, selenium, tellurium, antimony, tin, lead and arsenic in anode sludges, *Natl. Inst. Metall. Repub. S. Afr. Rep.,* No. 1771, 1975.

93. **Acatini, C., De Berman, S. N., Colombo, O., and Fondo, O.,** Determination of silver, copper, lead, tin, antimony, iron, calcium, zinc, magnesium, potassium and manganese in canned tomatoes by atomic absorption spectrometry, *Revta Asoc. Bioquim. Argent.,* 40, 175, 1975 (in Spanish).

94. **Langmyhr, F. J., Lind, T., and Jonsen, J.,** Atomic absorption spectrometric determination of manganese, silver and zinc in dental material by atomization directly from the solid state, *Anal. Chim. Acta,* 80, 297, 1975.

95. **McElhaney, R. J.,** Determination of gold, silver and cobalt in aluminum by flameless atomic absorption spectroscopy, *J. Radioanal. Chem.,* 32, 99, 1976.

96. **Shaburova, V. P., Yudelevich, I. G., Seryakova, I. V., and Zolotov, Y. A.,** Extraction-atomic absorption determination of copper, silver and thallium in some metal halides, *Zh. Anal. Khim.,* 31, 255, 1976.

97. **Terashima, S.,** Determination of microamounts of silver in standard silicates by atomic absorption spectrometry with carbon tube atomizer, *Bunseki Kagaku,* 25, 279, 1976.

98. **Vall, G. A., Usol'seva, M. V., Yudelevich, I. G., Seryakova, I. V., and Zolotov, Y. A.,** Atomic absorption determination of silver after its solvent extraction with diphenylthiourea, *Zh. Anal. Khim.,* 31, 27, 1976.

99. **Hamner, R. M., Lechak, D. L., and Greenberg, P.,** Determination of silver, arsenic, bismuth, antimony, selenium, and tellurium in chromium metal with flameless atomic absorption spectroscopy, *At. Absorpt. Newsl.,* 15, 122, 1976.

100. **Henrion, G. and Raguse, H. D.,** Method of internal standardization for flame spectrometric determination of silver and chromium, *Chem. Anal.,* 21, 281, 1976.

101. **Khlebnikova, A. A. and Torgov, V. G.,** Atomic absorption determination of silver using prior solvent extraction with petroleum sulfides, *Zh. Anal. Khim.,* 31, 1090, 1976.

102. **Tomi, B. and Manoliu, C.,** Determination of silver in artificial zeolites by atomic absorption spectrophotometry, *Rev. Chim.,* 27, 231, 1976.

103. **Royal, S. J. and Mallet, R. C.,** Rapid atomic absorption method for determination of silver in sulfide ores and concentrates, *Natl. Inst. Metall. Repub. S. Afr. Rep.,* No. 1797, 1976.

104. **Wunderlich, E. and Burghardt, M.,** Determination of traces of silver in pure copper by atomic absorption spectrophotometry, *Z. Anal. Chem.,* 281, 299, 1976.

105. **Halasz, A. and Polyak, K.,** Processes in the graphite tube, anionic and matrix effects in flameless atomic absorption analysis. I. Determination of silver, *Acta Chim. Acad. Sci. Hung.,* 91, 261, 1976.

106. **Bazhov, A. S. and Sokolova, E. A.,** Solvent extraction-atomic absorption determination of silver and gold, *Zh. Anal. Khim.,* 32, 65, 1977.

107. **Tindall, F. M.,** Hints of chemical analysis. I. Tungsten determination by atomic absorption spectrophotometry. II. Revised notes on gold and silver determination by atomic absorption spectrophotometry. III. Mercury and copper concentrates by atomic absorption spectrophotometry, *At. Absorpt. Newsl.,* 16, 37, 1977.

108. **Bea-Barredo, F., Polo-Polo, C., and Polo Diez, L.,** The simultaneous determination of gold, silver and cadmium at ppb levels in silicate rocks by atomic absorption spectrometry with electrothermal atomization, *Anal. Chim. Acta,* 94, 283, 1977.

109. **Aihara, M. and Kiboku, M.,** Determination of silver by atomic absorption spectrophotometry following extraction with potassium xanthate-methyl isobutyl ketone, *Bunseki Kagaku,* 26, 559, 1977.

110. **Bogdanova, V. I. and Dobretsova, I. L.,** Behavior of silver in a flameless atomic absorption determination, *Zh. Anal. Khim.,* 32, 1717, 1977.

111. **Pchelintseva, N. F.,** Atomic absorption determination of silver in rocks, *Zavod. Lab.,* 43, 693, 1977.

112. **Govindaraju, K., Morel, J., and L'ttomel, N.,** Solid sampling atomic absorption determination of silver in silicate rock reference samples. Application to a homogeneity study of silver in a one-ton- two-mica granite reference sample, *Geostandards Newsl.,* 1, 137, 1977.

113. **Pjatnicki, I. V. and Pilipjuk, J. P.,** Atomic absorption determination of silver by extraction with a pyrrdine-containing solution of cinnamic acid in chloroform, *Ukr. Chim. Zh.,* 43, 639, 1977.

114. **Ryabinin, A. I. and Lazareva, E. A.,** Solvent extraction-atomic absorption determination of copper, silver and cadmium in Black sea water, *Zh. Anal. Khim.,* 33, 298, 1978.

115. **Samchuk, A. I. and Kovtun, G. P.,** Solvent extraction-atomic absorption determination of silver in rocks and minerals, *Zh. Anal. Khim.,* 33, 1924, 1978.

116. **Watling, R. J.,** The use of a slotted tube for the determination of lead, zinc, cadmium, bismuth, cobalt, manganese and silver by atomic absorption spectrometry, *Anal. Chim. Acta,* 97, 395, 1978.

117. **Agapova, T. E., Polyakova, V. A., and Sedykh, E. M.,** Atomic absorption determination of silver in non-ferrous metallurgy products, *Zh. Anal. Khim.,* 33, 1285, 1978.

118. **Aziz-Alrahman, A. M. and Headridge, J. B.,** Determination of silver in irons and steels by atomic absorption spectrometry with an induction furnace. Direct analysis of solid samples, *Talanta,* 25, 413, 1978.

119. **Elson, C. M., Dostal, J., Hynes, D. L., and de Albuquerque, C. A. R.,** Silver, cadmium and lead contents of some rock reference samples, *Geostandards Newsl.,* 11, 121, 1978.

120. **Howlett, C. and Taylor, A.,** Measurement of silver in blood by atomic absorption spectrophotometry using the micro cup technique, *Analyst (London),* 103, 916, 1978.

121. **Shaeffer, J. D., Mulvey, G., and Skogerboe, R. K.,** Determination of silver in precipitation by furnace atomic absorption spectrometry, *Anal. Chem.,* 50, 1239, 1978.

122. **Sukhanova, G. B.,** Mechanism of the population of lightly excited states of copper and silver in a hollow cathode discharge, *Izv. Vysskh. Uchebn. Zavod. Fiz.,* 21, 105, 1978.

123. **Viets, J. G.,** Determination of silver, bismuth, cadmium, lead, copper and zinc in geologic materials by atomic absorption spectrometry with tricaprylyl methyl ammonium chloride, *Anal. Chem.,* 50, 1097, 1978.

124. **Prudnikov, E. D., Kolosova, L. P., Kalachev, V. K., Shapkina, Y. S., Novatskaya, N. V., and Bychkov, Y. A.,** Pulse atomic absorption determination of trace amounts of platinum metals, gold and silver in natural materials, *Zh. Anal. Khim.,* 33, 468, 1978.

125. **Aleksandrov, S., Gyulmezova, G., and Sanabria, J.,** A study on the possibility for emission spectral and atomic absorption determination of silver, gold and palladium in anode slime, *Dokl. Bolg. Akad. Nauk,* 31, 73, 1978.

126. **Fishkova, N. L. and Vilenkin, V. A.,** Application of an HGA-74 graphite atomizer to atomic absorption determination of gold, silver, platinum and palladium in solutions of complicated compositions, *Zh. Anal. Khim.,* 33, 897, 1978 (in Russian).

127. **Erkovich, G. E., Yakovuk, V. I., and Malykh, V. D.,** Atomic absorption determination of impurities in silver-copper alloys, *Zavod. Lab.,* 44, 1480, 1978.

128. **Kujirai, O., Kobayashi, T. and Sudo, E.,** Determination of sub-parts per million level of silver in heat resisting alloys by graphite furnace atomic absorption spectrometry, *Trans. Jpn. Inst. Met.,* 19, 159, 1978.

129. **Karmanova, N. G. and Pogrebnyak, Y. F.,** Atomic absorption determination of silver in sulfide ores, *Zavod. Lab.,* 45, 124, 1979.

130. **Skorko-Trybulova, Z., Boguszewska, Z., and Rozanska, B.,** Determination of silver in copper ores by atomic absorption spectrophotometry after extraction separation with triphenyl phosphate, *Mikrochim. Acta,* 1, 151, 1979.

131. **Armannsson, H.,** Dithizone extraction and flame atomic absorption spectrometry for the determination of cadmium, zinc, lead, copper, nickel and silver in sea water and biological tissues, *Anal. Chim. Acta,* 110, 21, 1979.

132. **Arpadjan, S. and Alexandrova, I.,** Determination of trace elements (lead, bismuth, copper, nickel and silver) in high purity tin by flame atomic absorption spectrophotometry, *Z. Anal. Chem.,* 298, 159, 1979.

133. **Aruscavage, P. A. and Campbell, E. Y.,** The determination of silver in silicate rocks by electrothermal atomic absorption spectrometry, *Anal. Chim. Acta,* 109, 171, 1979.

134. **Ward, N. I., Roberts, E., and Brooks, R. R.,** Silver uptake by seedlings of loliumperenne and trifolium repens, *N.Z. J. Sci.,* 22, 129, 1979.

135. **Konovalov, G. F. and Gur'eva, M. P.,** Extraction-atomic absorption determination of silver and gold, *Zavod. Lab.,* 45, 196, 1979.

136. **Forrester, J. E., Lehecka, V., Johnston, J. R., and Ott, W. L.,** Direct determination of trace quantities of antimony, arsenic, bismuth, cadmium, lead, selenium, silver, tellurium and thallium in high purity nickel by electrothermal atomic absorption spectrometry, *At. Absorpt. Newsl.,* 18, 73, 1979.

137. **Katskov, D. A. and Grinshtein, I. L.,** Study of the chemical interaction of copper, gold and silver with carbon by an atomic absorption method using an electrothermal atomizer, *Zh. Prikl. Spektrosk.,* 30, 787, 1979.

138. **Prudnikov, E. D. and Rantsev, A. A.,** Atomic absorption determination of gold and silver using a flameless atomizer in an air atmosphere, *Zh. Anal. Khim.,* 34, 1725, 1979.

139. **Brandvold, L. A.,** Reliability of gold and silver analyses by commercial labs in the southwest, *N. M. Geol.,* 1, 11, 1979.

140. **Kantor, T., Bezur, L., Pungor, E., Fodor, P., Nagy-Balogh, J., and Heincz, G.,** Determination of the thickness of silver, gold and nickel layers by a laser microprobe and flame atomic absorption technique, *Spectrochim. Acta,* 34b, 341, 1979.

141. **Jackwerth, E., Hoehn, R., and Musaick, K.,** Preconcentration of traces of silver, gold, bismuth, copper and palladium from pure lead by partial precipitation of the matrix with sodium borohydride as a reducing agent, *Z. Anal. Chem.,* 299, 362, 1979.

142. **Carles, J.,** Determination of traces of silver in capvern waters, *Presse Therm. Clim.,* 116, 117, 1979.

143. **Baeckman, S. and Karlsson, R. W.,** Determination of lead, bismuth, zinc, silver and antimony in steel and nickel base alloys by atomic absorption spectrophotometry using direct atomization of solid samples in a graphite furnace, *Analyst (London),* 104, 1017, 1979.

144. **Kitagawa, K., Shigeyasu, T., and Takeuchi, T.,** Application of the atomic faraday effect to the trace determination of elements (cadmium, silver and copper). Effect of the hyperfine structure on the Zeeman splitting and line crossing, *Spectrochim. Acta,* 34b, 389, 1979.

145. **Marinescu, I. and Tamas, M.,** Phytochemical study on poplar buds as possible propolis source. Some aspects of trace elements contained by propolis and poplar buds, (Ag), *Apic. Rom.,* 54, 14, 1979.

146. **Krasiejiko, M., Marczenko, Z., and Kowalski, T.,** Trace silver determination in high purity platinum using dithizone and an atomic absorption method after extraction with triphenyl phosphine, *Chem. Anal.,* 24, 1037, 1979.

147. **Baker, A. A., Headridge, J. B., and Nicholson, R. A.,** Determination of silver and thallium in nickel base alloys by atomic absorption spectrometry with introduction of solid samples into an induction furnace, *Anal. Chim. Acta,* 113, 47, 1980.

148. **Kantor, T., Bezur, L., Pungor, E., Fodor, P., Nagy-Balogh, J., and Heincz, G.,** Determination of thickness of silver, gold and nickel layers by a laser microprobe and flame atomic absorption technique, *Magy. Kem. Lapja,* 35, 266, 1980.

149. **Tsukuhara, I. and Tanaka, M.,** Determination of silver in copper and lead metals and alloys and in zinc and selenium by atomic absorption spectrometry after separation by extraction of the tris-*n*-octyl methyl ammonium-silver bromide complex, *Talanta,* 27, 237, 1980.

150. **Vall, G. A., Pobdubnaya, L. P., Vanifatova, N. G., Yudelevich, I. G., and Zolotov, Y. A.,** Atomic absorption and spectrophotometric determination of silver in geological samples after its solvent extraction by *o*-isopropyl N-methyl thiocarbamate, *Zh. Anal. Khim.,* 35, 1980.

151. **Colella, M. B., Siggia, S., and Barnes, R. M.,** Poly (acrylamidoxime)resin for determination of trace metals in natural waters, *Anal. Chem.,* 52, 2347, 1980.

152. **Eames, J. C. and Matousek, J. P.,** Determination of silver in silicate rocks by furnace atomic absorption spectrometry, *Anal. Chem.,* 52, 1980.

153. **Jackson, F. J., Read, J. I., and Lucas, B. E.,** Determination of total chromium, cobalt and silver in foodstuffs by flame atomic absorption spectrophotometry, *Analyst,* 105, 359, 1980.

154. **Kozlicka, M., Malusecka, M., Jedrzejewska, H., Kubicka, M., and Romanska, M.,** Effects of precision of analytical method on estimation of silver balance in flotation process of copper ores, *Erzmetall,* 33, 282, 1980.

155. **Manoliu, C., Popescu, O., Balasa, T., Tamas, V., and Georgescu, I.,** Study on determination of metal traces in solvents, *Rev. Chim.,* 31, 291, 1980.

156. **Ni, Z., Jin, L., and Wu, D.,** Application of graphite furnace atomic absorption to the determination of trace silver in rain, *Huang Kexue,* 1, 48, 1980 (in Chinese).

157. **Stein, V. B. and McClellan, B.,** Enhancement of atomic absorption sensitivity for cadmium, manganese, nickel and silver and determination in submicrogram quantities of cadmium and nickel in environmental samples, *Environ. Sci. Technol.,* 14, 872, 1980.

158. **Sterritt, R. M. and Lester, J. N.,** Determination of silver, cobalt, manganese, molybdenum and tin in sewage sludge by a rapid electrothermal atomic absorption spectroscopic method, *Analyst (London),* 105, 616, 1980.

159. **Posta, J.,** Determination of the optimum parameters of a combined atomic absorption spectrophotometric plus arc-flame method for the analysis of floating dust samples, *Hung. Sci. Instrum.,* 47, 33, 1980 (in Hungarian).

160. **Hoshino, Y., Utsunomiya, T., and Fukui, K.,** Graphite furnace atomic absorption spectrometry utilizing selective concentration onto tungsten wire, *Eng. Mater. Tokyo Inst. Technol.,* 5, 109, 1980.

161. **Fink, L. K., Jr., Harris, A. B., and Schick, L. L.,** Trace Metals in Suspended Particulates, Biota and Sediments of the St. Croix, Narraguagus and Union Estuaries and the Goose Cove Region of Penobscot Bay, Rep. W80-04809, OWRT-A-041-ME (!), 1980; available from the NTIS, Springfield, Va.; *Govt. Rep. Announce. Index (U.S.),* 80, 3372, 1980.

162. **Warburton, J. A., Molenar, J. V., Owens, M. S., and Anderson, A.,** Heavy metal enrichment in Antarctic precipitation and near surface snow, *Pure Appl. Geophys.,* 118, 1130, 1980.

163. **Sourova, J. and Capkova, A.,** Determination of trace elements in water with a high iron content by AAS, *Vodni. Hospod.,* 30b, 133, 1980 (in Czech).

164. **Sedykh, E. M., Belyaev, Y. I., and Sorokina, E. V.,** Elimination of matrix effects in electrothermal atomic absorption determination of silver, lead, cobalt, nickel, and tellurium in samples of complicated composition, *Zh. Anal. Khim.,* 35, 2348, 1980 (in Russian).

165. **Sedykh, E. M., Belyaev, Y. I., and Sorokina, E. V.,** Matrix effect during electrothermal atomic absorption determination of silver, tellurium, lead, cobalt and nickel in materials of complex compositions, *Zh. Anal. Khim.,* 35, 2162, 1980 (in Russian).

166. **Ol'khovich, P. F. and Prishchep, N. N.,** Determining the content of total and ionic silver in "Kievskaya" mineral water, *Khim. Tekhnol. Vody,* 2, 141, 1980.

167. **Hughes, M. J.,** Analysis of Roman tin and pewter ingots, Occas. Pap.- British Museum (Aspects early Metall.), 17, 41, 1980.

168. **Oreshkin, V. N., Belyaev, Y. I., Tatsii, Y. G., and Vnukovskaya, G. L.,** Direct simultaneous determination of cadmium, lead and silver in sea, river and eolian suspended matter by flameless atomic absorption, *Okeanologiya,* 20, 736, 1980 (in Russian).

169. **Lindh, U., Brune, D., Nordberg, G., and Wester, P. O.,** Levels of Sb, As, Cd, Pb, Hg, Se, Ag, Sn and Zn in bone tissue of industrially exposed workers, *Sci. Total Environ.,* 16, 109, 1980.

170. **Kersey, A. D., Dawson, J. B., and Ellis, D. J.,** Contributions of magnetically induced dichroism and the resonant voigt effect to the detection of silver by atomic spectroscopy, *Spectrochim. Acta,* 35b, 865, 1980.

171. **Jedrzejewska, H., Kozlicka, M., and Malusecka, M.,** Estimation of precision and accuracy of silver determination in lean copper ores by atomic absorption, *Chem. Anal.,* 25, 809, 1980.

172. **Ni, Z., Chin, L., and Wu, T.,** Application of graphite furnace atomic absorption in the determination of trace silver in rain, *Ching K'o Hsueh,* 1, 48, 1981 (in Chinese).

173. **Kimura, M. and Kawanami, K.,** Separation and preconcentration of trace amounts of several metals in sodium perchlorate using activated carbon as a collector, *Nippon Kagaku Kaishi,* 1, 1, 1981 (in Japanese).

174. **Malykh, V. D., Erkovich, G. E., and Yakovuk, V. I.,** Injection method for dispensing solutions into a flame during atomic absorption determination of impurities in pure gold and silver, *Zh. Anal. Khim.,* 36, 1730, 1981.

175. **Rains, T. C.,** Determination of aluminum, barium, calcium, lead, magnesium and silver in ferrous alloys by atomic emission and atomic absorption spectrometry, *ASTM STP,* 747, 43, 1981.

176. **Brooks, R. R., Holzbecher, J., Ryan, D. E., Zhang, H. F., and Chatterjee, A. K.,** A rapid method for the determination of gold and silver in sulfide ores and rocks, *At. Spectrosc.,* 2, 151, 1981.

177. **Moreiskaya, L. V., Nemodruk, A. A., Simonova, E., and Raevich, T. L.,** Stoichiometry in the atomic absorption spectrometric method. Determination of high component content, *Zavod. Lab.,* 47, 38, 1981.

178. **Adelman, H., Jenniss, S. W., and Katz, S. A.,** Interlaboratory analysis of sewage sludge, *Am. Lab.,* 31, 1981.

179. **Bye, R.,** Improvement of the sensitivity of the "sampling boat" technique in atomic absorption spectrometry. Determination of silver, lead and cadmium, *Z. Anal. Chem.,* 306, 30, 1981.

180. **Luail, N., Bol'shova, T. A., Alimarin, I. P., and Zlomanova, G. G.,** Use of extraction chromatography for concentration of trace silver with subsequent atomic absorption determination, *Vestn. Mosk. Univ. Ser. 2: Khim.,* 22, 280, 1981 (in Russian).

181. **Takahashi, Y., Shiozawa, Y., Kawai, T., Shimizu, K., Kato, H., and Hasegawa, J.,** Elements dissolved from three types of amalgams in synthetic saliva, *Aichi Gakuin Daigaku Shigakkaishi,* 19, 107, 1981.

182. **Kawahara, H., Yamada, T., Nakamura, M., Tomoda, T., Kobayashi, H., Saijo, A., Kawata, Y., and Hikari, S.,** Solubility of metal components into tissue culture medium from dental amalgam, *Shika Rikogaku Zasshi,* 22, 285, 1981.

183. **Teliya, N. M., Akimov, V. K., Dolidze, L. S., and Ivanov, V. K.,** Extraction of silver cyanide complexes and their determination by atomic absorption spectroscopy, *Soobshch. Akad. Nauk Gruz. SSSR,* 103, 593, 1981.

184. **Monteil, A. and Welte, B.,** Comparison of the different methods of attack during the determination of micropollutants in the sediments, Congr. Mediterr. Ing. Quim. (Actas), 2nd C20-1/C20-13, F.O.I.M., Barcelona, Spain, 1981 (in French).

185. **Stryjewska, E. and Just, B.,** Voltammetric determination of trace amounts of silver in waste samples with application of a rotating paste electrode (a comparison), *Chem. Anal. (Warsaw),* 26, 501, 1981 (in Polish).

186. **Khrapai, V. P., Provodenko, L. B., Kharitonova, V. A., and Babayants, T. A.,** Determination of trace impurities in high-purity gold and silver by electrothermal atomic absorption spectrometry, *Zavod. Lab.,* 47, 33, 1981.

187. **Yudelevich, I. G. and Startseva, E. A.,** Extraction-atomic absorption methods for the determination of gold and silver, *Zavod. Lab.,* 47, 24, 1981.

188. **Young, R. S.,** Analysis of nickel refinery slimes and residues, *Talanta,* 28, 25, 1981.

189. **Massee, R. and Maessen, F. J. M. J.,** Losses of silver, arsenic, cadmium, selenium and zinc traces from distilled water and artificial sea water by sorption on various container surfaces, *Anal. Chim. Acta,* 127, 181, 1981.

190. **Varma, A.,** Analytical techniques for corrosion studies on noble metals, *Talanta,* 28, 701, 1981.

191. **Salmon, S. G. and Holcombe, J. A.,** Alteration of metal release mechanisms in graphite furnace atomizers by chemosorbed oxygen, *Anal. Chem.,* 54, 630, 1982.

192. **Katskov, D. A. and Grinshtein, I. L.,** Formation of copper, silver and calcium acetylenides in graphite furnaces from atomic absorption analysis, *Zh. Prikl. Spektrosk.,* 36, 181, 1982.

193. **Jackwerth, E. and Salewski, S.,** Contribution to the multielement preconcentration from pure cadmium, *Z. Anal. Chem.,* 310, 108, 1982.

194. **Dillon, J. J., Hilderbrand, D. C., and Groon, K. S.,** APDC-MIBK extraction for the determination of trace amounts of weak-acid soluble silver in marine sediments by atomic absorption spectrophotometry, *At. Spectrosc.,* 3, 66, 1982.

195. **Balaes, G. E., Dixon, K., Russell, G. M., and Wall, G. J.,** Analysis of activated carbon as used in the carbon-in-pulp process for gold and eight other constituents, *S. Afr. J. Chem.,* 35, 4, 1982.

196. **Ohta, K., Smith, B. W., Suzuki, M., and Winefordner, J. D.,** The determination of atom vapor diffusion coefficients by high temperature gas chromatography with atomic absorption detection, *Spectrochim. Acta,* 37b, 343, 1982.

197. **Andronov, Y. G., Efimov, V. K., Polyakov, S. I., and Cherkason, V. A.,** Highly selective rapid method for the determination of silver in chemical-photographic industrial solutions, *Prikl. Fotogr. Kinematogr.,* 27, 129, 1982.

198. **Kauffman, R. E., Saba, C. S., Rhine, W. E., and Eisentraut, K. J.,** Quantitative multielement determination of metallic wear species in lubricating oils and hydraulic fluids, *Anal. Chem.,* 54, 975, 1982.

199. **Sickles, D. W., McLendon, R. E., and Rosenquist, T. H.,** Alternative method for quantitative anzyme histochemistry of muscle fibers. Application of photographic densitometry combined with atomic absorption spectrophotometry, *Histochemistry,* 73, 577, 1982.

200. **Harsanyi, E. G., Toth, K., Polos, L., and Pungor, E.,** Adsorption phenomena of silver iodide based ion selective electrodes, *Anal. Chem.,* 54, 1094, 1982.

AUTHOR INDEX

Super, A. B., 61
Suzuki, M., 196
Svehla, G., 8
Takahashi, M., 45
Takahashi, Y., 181
Takeuchi, T., 17, 144
Talalaev, B. M., 28, 39
Tamas, M., 145, 155
Tanaka, M., 149
Tatsii, Y. G., 168
Taylor, A., 120
Teliya, N. M., 183
Terashima, S., 97
Thomassen, Y., 78
Tindall, F. M., 11, 16, 107
Tomi, B., 102
Tomoda, T., 182
Torgov, V. G., 101
Toth, K., 200
Tsukahara, I., 45, 149
Uchida, T., 90
Usol'teva, M. V., 67, 98
Utsunomiya, T., 160
Vall, G. A., 64, 73, 98, 150
Van Dinh, P., 74
Vanifatova, N. G., 150
Varma, A., 190
Vasilev, E. N., 28
Venghiattis, A., 18
Verbeek, F., 59, 72

Viets, J. G., 123
Vilenkin, V. A., 126
Vnukovskaya, G. L., 81, 168
Wagenaar, H. C., 89
Wall, G. J., 195
Walton, G., 62
Warburton, J. A., 29, 162
Watling, R. J., 116
Welte, B., 184
West, P. K., 20
West, P. W., 20
West, T. S., 6, 24, 52
Wester, P. O., 169
Willis, J. B., 47
Wilson, L., 7
Winefordner, J. D., 55, 196
Wohlers, C. C., 83
Wold, L. T., 75
Woodriff, R., 61
Wu, D., 156
Wunderlich, E., 104
Yakovuk, V. I., 127, 174
Yamada, T., 182
Yanagisawa, M., 17
Young, R. S., 188
Yudelevich, I. G., 64, 73, 96, 98, 150, 187
Zdorov, E. P., 82
Zhang, H. F., 176
Zlomanova, G. G., 180
Zolotov, Y. A., 96, 98, 150

SUBJECT INDEX (SILVER)

Determination, 5, 6, 8, 16, 17, 22, 35, 36, 41, 55, 71, 72, 87, 90, 98, 100, 101, 105, 106, 109, 113, 116, 121, 135, 137, 138, 140, 148, 179, 180, 187
Determination in
air, 26, 74
alloys, 7, 13, 45, 50, 76, 86, 127, 128, 143, 147, 175
aluminum, 7, 45, 50, 69, 76, 95
ammonia solutions, 91
anodes, 33
antimony, 64
artificial zeolites, 102
biology, 88, 120, 131, 169, 181, 182, 199
blood, 120
cadmium, 63, 193
calcareous samples, 84
chromium, 99
commercial labs, 139
commodities, 32
complex solutions, 126, 165
concentrates, 1, 2, 33, 43, 51, 103
copper, 33, 45, 51, 56, 66, 104, 130, 149, 154, 171
corrosion, 190
dental material, 94, 182

drills, 57
dust, 159
electrolytes, 37, 38
foodstuffs, 32, 153
geological samples, 53, 123, 150
gold, 3, 51, 174, 186
halides, 96
hydraulic fluids, 198
industrial solutions, 39, 197
iron, 69, 76, 79, 118
lead, 1, 2, 141, 149
magnesium, 69
marine organisms, 85
metal halides, 96
metallurgy, 43, 53, 117
minerals, 12, 13, 19, 60, 62, 70, 82, 115
natural material, 124
nickel, 30, 76, 86, 136, 143, 147
oils, 9, 15, 52, 198
ores, 10, 31, 43, 46, 48, 54, 58, 62, 73, 75, 103, 129, 130, 154, 171, 176
petroleum products, 4
pewter, 167
plant intermediates, 14
plant liquors, 37
plants, 134, 145

GOLD, Au (ATOMIC WEIGHT 197.967)

Gold, both relatively and absolutely one of the rare elements, has been an object of human cupidity for the whole course of civilization. It occurs as free metal in nature owing to its chemical resistance and as well as alloyed with silver, bismuth, and tellurium in ores, rocks, and minerals. It is also present in some river beds and seawater. Gold is used to manufacture coins, jewelry, and dental alloys. It is also used for plating and for toning the silver image in photography. Gold isotopes are used in medicine to treat cancer and other diseases.

Gold determination is required for evaluation of ores, rocks, and minerals. It is also determined in alloys, cyanide mill solutions, and plating baths, etc. Sample preparation is by dissolution in aqua regia (four parts hydrochloric acid to one part nitric acid) or in cyanide solutions or in molten oxidizing alkaline salts.

Large amounts of iron, copper, and calcium suppress the analytical signal, when gold is extracted with MIBK (methyl isobutyl ketone). Cyanide complexes also depress the signal. Background correction is necessary for samples containing high acid concentrations or organic solvents. Matrix matching and standard addition methods will also eliminate any metal interferences. Platinum and palladium do not interfere. Gold can be determined in low-temperature flames, but with low sensitivity.

Standard Solution

To prepare 1000 mg/ℓ gold solution, dissolve 0.1000 g of gold metal in 10 mℓ of aqua regia, heat to near dryness, dissolve the residue in 5 mℓ hydrochloric acid, cool, and dilute to 100 mℓ with deionized water.

Instrumental Parameters

Wavelength	242.8 nm
Slit width	0.7 nm
Light source	Hollow cathode lamp
Flame	Air-acetylene, oxidizing (lean, blue)
Sensitivity	0.3 mg/ℓ
Detection limit	0.01 mg/ℓ
Optimum range	2—20 mg/ℓ
Secondary wavelengths and their sensitivities	
267.6 nm	0.2 mg/ℓ
312.3 nm	90 mg/ℓ
274.8 nm	110 mg/ℓ

REFERENCES

1. **Greaves, M. C.,** Determination of gold and silver in solution by atomic absorption spectroscopy, *Nature (London),* 199, 552, 1963.
2. **Ginzburg, V. L., Livshits, D. W., and Satarina, G. I.,** Atomic absorption flame photometric determination of silver, gold, palladium, platinum and rhodium, *Zh. Anal. Khim.,* 19, 1089, 1964.
3. **Skewes, H. R.,** The determination of gold in mill cyanide solutions by atomic absorption spectroscopy, *Aust. Inst. Min. Metall.,* 211, 217, 1964.
4. **Khalifa, H., Svehla, G., and Erdey, L.,** Precision of the determination of copper and gold by atomic absorption spectrophotometry, *Talanta,* 12, 703, 1965.
5. **Simmons, E. C.,** Gold assay by atomic absorption spectrophotometry, *At. Absorpt. Newsl.,* 4, 281, 1965.
6. **Tindall, F. M.,** Silver and gold assay by atomic absorption spectrophotometry, *At. Absorpt. Newsl.,* 4, 339, 1965.
7. **Strelow, F. W. E., Feast, E. C., Mathews, P. M., Bothma, C. J. C., and Van Zyl, C. R.,** Determination of gold in cyanide waste solutions by solvent extraction and atomic absorption spectrometry, *Anal. Chem.,* 38, 115, 1966.
8. **Aswathanarayana, R. and Vishnoi, D. N.,** Quantitative determination of gold by AAS and study of inter-element absorption effect on gold analysis, *Chem. Age (India),* 17, 532, 1966.
9. **Butler, L. R. P., Brink, J. A., and Engelbrecht, S. A.,** Automatic atomic absorption method for assaying gold in mill cyanide solutions, *Trans. S. Afr. Inst. Min. Metall.,* 76, C188, 1967.
10. **Giraud, J. L.,** The Use of Organic Solvents in Flame Emission and AAS Applied to the Determination of Traces of Gold in Arsenious Cobalt Ores, Thesis, University of Lyon, 1967 (in French).
11. **Van Sickle, G. H. and Lakin, H. W.,** Atomic absorption method for determination of gold in large samples of geological materials, *U.S. Geol. Surv.,* No. 561, 1968.
12. **Tomsett, S. L.,** Determination of lithium, strontium, barium and gold in biological material by atomic absorption spectrophotometry, *Proc. Assoc. Clin. Biochem.,* 5, 125, 1968.
13. **Groenwald, T.,** Determination of gold (I) in cyanide solutions by solvent extraction and AAS, *Anal. Chem.,* 40, 863, 1968.
14. **Pollack, E. N. and Anderson, S. I.,** The determination of gold by AA spectroscopy, *Anal. Chim. Acta,* 41, 441, 1968.
15. **Van Rensburg, H. C. and Zeeman, P. B.,** The determination of gold, titanium, palladium and rhodium by atomic absorption spectrophotometry with an ultrasonic nebulizer and a multi-element high density hollow cathode lamp with selective modulation, *Anal. Chim. Acta,* 43, 173, 1968.
16. **Zeeman, P. B. and Brink, J. A.,** The use of a non-absorbing reference line in the simultaneous determination of platinum, rhodium, palladium and gold by atomic absorption spectroscopy, *Analyst (London),* 93, 388, 1968.
17. **Lorber, A., Cohen, R. L., Chang, C. C., and Anderson, H. E.,** Gold determination in biological fluids by atomic absorption spectrophotometry: application to chrysotherapy in rheumatoid arthritis patients, *Arthritis Rheumatoid,* 11, 170, 1968.
18. **Aldous, K. M., Dagnall, R. M., and West, T. S.,** Preparation and spectral characteristics of microwave excited electrodeless discharge tubes for palladium, silver, platinum and gold, *Analyst (London),* 94, 347, 1969.
19. **Chao, T. T.,** Determination of gold in water in the nanogram range by anion exchange and atomic absorption spectrometry, *Econ. Geol.,* 64, 287, 1969.
20. **Groenwald, T.,** Quantitative determination of gold in solutions by solvent extraction and atomic absorption spectrometry, *Anal. Chem.,* 41, 1012, 1969.
21. **Van Loon, J. C.,** Determination of platinum, palladium and gold in silver assay bead by atomic absorption spectrophotometry, *Z. Anal. Chem.,* 246, 122, 1969.
22. **Zlatkis, A., Bruening, W., and Bayer, E.,** Determination of gold in natural waters at the parts per billion level by chelation and atomic absorption spectrometry, *Anal. Chem.,* 41, 1692, 1969.
23. **Eroshevich, T. A., Kukushkin, U. N., and Makarov, D. F.,** Atomic absorption determination of platinum metals, gold and silver in copper-nickel slurries and products of their reprocessing, *Zh. Prikl. Khim.,* 42, 2174, 1969.
24. **Smart, H. T. and Campbell, O. J.,** Determination of gold in sodium aurothiomalate injection B.P. by atomic absorption spectroscopy, *Can. J. Pharm. Sci.,* 4, 73, 1969.
25. **Carlson, G. G. and Van Loon, J. C.,** A study of the determination of gold in solutions containing high concentrations of other salts, *At. Absorpt. Newsl.,* 9, 90, 1970.
26. **Kallmann, S. and Hobart, E. W.,** Determination of silver, gold and palladium by a combined fire assay atomic absorption procedure, *Talanta,* 17, 845, 1970.
27. **Michailova, T. P. and Rezepina, V. A.,** Determination of gold and silver in plant liquors and electrolytes by atomic absorption spectrophotometry, *Analyst (London),* 95, 769, 1970.

28. **Michailova, V. A., Baranov, S. V., Aleksandrov, V. V., Sasov, V. N., and Rezepina, V. A.,** Atomic absorption determination of gold in electrolytes, *Izv. Sib. Otd. Akad. Nauk SSSR Ser. Khim. Nauk,* 2, 107, 1970.

29. **Michailova, T. P., Rezepina, V. A., and Dolsenko, L. G.,** Atomic absorption method for determining gold and silver in electrolytes and technological solutions, *Zh. Anal. Khim.,* 25, 1477, 1970.

30. **Yudelevich, I. G., Vall, G. A., Torgov, V. G., and Korda, T. M.,** Extraction-atomic absorption determination of gold in solutions, *Zh. Anal. Khim.,* 25, 870, 1970.

31. **Ivanov, N. P., Demidov, A. A., Baranov, S. V., Mikhel'son, D. M., Kulashev, A. V., and Khyazav, V. V.,** Determination of gold by atomic absorption spectrophotometry, *Tr. Vses. Nauchno-Issled. Inst. Khim. Reakt. Osovo Chist. Khim. Veshchestv.,* 32, 167, 1970.

32. **Fishkova, N.,** Atomic absorption determination of platinum, palladium and gold in silver regulus of test melting, *Zavod. Lab.,* 36, 1461, 1970.

33. **Mallet, R. C.,** Review of techniques for the determination of gold and silver by atomic absorption spectroscopy, *Miner. Sci. Eng.,* 2, 128, 1970.

34. **Aggett, J. and West, T. S.,** Atomic absorption and fluorescence spectroscopy with a carbon filament atom reservoir. Determination of gold by atomic fluorescence and atomic absorption spectroscopy with an unclosed atom reservoir, *Anal. Chim. Acta,* 55, 349, 1971.

35. **Dunckley, J. V.,** Estimation of gold in serum by atomic absorption spectroscopy, *Clin. Chem. (Winston-Salem, N.C.),* 17, 992, 1971.

36. **Groenwald, T. and Jones, B. M.,** Determination of gold in solutions of thiourea, *Anal. Chem.,* 43, 1689, 1971.

37. **Hildon, M. A. and Sully, G. R.,** Determination of gold in the ppb and ppm range by atomic absorption spectrophotometry, *Anal. Chim. Acta,* 54, 245, 1971.

38. **Ichinose, N.,** Extraction and determination of metal salts with methyl isobutyl ketone. Atomic absorption spectrophotometric determination of gold in ores, *Bunseki Kagaku,* 20, 660, 1971.

39. **Kerber, J. D.,** The direct determination of gold in polyester fibers with the HGA-70 graphite, *At. Absorpt. Newsl.,* 10, 104, 1971.

40. **Uchida, T., Iida, C., Yashuhara, N., and Nakagawa, M.,** Determination of gold in biological materials by atomic absorption spectroscopy, *Anal. Lett.,* 4, 555, 1971.

41. **Willis, J. B.,** Atomization problems in atomic absorption spectroscopy. III. Absolute atomization efficiencies of sodium, copper, silver and gold in a meker-type air-acetylene flame, *Spectrochim. Acta,* 26b, 177, 1971.

42. **Fishkova, N. L. and Falkova, O. B.,** Atomic absorption analysis of solutions containing gold, *Tr. Isent. Nauchno-issled. Gornorazved. Inst. Tsvet. Redk. Blagorod. Metall.,* 97, 143, 1971.

43. **Malykh, V. D., Erkovich, G. E., Goncharova, N. N., and Shipitsin, S. A.,** Determination of gold and silver in cyanide solutions by atomic absorption, *Nauk Tr. Irkutzk. Gos. Nauchno-issled. Inst. Redk. Tsvet. Metall.,* 22, 111, 1971 (in Russian).

44. **Palmer, I. and Streichert, G.,** Co-precipitation of noble metals with tellurium (I), platinum, palladium, rhodium and gold, *Natl. Inst. Metall. Repub. S. Afr. Rep.,* No. 1273, 1971.

45. **Bratzel, M. P., Jr., Chakrabarti, C. L., Sturgeon, R. E., McIntyre, M. W., and Agemian, H.,** Determination of gold and silver in parts per billion or lower levels in geological metallurgical samples by atomic absorption spectrometry with a carbon rod atomizer, *Anal. Chem.,* 44, 372, 1972.

46. **Balazs, N. D. H., Pole, D. J., and Masarei, J. R.,** Determination of gold in body fluids by atomic absorption spectrophotometry, *Clin. Chim. Acta,* 40, 213, 1972.

47. **Fishkova, N. L., Falkova, O. B., and Meshalkina, R. D.,** Atomic absorption analysis of solutions containing gold, *Zh. Anal. Khim.,* 27, 1916, 1972.

48. **Galanova, A. P., Pronin, V. A., Vall, G. A., Yudelevich, I. G., and Gilbert, E. N.,** Extraction and atomic absorption method for determining gold in ores and their reprocessing products, *Zavod. Lab.,* 38, 646, 1972.

49. **Jager, H.,** The use of a slow-discharge lamp as a light source in the spectrometric analysis of gold, *Anal. Chim. Acta,* 58, 57, 1972.

50. **Bazhov, A. S. and Sokolova, E. A.,** Atomic absorption determination of gold in ores, *Zh. Anal. Khim.,* 27, 2442, 1972.

51. **Britske, M. E., Ioffe, V. P., Saveleva, A. N., Slabodenyuk, I. V., and Khirulina, N. P.,** Determination of silver and gold in copper industry products by atomic absorption, *Zavod. Lab.,* 38, 1458, 1972.

52. **Yudelevich, I. G. and Vilyugina, M. D.,** Determination of palladium, platinum, rhodium and gold by an atomic absorption method, *Izv. Sib. Otd. Akad. Nauk SSSR Ser. Khim. Nauk,* 166, 1972.

53. **Beleva-Naumova, S. and Dancheva, R.,** Atomic absorption determination of gold, *Rudodobiv,* 27, 25, 1972 (in Bulgarian).

54. **Steele, T. W., Mallet, R. C., Pearton, D. C. G., and Erving, E. J.,** Determination by atomic absorption spectrophotometry of platinum, palladium, rhodium, ruthenium, gold, silver and lead in drills, *Lab. Meth.,* No. 78/4, 1972.

55. **Adriaenssens, E. and Verbeek, F.,** Atomic absorption spectrophotometry of silver, gold, palladium and platinum in a potassium cyanide medium, *At. Absorpt. Newsl.,* 12, 57, 1973.

56. **Aggett, J.,** Determination of gold in serum by atomic absorption spectrometry with a carbon filament atom reservoir, *Anal. Chim. Acta,* 63, 473, 1973.

57. **Harth, M., Haines, D. S. M., and Bondy, D. C.,** A simple method for the determination of gold in serum, blood and urine by atomic absorption spectroscopy, *Am. J. Clin. Pathol.,* 59, 423, 1973.

58. **Sen Gupta, J. G.,** Determination of gold, platinum group metals and some common metals in native silver by atomic absorption spectrometry, *Anal. Chim. Acta,* 63, 19, 1973.

59. **Jackwerth, E., Hoehn, R., and Koos, K.,** Trace enrichment by partial dissolution of the matrix in presence of mercury. Determination of bismuth, copper, lead, nickel, silver, gold and palladium in high purity cadmium by atomic absorption spectrometry, *Z. Anal. Chem.,* 264, 1, 1973 (in German).

60. **Dietz, A. A. and Rubinstein, H. M.,** Serum gold. I. Estimation by atomic absorption spectroscopy, *Ann. Rheum. Dis.,* 32, 124, 1973.

61. **Dunckley, J. V.,** Estimation of gold in urine by atomic absorption spectroscopy, *Clin. Chem. (Winston-Salem, N.C.),* 19, 1081, 1973.

62. **Galanova, A. P., Kudryavina, A. K., Pronin, V. A., Yudelevich, I. G., Vall, G. A., and Gilbert, E. W.,** Extraction-atomic absorption determination of gold and silver in antimony containing products, *Izv. Sib. Otd. Akad. Nauk SSSR Ser. Khim. Nauk,* 89, 1973.

63. **Kreimer, S. E., Rakhlina, M. L., Lomekhov, A. S., Mikhailov, P. M., and Khazov, V. S.,** Chemical atomic absorption determination of palladium, gold and platinum, *Zavod. Lab.,* 39, 947, 1973.

64. **Rubinstein, H. M. and Dietz, A. A.,** Serum gold. II. Levels in rheumatoid arthritis, *Ann. Rheum. Dis.,* 32, 128, 1973.

65. **Schaefer, C. and Vomhof, D. W.,** Determination of gold in jewellery alloys by atomic absorption spectrometry, *At. Absorpt. Newsl.,* 12, 133, 1973.

66. **Sen Gupta, J. G.,** A review of the methods for the determination of platinum group metals, silver and gold by atomic absorption spectroscopy, *Miner. Sci. Eng.,* 5, 207, 1973.

67. **Yudelevich, I. G. and Vall, G. A.,** Determination of gold in organic solvents by atomic absorption method, *Zh. Anal. Khim.,* 28, 1076, 1973.

68. **Makarov, D. F., Kukushkin, Yu. N., and Eroshevich, T. A.,** Atomic absorption determination of gold and silver in black and cathodic copper, *Zh. Prikl. Khim. (Leningrad),* 46, 656, 1973 (in Russian).

69. **Pronin, V. A., Galanova, A. P., Kudryavina, A. K., Shastina, Z. N., Apolitskii, V. N., and Usol'teva, M. V.,** Stability of standard solutions of gold, silver and platinum used in atomic absorption analysis, *Zh. Anal. Khim.,* 28, 2328, 1973.

70. **Serebryanyi, B. L., Fishkova, N. L., Petrukhin, O. N., and Rakoviskii, E. E.,** Atomic absorption determination of gold after its extraction by triphenylphosphine from hydrochloric acid and cyanide solutions, *Zh. Anal. Khim.,* 28, 2333, 1973.

71. **Mizuike, A., Fukuda, K., and Sakamoto, T.,** Rapid collection of traces of silver (or gold or palladium) with mercury in an ultrasonic field, *Bull. Chem. Soc. Jpn.,* 46, 3596, 1973.

72. **Bowditch, D. C.,** Comparative study of three analytical procedures for the collection and determination of gold and platinoids in precious metal bearing ores, *Aust. Miner. Dev. Lab. Bull.,* 15, 71, 1973.

73. **Shelkovnikova, O. S., Chuchalin, L. K., Vall, G. A., Korda, T. M., and Yudelevich, I. G.,** Extraction of gold (III) by tributyl phosphate from chloride and its atomic absorption determination, *Zh. Anal. Khim.,* 28, 2147, 1973.

74. **Vall, G. A., Usol'tseva, M. V., Pronin, V. A., Shipitshin, S. A., Skudaev, U. D., and Yudelevich, I. G.,** Extraction-atomic absorption determination of gold in lead products using a furnace flame atomizer, *Izv. Sib. Otd. Akad. Nauk SSSR Ser. Khim. Nauk,* 5, 71, 1974.

75. **Eckelmans, V., Graauwmans, E., and De Jaegere, S.,** Mutual interference of gold, platinum and palladium in atomic absorption spectroscopy, *Talanta,* 21, 715, 1974.

76. **Maessen, F. J. M. J. and Posma, F. D.,** Direct determination of gold, cobalt and lithium in blood plasma using the minimassmann carbon atomizer, *Anal. Chem.,* 46, 1445, 1974.

77. **Makarov, D. F., Kukushkin, Y. N., and Eroshevich, T. A.,** Determination of gold, platinum, palladium and rhodium in a silver collector by an atomic absorption method, *Zh. Prikl. Khim.,* 47, 1212, 1974.

78. **Fishkova, N. L.,** Determination of the platinum metals, gold and silver by atomic absorption spectrophotometry, *Zh. Anal. Khim.,* 29, 2121, 1974.

79. **Hovorka, D. and Jaros, M.,** Gold abundances in west carpathian ultramatic rocks, *Geol. Sb. (Bratislava),* 25, 355, 1974.

80. **Men'shikov, V. I., Malykh, V. D., and Shestakova, T. D.,** Direct determination of gold in solids by an atomic absorption method, *Zh. Anal. Khim.,* 29, 2132, 1974.

81. **Strong, B. and Murray-Smith, R.,** Determination of gold in copper-bearing sulfide ores and metallurgical flotation products by atomic absorption spectrometry, *Talanta,* 21, 1253, 1974.

82. **Adriaenssens, E. and Verbeek, F.,** The determination of silver, gold, palladium, platinum by a combined fire-assay atomic absorption source, *At. Absorpt. Newsl.,* 13, 141, 1974.

83. **Bogoslovskaya, M. N., Birger, G. I., Bruk, B. S., Gromova, T. I., and Satarina, G. I.,** Atomic absorption determination of gold in ores and tailings, *Zavod. Lab.,* 41, 683, 1975.

84. **Fishkova, N. L., Zdorova, E. P., and Popova, N. N.,** Determination of low gold and silver contents in mineral raw materials by the assay-atomic absorption method, *Zh. Anal. Khim.,* 30, 806, 1975.

85. **Musha, S. and Takahashi, Y.,** Enrichment of trace amounts of gold in water utilizing the coagulation of soybean protein and its determination by atomic absorption spectrometry and emission spectrography. II. Enrichment of trace metals utilizing the coagulation of soybean protein, *Bunseki Kagaku,* 24, 395, 1975.

86. **Mihalka, S. and Resmann, A.,** Comparing error in the determination of gold in ores by Docimastic, spectrophotometric and atomic absorption methods, *Rev. Chim.,* 26, 875, 1975.

87. **Jackwerth, E. and Willmer, P. G.,** Separation of traces of gold and palladium from high purity cadmium, indium, nickel, lead and zinc with subsequent determination by graphite tube atomization, *Talanta,* 23, 197, 1976.

88. **Moloughney, P. E. and Faye, G. H.,** A rapid assay atomic absorption method for the determination of platinum palladium and gold in ores and concentrates: a modification of the tin collection scheme, *Talanta,* 23, 377, 1976.

89. **Bazhov, A. S., Sokolova, E. A., and Shcherbov, D. P.,** Extraction-atomic absorption determination of low gold content in geological samples, *Zh. Anal. Khim.,* 31, 1098, 1976.

90. **Das, N. R. and Bhattacharya, S. N.,** Solvent extraction of gold, *Talanta,* 23, 535, 1976.

91. **Kamel, H., Brown, D. H., Ottaway, J. M., and Smith, W. E.,** Determination of gold in blood fractions by atomic absorption spectrometry using carbon rod and carbon furnace atomization, *Analyst (London),* 101, 790, 1976.

92. **Machiroux, R. and Anh, D. T. K.,** Application of flameless atomic absorption spectrometry to the determination of gold in raw sulfide ores of zinc and lead, *Anal. Chim. Acta,* 86, 35, 1976 (in French).

93. **McElhaney, R. J.,** Determination of gold, silver and cobalt in aluminum by flameless atomic absorption spectroscopy, *J. Radioanal. Chem.,* 32, 99, 1976.

94. **Singhinolfi, G. P. and Santos, A. M.,** Determination of gold in geological samples at parts per million levels by flameless atomic absorption spectroscopy, *Mikrochim. Acta,* 2, 33, 1976.

95. **Sukiman, S.,** Determination of gold in ores by atomic absorption spectrometry after chromatographic separation, *Anal. Chim. Acta,* 84, 419, 1976.

96. **Chowdhury, A. N. and Das, A. K.,** Determination of gold in rocks and minerals by atomic absorption spectrophotometry, *Indian J. Technol.,* 14, 353, 1976.

97. **Gurin, P. A. and Chechuli, L. I.,** Determination of low concentrations of gold by a combined chemical and atomic absorption method, *Nauchn. Tr. Irkutsk. Gos. Nauchno-Issled. Inst. Redk. Tsvetn. Met.,* 28, 54, 1976.

98. **Bazhov, A. S. and Sokolova, E. A.,** Solvent extraction-atomic absorption determination of silver and gold, *Zh. Anal. Khim.,* 32, 65, 1977.

99. **Moloughney, P. E.,** An abbreviated fire assay atomic absorption method for the determination of gold and silver in ores and concentrates, *Talanta,* 24, 135, 1977.

100. **Rubeska, I., Koreckova, J., and Weiss, D.,** The determination of gold and palladium in geological materials by atomic absorption after extraction with dibutyl sulfide, *At. Absorpt. Newsl.,* 16, 1, 1977.

101. **Tindall, F. M.,** Hints of chemical analysis. I. Tungsten determination by atomic absorption spectrophotometry. II. Revised notes on gold and silver determination by atomic absorption spectrometry. III. Mercury in copper concentrates by atomic spectrophotometry, *At. Absorpt. Newsl.,* 16, 37, 1977.

102. **Dunckley, J. V. and Staynes, F. A.,** Estimation of gold in urine by flameless atomic absorption spectroscopy, *Ann. Clin. Biochem.,* 14, 53, 1977.

103. **Fishkova, N. L.,** Atomic absorption determination of platinum, palladium and gold in ores of complex composition, *Zh. Anal. Khim.,* 32, 1776, 1977.

104. **Kamel, H., Brown, D. H., Ottaway, J. M., and Smith, W. E.,** Determination of gold in tissue by carbon furnace atomic absorption spectrometry, *Talanta,* 24, 309, 1977.

105. **Men'shikov, V. I., Khlebnikova, A. A., Tsykhanskii, V. D., and Malykh, V. D.,** Atomic absorption determination of gold micro amounts in geochemical sample by using pulsed atomizers, *Zh. Anal. Khim.,* 32, 954, 1977.

106. **Schattenkichner, M. and Grobenski, Z.,** The measurement of gold in blood and urine by atomic absorption in the treatment of rheumatoid arthritis, *At. Absorpt. Newsl.,* 16, 84, 1977.

107. **Torgov, V. G. and Khlebnikova, A. A.,** Atomic absorption determination of gold in a flame and a flameless graphite atomizer with prior solvent extraction with petroleum sulfides, *Zh. Anal. Khim.,* 32, 960, 1977.

108. **Ward, R. J., Danpure, C. J., and Fyfe, D. A.,** Determination of gold in plasma and plasma fractions by atomic absorption spectrometry and by neutron activation analysis, *Clin. Chim. Acta,* 81, 87, 1977.

109. **Bea-Barredo, F., Polo-Polo, C., and Polo-Diez, L.,** The simultaneous determination of gold, silver and cadmium at ppb levels in silicate rocks by atomic absorption spectrometry with electrothermal atomization, *Anal. Chim. Acta,* 94, 283, 1977.

110. **Belskii, N. K., Nebol'sina, L. A., Yuzko, M. I., Fomina, T. A., Shubochkina, E. F., and Shubochkina, L. K.,** Atomic absorption determination of gold, palladium, platinum and rhodium in silver alloys, *Zh. Anal. Khim.,* 33, 336, 1978.

111. **Barrett, M. J., DeFries, R., and Henderson, W. M.,** Rapid determination of gold in whole blood of arthritis patients using flameless atomic absorption spectrophotometry, *J. Pharm. Sci.,* 67, 332, 1978.

112. **Prudnikov, E. D., Kolosova, L. P., Kalachev, V. K., Shapkin, Y. S., Novatskaya, N. V., and Bychkov, Y. A.,** Pulse atomic absorption determination of trace amounts of platinum metals, gold and silver in natural materials, *Zh. Anal. Khim.,* 33, 468, 1978.

113. **Aleksandrov, S., Gyulmezova, G., and Sanabria, J.,** A study on the possibility for emission spectral and atomic absorption determination of silver, gold and palladium in anode film, *Dokl. Bolg. Akad. Nauk,* 31, 73, 1978.

114. **Fishkova, N. L. and Vilenkin, V. A.,** Application of an HGA-74 graphite atomizer to atomic absorption determination of gold, silver, platinum and palladium in solutions of complicated composition, *Zh. Anal. Khim.,* 33, 897, 1978 (in Russian).

115. **Kahn, N. and Van Loon, J. C.,** Direct atomic absorption spectrophotometric analysis of anion complexes of platinum and gold after their concentration and separation from aqueous solutions by anion exchange chromatography, *Anal. Lett.,* A11, 991, 1978.

116. **Robert, R. V. D. and Mallet, R. C.,** Measurements of trace amounts of gold in solution by atomic absorption spectrophotometry and carbon rod atomization, *Natl. Inst. Metall. Repub. S. Afr. Rep.,* No. 1948, 1978.

117. **Konovalov, G. F. and Gur'eva, M. P.,** Extraction-atomic absorption determination of silver and gold, *Zavod. Lab.,* 45, 196, 1979.

118. **Parkes, A. and Murray-Smith, R.,** A rapid method for the determination of gold and palladium in soils and rocks, *At. Absorpt. Newsl.,* 18, 57, 1979.

119. **Wawschinek, O. and Rainer, F.,** Determination of gold in serum and urine by atomic absorption after extraction with dimorpholinethiuramdisulfide in methyl isobutyl ketone, *At. Absorpt. Newsl.,* 18, 50, 1979.

120. **Gilbert, E. N., Androsova, N. N., and Badmaeva, Z. O.,** Flameless solvent extraction-atomic absorption determination of gold and palladium in molybdenites, *Zh. Anal. Khim.,* 34, 1150, 1979.

121. **Hall, S. H.,** A rapid method for gold extraction using MIBK, *At. Absorpt. Newsl.,* 18, 126, 1979.

122. **Katskov, D. A. and Grinshtein, I. L.,** Study of the chemical interaction of copper, gold and silver with carbon by an atomic absorption method using an electrothermal atomizer, *Zh. Prikl. Spektrosk.,* 30, 787, 1979.

123. **Prudnikov, E. D. and Rantsev, A. A.,** Atomic absorption determination of gold and silver using a flameless atomizer in an air atmosphere, *Zh. Anal. Khim.,* 34, 1725, 1979.

124. **Savel'eva, A. N. and Agpova, T. E.,** Effect of sample composition on determination of gold by flameless atomic absorption, *Zh. Anal. Khim.,* 34, 1733, 1979.

125. **Shestopalova, L. F. and Shestakova, T. D.,** Possibility of quantitative determination of gold in its solutions by an atomic absorption method, *Kolloidn. Zh.,* 41, 615, 1979.

126. **Wawschinek, O.,** Determination of copper and gold in serum by flameless atomic absorption, *Mikrochim. Acta,* 2, 111, 1979.

127. **Brandvold, L. A.,** Reliability of gold and silver analyses by commercial labs in the southwest, *N.M. Geol.,* 1, 11, 1979.

128. **Kantor, T., Bezur, L., Pungor, E., Fodor, P., Nagy-Balogh, J., and Heincz, G.,** Determination of the thickness of silver, gold and nickel layers by a laser microprobe and flame atomic absorption technique, *Spectrochim. Acta,* 34b, 341, 1979.

129. **Dunckley, J. V., Grennan, D. M., and Palmer, D. G.,** Estimation of serum and urine gold by atomic absorption spectroscopy in rheumatoid patients receiving gold therapy, *J. Anal. Toxicol.,* 3, 242, 1979.

130. **Jackwerth, E., Hoehn, R., and Musaick, K.,** Preconcentration of traces of silver, gold, bismuth, copper and paladium from pure lead by partial precipitation of the matrix with sodium borohydride as a reducing agent, *Z. Anal. Chem.,* 299, 362, 1979.

131. **Guerra, R.,** Rapid determination of gold and copper in lead/tin-based solder, *At. Spectrosc.,* 1, 58, 1980.

132. **Haddon, M. J. and Pantony, D. A.,** Specific solvent extraction method for the determination of gold in ores and products, *Analyst (London),* 105, 371, 1980.

133. **Kantor, T., Bezur, L., Pungoe, E., Fodor, P., Nagy-Balogh, J., and Heincz, G.,** Determination of thickness of silver, gold and nickel layers by a laser microprobe and flame atomic absorption technique, *Magy. Kem. Lapja,* 35, 266, 1980.

134. **Meier, A. L.,** Flameless atomic absorption determination of gold in geological materials, *J. Geochem. Explor.,* 13, 77, 1980.

135. **Melethil, S., Poklis, A., and Sagar, V. A.,** Binding of gold to bovine serum albumin using flameless atomic absorption, *J. Pharm. Sci.,* 69, 585, 1980.

136. **Tsukahara, I. and Tanaka, M.,** Determination of gold in silver, copper, lead, selenium and anode slime by atomic absorption spectrometry, *Talanta,* 27, 655, 1980.

137. **Young, R. S.,** Analysis for gold, *Gold Bull.,* 13, 9, 1980.

138. **Balaes, G. and Robert, R. V. D.,** Analysis by atomic absorption spectrophotometry of activated charcoal, *Natl. Inst. Metall. Repub. S. Afr. Rep.,* No. 2060, 1980.

139. **Brandt, P. J., Van Dalen, J. H., and Wessels, F. W.,** Automatic determination of trace amounts in gold-plant barren solutions, *J. S. Afr. Inst. Min. Metall.,* 80, 197, 1980.

140. **Cambel, B., Stresko, V., and Skerencakova, O.,** Content of gold in pyrites of various genesis, *Geol. Zb.,* 31, 139, 1980.

141. **Lo, T. S., Chou, L., Tu, C. C., and Liu, H. L.,** Plastic foam adsorption and thiourea desorption in atomic absorption determination of gold, *Fen Hsi Hua Hsueh,* 8, 177, 1980 (in Chinese).

142. **Royal, S. J.,** Aerosol deposition and carbon rod atomization of gold, *Natl. Inst. Metall. Repub. S. Afr. Rep.,* No. 2063, 1980.

143. **Royal, S. J., Robert, R. V. D., Ormrod, G. T. M., and Mallet, R. C.,** On-line monitoring of gold in barren solutions, *Natl. Inst. Metall. Repub. S. Afr. Rep.,* No. 2064, 1980.

144. **Clark, J. R. and Viets, J. G.,** Multielement extraction system for the determination of 18 trace elements in geochemical samples, *Anal. Chem.,* 53, 61, 1981.

145. **Clark, J. R. and Viets, J. G.,** Back-extraction of trace elements from organometallic halide extracts for determination of flameless atomic absorption spectrometry, *Anal. Chem.,* 53, 65, 1981.

146. **Kontas, E.,** Rapid determination of gold by flameless atomic absorption spectrometry in the ppb and ppm ranges without organic solvent extraction, *At. Spectrosc.,* 2, 59, 1981.

147. **Korda, T. M., Zelentsova, L. V., and Yudelevich, I. G.,** Solvent selection for organic sulfides for extraction-atomic absorption determination of gold and palladium with electrothermal atomization, *Zh. Anal. Khim.,* 36, 86, 1981 (in Russian).

148. **Hernandez, M. J., Carabias Martinez, R., and Hernandez Hernandez, P.,** Electrochemical preconcentration of gold traces and post determination by atomic absorption spectrometry, *Afinidad,* 38, 28, 1981.

149. **Brooks, R. R., Holzbecher, J., Ryan, D. E., Zhang, H. F., and Chatterjee, A. K.,** A rapid method for the determination of gold and silver in sulfide ores and rocks, *At. Spectrosc.,* 2, 151, 1981.

150. **Siems, D. F., Meier, A. L., and Lesure, F. G.,** Analyses and description of geochemical samples, Shining Rock Wilderness, Haywood County, North Carolina, *U.S. Geol. Surv. Open-File Rep.,* 81-593, 1, 1981.

151. **Brooks, R. R., Chatterjee, A. K., Amulya, K., and Ryan, D. E.,** Determination of gold in natural waters at the parts-per-trillion (pg cm^{-3}) level, *Chem. Geol.,* 33, 163, 1981.

152. **Yang, J., Ye, S., Gui, F., and Wang, J.,** Use of octyl disulfide in the analysis of noble metals. I. Octyl disulfide-xylene extraction and atomic absorption spectrometric determination of small amounts of gold and palladium in ores, *Fen Hsi Hua Hsueh,* 9, 323, 1981 (in Chinese).

153. **Diamantatos, A.,** A solvent extraction scheme for the determination of platinum, palladium, rhodium, iridium and gold in platiniferrous materials, *Anal. Chim. Acta,* 131, 53, 1981.

154. **Kao, S. G., Wei, L., Chang, C. P., and Hsu, N. N.,** Preparation of gold-1 wt% beryllium master alloy, *Kuang Yeh,* 25, 125, 1981 (in Chinese).

155. **Kovarskii, N. Y., Kovekovdova, L. T., Pryazhevskaya, I. S., Belenskii, V. S., Shapovalov, E. N., and Popkova, S. M.,** Preconcentration of trace elements from sea water by electro-deposited magnesium hydroxide, *Zh. Anal. Khim.,* 36, 2264, 1981.

156. **Saveleva, A. N. and Agpova, T. E.,** Determination of gold, *Otkrytiya Izobret. Prom. Obraztsy Tovarnye Znaki,* 35, 99, 1981.

157. **Aihara, M. and Kiboku, M.,** Extraction and atomic absorption spectrophotometry of gold by using potassium xanthates, *Bunseki Kagaku,* 30, 394, 1981.

158. **Futekov, L., Dobreva, D., and Specker, H.,** AAS determination of gold in the ppm to ppb range in selenium and ores after enrichment by extraction and adsorption, *Z. Anal. Chem.,* 306, 381, 1981.

159. **Petrosyn, R. A., Eginyan, O. S., Boyadzhyan, V. K., and Khachatryan, R. M.,** Atomic absorption determination of palladium and gold in a silicon dioxide-based catalyst, *Arm. Khim. Zh.,* 34, 209, 1981.

160. **Turkall, R. M. and Bianchine, R.,** Determination of gold in tissue and feces by atomic absorption spectrophotometry using carbon rod atomization, *Analyst (London),* 106, 1096, 1981.

161. **Young, R. S.,** Analysis of nickel refinery slimes and residues, *Talanta,* 28, 25, 1981.

162. **Bazhov, A. S. and Soklova, E. A.,** Gold Determination, USSR SU 880,985 (CI.COIG7/00), 1981; Appl. 2,843,784, 1979; from *Otkrytiya Izobret. Prom. Obraztsy Tovarnye Znaki,* 42, 118, 1981.

163. **Campbell, W. L.,** Manganese dioxide causes spurious gold values in flame atomic absorption readings from hydrogen bromide-bromine digestions, *J. Geochem. Explor.,* 15, 613, 1981.

164. **Varma, A.,** Analytical techniques for corrosion studies on noble metals, *Talanta,* 28, 701, 1981.

165. **Zhang, Z. and Lin, J.,** Determination of trace gold in geological samples by graphite furnace atomic absorption spectrometry, *Fenxi Huaxue,* 9, 703, 1981 (in Chinese).

166. **Demchenko, V. Y., Prischep, N. N., and Pilipenko, A. T.,** Use of pyrazolone-5-(4-azo-1')-2-naphthol for the extraction-atomic absorption determination of gold (III), *Zavod. Lab.,* 47, 5, 1981.

167. **Yudelevich, I. G. and Startseva, E. A.,** Extraction-atomic absorption methods for the determination of gold and silver, *Zavod. Lab.,* 47, 24, 1981.

168. **Anoshin, G. N., Lepezin, G. G., Mel'gunov, S. V., Mirievskaya, O. S., and Tsimbalist, V. G.,** Behavior of gold and silver during progressive metamorphism (as illustrated by the metamorphic complexes of the Altai-Sayan folded region of USSR), *Dokl. Nauk USSR,* 262, 1477, 1982 (in Russian).

169. **Rodgers, A. I. A., Brown, D. H., Smith, W. E., Lewis, D., and Capell, H. A.,** Distribution of gold in blood following administration of myocrisin and auranofin, *Anal. Proc. (London),* 19, 87, 1982.

170. **Louail, N., Bol'shova, T. A., and Alimarin, I. P.,** Extraction chromatographic preconcentration of gold from a mixture of elements by using *o*-isopropyl *N*-methylthiocarbamate, *Vestn. Mosk. Univ. Ser. 2, Khim.,* 23, 138, 1982.

171. **Johnsen, A. C., Wibetoe, G., Langmyhr, F. J., and Aaseth, J.,** Atomic absorption spectrometric determination of the total content and distribution of copper and gold in synovial fluid from patients with rheumatoid arthritis, *Anal. Chim. Acta,* 135, 243, 1982.

172. **Balaes, G. E., Dixon, K., Russell, G. M., and Wall, G. J.,** Analysis of activated carbon as used in the carbon-in-pulp process, for gold and eight other constituents, *S. Afr. J. Chem.,* 35, 4, 1982.

173. **Sharma, R. P.,** A microanalytical method for the analysis of gold in biological media by flameless atomic absorption spectroscopy, *Ther. Drug Monit.,* 4, 219, 1982.

174. **Anoshin, G. N., Volynets, O. N., and Flerov, G. B.,** Geochemistry of gold and silver in basalts of the Great Fissure Tolbachik Eruption (USSR) 1975-76, *Dokl. Akad. Nauk SSSR,* 264, 195, 1982.

175. **Werbicki, J. J., Jr.,** How much gold?, *Prod. Finish. (London),* 46, 42, 1982.

176. **Gabriel, N. E. and Law, H. H.,** New analytical procedure for determining the gold content of anion exchange resins, *Annu. Tech. Conf., Proc. Am. Electroplat. Soc.,* 69th, Paper A-3, American Electroplating Society, Winter Park, Fla., 1982.

AUTHOR INDEX

Aaseth, J., 171
Adriaenssens, E., 55, 82
Agemian, H., 45
Aggett, J., 34, 56
Agpova, T. E., 124, 156
Aihara, M., 157
Aldous, K. M., 18
Aleksandrov, S., 113
Aleksandrov, V. V., 28
Alimarin, I. P., 170
Amulya, K., 151
Anderson, H. E., 17
Anderson, S. I., 14
Androsova, N. N., 120
Anh, D. T. K., 92
Anoshin, G. N., 168, 174
Apolitskii, V. N., 69
Aswathanarayana, R., 8
Badmaeva, Z. O., 120
Balaes, G. E., 138, 172
Balasz, N. D. H., 46
Baranov, S. V., 28, 31
Barrett, M. J., 111
Bayer, E., 22
Bazhov, A. S., 50, 89, 98, 162
Bea-Barredo, F., 109
Belenskii, V. S., 155
Belskii, N. K., 110
Beleva-Naumova, S., 53
Bezur, L., 128, 133
Bhattacharya, S. N., 90
Bianchine, R.,160
Birger, G. I., 83
Bogoslovskaya, M. N., 83
Bol'shova, T. A., 170
Bondy, D. C., 57
Bothma, C. J. C., 7
Boyadzhyan, V. K., 159
Bowditch, D. C., 72
Brandt, P. J., 139
Brandvold, L. A., 127
Bratzel, M. P., 45
Brink, J. A., 9, 16
Britske, M. E., 51
Brooks, R. R., 149, 151
Brown, D. H., 91, 104, 169
Bruening, W., 22
Bruk, B. S., 83
Butler, L. R. P., 9
Bychkov, Y. A., 112
Cambel, B.,140
Campbell, O. J., 24
Campbell, W. L., 163
Capell, H. A., 169
Carabias Martinez, R., 148
Carlson, G. G., 25
Chakrabarti, C. L., 45
Chang, C. C., 17
Chang, C. P., 154

Chao, T. T., 19
Chatterjee, A. K., 149, 151
Chechuli, L. I., 97
Chou, L., 141
Chowdhury, A. N., 96
Chuachalin, L. K., 73
Clark, J. R., 144, 145
Cohen, R. L., 17
Dagnall, R. M., 18
Dancheva, R., 53
Danpure, C. J., 108
Das, A. K., 96
Das, N. R., 90
DeFries, R., 111
De Jaegere, S., 75
Demchenko, V. Y., 166
Demidov, A. A., 31
Diamantatos, A., 153
Dietz, A. A., 60, 64
Dixon, K., 172
Dobreva, D., 158
Dolsenko, L. G., 29
Dunckley, J. V., 35, 102
Eckelmans, V., 75
Eginyan, O. S., 159
Engelbrecht, S. A., 9
Erdey, L., 4
Erkovich, G. E., 43
Eroshevich, T. A., 23, 68, 77
Falkova, O. B., 42, 47
Faye, G. H., 88
Feast, E. C., 7
Fishkova, N. L., 32, 42, 47, 70, 78, 84,103, 114
Flerov, G. B., 174
Fodor, P., 128, 133
Fomina, T. A., 110
Fukuda, K., 71
Futekov, L., 158
Fyfe, D. A., 108
Gabriel, N. E., 176
Galanova, A. P., 48, 62, 69
Gilbert, E. N., 48, 62, 120
Ginzburg, V. L., 2
Giraud, J. L., 10
Goncharova, N. N., 43
Graauwmans, E., 75
Greaves, M. C., 1
Grennan, D. M., 129
Grinshtein, I. L., 122
Grobenski, Z., 106
Groenwald, T., 13, 20, 36
Guerra, R., 131
Gui, F., 152
Gureva, M. P., 117
Gurin, P. A., 97
Gyulmezova, G., 113
Haddon, M. J., 132
Haines, D. S. M., 57
Hall, S. H., 121

SUBJECT INDEX (GOLD)

Group IIB Elements — Zn, Cd, and Hg

ZINC, Zn (ATOMIC WEIGHT 65.38)

Zinc is found in nature only in the combined form in ores and minerals. It is one of the most useful of all the metals. It is mostly used in metallurgical industries, such as in the manufacture of brass, commercial bronze, German silver, soft solder, aluminum solder, and die casting and typewriter metals. It is also used in rubber goods, automotive, electrical, and hardware industries. It is extensively used for galvanization as a protective coating on iron surfaces to prevent corrosion. Zinc compounds are used in cosmetics, floor coverings, ink, plastics, paints and pigments, soap, pharmaceuticals, textiles, and storage batteries, etc. Zinc oxide is used in glazes, enamels, dusting powders, and ointments. Zinc chloride is used as a wood preservative.

Determination of zinc is required in the evaluation of buying and selling of the ores, galvanizing plants, foundries, brass mills, blast furnaces, brass, alloys, polluted water, sewage, industrial wastes, as well as the above-mentioned industries.

The dissolution procedure for zinc will largely depend on the nature of the material and the quantity of zinc present in the sample. Samples can be dissolved by treatment with hydrochloric acid, nitric acid, or sulfuric acid. No interferences have been reported in the AA method.

Standard Solution

To prepare 1000 mg/ℓ of zinc solution, dissolve 1.0000 g of zinc metal in minimum volume of 1:1 hydrochloric acid and dilute to 1 ℓ with 1% (v/v) hydrochloric acid.

Instrumental Parameters

Wavelength	213.9 nm
Slit	0.7 nm
Light source	Hollow cathode lamp
Flame	Air-acetylene, oxidizing (lean, blue)
Sensitivity	0.01 mg/ℓ
Detection limit	0.002 mg/ℓ
Optimum range	1 mg/ℓ
Secondary wavelength and its sensitivity	
307.6 nm	68.0—79.0 mg/ℓ

REFERENCES

1. **David, D. J.,** The determination of zinc and other elements in plants by atomic absorption spectroscopy, *Analyst (London),* 83, 655, 1958.
2. **Gidley, J. A. F. and Jones, J. T.,** The determination of zinc in metallurgical materials by atomic absorption spectrophotometry, *Analyst (London),* 85, 249, 1960.
3. **Allen, J. E.,** The determination of zinc in agriculture materials by atomic absorption spectroscopy, *Analyst (London),* 86, 530, 1961.
4. **Finkelstein, N. P. and Jansen, A. V.,** An investigation of the reported interference by halogen acids in the determination of zinc by atomic absorption spectrophotometry, *S. Afr. Ind. Chem.,* June 1961.
5. **Strasheim, A. and Verster, F.,** The determination of copper and zinc in human tissues using atomic absorption spectroscopy, *Tydskr. Natuurwet.,* 1, 197, 1961 (in Afrikaans).
6. **Schuler, V. C. O., Jansen, A. V., and James, G. S.,** The development of atomic absorption methods for the determination of silver, copper, iron, lead and zinc in high purity gold and the role of organic additives, *J. S. Afr. Inst. Min. Metall.,* 62, 790, 1962.
7. **Willis, J. B.,** Determination of lead and other heavy metals in urine by atomic absorption spectroscopy, *Anal. Chem.,* 34, 614, 1962.
8. **Zeeman, P. B. and Butler, L. R. P.,** The determination of lead, copper and zinc in wines by atomic absorption spectroscopy, *Appl. Spectrosc.,* 16, 20, 1962.
9. **Gerstenfeldt, H.,** Determination of zinc in vegetables and animal substances, fertilizers and soils by absorption flame photometry, *Landwirtsch. Forsch.,* 15, 64, 1962 (in German).
10. **Wallace, F. J.,** The determination of zinc in metallurgical materials by atomic absorption spectroscopy, *Hilger J.,* 7, 39, 1962.
11. **Erdey, L., Svehla, G., and Koltai, L.,** The accuracy of zinc determination by atomic absorption methods, *Talanta,* 10, 531, 1963.
12. **Honegger, N.,** Serum zinc analysis by absorption flame photometry, *Arztl. Lab.,* 2, 41, 1963 (in German).
13. **Parker, H. E.,** Magnesium, calcium and zinc in animal nutrition, *At. Absorpt. Newsl.,* No. 13, May 1963.
14. **Musha, S., Munemori, M., and Nakanishi, Y.,** Determination of zinc, lead and calcium in poly (vinylchloride) by atomic absorption spectrophotometry, *Bunseki Kagaku,* 13, 330, 1964.
15. **Belt, C. B., Jr.,** Atomic absorption spectrophotometry and the analysis of silicate rocks for copper and zinc, *Econ. Geol.,* 59, 240, 1964.
16. **Belt, C. B., Jr.,** The determination of copper and zinc in Bayer process liquor, *At. Absorpt. Newsl.,* No. 19, March 1964.
17. **Fuwa, K., Pulido, P., McKay, R., and Vallee, B. L.,** Determination of zinc in biological materials by atomic absorption spectrophotometry, *Anal. Chem.,* 36, 2407, 1964.
18. **Nikolaev, G. I.,** Atomic absorption method for zinc determination in metals and alloys, *Zh. Anal. Khim.,* 19, 63, 1964 (in Russian).
19. **Shafto, R. G.,** The determination of copper, iron, lead and zinc in nickel plating solutions by atomic absorption, *At. Absorpt. Newsl.,* 3, 115, 1964.
20. **Buchanan, J. R. and Muraoka, T. T.,** Determination of zinc and manganese in tree leaves by atomic absorption spectroscopy, *At. Absorpt. Newsl.,* No. 24, September 1964.
21. **Billings, G. K. and Harriss, R. C.,** Cation analysis of marine waters by atomic absorption spectrometry. Gulf of Mexico coastal waters, *Tex. J. Sci.,* 17, 129, 1965.
22. **Biechler, D. G.,** Determination of trace copper, lead, zinc, cadmium, nickel and iron in industrial waste waters by atomic absorption spectrometry after ion exchange concentration on DOWEX A-I, *Anal. Chem.,* 37, 1054, 1965.
23. **Bordonali, C., Biancifiori, M. A., and Besazza, G.,** Determination of traces of metals in sodium by atomic absorption (spectrometry). Determination of copper, manganese, iron, lead, nickel and zinc, *Chim. Ind. (Milan),* 47, 397, 1965.
24. **Farrar, B.,** Determination of copper and zinc in ore samples and lead base alloys, *At. Absorpt. Newsl.,* 4, 325, 1965.
25. **Herrmann, R. and Neu, W.,** Magnesium and zinc in chloroplasts, *Experientia,* 21, 436, 1965.
26. **Nikolaev, G. I.,** Atomic absorption method for the determination of aluminum and zinc admixtures in solid samples of refractory metals, *Zh. Anal. Khim.,* 20, 445, 1965.
27. **Passmore, W. and Adams, P. B.,** Determination of iron and zinc in glass by atomic absorption spectrophotometry, *At. Absorpt. Newsl.,* 4, 237, 1965.
28. **Rousselet, P. and Girard, M. L.,** Use of atomic absorption for micro determination of copper and zinc in biological media, *C. R.,* 260, 3780, 1965 (in French).
29. **Sprague, S. and Slavin, W.,** Determination of iron, copper and zinc in blood serum by an atomic absorption method requiring only dilution, *At. Absorpt. Newsl.,* 4, 228, 1965.

30. **Vollmer, J., Sebens, C., and Slavin, W.,** Improved zinc hollow cathode lamps, *At. Absorpt. Newsl.,* 4, 306, 1965.
31. **Burrell, D. C.,** The determination of zinc in silicates by atomic absorption spectroscopy, *Nor. Geol. Tidsskr.,* 45, 21, 1965 (in English).
32. **Bradfield, E. G. and Spencer, D.,** Leaf analysis as a guide to the nutrition of fruit crops. VI. Determination of magnesium, zinc and copper by atomic absorption spectroscopy, *J. Sci. Food Agric.,* 16, 35, 1965.
33. **Prasad, A. S., Oberleas, D., and Halsted, J. A.,** Determination of zinc in biological fluids by atomic absorption spectrophotometry in normal and cirrhotic subjects, *J. Lab. Clin. Med.,* 66, 508, 1965.
34. **Bradfield, E. G. and Osborne, M.,** Application of atomic absorption spectroscopy to the determination of Zn, Fe, Cu, Mn and Pb in bottled ciders, *Long Ashton Agric. Hortic. Res. Stn. Univ. Bristol Annu. Rep.,* 157, 1965.
35. **Burrell, D. C.,** The geochemistry and origin of amphibolites from Bamble, South Norway. I. The determination of zinc in amphibolites by atomic absorption spectroscopy, *Nor. Geol. Tidsskr.,* 45, 24, 1965.
36. **Doerffel, K. and Nitzsche, U.,** Determination of zinc by atomic absorption spectroscopy, *Wiss. Z. Tech. Hochsch. Chem. Leuna-Merseburg,* 7, 9, 1965 (in German).
37. **Dumanski, J.,** Adaptation and modification in determining copper, zinc and manganese in plant materials by atomic absorption spectroscopy, *Rocz. Nauk Roln. Ser. A,* 90, 431, 1965 (in Polish).
38. **Calkins, R. C.,** The determination of manganese and zinc in aluminous materials, *Appl. Spectrosc.,* 20, 146, 1966.
39. **Chang, T. L., Gover, T. A., and Harrison, W. W.,** Determination of magnesium and zinc in human brain tissue by atomic absorption spectroscopy, *Anal. Chim. Acta,* 34, 17, 1966.
40. **Kahnke, M. J.,** Atomic absorption spectrophotometry applied to the determination of zinc in formalinized human tissue, *At. Absorpt. Newsl.,* 5, 7, 1966.
41. **Price, C. A. and Quigley, J. W.,** A method for determining quantitative zinc requirements for growth, *Soil Sci.,* 101, 11, 1966.
42. **Berry, R. K., Bell, M. C., and Wright, P. L.,** Influence of dietary calcium, zinc and oil upon in-vitro uptake of zinc-65 by porcine blood cells, *J. Nutr.,* 88, 284, 1966.
43. **Cheek, D. B., Powell, G. K., Reba, R., and Feldman, M.,** Manganese, copper and zinc in rat muscle and liver cells and in thyroid and pituitary insufficiency, *Bull. Johns Hopkins Hosp.,* 118, 338, 1966.
44. **Ezell, J. B., Jr.,** Determination of zinc in aluminous materials by atomic absorption spectroscopy, *At. Absorpt. Newsl.,* 5, 122, 1966.
45. **Fishman, M. J.,** The use of atomic absorption for analysis of natural waters, *At. Absorpt. Newsl.,* 5, 102, 1966.
46. **Fixman, M. and Boughton, L.,** Mineral assay for silver, zinc and cadmium, *At. Absorpt. Newsl.,* 5, 33, 1966.
47. **Gaumer, M. W., Sprague, S., and Slavin, W.,** An automated procedure for the determination of trace metals by atomic absorption spectroscopy, *At. Absorpt. Newsl.,* 5, 58, 1966.
48. **Mansell, R. E., Emmel, H. W., and McLaughlin, F. L.,** Analysis of magnesium and aluminum alloys by atomic absorption spectroscopy, *Appl. Spectrosc.,* 20, 231, 1966.
49. **Nielson, F. H., Sunde, M. L., and Holkstra, W. G.,** Effect of dietary amino acid source on the zinc-deficiency syndrome in the chick, *J. Nutr.,* 89, 24, 1966.
50. **Sattur, T. W.,** Routine atomic absorption analysis on nonferrous alloys and plant intermediates, *At. Absorpt. Newsl.,* 5, 37, 1966.
51. **Bell, G. F.,** The analysis of aluminum alloys by means of atomic absorption spectrophotometry, *At. Absorpt. Newsl.,* 5, 73, 1966.
52. **Simpson, G. R. and Blay, R. A.,** Rapid method for determination of the metals, copper, zinc, tin, iron and calcium in foodstuffs by atomic absorption spectroscopy, *Food Trade Rev.,* August 1966.
53. **Imai, T.,** The determination of zinc with atomic absorption, *Jpn. Analyst,* 16, 900, 1967 (in Japanese).
54. **Bell, G. F.,** On the effect of copper on the determination of zinc in aluminum alloys, *At. Absorpt. Newsl.,* 6, 18, 1967.
55. **Parker, M. W., Humoller, F. L., and Mahler, D. J.,** Determination of copper and zinc in biological material, *Clin. Chem. (Winston-Salem, N.C.),* 13, 40, 1967.
56. **Scott, T. C., Roberts, E. D., and Cain, D. A.,** Determination of minor constituents in ferrous materials by atomic absorption spectrophotometry, *At. Absorpt. Newsl.,* 6, 1, 1967.
57. **Girard, M. L. and Rousselet, F.,** Atomic absorption applied to biology. V. Progress made in the microdetermination of copper and zinc, *Ann. Pharm. Fr.,* 25, 353, 1967 (in French).
58. **Hanig, R. C. and Aprison, M. H.,** Determination of calcium, copper, iron, magnesium, manganese, potassium, sodium, zinc and chloride concentrations in several brain areas, *Anal. Biochem.,* 21, 169, 1967.
59. **Prasad, A. S., Oberleas, D., Wolf, P., and Horwitz, J. P.,** Studies on zinc deficiency: changes in trace elements and enzyme activities in tissues of zinc-deficient rats, *J. Clin. Invest.,* 46, 549, 1967.

60. **Ramakrishna, T. V., Robinson, J. W., and West, P. W.,** Determination of copper, cadmium and zinc by atomic absorption spectroscopy, *Anal. Chim. Acta,* 37, 20, 1967.

61. **Abbey, S.,** Analysis of rocks and minerals by atomic absorption spectroscopy. I. Determination of magnesium, lithium, zinc and iron, *Geol. Surv. Can. Pap.,* 67, Ottawa, Ont., 1967.

62. **Brierley, G. P. and Knight, V. A.,** Ion transport by heart mitochondris. X. The uptake and release of Zn^{+2} and its relation to the energy-linked accumulation of magnesium, *Biochemistry,* 6, 3892, 1967.

63. **Ramirez-Munoz, J.,** Determination of zinc in vegetable materials by atomic absorption flame photometry, *An. Edafol. Agrobiol. (Madrid),* 26, 1211, 1967 (in Spanish).

64. **Settlemire, C. T. and Matrone, G.,** In vivo interference of zinc with ferritin iron in the rat, *J. Nutr.,* 92, 153, 1967.

65. **Settlemire, C. T. and Matrone, G.,** In vivo effect of zinc on iron turnover in rats and life span of the erythrocyte, *J. Nutr.,* 92, 159, 1967.

66. **Reinhold, J. G., Pascoe, E., and Kfoury, G. A.,** Capillary diameter and rate of aspiration as factors affecting the accuracy of zinc analysis in serum by atomic absorption spectrophotometry, *Anal. Biochem.,* 25, 557, 1968.

67. **Sato, T., Motoyama, Y., and Uchida, K.,** Behaviour of inorganic ions caused by position processes. Determination of zinc in deposited film by atomic absorption spectroscopy, *Shikizai Kyokaishi,* 41, 554, 1968.

68. **Amati, A., Rostelli, R., and Minguzzi, A.,** Determination of iron, copper, zinc and manganese in must and wine by atomic absorption spectrophotometry, *Ind. Agrar.,* 6, 1630, 1968 (in Italian).

69. **Endo, Y., Hata, T., and Nakahara, Y.,** Atomic absorption determination of calcium, magnesium, manganese, copper, zinc and aluminum in iron ores, *Jpn. Analyst,* 17, 679, 1968 (in Japanese).

70. **Matsumoto, H., Tsunematsu, K., and Shiraishi, T.,** Determination of zinc in biological materials by atomic absorption spectrometry. I. Experimental conditions and application to human serum, *Jpn. Analyst,* 17, 703, 1968 (in Japanese).

71. **Hackley, B. M., Smith, J. C., and Halsted, J. A.,** A simplified method for plasma zinc determination by atomic absorption spectrophotometry, *Clin. Chem. (Winston-Salem, N.C.),* 14, 1, 1968.

72. **Olson, A. D. and Hamlin, W. B.,** Serum copper and zinc by atomic absorption spectrophotometry, *At. Absorpt. Newsl.,* 7, 69, 1968.

73. **Roach, A. G., Sanderson, P., and Williams, D. R.,** Determination of trace amounts of copper, zinc and magnesium in animal feeds by atomic absorption spectrophotometry, *Analyst (London),* 93, 42, 1968.

74. **Rogers, G. R.,** Collaborative study of atomic absorption spectrophotometric method for determining zinc in foods, *J. Assoc. Off. Anal. Chem.,* 51, 1042, 1968.

75. **Rosner, F. and Gorfien, P. C.,** Erythrocyte and plasma zinc and magnesium levels in health and disease, *J. Lab. Clin. Med.,* 72, 213, 1968.

76. **Woodburg, J., Lyons, K., Corretta, R., Hahn, A., and Sullivan, J. F.,** Cerebrospinal fluid and serum levels of magnesium, zinc and calcium in man, *Neurology,* 18, 700, 1968.

77. **Olwa, K., Kimura, T., Makino, H., and Okuda, M.,** Determination of zinc in blood serum and red blood cells by atomic absorption spectrometry, *Jpn. Analyst,* 17, 810, 1968 (in Japanese).

78. **Davies, I. J. T., Musa, M., and Dormandy, T. L.,** Measurement of plasma zinc by atomic absorption, *J. Clin. Pathol.,* 21, 359, 1968.

79. **Girard, M. L.,** Determination of copper and zinc by atomic absorption, *Clin. Chim. Acta,* 20, 243, 1968 (in French).

80. **Harrison, W. W., Netsky, M. C., and Brown, M. D.,** Trace elements in human brain: copper, zinc, iron and magnesium, *Clin. Chim. Acta,* 21, 55, 1968.

81. **Premi, P. R. and Cornfield, A. H.,** Determination of total copper, zinc, iron, manganese and chromium in plant materials and organic residues by extraction with hydrochloric acid followed by atomic absorption spectroscopy, *Spectrovision,* 19, 15, 1968.

82. **Pemsler, J. P. and Rapperport, E. J.,** Thermodynamic activity measurements using atomic absorption copper-zinc, *Trans. Met. Soc. AIME,* 245, 1395, 1969.

83. **Netsky, M. G., Harrison, W. W., Brown, E., and Benson, C.,** Tissue zinc and human disease relation of zinc content of kidney, liver and lung to atherosclerosis and hypertension, *Am. J. Clin. Pathol.,* 51, 358, 1969.

84. **Bailey, B. W. and Rankin, J. M.,** The use of organic liquids as fuels in flame spectroscopy, *Spectrosc. Lett.,* 2, 159, 1969.

85. **Donaldson, E. M., Charette, D. J., and Rolko, V. H. E.,** Determination of cobalt and zinc in high-purity niobium, tantalum, molybdenum and tungsten metals by atomic absorption spectrophotometry after separation by extraction, *Talanta,* 16, 1305, 1969.

86. **Lundy, R. G. and Watje, W. F.,** Determination of aluminum, iron, chromium, nickel and zinc in inhibited red fuming nitric acid by atomic absorption spectrophotometry, *At. Absorpt. Newsl.,* 8, 124, 1969.

87. **L'vov, B. V. and Khartsyzov, A. D.,** Atomic absorption spectral analysis of radioactive preparations of zinc and cadmium, *Zh. Anal. Khim.,* 24, 936, 1969.
88. **McCracken, J. D., Vecchione, M. C., and Longo, S. L.,** Ion exchange separation of traces of tin, cadmium and zinc from copper and their determination by atomic absorption spectrophotometry, *At. Absorpt. Newsl.,* 8, 102, 1969.
89. **Sato, T., Motoyama, Y., and Ohe, O.,** Determination of zinc in electrodeposited films by atomic absorption spectrometry, *J. Paint Technol.,* 41, 438, 1969.
90. **Spencer, D. W. and Brewer, P. G.,** The distribution of copper, zinc and nickel in seawater by the gulf of Maine and the Sargasso sea, *Geochim. Cosmochim. Acta,* 33, 325, 1969.
91. **Racker, E. T.,** Chloric acid digestion in the determination of trace metals (iron, zinc and copper) in brain and hair by atomic absorption spectrophotometry, *Clin. Chim. Acta,* 24, 233, 1969.
92. **Burrell, D. C. and Wood, G. G.,** Direct determination of zinc in sea water by atomic absorption spectrophotometry, *Anal. Chim. Acta,* 48, 45, 1969.
93. **Emmermann, R. and Luecke, W.,** Determination of trace amounts of lead, zinc and silver in soil samples by atomic absorption spectrometry using a tantalum sampling boat, *Z. Anal. Chim.,* 248, 325, 1969.
94. **Dawson, J. B. and Walker, B. E.,** Direct determination of zinc in whole blood, plasma and urine by atomic absorption spectroscopy, *Clin. Chim. Acta,* 26, 465, 1969.
95. **Osada, H.,** Determination of metals in foods by atomic absorption spectrophotometry. Determination of zinc, *Eiya To Shokuryo,* 22, 552, 1969.
96. **Spielholtz, G. I. and Toralballa, G. C.,** The determination of zinc in crystalline insulin and in certain insulin preparations by atomic absorption spectroscopy, *Analyst (London),* 94, 1072, 1969.
97. **Tusl, J. and Krska, M.,** Determination of magnesium, iron, manganese, copper and zinc in foods by atomic absorption spectrophotometry, *Prum. Potravin,* 21, 119, 1970 (in Czech).
98. **Miksovsky, M. and Moldan, B.,** Determination of copper, zinc, manganese and iron in silicate samples by atomic absorption spectroscopy, *Chem. Zvesti,* 24, 128, 1970.
99. **Mikhailova, T. P.,** Atomic absorption for determining zinc in a zinc chloride-potassium chloride-lithium chloride melt, *Zavod Lab.,* 36, 1346, 1970.
100. **Ali, S. A. and Burrell, D. C.,** Determination of zinc in interstitial waters by atomic absorption spectrophotometry, *Pak. J. Sci. Ind. Res.,* 12, 506, 1970.
101. **Gandrud, B. and Marshall, J. C.,** The determination of arsenic, lead, nickel and zinc in copper by atomic absorption spectrophotometry, *Appl. Spectrosc.,* 24, 367, 1970.
102. **Kazimierczyk, S., Michalewska, M., and Jarosz, R.,** Determination of silver, copper and zinc by atomic absorption spectroscopy, *Chem. Anal. (Warsaw),* 15, 553, 1970.
103. **Knauer, G. A.,** The determination of magnesium, manganese, iron, copper and zinc in marine shrimp, *Analyst (London),* 95, 476, 1970.
104. **Lerner, L. A. and Ivanov, D. N.,** Determination of total amount of zinc, copper, cobalt and manganese in soils by atomic absorption spectrophotometry, *Agrokhimiya,* 3, 133, 1970.
105. **Nix, J. and Goodwin, T.,** The simultaneous extraction of iron, manganese, copper, cobalt, nickel, chromium, lead and zinc from natural water for determination by atomic absorption spectroscopy, *At. Absorpt. Newsl.,* 9, 119, 1970.
106. **Peterson, G. E. and Kahn, H. L.,** The determination of barium, calcium and zinc in additives and lubricating oils using atomic absorption spectrophotometry, *At. Absorpt. Newsl.,* 9, 71, 1970.
107. **Rudnevskii, N. K., Demarin, V. T., and Nyrkova, O. A.,** Use of atomic absorption spectra for determining excess quantities of selenium and zinc in zinc-selenide, *Zh. Prikl. Spektrosk.,* 12, 156, 1970.
108. **Toma, O. and Crisan, T.,** Atomic absorption and emission spectroscopic method for the determination of iron, zinc and sodium in alumina, *Rev. Chim. (Bucharest),* 21, 334, 1970.
109. **Tusl, J.,** Determination of zinc in stock and mixed feeds by EDTA extraction and atomic absorption spectrophotometry, *J. Assoc. Off. Anal. Chem.,* 53, 1190, 1970.
110. **Tusl, J. and Nemeskal, S.,** Determination of zinc in stock feeds by atomic absorption spectrophotometry, *Chem. Listy,* 64, 496, 1970.
111. **Ure, A. M. and Berrow, M. L.,** Analysis of EDTA extracts of soils for copper, zinc and manganese by atomic absorption spectrophotometry with a mechanically separated flame, *Anal. Chim. Acta,* 52, 243, 1970.
112. **Zook, E., Greene, F., and Morris, E.,** Nutrient composition of selected wheats and wheat products. Distribution of manganese, copper, nickel, zinc, magnesium, lead, tin, cadmium, chromium and selenium as determined by atomic absorption spectroscopy and calorimetry, *Cereal Chem.,* 47, 720, 1970.
113. **Yamamoto, Y., Kumamaru, T., Hayashi, Y., and Kanke, M.,** Determination of a trace amount of cadmium, zinc, lead and copper in water by atomic absorption spectrophotometry combined with solvent extraction, *Bunseki Kagaku,* 20, 347,1971.
114. **Ghai, O. P. and Vasuki, K.,** Plasma zinc concentration in Indian childhood cirrhosis, *Indian Pediatr.,* 8, 309, 1971.

115. **Kanke, M., Hayashi, Y., Kumamaru, T., and Yamamoto, Y.,** Determination of ppb levels of cadmium, zinc, lead and copper extracted as their dithizonates into nitrobenzene by atomic absorption spectrophotometry, *Nippon Kagaku Zasshi,* 92, 983, 1971.

116. **Krasilshchik, V. Z. and Yakovleva, A. E.,** Analysis of oxides of nickel and cobalt (for silver, copper, lead, zinc, bismuth and cadmium by spectrographic excitation) in a discharge tube with a hollow cathode, *Zavod. Lab.,* 37, 181, 1971 (in Russian).

117. **Muradova, O. N. and Muradov, V. G.,** Determination of the absolute concentration of zinc atoms in the gas phase by atomic absorption, *Izv. Vyssh. Ucheb. Zaved.,* 14, 123, 1971.

118. **Nemets, A. M. and Nikolaev, G. I.,** Determination of thermodynamic characteristics of copper-zinc alloys by an atomic absorption method, *Zh. Prikl. Spektrosk.,* 15, 583, 1971.

119. **Boawn, L. C. and Rasmussen, P. E.,** Crop response to excessive zinc fertilization of alkaline soil, *Agron. J.,* 63, 874, 1971.

120. **Dancheva, R.,** Determination of zinc by an atomic absorption method, *Khim. Ind. (Sofia),* 43, 176, 1971.

121. **Follett, H. H. and Lindsay, W. L.,** Changes in DTPA-extractable zinc, iron, manganese and copper in soils following fertilization, *Soil Sci. Soc. Am. Proc.,* 35, 600, 1971.

122. **Cambrell, J. W.,** The determination of lead and zinc as principal constituents by atomic absorption spectroscopy, *At. Absorpt. Newsl.,* 10, 81, 1971.

123. **Ghanbari, H. A. and Mameesh, M. S.,** Iron, zinc, manganese and copper content of semidwarf wheat varieties grown under different agronomic conditions, *J. Am. Assoc. Cereal Chem.,* 48, 411, 1971.

124. **Ishii, T. and Musha, S.,** Atomic absorption spectrophotometric analysis of zinc in vulcanizates with ethyl acetoacetates as metal solubility reagent, *Bunseki Kagaku,* 20, 489, 1971.

125. **James, B. E. and Macmahon, R. A.,** An effect of trichloroacetic acid on the determination of zinc by atomic absorption spectroscopy, *Clin. Chim. Acta,* 32, 307, 1971.

126. **Jones, C. E. and Pracht, L. E.,** Collaborative study of a zincon ion exchange method for the quantitative determination of zinc in fertilizers, *J. Assoc. Offic. Anal. Chem.,* 54, 790, 1971.

127. **Kashigi, M. and Oshima, S.,** Determination of calcium, barium and zinc in lubricating oils by atomic absorption spectrophotometry, *Anal. Chim. Acta,* 55, 436, 1971.

128. **Kuroha, T., Tsukahara, I., and Shibuya, S.,** Determination of microamounts of cadmium and zinc in copper, nickel, aluminum and uranium metals by solvent extraction-atomic absorption spectrophotometry, *Bunseki Kagaku,* 20, 1137, 1971.

129. **Lerner, L. A., Orlova, L. P., and Ivanov, D. N.,** Use of a rapid method of sample decompositions during the atomic absorption determination of the total copper, zinc and manganese content in soils, *Agrokhimya,* 5, 138, 1971.

130. **Luecke, W. and Emmermann, R.,** The application of the boat technique for lead, zinc, silver and cadmium in soil samples, *At. Absorpt. Newsl.,* 10, 45, 1971.

131. **Maher, D. J., Walsh, J. R., and Haynie, G. D.,** Magnesium, zinc and copper in dialysis patients, *Am. J. Clin. Pathol.,* 56, 17, 1971.

132. **Matousek, J. P. and Stevens, B. J.,** Biological applications of the carbon rod atomizer in atomic absorption spectroscopy. Preliminary studies on magnesium, iron, copper, lead and zinc in blood and plasma, *Clin. Chem. (Winston-Salem, N.C.),* 17, 363, 1971.

133. **Meret, S. and Henkin, R. I.,** Simultaneous direct estimation by atomic absorption spectrophotometry of copper and zinc in serum, urine and cerebrospinal fluid, *Clin. Chem. (Winston-Salem, N.C.),* 17, 369, 1971.

134. **Murthy, G. K., Rhea, V., and Peeler, J. T.,** Levels of antimony, cadmium, chromium, cobalt, manganese and zinc in institutional total diets, *Environ. Sci. Technol.,* 5, 436, 1971.

135. **Nikolaev, G. I. and Nemets, A. M.,** Use of atomic absorption spectroscopy for determining saturated vapor pressure and the heat of sublimation of zinc, *Zh. Prikl. Spektrosk.,* 15, 23, 1971.

136. **Obolenskaya, L. I., Kamenskaya, L. S., and Buzyukina, V. V.,** Atomic absorption method for determining zinc, manganese, iron and magnesium in plants and soil, *Agrokhimiya,* 1, 136, 1971.

137. **Orren, M. J.,** Determination of copper, zinc, iron, manganese, potassium, lithium and rubidium in sea water by atomic absorption spectrophotometry, *J. S. Afr. Inst.,* 24, 96, 1971.

138. **Rosman, K., Jr. and Jeffery, P. M.,** Determination of zinc in standard reference materials by isotope dilution and atomic absorption analysis, *Chem. Geol.,* 8, 25, 1971.

139. **Rossi, N. and Beauchamp, E. G.,** Influence of relative humidity and associated anion on the absorption of Mn and Zn by soybean leaves, *Agron. J.,* 63, 860, 1971.

140. **Thomas, B. G.,** Determination of silver, lead and zinc in high grade ores, *At. Absorpt. Newsl.,* 10, 73, 1971.

141. **Bergner, K. C. and Lang, B.,** The determination of iron, copper, zinc, manganese and cadmium in grape juice and wine by atomic absorption spectrophotometry, *Dtsch. Lebensmitt. Rdsch.,* 67, 121, 1971 (in German).

142. **Bohnstedt, U.,** Determination of traces of lead, bismuth, zinc and cadmium in alloys. An example of the use of combined analytical procedures, *DEW Tech. Ber.,* 11, 101, 1971 (in German).

143. **Fratkin, Z. G., Sokolov, A. A., and Bogdanova, N. N.,** Atomic absorption determination of zinc, copper, nickel and cobalt in compounds of phosphorus (used as fertilizers), *Tr. Leningr. Nauchno Issled. Proekt. Inst. Osnov. Khim. Prom.,* 4, 283, 1971 (in Russian).

144. **Higgins, M. L. and Pickering, W. F.,** The precision of determinations of zinc content of food, *Talanta,* 18, 986, 1971.

145. **Toma, O. and Crisan, T.,** Determination of copper, tin, lead, zinc, nickel and iron in bronze by atomic absorption spectrophotometry, *Metallurgia,* 23, 715, 1971 (in Romanian).

146. **Romano Gallo, J., Bataglia, O. C., and Nayme Miguel, P. T.,** Determination of copper, iron, manganese and zinc in a bulk plant extract by atomic absorption spectrophotometry, *Bragantia,* 30, 155, 1971 (in Portuguese).

147. **El-Hinnawi, E.,** Determination of zinc in geological materials via atomic absorption spectrometry, *Neues Jahrb. Mineral. Monatsh.,* 1972, 299, 1972.

148. **Everson, R. J.,** Zinc contamination from rubber products, *At. Absorpt. Newsl.,* 11, 130, 1972.

149. **Fell, G. S., Canning, E., Husain, S. L., and Scott, R.,** Copper and zinc in human health and disease, in *Trace Substance in Environmental Health,* V. Hemphill, D. D., Ed., University of Missouri, Columbia, 1972.

150. **Giordano, P. M. and Mortvedt, J. J.,** Rice response to Zn in flooded and nonflooded soil, *Agron. J.,* 64, 521, 1972.

151. **Guttenberger, J. and Marold, M.,** Direct determination of the additive elements calcium, barium and zinc in lubricating oils. Solutions of inorganic salts in dimethyl sufoxide as standards in atomic absorption spectroscopy, *Z. Anal. Chem.,* 262, 102, 1972.

152. **Hag, A. U. and Miller, M. H.,** Prediction of available soil Zn, Cu and Mn using chemical extractants, *Agron. J.,* 64, 779, 1972.

153. **Rosner, F. and Lee, S. L.,** Zinc and magnesium content of leukocyte alkaline phosphatase isoenzymes, *J. Lab. Clin. Med.,* 79, 228, 1972.

154. **Pekarek, R. S., Beisel, W. R., Bartelloni, P. J., and Bostian, K. A.,** Determination of serum zinc concentrations in normal adult subjects by atomic absorption spectrophotometry, *Am. J. Clin. Pathol.,* 57, 506, 1972.

155. **Marinho, M. L. and Igue, K.,** Factors affecting zinc absorption by corn from volcanic ash soil, *Agron. J.,* 64, 3, 1972.

156. **Kubota, N. and Imai, T.,** Silver plating obtained from cyanide baths. Determination of copper and iron in silver plating solutions by atomic absorption, *Kinzoku Hyomen Gijutsu,* 23, 95, 1972.

157. **Lindeman, R. D., Bottomley, R. G., Cornelison, R. L., Jr., and Jacobs, L. A.,** Influence of accute tissue injury on zinc metabolism in man, *J. Lab. Clin. Med.,* 79, 452, 1972.

158. **Alley, M. M., Martens, D. C., Schnappinger, M. G., Jr., and Hawkins, G. W.,** Field calibration of soil tests for available zinc, *Soil Sci. Soc. Am. Proc.,* 36, 621, 1972.

159. **Norvell, W. A. and Lindsay, W. L.,** Reactions of DTPA chelates of iron, zinc, copper and manganese with soils, *Soil Sci. Soc. Am. Proc.,* 36, 778, 1972.

160. **Rudnevskii, N. K. and Demarian, V. T.,** Atomic absorption determination of zinc oxide impurities of zinc tungstate, *Tr. Khim. Khim. Tekhnol.,* 1971, 106, 1972.

161. **Salvesen, B. and Aaroe, B.,** Determination of zinc in pharmaceuticals (and medicated feeds) by atomic absorption spectrophotometry, *Medd. Nor. Farm. Selsk.,* 34, 9, 1972.

162. **Smith, D. L. and Schrenk, W. G.,** Application of atomic absorption spectroscopy to plant analysis. I. Comparison of zinc and manganese analysis with official AOAC calorimetric methods, *J. Assoc. Off. Anal. Chem.,* 55, 699, 1972.

163. **Heinemann, G.,** Iron, copper and zinc analysis by atomic absorption spectrophotometry, *Z. Klin. Chem. Klin. Biochem.,* 10, 467, 1972.

164. **Kidani, Y., Noji, M., and Koike, H.,** Quantitative determination of copper, nickel, cobalt and zinc in metal complexes by atomic absorption spectrometry, *Bunseki Kagaku,* 21, 1652, 1972.

165. **Toma, O. and Crisan, T.,** The determination by atomic absorption spectrophotometry of traces of iron, zinc, sodium and magnesium in ABS (acrylonitrile-butadiene-styrene copolymer), *Chim. Anal.,* 2, 25, 1972 (in Romanian).

166. **Toma, O. and Crisan, T.,** Atomic absorption determination of zinc and traces of iron, lead, sodium, nickel, calcium and magnesium in saturated zinc chloride solutions, *Rev. Chim.,* 23, 189, 1972 (in Romanian).

167. **Toma, O. and Crisan, T.,** Determination of the concentration of zinc and traces of iron, lead, sodium, nickel, calcium and magnesium in saturated zinc chloride solution with atomic absorption spectrophotometry, *Chim. Anal.,* 2, 189, 1972.

168. **Krema, V. and Sommer, L.**, Contribution to the atomic absorption spectrophotometry of zinc in aqueous solutions, *Scr. Chem.*, 2, 55, 1972.

169. **Ishizuka, T., Sunahara, H., and Tanaka, K.**, Determination of copper, iron, lead and zinc in yttrium oxide and yttrium oxide sulfide by atomic absorption spectrometry, *Bunseki Kagaku*, 21, 847, 1972.

170. **Maas, E. V., Ogata, G., and Garber, M. J.**, Influence of salinity on Fe, Mn and Zn uptake by plants, *Agron. J.*, 64, 793, 1972.

171. **Marks, G. E., Moore, C. E., Kanabrocki, E. L., Oester, Y. T., and Kaplan, E.**, Determination of trace elements in human tissue. I. Cd, Fe, Zn, Mg and Ca, *Appl. Spectrosc.*, 26, 523, 1972.

172. **Mills, C. F. and Dalgarno, A. C.**, Copper and zinc status of ewes and lambs receiving increased dietary concentrations of cadmium, *Nature (London)*, 239, 171, 1972.

173. **Moody, R. R. and Taylor, R. B.**, Determination of zinc in pharmaceutical preparations by atomic absorption spectrophotometry, *J. Pharm. Pharmacol.*, 24, 848, 1972.

174. **Nassi, L., Poggini, G., Vechhi, C., and Galvan, P.**, Atomic absorption spectrophotometry in the determination of total free and bound zinc in human colostrum and milk, *Boll. Soc. Ital. Biol. Sper.*, 48, 86, 1972.

175. **Kurz, D., Roach, J., and Eyring, E. J.**, Direct determination of serum zinc and copper by atomic absorption spectrophotometry, *Biochem. Med.*, 6, 274, 1972.

176. **Ishizuka, T., Sunahara, H., and Tanaka, K.**, Determination of copper, iron, lead and zinc in yttrium oxide and yttrium oxide sulfide by atomic absorption spectrometry, *Nagoya Kogyo Gijutsu Shikensho Hokoku*, 21, 385, 1972.

177. **Piscator, M. and Lind, B.**, Cadmium, zinc, copper and lead in renal cortex, *Arch. Environ. Health*, 24, 426, 1972.

178. **Lindeman, R. D., Yunice, A. A., Baxter, D. J., Miller, L. R., and Nordquist, J.**, Myocardial zinc metabolism in experimental myocardial infarction, *J. Lab. Clin. Med.*, 81, 194, 1973.

179. **Murthy, L., Menden, E. E., Eller, P. M., and Petering, H. G.**, Atomic absorption determination of zinc, copper, cadmium and lead in tissues solubilized by aqueous tetra methyl ammonium hydroxide, *Anal. Biochem.*, 53, 365, 1973.

180. **Nakagawa, R.**, Atomic absorption analysis of zinc, lead and cadmium in hot spring waters, *Nippon Kagaku Kaishi*, 514, 1973.

181. **Buchauer, M. J.**, Contamination of soil and vegetation near a zinc smelter by zinc, cadmium, copper and lead, *Environ. Sci. Technol.*, 7, 131, 1973.

182. **Chaube, S., Nishimura, H., and Swinyard, C. A.**, Zinc and cadmium in normal human embryos and fetuses. Analysis by atomic absorption spectrophotometry, *Arch. Environ. Health*, 26, 237, 1973.

183. **Clark, D., Dagnall, R. M., and West, T. S.**, Atomic absorption determination of zinc with a graphite furnace, *Anal. Chim. Acta*, 63, 11, 1973.

184. **Cohen, I. K., Schechter, P. J., and Henkin, R. L.**, Hypogeusia, anorexis and altered zinc metabolism following thermal burn, *JAMA*, 223, 914, 1973.

185. **Dittrich, K. and Zeppan, W.**, Atomic absorption spectrophotometric determination of zinc in gallium arsenide and gallium aluminum arsenide, *Talanta*, 20, 126, 1973 (in German).

186. **Dong, A. E.**, A study of the effect of pH on the determination of zinc by atomic absorption spectrophotometry, *Appl. Spectrosc.*, 27, 124, 1973.

187. **Reusomann, G. and Westphalen, J.**, Automated determination of lead, zinc and cadmium in vegetable material, *Staub Reinhalt. Luft*, 33, 435, 1973 (in German).

188. **Dybczynska, I., Fijalkowski, J., Chruscinska, T., and Myszka, E.**, Determination of copper, zinc and iron in platinum-rhodium and platinum-indium alloys by atomic absorption spectroscopy, *Chem. Anal. (Warsaw)*, 18, 169, 1973.

189. **Kelly, W. R. and Moore, C. B.**, Iron spectral interference in the determination of zinc by atomic absorption spectrometry, *Anal. Chem.*, 45, 1274, 1973.

190. **Kurz, D., Roach, J., and Eyring, E. J.**, Determination of zinc by flameless atomic absorption spectrophotometry, *Anal. Biochem.*, 53, 586, 1973.

191. **Nakahara, T., Munemori, M., and Musha, S.**, Atomic absorption spectrophotometric determination of zinc in premixed inert gas (entrained air) - hydrogen flame and the measurement of the flame temperature, *Bull. Chem. Soc. Jpn.*, 46, 639, 1973.

192. **Facchinetti, M., Grassi, R. L., and Diez, A. L.**, Atomic absorption spectrophotometry of extractable copper, zinc, manganese and iron in soils, *Agrochimica*, 17, 413, 1973.

193. **Jensen, R. F. and Kaehler, M.**, A simultaneous determination of zinc and cadmium, *Anal. Chim. Acta*, 67, 465, 1973.

194. **Kaplan, P. D., Blackstone, M., and Richdale, N.**, Direct determination of cadmium, nickel and zinc in rat lungs by atomic absorption spectrophotometry, *Arch. Environ. Health*, 27, 387, 1973.

195. **Machata, G. and Bindes, R.**, Determination of traces of lead, thallium, zinc and cadmium in biological material by flameless atomic absorption spectrophotometry, *Z. Rechtsmed.*, 73, 29, 1973 (in German).

196. **Olzhataev, B. A., Koifman, M. D., Shugurov, E. V., and Nedugova, G. A.,** Development of methods for the atomic absorption determination of zinc in products of the flotation beneficiation of lead-zinc ores, *Izv. Akad. Nauk. Uzb. SSR Ser. Fiz. Mat. Nauk,* 17, 66, 1973.

197. **Szivos, K., Polos, L., Bezur, L., and Pungor, E.,** Comparative studies of (methods for) determination of the zinc content of insulin preparations and crystalline insulin, *Acta. Pharm. Hung.,* 43, 90, 1973 (in Hungarian).

198. **Okusu, H., Ueda, Y., Ota, K., and Kawano, K.,** Determination of cadmium, zinc, copper and lead in waste water by atomic absorption spectrometry, *Bunseki Kagaku,* 22, 84, 1973.

199. **Steele, T. H.,** Dissocation of zinc excretion from other cations in man, *J. Lab. Clin. Med.,* 81, 205, 1973.

200. **Sorenson, J. R. J., Levin, L. S., and Petering, H. G.,** Cadmium, copper, lead, mercury and zinc concentrations in the hair of individuals living in the United States, *Interface,* 2, 17, 1973.

201. **Smeyers-Verbeke, J., Massart, D. L., Versieck, J., and Speecke, A.,** The determination of copper and zinc in biological materials. A comparison of atomic absorption with spectrophotometry and neutron activation, *Clin. Chim. Acta,* 44, 243, 1973.

202. **Baker, D. E.,** A new approach to soil testing. II. Ionic equilibria involving H, K, Ca, Mg, Mn, Fe, Cu, Zn, Na, P and S, *Soil Sci. Soc. Am. Proc.,* 37, 537, 1973.

203. **Fernandez-Madrid, F., Prasad, A. S., and Oberleas, D.,** Effect of zinc deficiency on nucleic acids, collagen and noncollageneous protein of the connective tissue, *J. Lab. Clin. Med.,* 82, 951, 1973.

204. **Gish, C. D. and Christensen, R. E.,** Cadmium, nickel, lead and zinc in earthworms from roadside soil, *Environ. Sci. Technol.,* 7, 1060, 1973.

205. **Hohnadel, D. C., Sunderman, F. W., Jr., Nechay, M. W., and McNeeley, M. D.,** Atomic absorption spectrometry of nickel, copper, zinc and lead in sweat collected from healthy subjects during sauna bathing, *Clin. Chem. (Winston-Salem, N.C.),* 19, 1288, 1973.

206. **Johnson, D. L., West, T. S., and Dagnall, R. M.,** Comparative study of the determination of zinc and molybdenum by atomic absorption spectrometry with a carbon filament atom reservoir, *Anal. Chim. Acta,* 66, 171, 1973.

207. **Lagerwerff, J. V., Brower, D. L., and Biersdorf, G. T.,** Accumulation of cadmium, copper, lead and zinc in soil and vegetation in the proximity of a smelter, in *Trace Substances in Environmental Health,* VI. Hemphill, D. D., Ed., University of Missouri, Columbia, 1973, 71.

208. **Mikhailova, T. P., Klimashova, B. P., and Babeuva, L. V.,** Determination of zinc, cadmium, copper and iron in aluminum and gallium nitrides using an atomic absorption method, *Izv. Sib. Otd. Akad. Nauk SSSR Ser. Khim. Nauk,* 135, 1973.

209. **Mizuno, T.,** Determination of traces of iron, copper, zinc, nickel and cobalt in water by atomic absorption spectrometry using PAN - MIBK (1-[2-Pyridyl azo]-2-naphthol) - (methyl isobutyl ketone) extraction, *Nippon Kagaku Kaishi,* 10, 1904, 1973.

210. **Osiname, O. A., Kang, B. T., Schulte, E. E., and Gorey, R. B.,** Zinc response of maize (*Zea mays* L.) grown on sandy inceptisols in Western Nigeria, *Agron. J.,* 65, 875, 1973.

211. **Sorenson, J. R. J., Melby, E. G., Nord, P. J., and Petering, H. G.,** Interferences in the determination of metallic elements in human hair. Evaluation of zinc, copper, lead and cadmium using atomic absorption spectrophotometry, *Arch. Environ. Health,* 27, 36, 1973.

212. **Toma, O. and Crisan, T.,** Determination of iron, zinc, chromium and sodium in catalysts by atomic absorption spectrometry, *Rev. Chim.,* 24, 555, 1973.

213. **Warncke, D. D. and Barber, S. A.,** Diffusion of zinc in soils. III. Relation to zinc absorption isotherms, *Soil Sci. Soc. Am. Proc.,* 37, 355, 1973.

214. **Osiname, O. A., Schulte, E. E., and Corey, R. R.,** Tests for available copper and zinc in soils of Western Nigeria, *J. Sci. Food Agric.,* 24, 1341, 1973.

215. **Peaston, R. A.,** Determination of copper and zinc in plasma and urine by atomic absorption spectrophotometry, *Med. Lab. Technol.,* 30, 249, 1973.

216. **Harms, U.,** Determination of transition metals, Mn, Fe, Co, Cu and Zn in river fish with the help of X-ray fluorescence and flameless atomic absorption, *Mitteilungsbl. GDCh. Fachgruppe Lebensmittelchem. Gerichtl. Chem.,* 27, 271, 1973.

217. **Koester, H. M.,** The atomic absorption determination of copper (II), iron (III) and zinc in silicate rocks after separation on chloride-form Dowex I-X8 or amberlite CG 400-I ion exchanger, *Neues Jahrb. Miner. Abh.,* 119, 145, 1973 (in German).

218. **Hahn, C.,** Determination of lead, cadmium and zinc in fired glazes and engobe decorations using atomic absorption spectroscopy, *Silikat J.,* 12, 38, 1973.

219. **Brockhaus, A.,** Quantitative detection of lead, cadmium and zinc in blood and organs, *Staub-Reinhalt. Luft,* 33, 437, 1973 (in German).

220. **Chao, T. T. and Sanzolone, R. F.,** The atomic absorption spectrophotometric determination of microgram levels of cobalt, nickel, copper, lead and zinc in soil and sediment extracts containing large amounts of manganese and iron, *J. Res. U.S. Geol. Surv.,* 1, 681, 1973.

221. **Hopp, H. U.,** Atomic absorption spectrophotometric determination of zinc, calcium, barium and magnesium in mineral oil products, *Erdoel Kohle Erdas Petrochem. Brennst. Chem.*, 27, 435, 1974.

222. **Jensen, F. O., Dolezal, J., and Langmyhr, F. J.,** Atomic absorption spectrometric determination of cadmium, lead and zinc in salts or salt solutions by hanging mercury drop electrodeposition and atomization in a graphite furnace, *Anal. Chim. Acta*, 72, 245, 1974.

223. **Langmyhr, F. J., Stuberg, J. R., Thomassen, Y., Hanssen, J. E., and Dolezal, J.,** Atomic absorption spectrometric determination of cadmium, lead, silver, thallium and zinc in silicate rocks by direct atomization from the solid state, *Anal. Chim. Acta*, 71, 35, 1974.

224. **Baker, A. S. and Smith, R. L.,** Preparation of solutions for atomic absorption analysis of iron, manganese, zinc and copper in plant tissue, *J. Agric. Food. Chem.*, 22, 103, 1974.

225. **Bentley, G. E. and Parsons, M. L.,** A further study of the effect of pH on the atomic absorption analysis of zinc, *Appl. Spectrosc.*, 28, 71, 1974.

226. **Brady, D. V., Montalvo, J. G., Jr., Glowacki, G., and Piscinotta, A.,** Direct determination of zinc in sea-bottom sediments by carbon tube atomic absorption, *Anal. Chim. Acta*, 70, 448, 1974.

227. **Hartley, T. F., Dawson, J. B., and Hodgkinson, A.,** Simultaneous measurement of Na, K, Ca, Mg, Cu and Zn balances in man, *Clin. Chim. Acta*, 52, 321, 1974.

228. **Holmes, C. W., Slade, E. A., and McLerran, C. J.,** Migration and redistribution of zinc and cadmium in marine esturine system, *Environ. Sci. Technol.*, 8, 255, 1974.

229. **Loffe, E. S., Prisenko, V. S., and Romazanova, I. I.,** Atomic absorption determination of zinc in hydrometallurgical conversion solutions, *Zavod. Lab.*, 40, 358, 1974.

230. **Smeyers-Verbeke, J., Defrise-Gussenhoven, E., Ebinger, G., Loewenthal, A., and Massart, D. L.,** Distribution of Cu and Zn in human brain tissue, *Clin. Chim. Acta*, 51, 309, 1974.

231. **Arroyo, M. and Palenque, E.,** Analytical determination of zinc in biological fluids by atomic absorption spectrophotometry, *Rev. Clin. Esp.*, 133, 211, 1974 (in Spanish).

232. **Brachaczek, W. W., Butler, J. W., and Pierson, W. R.,** Acid enhancement of zinc and lead atomic absorption signals, *Appl. Spectrosc.*, 28, 585, 1974.

233. **Diez, A. L., Grassi, R. L., and Facchinetti, M.,** Determination of copper, zinc, manganese and iron in garlic plants by atomic absorption spectrophotometry, *Agrochimica*, 18, 128, 1974.

234. **Falchuk, K. H., Evenson, M., and Vallee, B. L.,** Multichannel atomic absorption instrument. Simultaneous analysis of zinc, copper and cadmium in biological materials, *Anal. Biochem.*, 62, 255, 1974.

235. **Fodor, P., Polos, L., Bezur, L., and Pungor, E.,** Determination of calcium, mangesium, iron and zinc in protein concentrates by atomic absorption spectrometry, *Period. Polytech. Chem. Eng.*, 18, 125, 1974.

236. **Jones, M. H. and Woodcock, J. T.,** On stream atomic absorption determination of zinc and manganese in flotation liquors containing calcium sulfate, *Anal. Chim. Acta*, 69, 275, 1974.

237. **Komarek, J., Horak, J., and Sommer, L.,** Atomic absorption of zinc after extraction of its chelate with 1-(2-pyridyl azo)-2-naphthol, *Collect. Czech. Chem. Commun.*, 39, 92, 1974.

238. **Kubota, J., Mills, E. L., and Oglesby, R. T.,** Lead, Cd, Zn, Cu and Co in streams and lake waters of Cayuga lake Basin, New York, *Environ. Sci. Technol.*, 8, 243, 1974.

239. **Lee, K. T. and Jacob, E.,** Determination of serum iron and zinc by atomic absorption spectroscopy, *Mikrochim. Acta*, 1, 65, 1974.

240. **McBean, L. D., Smith, J. C., Jr., Berne, B. H., and Halsted, J. A.,** Serum zinc and ALPHA$_2$ macroglobulin concentration in myocardial infarction, decubitus ulcer, multiple myeloma, prostatic carcinoma, Down's Syndrome and nephrotic syndrome, *Clin. Chim. Acta*, 50, 43, 1974.

241. **Mikhailova, T. P. and Slyadneva, N. A.,** Atomic absorption determination of zinc and cadmium in industrial solutions, *Zh. Anal. Khim.*, 29, 1912, 1974.

242. **Weiner, J.,** Determination of zinc in beer by atomic absorption spectroscopy, *J. Inst. Brew. London*, 80, 486, 1974.

243. **Van Raaphorst, J. C., Van Weers, A. W., and Haremaker, H. M.,** Loss of zinc and cobalt during dry ashing of biological material, *Analyst (London)*, 99, 523, 1974.

244. **Throneberry, G. O.,** Phosphorus and zinc measurements in kjeldahl digests, *Anal. Biochem.*, 60, 358, 1974.

245. **Talmi, Y.,** Determination of zinc and cadmium in environmentally based samples by the radio-frequency spectrometric source, *Anal. Chem.*, 46, 1005, 1974.

246. **Bergman, B., Sjostrom, R., and Wing, K. R.,** Variation with age of tissue zinc concentrations in albino rats determined by atomic absorption spectrophotometry, *Acta Physiol. Scand.*, 92, 440, 1974.

247. **Brovko, I. A., Nazarov, S. N., and Rish, M. A.,** Atomic absorption determination of zinc, cadmium, cobalt, copper and nickel after their extraction concentration in the di-phenyl-carbazone-pyridine-toluene system, *Zh. Anal. Khim.*, 29, 2387, 1974.

248. **Fletcher, G. E. and Collins, A. G.,** Atomic absorption methods of analysis of oilfield brines: barium, calcium, copper, iron, lead, lithium, magnesium, manganese, potassium, sodium, strontium and zinc, *Rep. Invest. U.S. Bur. Mines*, RI 7861, 1974.

249. **Lofberg, R. T. and Levri, E. A.**, Analysis of copper and zinc in hemolyzed serum samples, *Anal. Lett.*, 7, 775, 1974.

250. **Oelschlagger, W. and Lantzsch, H. J.**, Sources of error in the determination of zinc by means of atomic absorption spectrophotometry, *Forschungsdiest*, 27, 31, 1974.

251. **Reith, J. F., Engelsma, J., and Van Ditmarsch, M.**, Lead and zinc contents of food and diets in the Netherlands, *Z. Lebensm. Unters. Forsch.*, 156, 271, 1974.

252. **Sinha, R. C. P. and Banerjee, B. K.**, Interferences in estimation of trace amounts of cobalt, copper and zinc in soils by atomic absorption spectrophotometry, *Technol. Sindri*, 11, 263, 1974.

253. **Oelschlager, W. and Schmidt, S.**, Determination of zinc in vegetable and animal materials and in mineral fodders using atomic absorption spectrophotometry, *Landwirtsch. Forsch.*, 27, 85, 1974 (in German).

254. **Beyer, M. E. and Bond, A. M.**, Simultaneous determination of cadmium, copper, lead and zinc in lead and zinc concentrates by a.c. polarographic methods. Comparison with atomic absorption spectrometry, *Anal. Chim. Acta*, 75, 409, 1975.

255. **Burch, R. E., Hahn, H. K. J., and Sullivan, J. F.**, Newer aspects of the roles of zinc, manganese and copper in human nutrition, *Clin. Chem. (Winston-Salem, N.C.)*, 21, 501, 1975.

256. **Burch, R. E., Williams, R. V., Hahn, H. K. J., Jetton, M. M., and Sullivan, J. F.**, Serum and tissue enzyme activity and trace element content in response to zinc deficiency in the pig, *Clin. Chem. (Winston-Salem, N.C.)*, 21, 568, 1975.

257. **Chooi, M. K., Todd, J. K., and Boyd, N. D.**, Effect of carbon cup aging on plasma zinc determination by flameless atomic absorption spectrometry, *Clin. Chem. (Winston-Salem, N.C.)*, 21, 632, 1975.

258. **Dittrich, K. and Zeppan, W.**, Determination of zinc in hydrochloric acid, gallium arsenide and gallium aluminum arsenide using flameless atomic absorption, *Talanta*, 22, 299, 1975.

259. **Evenson, M. A. and Anderson, C. T., Jr.**, Ultramicro analysis for copper, cadmium and zinc in human liver tissue by use of atomic absorption spectrophotometry, *Clin. Chem. (Winston-Salem, N.C.)*, 21, 537, 1975.

260. **Hermann, P.**, Determination of lead and zinc in powders by atomic absorption, *Spectrochim. Acta*, 30b, 15, 1975 (in German).

261. **Jacob, R. A. and Klevay, L. M.**, Determination of trace amounts of copper and zinc in edible fats and oils by acid extraction and atomic absorption spectrophotometry, *Anal. Chem.*, 47, 741, 1975.

262. **Korkisch, J., Goedl, L., and Gross, H.**, Use of ion exchange for the determination of trace elements in natural waters. VI. Zinc, *Talanta*, 22, 289, 1975.

263. **Momcilovic, B., Belonje, B., and Shah, B. G.**, Effect of the matrix of the standard on results of atomic absorption spectrophotometry of zinc in serum, *Clin. Chem. (Winston-Salem, N.C.)*, 21, 588, 1975.

264. **Pekarek, R. S., Kluge, R. M., DuPont, H. L., Wannemacher, R. W., Jr., Hornick, R. B., Bostian, K. A., and Beisel, W. R.**, Serum zinc, iron and copper concentrations during typhoid fever in man. Effect of chloramphenicol therapy, *Clin. Chem. (Winston-Salem, N.C.)*, 21, 528, 1975.

265. **Prasad, A. S., Schoomaker, E. B., Ortega, J., Brewer, G. J., Oberleas, D., and Oelshlegel, F. J., Jr.**, Zinc deficiency in sickle cell disease, *Clin. Chem. (Winston-Salem, N.C.)*, 21, 582, 1975.

266. **Vasilades, J. and Sahawneh, T.**, Effect of diphenylhydantoin on serum copper, zinc and magnesium, *Clin. Chem. (Winston-Salem, N.C.)*, 21, 637, 1975.

267. **Wang, J. and Pierson, R. N., Jr.**, Distribution of zinc in skeletal muscle and liver tissue in normal and dietary controlled alcoholic rats, *J. Lab. Clin. Med.*, 85, 50, 1975.

268. **Carr, G. and Wilkinson, A. W.**, Zinc and copper urinary excretions in children with burns and scalds, *Clin. Chem. Acta*, 61, 199, 1975.

269. **Greig, R. A.**, Comparison of atomic absorption and neutron activation analysis for the determination of silver, chromium and zinc in various marine organisms, *Anal. Chem.*, 47, 1682, 1975.

270. **Greiner, A. C., Chan, S. C., and Nicolson, G. A.**, Determination of calcium, copper, magnesium and zinc content of identical areas in human cerebral hemispheres of normals, *Clin. Chim. Acta*, 61, 335, 1975.

271. **Greiner, A. C., Chan, S. C., and Nicolson, G. A.**, Human brain contents of calcium, copper, magnesium and zinc in some neurological pathologies, *Clin. Chim. Acta*, 64, 211, 1975.

272. **Henkin, R. I., Mueller, C. W., and Wolf, R. O.**, Estimation of zinc concentration of parotid saliva by flameless atomic absorption spectrophotometry in normal subjects and in patients with idiopathic hypogeusia, *J. Lab. Clin. Med.*, 86, 175, 1975.

273. **Kirk, M., Perry, E. G., and Arritt, J. M.**, Separation and atomic absorption measurement of trace amounts of lead, silver, zinc, bismuth and cadmium in high nickel alloys, *Anal. Chim. Acta*, 80, 163, 1975.

274. **Lukasiewicz, R. J. and Buell, B. E.**, Zinc determination in lubricating oils and additives by atomic absorption spectrometry, *Anal. Chem.*, 47, 1673, 1975.

275. **Luzar, O. and Sliva, V.**, Determination of calcium oxide, magnesium oxide, alumina, silica, iron, copper, zinc, lead, cadmium, sodium oxide and potassium oxide in iron ores and agglomerates, *Hutn. Listy*, 30, 55, 1975 (in Czech).

276. **Machata, G.,** The normal level of cadmium, copper and zinc in blood of the Viennese population, *Wien. Klin. Wochenschr.,* 87, 484, 1975 (in German).

277. **Pomeroy, R. K., Drikitis, N., and Koga, Y.,** Determination of zinc in hair using atomic absorption spectroscopy, *J. Chem. Educ.,* 52, 544, 1975.

278. **Touillon, C., Bansillon, V., Vellon, J. J., Badinand, A., and Comtet, J. J.,** Study of the zinc levels in serum and erythrocytes in burn patients, *Clin. Chim. Acta,* 63, 115, 1975.

279. **Iosof, V., Mihalka, S., and Colios, E.,** Determination of copper, lead, zinc in mineral products by atomic absorption spectrophotometry, *Rev. Chim.,* 26, 680, 1975.

280. **LeBihan, A. and Courtot-Coupez, J.,** Direct determination of zinc in sea water using flameless atomic absorption spectrophotometry, *Analusis,* 3, 559, 1975.

281. **Langmyhr, F. J., Lind, T., and Jonsen, J.,** Atomic absorption spectrometric determination of manganese, silver and zinc in dental material by atomization directly from the solid state, *Anal. Chim. Acta,* 80, 297, 1975.

282. **Schnappinger, M. G., Jr., Martens, D. C., and Plank, C. O.,** Zinc availability as influenced by application of fly ash to soil, *Environ. Sci. Technol.,* 9, 258, 1975.

283. **Acatini, C., deBerman, S. N., Colombo, O., and Pondo, O.,** Determination of silver, copper, lead, tin, antimony, iron, calcium, zinc, magnesium, potassium and manganese in canned tomatoes by atomic absorption spectrophotometry, *Rev. Asoc. Bioquim Argent.,* 40, 175, 1975 (in Spanish).

284. **Bulas, E., Daescu, A., and Basceanu, M.,** Indirect determination of zinc in the presence of aluminum, iron, calcium and magnesium, *Rev. Chim.,* 26, 689, 1975 (in Romanian).

285. **El-Kinawy, S. A., Walash, M. I., Abou-Bakr, M. S., and Diaie, I. Z.,** Determination of minerals in commercial pharmaceutical preparations by atomic absorption spectrophotometry. I. Determination of zinc, iron and magnesium in some multivitamin and mineral preparations, *J. Drug Res. Egypt,* 7, 151, 1975.

286. **Guido, O. O. and Amaya, C.,** Direct determination of calcium, cobalt and zinc in nuclear-grade uranium oxide by atomic absorption, *Inf. Com. Nac. Energ. Atom. Argentina,* 392, 1975 (in Spanish).

287. **Postel, W., Georg, A., Drawert, F., and Guvenc, U.,** Determination of trace elements in foods by atomic absorption spectrophotometry. IV. Zinc in beer, *Brauwissenschaft,* 28, 301, 1975 (in German).

288. **Vondenhoff, T.,** Determination of lead, cadmium, copper and zinc in plant and animal material by atomic absorption in a flame and in a graphite tube after sample decomposition by the Schoniger technique, *Mitteilungsbl. GDCh. Fachgruppe Lebensmittelchem. Gerichtl. Chem.,* 29, 341, 1975 (in German).

289. **Rudnevskii, N. K., Demarian, V. T., Molyanov, A. I., and Sklyumina, L. V.,** Effect of heating the aerosol on sensitivity of atomic absorption determination of certain elements (copper, iron, magnesium and zinc), *Tr. Khim. Khim. Technol.,* 1, 108, 1975 (in Russian).

290. **Rudnevskii, N. K., Demarian, V. T., Sklemina, L. V., and Tumanova, A. N.,** Influence of some organic solvents on sensitivity of atomic absorption determination of iron, cadmium, copper, magnesium, sodium, vanadium, selenium and zinc, *Tr. Khim. Khim. Technol.,* 1, 106, 1975.

291. **Berndt, H. and Jackwerth, E.,** Automated injection method for dispensing small volume samples in flame atomic absorption. Multielement analysis of high purity aluminum and determination of copper, iron and zinc in serum, *At. Absorpt. Newsl.,* 15, 109, 1976.

292. **Bezur, L., Silvosh, K., Popova, S., and Pungor, E.,** Atomic absorption determination of copper, manganese, zinc, iron, potassium, sodium, calcium and magnesium in some animal foods, *Khim. Ind.,* 48, 204, 1976.

293. **Christensen, J. J., Hearty, P. A., and Izatt, R. M.,** Determination of chromium in several proposed standard samples and of zinc and chromium in wheat milling and beet sugar refining samples, *J. Agric. Food Chem.,* 24, 811, 1976.

294. **Koizumi, H. and Yasuda, K.,** Determination of lead, cadmium and zinc using the Zeeman effect in atomic absorption spectrometry, *Anal. Chem.,* 48, 1178, 1976.

295. **Korkisch, J. and Huebner, H.,** Atomic absorption spectrophotometric determination of cobalt, copper, manganese and zinc in multivitamin preparations after separation by means of anion exchange, *Mikrochim. Acta,* 2, 311, 1976.

296. **Ohki, K., Anderson, O. E., and Jones, L. S.,** Residual Zn availability related to growth and nutrient gradients in corn, *Agron. J.,* 68, 612, 1976.

297. **Roberts, A. H. C., Turner, M. A., and Syers, J. K.,** Simultaneous extraction and concentration of cadmium and zinc from soil extracts, *Analyst (London),* 101, 574, 1976.

298. **Sanzolone, R. P. and Chao, T. T.,** Atomic absorption spectrometric determination of copper, zinc, and lead in geological materials, *Anal. Chim. Acta,* 86, 163, 1976.

299. **Satake, M., Asano, T., Takagi, Y., and Yonekubo, T.,** Determination of copper, zinc, lead, cadmium and manganese in brackish and coastal waters by the combination of chelating ion exchange separation and atomic absorption spectrophotometry, *Nippon Kagaku Kaishi,* 5, 762, 1976.

300. **Sohler, A., Wolcott, P., and Pfeiffer, C. C.,** Determination of zinc in fingernails by non-flame atomic absorption spectroscopy, *Clin. Chim. Acta,* 70, 391, 1976.

301. **Tomi, B. and Manoliu, C.,** Atomic absorption spectrophotometric determination of zinc deposited on active carbon supports, *Rev. Chim.,* 27, 704, 1976.

302. **Wuyts, L., Smeyers-Verbeke, J., and Massart, D. L.,** Atomic absorption spectrophotometry of copper and zinc in human brain tissue. A critical investigation of two digestion techniques, *Clin. Chim. Acta,* 72, 405, 1976.

303. **Lafargue, P., Couture, J. C., Monteil, R., Guilband, J., and Saliou, L.,** Evolution of serum copper and zinc levels in burn patients, *Clin. Chim. Acta,* 66, 181, 1976 (in French).

304. **McGary, E. D. and Young, B. E.,** Quantitative determination of zinc, iron, calcium and phosphorus in the total diet market basket by atomic absorption and calorimetric spectrophotometry, *J. Agric. Food Chem.,* 24, 539, 1976.

305. **Morris, N. M., Clarke, M. A., Tripp, V. W., and Carpenter, F. G.,** Determination of lead, cadmium and zinc in sugar, *J. Agric. Food Chem.,* 24, 45, 1976.

306. **Mzhel'skaya, T. I.,** Determination of copper, iron and zinc content in blood serum using the atomic absorption spectrophotometer "SPEKTR-I", *Lab. Delo,* 4, 229, 1976.

307. **Michie, D. D., Bell, N. H., and Wirth, F. H.,** Technique of handling of neonatal blood samples for zinc analysis, *Am. J. Med. Technol.,* 42, 424, 1976.

308. **Wall, G.,** Determination by atomic absorption spectrophotometry of copper, lead and zinc in sulfide concentrates, *Rep. Natl. Inst. Metall. Repub. S. Afr. Rep.,* No. 1798, 1976.

309. **Hoffman, H. E., Nathansen, N., and Altmann, F.,** Determination of calcium, zinc and lead in unused lubricating oils with the AAS-I atomic absorption spectrophotometer, *Jenaer Rundsch.,* 21, 302, 1976 (in German).

310. **Wunderlich, E. and Hädeler, W.,** Determination of traces of zinc, cadmium and bismuth in pure copper by AAS, *Z. Anal. Chem.,* 281, 300, 1976 (in German).

311. **Inoue, S., Yotsuyanagi, T., Sasaki, M., and Aomura, K.,** Atomic absorption method for the determination of trace amounts of cadmium, copper, lead and zinc by ion-pair extraction of dimercaptomaleonitrile complexes, *Bunseki Kagaku,* 26, 550, 1977.

312. **Kang, H. K., Harvey, P. W., Valentine, J. L., and Swendseid, M. E.,** Zinc, iron, copper and magnesium concentrations in tissues of rats fed various amounts of zinc, *Clin. Chem. (Winston-Salem, N.C.),* 23, 1834, 1977.

313. **Mosescu, N., Kalmutchi, G., and Badea, S.,** Zinc determination by atomic absorption of mineral oil in additives, *Rev. Chim.,* 28, 373, 1977.

314. **Stevens, M. D., Mackenzie, W. F., and Anand, V. D.,** A simplified method for determination of zinc in whole blood, plasma and erythrocytes by atomic absorption spectrophotometry, *Biochem. Med.,* 18, 158, 1977.

315. **Zawadzko, H., Baralkiewiez, D., and Elbanowska, H.,** Flame atomic absorption spectrometric determination of cobalt, cadmium, lead, nickel, copper and zinc in natural waters, *Chem. Anal.,* 22, 913, 1977.

316. **Berenguer-Navarro, V. and Hernandez-Mendez, J.,** Analysis by atomic absorption using emulsion formations. I. Determination of zinc in lubricant oils, *Quim. Anal.,* 31, 81, 1977 (in Spanish).

317. **Del-Castiho, P. and Herber, R. F.,** The rapid determination of cadmium, lead, copper and zinc in whole blood by atomic absorption spectrometry with electrothermal atomization. Improvements in precision with a peak-shape monitoring device, *Anal. Chim. Acta,* 94, 269, 1977.

318. **Djulgerova, R. B.,** Spectroscopical effects arising under application of pulse supply to zinc hollow cathode discharge, *Spectrosc. Lett.,* 10, 727, 1977.

319. **Harada, T., Fujiwara, K., and Fuwa, K.,** Problems in determination of zinc at sub-picogram amounts by carbon rod atomic absorption spectrophotometry, *Bunseki Kagaku,* 26, 877, 1977.

320. **Capar, S. G.,** Atomic absorption spectrophotometric determination of lead, cadmium, zinc and copper in clams and oysters: collaborative study, *J. Assoc. Off. Anal. Chem.,* 60, 1400, 1977.

321. **Petering, H. G., Murthy, L., and O'Flaherty, E.,** Influence of dietary copper and zinc on rat lipid metabolism, *J. Agric. Food Chem.,* 25, 1105, 1977.

322. **Lorber, K. and Mueller, K.,** Atomic absorption spectrometry determination of copper and zinc in high purity tungsten and molybdenum compounds after column extraction chromatographic enrichment with dithizone in *O*-dichlorobenzene, *Mikrochim. Acta,* 2, 5, 1977.

323. **Maurer, J.,** Extraction method for the simultaneous determination of sodium, potassium, calcium, magnesium, iron, copper, zinc and manganese in organic material using atomic absorption spectrophotometry, *Z. Lebensm. Unters. Forsch.,* 165, 1, 1977.

324. **Menden, E. E., Brockman, D., Choudhury, H., and Petering, H. G.,** Dry ashing of animal tissues for atomic absorption spectrometric determination of zinc, copper, cadmium, lead, iron, manganese, magnesium and calcium, *Anal. Chem.,* 49, 1644, 1977.

325. **Pandey, L. P., Ghose, A., and Dasgupta, P.,** Determination of zinc, silver, copper, iron and antimony in lead metal by atomic absorption spectrophotometry, *J. Inst. Chem.,* 49, 35, 1977.

326. **Putov, I., Popova, S., and Brashnarova, A.,** Determination of copper, nickel, cobalt, iron, zinc and manganese in rubber stocks and adhesives by an atomic absorption spectrophotometric method, *Khim. Ind.,* 49, 16, 1977.

327. **Yasuda, S. and Kakiyama, N.,** Vaporization of atoms and molecules during heating of cadmium, lead and zinc salts in a carbon tube atomizer, *Anal. Chim. Acta,* 89, 369, 1977.

328. **Tzouwara-Karayanni, S. M.,** Determination of ionic and organically bound zinc in the sperm of the sea urchin by atomic absorption spectrophotometry, *Microchem. J.,* 22, 259, 1977.

329. **Butrimovitz, G. P. and Purdy, W. C.,** The determination of zinc in blood plasma by atomic absorption spectrometry, *Anal. Chim. Acta,* 94, 63, 1977.

330. **Campbell, W. C. and Ottaway, J. M.,** Direct determination of cadmium and zinc in sea water by carbon furnace atomic absorption spectrometry, *Analyst (London),* 102, 495, 1977.

331. **Armannsson, H.,** The use of dithizone extraction and atomic absorption spectrometry for the determination of cadmium, zinc, copper, nickel and cobalt in rocks and sediments, *Anal. Chim. Acta,* 88, 89,1977.

332. **Bogden, J. D., Troiano, R. A., and Joselow, M. M.,** Copper, zinc, magnesium and calcium in plasma and cerebrospinal fluid of patients with neurological diseases, *Clin. Chem. (Winston-Salem, N.C.),* 23, 485, 1977.

333. **Bower, N. W. and Ingle, J. D.,** Precision of flame atomic absorption measurements of arsenic, cadmium, calcium, copper, iron, magnesium, molybdenum, sodium and zinc, *Anal. Chem.,* 49, 574, 1977.

334. **Hinesly, T. D., Jones, R. L., Ziegler, E. L., and Tyler, J. J.,** Effects of annual and accumulative applications of sewage sludge on assimilation of zinc and cadmium by corn (*Zea mays* L.), *Environ. Sci. Technol.,* 11, 182, 1977.

335. **Pakalns, P. and Farrar, Y. J.,** The effect of surfactants on the extraction-atomic absorption spectrophotometry. Determination of copper, iron, manganese, lead, nickel, zinc, cadmium and cobalt, *Water Res.,* 11, 145, 1977.

336. **Ishii, T. and Ohnishi, R.,** Pretreatment for the determination of zinc and nickel in coal and coke by atomic absorption spectrophotometry, *Bunseki Kagaku,* 26, 874, 1977.

337. **Pellerin, F. and Goulle, J. P.,** Detection and rapid determination of cadmium, copper, lead and zinc in dyes and antioxidants authorized for use in drugs and foodstuffs, *Ann. Pharm. Fr.,* 35, 189, 1977.

338. **Zakhariya, A. N. and Olenovich, N. L.,** Atomic absorption determination of zinc in copper alloys, *Zh. Prikl. Spektrosk.,* 27, 792, 1977.

339. **Watling, R. J.,** The use of a slotted tube for the determination of lead, zinc, cadmium, bismuth, cobalt, manganese and silver by atomic absorption spectrometry, *Anal. Chim. Acta,* 97, 395, 1978.

340. **Arpadjan, S. and Kachov, I.,** Atomic absorption determination of cadmium, chromium and zinc in blood serum, *Zentralbl. Pharm. Pharmakother. Laboratoriumsdiagn.,* 117, 237, 1978 (in German).

341. **Bogden, J. D. and Troiano, R. A.,** Plasma calcium, copper, magnesium and zinc concentrations in patients with the alcohol withdrawal syndrome, *Clin. Chem. (Winston-Salem, N.C.),* 24, 1553, 1978.

342. **Boiteau, H. L. and Metayer, C.,** Microdetermination of lead, cadmium, zinc and tin in biological materials by atomic absorption spectrometry after mineralization and extraction, *Analusis,* 6, 350, 1978.

343. **Fudagawa, N. and Kawase, A.,** Determination of copper and zinc in tea by atomic absorption spectrometry, *Bunseki Kagaku,* 27, 353, 1978.

344. **Hoenig, M. and Vanderstappen, R.,** Determination of cadmium, copper, lead, zinc and manganese in plants by flame atomic absorption spectroscopy: mineralization effects, *Analusis,* 6, 312, 1978.

345. **Ellis, L.,** Serum zinc levels and urinary zinc excretion in patients with renal transplants, *Clin. Chim. Acta,* 82, 105, 1978.

346. **Geladi, P. and Adams, F.,** The determination of cadmium, copper, iron, lead and zinc in aerosols by atomic absorption spectrometry, *Anal. Chim. Acta,* 96, 229, 1978.

347. **Glaeser, E.,** Atomic absorption spectroscopic determination of iron, copper, zinc, cadmium, manganese and cobalt in waters by the "injection method" after an extractive concentration, *Acta Hydrochim. Hydrobiol.,* 6, 83, 1978.

348. **Ishino, I., Matsumae, H., Shibata, K., Ariga, N., and Goshima, F.,** Determination of copper, cadmium and zinc in bone by atomic absorption spectrophotometry, *Bunseki Kagaku,* 27, 232, 1978.

349. **Kelson, J. R. and Shamberger, R. J.,** Methods compared for determining zinc in serum by flame atomic absorption spectroscopy, *Clin. Chem. (Winston-Salem, N.C.),* 24, 240, 1978.

350. **Stoveland, S., Astruc, M., Perry, R., and Lester, J. N.,** Rapid flameless atomic absorption analysis of the metallic content of sewage sludges. II. Chromium, nickel and zinc, *Total Environ.,* 9, 263, 1978.

351. **Mazzucotelli, A., Galli, M., Benassi, E., Loeb, C., Ottonello, G. A., and Tanganelli, P.,** Atomic absorption microdetermination of zinc in cerebrospinal fluid by ion exchange chromatography and electrothermal atomization, *Analyst (London),* 103, 863, 1978.

352. **Urquhart, N.,** Zinc contamination in a trace element control serum, *Clin. Chem. (Winston-Salem, N.C.),* 24, 1652, 1978.

353. **Viets, J. G.,** Determination of silver, bismuth, cadmium, copper, lead and zinc in geologic materials by atomic absorption spectrometry with tricaprylyl methyl ammonium chloride, *Anal. Chem.,* 50, 1097, 1978.

354. **Hioki, T., Dokiya, Y., Notsu, K., and Fuwa, K.,** Analytical condition for the determination of copper, zinc and nickel in silicates by atomic absorption spectrometry, *Bunseki Kagaku,* 27, 48t, 1978.

355. **Kantor, T., Fodor, P., and Pungor, E.,** Determination of traces of lead, cadmium and zinc in copper by an arc-nebulization and flame atomic absorption technique, *Anal. Chim. Acta,* 102, 15, 1978.

356. **Pilipyuk, Y. S., Pyatnitskii, I. V., and Kovalenko, A. A.,** Atomic absorption determination of zinc and cadmium in chloroform extracts of their amine complexes with cinnamic acid, *Ukr. Khim. Zh.,* 44, 1206, 1978 (in Russian).

357. **Mckown, M. M., Tschrin, C. R., and Lee, P. P. F.,** Investigation of Matrix Interference for Atomic Absorption Spectroscopy of Sediments, Rep. No. EPA-600/7-78/085, U.S. Environmental Protection Agency, Office of Research and Development, Corvallis, Ore., 1978.

358. **Terashima, S.,** Atomic absorption determination of manganese, iron, copper, nickel, cobalt, lead, zinc, silicon, aluminum, calcium, magnesium, sodium, potassium, titanium and strontium in manganese nodules, *Chichtsu Chosasho Geppo,* 29, 401, 1978 (in Japanese).

359. **Schuetze, I. and Mueller, W.,** Determination of trace metals in dietary fats and emulsifiers by flameless atomic absorption spectrometry. I. Determination of bound heavy metals, copper, nickel, zinc, lead and cadmium in dietary fats, *Nahrung,* 22, 777, 1978 (in German).

360. **Berndt, H. and Jackwerth, E.,** Mechanized micromethod ("injection method") of flame photometry (atomic absorption-atomic emission) for determination of serum electrolytes and trace elements (iron, copper, zinc). Determination of lithium, sodium, potassium, magnesium, calcium with a mechanized flame spectrometric micromethod, *J. Clin. Chem. Clin. Biochem.,* 17, 71, 1979.

361. **Bruland, K. W., Franks, R. P., Knauer, G. A., and Martin, J. H.,** Sampling and analytical methods for the determination of copper, cadmium, zinc and nickel in the nanogram per liter in sea water, *Anal. Chim. Acta,* 105, 233, 1979.

362. **Meranger, J. C., Subramanian, K. S., and Chalifoux, C.,** A national survey for cadmium, chromium, copper, lead, zinc, calcium and magnesium in Canadian drinking water supplies, *Environ. Sci. Technol.,* 13, 707, 1979.

363. **Okuneva, G. A., Gibalo, I. M., Kuzyakov, Y. Y., Zheleznova, A. A., and Shashova, M. V.,** Effect of cations during atomic absorption determination of iron, manganese, nickel, lead and zinc in copper-based alloys, *Zh. Anal. Khim.,* 34, 661, 1979.

364. **Vratkovskaya, S. V. and Pogrebnyak, Y. F.,** Determination of copper, lead and zinc in slightly mineralized water by flame atomic absorption spectrophotometry, *Zh. Anal. Khim.,* 34, 759, 1979.

365. **Wittman, Z.,** Direct determination of calcium, magnesium and zinc in lubricating oils and additives by atomic absorption spectrometry using a mixed solvent system, *Analyst (London),* 104, 156, 1979.

366. **Adler, J. F. and Bucklow, P. L.,** Determination of copper, zinc and chromium in carbon cloth by atomic absorption spectrometry without sample ashing, *At. Absorpt. Newsl.,* 18, 123, 1979.

367. **Arpadjan, S. and Alexandrova, I.,** Determination of trace elements (lead, bismuth, copper, zinc and silver) in high-purity tin by flame atomic absorption spectrophotometry, *Z. Anal. Chem.,* 298, 159, 1979.

368. **Armannsson, H.,** Dithizone extraction and flame atomic absorption spectrometry for the determination of cadmium, zinc, lead, copper, nickel, cobalt and silver in sea water and biological tissues, *Anal. Chim. Acta,* 110, 21, 1979.

369. **Berndt, H. and Jackwerth, E.,** Determination of iron, copper and zinc by a mechanized flame photometric micromethod. Mechanized micromethod ("injection method") of flame photometry (atomic absorption-atomic emission) for the determination of serum electrolytes and trace elements. II, *J. Clin. Chem. Clin. Biochem.,* 17, 489, 1979.

370. **Bozsai, G. and Csanady, M.,** Systematic investigations on the heavy metal pollution (cadmium, lead, copper, zinc, chromium and barium) of drinking water using atomic absorption spectrometric methods, *Z. Anal. Chem.,* 297, 370, 1979.

371. **Kumina, D. M. and Karyakin, A. V.,** Reduction of limits of detection for trace amounts of magnesium, zinc and copper in atomic absorption analysis using mixtures of organic solvents, *Zh. Anal. Khim.,* 34, 1411, 1979.

372. **Langmyhr, F. J., Eyde, B., and Jonsen, J.,** Determination of the total content and distribution of cadmium, copper and zinc in human parotid saliva, *Anal. Chim. Acta,* 107, 211, 1979.

373. **Smith, J. C., Butrimovitz, G. P., and Purdy, W. C.,** Direct measurement of zinc in plasma by atomic absorption spectroscopy, *Clin. Chem. (Winston-Salem, N.C.),* 25, 1487, 1979.

374. **Baeckman, S. and Karlsson, R. W.,** Determination of lead, bismuth, zinc, silver and antimony in steel and nickel-base alloys by atomic absorption spectrophotometry using direct atomization of solid samples in a graphite furnace, *Analyst (London),* 104, 1017, 1979.

375. **Capar, S. G. and Gould, J. H.,** Lead, fluoride and other elements in bonemeal supplements, *J. Assoc. Off. Anal. Chem.,* 62, 1054, 1979.

376. **Boyer, K. W., Capar, S. G., Jones, J. W., Suddendorf, R. F., and Forwalter, J.,** Multielement/trace analysis identifies 48 elements in foods. Simultaneous element analysis and data reduction reaches parts per billion, *Food Process,* 40, 72, 1979.

377. **Dogru, U., Arcasoy, A., and Cavdar, A. E.,** Zinc levels of plasma, erythrocyte, hair and urine in homozygote beta-thalassemia, *Acta Haematol.,* 62, 41,1979.

378. **Flanjak, J. and Lee, H. Y.,** Trace metal content of livers and kidneys of cattle, *J. Sci. Food Agric.,* 30, 503, 1979.

379. **Komarek, J. B., Havel, J., and Sommer, L.,** The use of chelates of copper, nickel, cobalt, cadmium and zinc with heterocyclic azo dyes in the AAS determination of these elements, *Collect. Czech. Chem. Commun.,* 44, 3241, 1979.

380. **Lambert, J. B., Szpunar, C. B., and Buikstra, J. E.,** Chemical analysis of excavated human bone from middle and late Woodland sites, *Archaeometry,* 21, 115, 1979.

381. **Sturgeon, H. E., Berman, S. S., Desauliners, A., and Russell, D. S.,** Determination of iron, manganese and zinc in seawater by graphite furnace atomic absorption spectrometry, *Anal. Chem.,* 51, 2364, 1979.

382. **Baucells, M., Lacort, G., and Roura, M.,** Determination of major and trace elements in marine sediments of Costa Brava by emission and atomic absorption spectroscopy, *Invest. Pesq.,* 43, 551, 1979.

383. **Berta, E. and Olah, J.,** Survey of heavy metal content of wastewater sludges in Hungary, *Hidrol. Kozl.,* 59, 469, 1979.

384. **Branca, P. and Maina, E.,** Contents of lead, copper, zinc, manganese and iron in products preserved in tin-plated containers, *Boll. Chim. Unione Ital. Lab. Prov. Parte Sci.,* 5, 593, 1979.

385. **Burdin, K. S., Krupina, M., and Savel'ev, I. B.,** Mollusks of the genus *Mytilus* as possible indicators of the concentration of heavy and transition metals in the marine environment, *Okeanologiya,* 19, 1038, 1979.

386. **Dolake, D. A. and Sievering, H.,** Trace element loading of Southern Lake Michigan by dry deposition of atmospheric aerosol, *Water Air Soil Pollut.,* 12, 485, 1979.

387. **El-Enany, F. F., Mahmoud, K. P., and Varma, M. M.,** Single organic extraction for determination of ten heavy metals in sea water, *J. Water Pollut. Control Fed.,* 51, 2545, 1979.

388. **Fujino, O., Matsin, M., and Shigematsu, T.,** Determination of trace metals in natural water by atomic absorption spectrometry. Preconcentration of metals by diethyldithiocarbamate/di-isobutyl ketone solvent extraction, *Mizu Shori Gijutsu,* 20, 201, 1979 (in Japanese).

389. **Korenev, A. P. and Konovalov, G. S.,** Optimization of conditions for determination of some metals (zinc and cadmium) in natural waters and in solutions by an atomic absorption method, *Gidrokhim. Mater.,* 73, 40, 1979.

390. **Marinescu, I. and Tamas, M.,** Phytochemical study on poplar buds as possible propolis source: some aspects of trace elements contained by propolis and poplar buds, *Apic. Rom.,* 54, 14, 1979.

391. **Price, W. J.,** New techniques of atomic absorption in chemical and biochemical analysis, *Sci. Ind.,* 12, 22, 1979.

392. **Williams, D. M.,** Trace metal determinations in blood obtained in evacuated collection tubes, *Clin. Chim. Acta,* 99, 23, 1979.

393. **Fricke, F. L., Robbins, W. B., and Caruso, J. A.,** Trace element analysis of food and beverages by atomic absorption spectrometry, *Prog. Anal. At. Spectrosc.,* 2, 185, 1979.

394. **Fonds, A. W., Kempf, T., Minderhoud, A., and Sonneborn, M.,** Heavy metal content of various kinds of water, *Reinhalt. Wasser,* 78, 1979.

395. **Anonymous,** Determination of Trace Metals in Liquid Coke-Oven Effluents by Atomic Absorption Spectrophotometry, Carbonization Res. Rep., 76, British Carbonization Research Association, Chesterfield, Derbyshire, England, 1979, 17.

396. **Ihida, M., Ishii, T., and Ohnishi, R.,** Precision on the determination of trace elements in coal, *Am. Chem. Soc. Div. Fuel Chem.,* 24, 262, 1979.

397. **Koppenaal, D. W., Schultz, H., Lett, R. G., Brown, F. R., Booher, H. B., Hattman, E. A., and Manahan, S. E.,** Trace element distributions in coal gasification products, *Prepr. Pap. Am. Chem. Soc. Div. Fuel Chem.,* 24, 299, 1979.

398. **Tsujii, K., Kuga, K., Murayama, S., and Yasuda, M.,** Evaluation of high-frequency discharge lamps for atomic absorption and atomic fluorescence spectrometry of cadmium, lead and zinc, *Anal. Chim. Acta,* 111, 103, 1979.

399. **Rockland, L. B., Wolf, W. R., Hahn, D. M., and Young, R.,** Estimation of zinc and copper in raw and cooked legumes: interlaboratory study of atomic absorption and X-ray fluorescence spectroscopy, *J. Food Sci.,* 44, 1711, 1979.

400. **Ueda, Y., Nakahara, M., Ishii, T., Suzuki, Y., and Suzuki, H.,** Amounts of trace elements in marine cephalopods, *J. Radiat. Res.,* 20, 338, 1979.

401. **Victor, A. H. and Strelow, F. W. E.,** Exchange chromatography and atomic absorption spectrometry, *Geostandards Newsl.,* 4, 217, 1980.

402. **Waughman, G. J. and Brett, T.,** Interference due to major elements during estimation of trace heavy metals in natural materials by atomic absorption spectrophotometry, *Environ. Res.,* 21, 385, 1980.

403. **Wesenberg, G. B. R., Fosso, G., Rasmussen, P., and Justesen, N. P. B.,** Effects of lead and cadmium uptake on zinc and copper levels in hard and soft tissues of rats, *Int. J. Environ. Stud.,* 15, 41, 1980.

404. **Abo-Rady, M. D. K.,** Aquatic macrophytes as indicator for heavy metal pollution in the river Seine, *Arch. Hydrobiol.,* 89, 387, 1980 (in German).

405. **Kelson, J. R.,** Selected method: direct measurement of zinc in plasma by atomic absorption spectroscopy comments, *Clin. Chem. (Winston-Salem, N.C.),* 26, 349, 1980.

406. **Chakrabarti, C. L., Wan, C. C., and Li, W. C.,** Direct determination of traces of copper, zinc, lead, cobalt, iron and cadmium in bovine liver by graphite furnace atomic absorption spectrometry using the solid sampling and the platform technique, *Spectrochim. Acta,* 35b, 93, 1980.

407. **Ladefoges, K.,** Determination of zinc in diet and feces by acid extraction and atomic absorption spectrophotometry, *Clin. Chim. Acta,* 100, 149, 1980.

408. **Petrov, I. L., Tsalev, D. L., and Barsev, A. I.,** Atomic absorption spectrometric determination of cadmium, cobalt, copper, manganese, nickel, lead and zinc in acetate soil extracts, *At. Spectrosc.,* 1, 47, 1980.

409. **Smith, J. C., Butrimovitz, G. P., and Purdy, W. C.,** Selected methods: direct measurement of zinc in plasma by atomic absorption spectroscopy. Reply to comments, *Clin. Chem. (Winston-Salem, N.C.),* 26, 350, 1980.

410. **Smith, R. G. and Windom, H. L.,** A solvent extraction technique for determining nanogram per liter concentrations of cadmium, copper, nickel and zinc in seawater, *Anal. Chim. Acta,* 113, 39, 1980.

411. **Adams, T. G., Atchison, G. J., and Vetter, R. J.** Impact of an industrially contaminated lake on heavy metal levels in its effluent stream, *Hydrobiologia,* 69, 187, 1980.

412. **Allen, K. G. D. and Klevay, L. M.,** Copper and zinc in selected breakfast cereals, *Nutr. Rep. Int.,* 22, 389, 1980.

413. **Arpadjan, S. and Stojanova, D.,** Application of detergents to the direct determination of iron, zinc and copper in milk by flame atomic absorption spectrophotometry, *Z. Anal. Chem.,* 302, 206, 1980.

414. **Bogden, J. D., Kemp, F. W., Troiano, R. A., Jortner, B. S., Timpone, C., and Giuliani, D.,** Effect of mercuric chloride and methyl mercury chloride exposure on tissue concentrations of six essential minerals, *Environ. Res.,* 21, 350, 1980.

415. **Borriello, R. and Sciaudone, G.,** Zinc, copper, iron and lead in bottled and canned beer by atomic absorption spectroscopy, *At. Spectrosc.,* 1, 131, 1980.

416. **Cavdar, A. E., Babacan, E., Arcasoy, A., and Ertem, U.,** Effect of nutrition on serum zinc concentration during pregnancy in Turkish women, *Am. J. Clin. Nutr.,* 33, 542, 1980.

417. **Amiard, J. C., Amiard-Triquet, C., Metayer, C., Marchant, J., and Ferre, R.,** Study of transfer of cadmium, lead, copper and zinc in esturine food chains. I. Study in the Loire estuary during summer, 1978, *Water Res.,* 14, 665, 1980.

418. **Chakrabarti, C. L., Wan, C. C., and Li, W. C.,** Atomic absorption spectrometric determination of cadmium, lead, zinc, copper, cobalt and iron in oyster tissue by direct atomization from the solid state using the graphite furnace platform technique, *Spectrochim. Acta,* 35b, 547, 1980.

419. **Clegg, J. B., Grainger, F., and Gale, I. G.,** Quantitative measurement of impurities in gallium arsenide, *J. Mater. Sci.,* 15, 747, 1980.

420. **Colella, M. B., Siggia, S., and Barnes, R. M.,** Poly (acrylamidoxime) resin for determination of trace metals in natural waters, *Anal. Chem.,* 52, 2347, 1980.

421. **Danielsson, L. G.,** Cadmium, cobalt, copper, iron, lead, nickel and zinc in Indian Ocean water, *Mar. Chem.,* 8, 199, 1980.

422. **Evans, W. H., Dellar, D., Lucas, B. E., Jackson, F. J., and Read, J. I.,** Observations on the determination of total copper, iron, manganese and zinc in food stuffs by flame atomic absorption spectrophotometry, *Analyst (London),* 105, 529, 1980.

423. **Grassi, R. L., Diez, A. L., and Acebal, S. G.,** Behavior of EDTA, CDTA and DTPA as agents for extraction of trace elements in the soil, *Agrochimica,* 24, 78, 1980.

424. **Holak, W.,** Analysis of foods for lead, cadmium, copper, zinc, arsenic and selenium using closed system sample digestion, *J. Assoc. Off. Anal. Chem.,* 63, 485, 1980.

425. **Horler, D. N. H., Barber, J., and Barringer, A. R.,** Multielemental study of plant surface particles in relation to geochemistry and biogeochemistry, *J. Geochem. Explor.,* 13, 41, 1980.

426. **Hubert, J., Candelaria, R. M., and Applegate, H. G.,** Determination of lead, zinc, cadmium and arsenic in environmental samples, *At. Spectrosc.,* 1, 90, 1980.

427. **Karlinskii, V. M., Bogomolova, G. G., and Kamf, G. G.,** Blood plasma and serum zinc levels in healthy people, *Lab. Delo,* 7, 441, 1980.

428. **Kiilerich, S., Christensen, M. S., Naestoff, J., and Christiansen, C.,** Determination of zinc in serum and urine by atomic absorption spectrophotometry: relationship between serum levels of zinc and proteins in 104 normal subjects, *Clin. Chim. Acta,* 105, 231, 1980.

429. **Ledl, G., Horak, O., and Janauer, G. A.,** Manganese, iron, copper and other heavy metals in Austrian rivers, *Oesterr. Abwasser Rundsch. OAR Inst.,* 25, 28, 1980.

430. **Lei, K. Y., Davis, M. W., Fang, M. M., and Young, L. G.,** Effect of peetin on zinc, copper and iron balances in humans, *Nutr. Rep. Int.,* 22, 459, 1980.

431. **Leissner, K. H., Fjelkegaard, B., and Tisell, L. E.,** Concentration and content of zinc in the human prostate, *Invest. Urol.,* 18, 30, 1980.

432. **Lekakis, J. and Kalofoutis, A.,** Zinc concentrations in serum as related to myocardial infarction, *Clin. Chem. (Winston-Salem, N.C.),* 26, 1660, 1980.

433. **Lepage, M.,** Atomic absorption spectrophotometry, *J. Am. Soc. Brew. Chem.,* 38, 116, 1980.

434. **Levi, S. and Purdy, W. G.,** Continuous sampling of dialysate for the atomic absorption spectrometric determination of copper and zinc from hemodialysis patients, *Anal. Chim. Acta,* 116, 375, 1980.

435. **L'vov, B. V., Orlov, N. A., and Mandrazhi, E. K.,** Theory of flame atomic absorption analysis. II. Mechanism of the pH effect on atomic absorption of zinc, *Zh. Anal. Khim.,* 35, 894, 1980.

436. **Magnusson, B. and Westerlund, S.,** Determination of cadmium, copper, iron, nickel, lead and zinc in Baltic sea water, *Mar. Chem.,* 8, 231, 1980.

437. **Manoliu, C., Popescu, O., Balasa, T., Tamas, V., and Georgescu, I.,** Study on determination of metal traces in solvents, *Rev. Chim.,* 31, 291, 1980.

438. **Michaelson, G., Ljunghall, K., and Danielson, B. G.,** Zinc in epidermis and dermis in healthy subjects, *Acta Derm. Venereol.,* 60, 295, 1980.

439. **Mosse, R. A. and Lisk, D. J.,** Elemental analysis of honeys from several nations, *Am. Bee J.,* 120, 592, 1980.

440. **Nadkarni, R. A.,** Multitechnique multielemental analysis of coal and flyash, *Anal. Chem.,* 52, 929, 1980.

441. **Pande, J. and Das, S. M.,** Metallic contents in water and sediments of Lake Nainital, India, *Water Air Soil Pollut.,* 13, 3, 1980.

442. **Pfueller, U., Fuchs, V., Golbs, S., Ebert, E., and Pfeifer, D.,** Solubilization to prepare samples for determination of heavy metals in biological objects by flameless atomic absorption spectrophotometry, *Arch. Exp. Veterinaermed.,* 34, 367, 1980.

443. **Schramel, P., Wolf, A., Seif, R., and Klose, B. J.,** New device for ashing of biological materials under pressure, *J. Anal. Chem.,* 302, 62, 1980.

444. **Segar, D. A. and Cantillo, A. Y.,** Determination of iron, manganese and zinc in sea water by graphite furnace atomic absorption spectrometry. Comments, *Anal. Chem.,* 52, 1766, 1980.

445. **Sinex, S. A., Cantillo, A. Y., and Helz, G. R.,** Accuracy of acid extraction methods for trace metals in sediments, *Anal. Chem.,* 52, 2342, 1980.

446. **Smith-Briggs, J. L.,** Pollution studies in the Clyde Sea area, *Anal. Proc.,* 17, 5, 1980.

447. **Sturgeon, R. E., Berman, S. S., Desauliners, A., and Russell, D. S.,** Determination of iron, manganese and zinc by graphite furnace atomic absorption spectrometry. Reply to comments, *Anal. Chem.,* 52, 1767, 1980.

448. **Stolzenberg, T. R. and Andren, A. W.,** Simple acid digestion method for the determination of ten elements in ambient aerosols by flame atomic absorption spectrometry, *Anal. Chim. Acta,* 11, 377, 1980.

449. **Tews, H.,** Zeeman splitting of ground and excited acceptor states in zinc telluride, *Solid State Commun.,* 34, 611, 1980.

450. **Tessier, A., Campbell, P. G. C., and Bisson, M.,** Trace metal speciation of USGS reference sample MAG-I, *Geostandards Newsl.,* 4, 145, 1980.

451. **Thompson, K. C. and Wagstaff, K.,** Simplified method for the determination of cadmium, chromium, copper, nickel, lead and zinc in sewage sludge using atomic absorption spectrophotometry, *Analyst (London),* 105, 883, 1980.

452. **Tschoepel, P., Kotz, L., Schultz, W., Veber, M., and Toelg, G.,** Causes and elimination of systematic errors in the determination of elements in aqueous solutions in the ng/mℓ and pg/mℓ range, *Z. Anal. Chem.,* 302, 1, 1980 (in German).

453. **Balaes, G. and Robert, R. V. D.,** Analysis by atomic absorption spectrophotometry of activated charcoal, *Natl. Inst. Metall. Repub. S. Afr. Rep.,* No. 2060, 1980.

454. **Brovko, I. A.,** Diphenyl carbazone as a reagent for extraction-atomic absorption determination of cadmium, cobalt, copper, manganese, nickel and zinc, *Zh. Anal. Khim.,* 35, 2095, 1980 (in German).

455. **Davison, W.,** Ultratrace analysis of soluble zinc, cadmium, copper and lead in Windermere Lake water using anodic stripping voltammetry and atomic absorption spectroscopy, *Freshwater Biol.,* 10, 223, 1980.

456. **Freeland-Graves, J. H., Ebangit, L. M., and Bodzy, P. W.,** Zinc and copper content of food used in vegetarian diets, *J. Am. Diet. Assoc.,* 77, 648, 1980.

457. **Hartmann, V. E. M. and Cavallini, Z. T. L.,** Determination of lead and zinc in Brazilian oils of sicilian lemon and orange limonene, *An. Acad. Bras. Cienc.,* 52, 311, 1980 (in Portuguese).

458. **Hoshino, Y., Utsunomiya, T., and Fukui, K.,** Graphite furnace atomic absorption spectrometry utilizing selective concentration onto tungsten wire, *Rep. Res. Lab. Eng. Mater. Tokyo Inst. Technol.,* 5, 105, 1980.

459. **Ivanova, E., Mareva, S., and Iordanov, N.,** Extraction-flame atomic absorption determination of micro traces of copper, zinc, nickel, cobalt, manganese and iron in some alkali salts, *Z. Anal. Chem.,* 303, 378, 1980.

460. **Itokawa, H., Watanabe, K., Tazaki, T., Hayashi, T., and Hayashi, H.** Quantitative analysis of metals in crude drugs, *Shoyakugaku Zasshi,* 34, 155, 1980 (in Japanese).

461. **Ishii, T., Hirano, S., Matsuba, M., and Koyanagi, T.,** Determination of trace elements in shellfish, *Nippon Suisan Bakkaishi,* 46, 1375, 1980 (in English).

462. **Kroma, V. and Komarek, J.,** Determination of zinc in body fluids with atomic absorption spectroscopy, *Chem. Listy,* 74, 770, 1980.

463. **Lindh, U., Brune, D., Nordberg, G., and Wester, P. O.,** Levels of antimony, arsenic, cadmium, copper, lead, mercury, selenium, silver, tin and zinc in bone tissue of industrially exposed workers, *Sci. Total Environ.,* 16, 109, 1980.

464. **Ludany, A., Kellermayer, M., and Jobst, K.,** Metal content of rat liver cell organelles, *Acta Biochim. Biophys. Acad. Sci. Hung.,* 15, 229, 1980.

465. **Maeda, T., Nakagawa, M., Kawakatsu, M., and Tanimoto, Y.,** Determination of metallic elements in serum and urine by flame and graphite furnace atomic absorption spectrophotometry, *Shimadzu Hyoron,* 37, 1, 1980 (in Japanese).

466. **Mendelson, R. A. and Huber, A. M.,** Effect of duration of alcohol administration on deposition of trace elements in the fetal rat, *Adv. Exp. Med. Biol.,* 132, 295, 1980.

467. **Ogihara, K., Seki, H., and Nagase, K.,** Comparison of the method of bottom sediment analysis and the method of the agricultural soil pollution prevention law, *Nagano-ken Eisei Kogai Kenkyusho Kenkyu Hokoku,* 2, 88, 1980 (in Japanese).

468. **Posta, J.,** Determination of the optimum parameters of a combined atomic absorption spectrophotometric plus arc flame method for the analysis of floating dust samples, *Hung. Sci. Instrum.,* 47, 33, 1980.

469. **Roguljic, A., Mikac-Davic, D., and Krusic, J.,** Copper, zinc and magnesium levels in healthy tissue and benign and malignant tumors of the uterus, *Period. Biol.,* 82, 213, 1980.

470. **Sheybal, I., Swiderska, K., Jurczyk, J., and Bec, H.,** Determination of sodium, potassium and zinc in iron ores and sinters by atomic absorption spectrometry, *Hutnik,* 47, 278, 1980 (in Polish).

471. **Stendal, H.,** Leaching studies for the determination of copper, zinc, lead, nickel and cobalt in geological materials by atomic absorption spectrophotometry, *Chem. Erde,* 39, 276, 1980.

472. **Tursunov, A. T., Brovko, I. A., and Nazarov, S. N.,** Use of furnace-flame system for the atomic absorption determination of some trace elements in soil extracts, *Uzb. Khim. Zh.,* 5, 5, 1980 (in Russian).

473. **Van Schalkwyk, S. J.,** Promotion of methylbenzylnitrosamine-induced esophageal cancer in rats by sub-clinical zinc deficiency, *Nutr. Rep. Int.,* 22, 891, 1980.

474. **Van Willis, W., El-Ahraf, A., Vinjamoori, D. V., and Aref, K.,** Analysis of animal feed ingredients and soil amendment products produced from cattle manure for selected trace metals using atomic absorption spectrophotometry, *J. Food Prot.,* 43, 834, 1980.

475. **Raj, K. P. S., Agrawal, Y. K., and Patel, M. R.,** Analysis of garlic for its metal contents, *J. Indian Chem. Soc.,* 27, 1121, 1980.

476. **Inhat, M., Gordon, A. D., Gaynor, J. D., Berman, S. S., Desauliners, A., Stoeppler, M., and Valenta, P.,** Interlaboratory analysis of natural fresh waters for Cu, Zn, Cd and Pb, *Int. J. Environ. Anal. Chem.,* 8, 259, 1980.

477. **Orpwood, B.,** Use of Chelating Ion-Exchange Resins for the Determination of Trace Metals in Drinking Water, Tech. Rep. TR.-Water Res. Cent. TR 153, Medmenham Laboratories, Medmenham, Marlow, Bucks., 35pp, 1980.

478. **Jiminez Torres, N. V., Castera, D. E., and Serra, J. A.,** Presence of essential trace elements in commercial solutions for parenteral nutrition, *Rev. Asoc. Esp. Farm. Hosp.,* 4, 87, 1980 (in Spanish).

479. **Fink, L. K., Jr., Harriss, A. B., and Schick, L. L.,** Trace Metals in Suspended Particulates, Biota and Sediments of the St. Croix, Narrguage and Union Estuaries and the Goose Cove Region of Penobscot Bay, Rep. W 80-04809, OWRT-A-041-ME (I) (1980); Available from the NTIS, Springfield, Va.; *Gov. Rep. Announce. Index (U.S.),* 80, 3372, 1980.

480. **Knight, M. J.,** Comparison of Four Digestion Procedures not Requiring Perchloric Acid for the Trace Element Analysis of Plant Material, Rep. ANL/LRP-TM-18, Argonne National Laboratory, Argonne, Ill., 1980; available from the NTIS, Springfield, Va.

481. **Jarkovsky, J., Plsko, E., and Stresko, V.,** Contribution to the analytical geochemistry of antimonites, *Acta Geol. Geogr. Univ. Comenianae Geol.,* 34, 71, 1980.

482. **Chakrabarti, C. L., Wan, C. C., and Li, W. C.,** Atomic absorption spectrometric determination of Cd, Pb, Zn, Cu, Co and Fe in oyster tissue by direct atomization from the solid state using the graphite furnace platform technique, *Spectrochim. Acta,* 35, 547, 1980.

483. **Caristi, C., Cimino, G., and Ziino, M.,** Heavy metal pollution. II. Determination of trace heavy elements in lemon, orange and mandarin juices by flameless atomic absorption spectrophotometry, *Essenz. Deriv. Agrum.,* 50, 165, 1980.

484. **Chebotareva, N. A. and Samokhvalov, S. G.,** Atomic absorption determination of Cu, Zn, Mn, Fe and Co in plant materials, *Byull. Pochv. In-ta Vaskhil.,* 24, 49, 1980.

485. **Das, A. K.,** AAS determination of trace metals in raw sewage, *Indian J. Environ. Health,* 22, 130, 1980.

486. **Acebal, G. S., Grossi, R. L., and Gutierrez, D. O.,** Behaviour of oxalic, citric and nitriloacetic acids as extractants of trace elements in the soil, *Agrochimica,* 24, 462, 1980.

487. **Freeland-Graves, J. H., Ebangit, M. L., and Bodzy, P. W.,** Zinc and copper content of foods used in vegetarian diets, *J. Am. Diet. Assoc.,* 77, 648, 1980.

488. **Levi, S. and Purdy, W. C.,** The AAS determination of Cu and Zn levels in the serum of hemodialysis patients, *Clin. Biochem.,* 13, 253, 1980.

489. **Motuzova, G. V. and Obukhov, A. I.,** Effect of conditions for soil decomposition on the results of the determination of trace nutrient content on it, *Biol. Nauk (Moscow),* 11, 87, 1980.

490. **Raghavan, T. N. V. and Agrawal, Y. K.** Trace elements in crude petroleum of North Gujarat, *Ind. J. Technol.,* 18, 422, 1980.

491. **Farag, R. S., El-Assar, S. T., Mostafa, M. A., and Abdel Rahim, E. A.,** Use of atomic absorption analysis of mineral content of certain animal hairs in relation to sex. I, *J. Drug Res.,* 12, 217, 1980.

492. **Baroccio, A. and Moauro, A.,** Determination of Some Elements in Vegetal Samples by Instrumental Neutron Activation. Comparison with the Data Obtained from Atomic Absorption, CNEN-RT/CHI (80) 15, Comitato Nazionale Energia Nucleare, Casaccio Nuclear Studies Center, Rome, Italy, 1980 (in Italian).

493. **Calixto, F. S. and Bauza, M.,** Determination of inorganic elements in the almond (*Prunus amygdalus*), *An. Bromatol.,* 32, 119, 1980.

494. **Clarke, J. R. and Viets, J. G.,** Multielement extraction system for the determination of 18 trace elements in geochemical samples, *Anal. Chem.,* 53, 61, 1981.

495. **Clarke, J. R. and Viets, J. G.,** Back extraction of trace elements from organimetallic halide extracts for determination by flameless atomic absorption spectrometry, *Anal. Chem.,* 53, 65, 1981.

496. **Iyengar, S. S., Martens, D. C., and Miller, P. W.,** Determination of copper and zinc in soil; extracts by atomic absorption spectrophotometry using APDC-MIBK solvent extraction, *Soil Sci.,* 131, 95, 1981.

497. **Laxen, D. P. and Harrison, R. M.,** Cleaning methods for polythene prior to the determination of trace metals in fresh water samples, *Anal. Chem.,* 53, 345, 1981.

498. **Meranger, J. C., Hollebone, B. R., and Blanchette, G. A.,** Effects of storage times, temperatures and container types on the accuracy of atomic absorption determinations of Cd, Cu, Hg, Pb and Zn in whole heparinized blood, *J. Anal. Toxicol.,* 5, 33, 1981.

499. **Meranger, J. C., Subramanian, K. S., and Chalifoux, C.,** Survey for cadmium, cobalt, chromium, copper, nickel, lead, zinc, calcium and magnesium in Canadian drinking water supplies, *J. Assoc. Off. Anal. Chem.,* 64, 44, 1981.

500. **Mohamed, N. and Fry, R. C.,** Slurry atomization direct atomic spectrochemical analysis of animal tissue, *Anal. Chem.,* 53, 450, 1981.

501. **Pakalns, P.,** Effect of surfactants on mixed chelate extraction-atomic absorption spectrophotometric determination of copper, nickel, iron, cobalt, cadmium, zinc and lead, *Water Res.,* 15, 7, 1981.

502. **Taylor, A. and Bryant, T. N.,** Comparison of procedures for determination of copper and zinc in serum by atomic absorption spectroscopy, *Clin. Chim. Acta,* 110, 83, 1981.

503. **Vieira, N. E. and Hansen, J. W.,** Zinc determined in 10-microliter serum or urine samples by flameless atomic absorption spectrometry, *Clin. Chem. (Winston-Salem, N.C.),* 27, 73, 1981.

504. **Clanet, F., Deloncle, R., and Popoff, G.,** Chelating resin catcher for capture, preconcentration and determination of toxic metal traces (Zn, Cd, Hg and Pb) in waters, *Water Res.,* 15, 591, 1981.

505. **Carleer, R., Francois, J. P., and Van Poucke, L. C.,** Determination of the main, minor and trace elements in lead/tin based solder by atomic absorption spectrophotometry, *Bull. Soc. Chim. Belg.,* 90, 397, 1981.

506. **Bewers, J. M., Doleziel, J., Yeats, P. A., and Barron, J. L.,** An intercalibration for trace metals in seawater, *Mar. Chem.,* 10, 173, 1981.

507. **Henrion, G., Gelbrecht, J., Hoffman, T., and Marquardt, D.,** AAS trace determination in tungsten after preconcentration on chelate resin Wofatit MC 50, *Z. Chem.,* 21, 192, 1981 (in German).

508. **Wandiga, S. O.,** The concentrations of zinc, copper, lead, manganese, nickel and fluoride in rivers and lakes of Kenya, *Sinet,* 3, 67, 1981.

509. **De Antonio, S. M., Katz, S. A., Scheiner, D. M., and Wood, J. D.,** Anatomical variations of trace metal levels in hair, *Anal. Proc. (London),* 18, 162, 1981.

510. **Engebretson, J. A. and Mason, W. H.,** Depletion of trace elements in mated male *Heliothis Virescens* and *Drosophila melanogaster, Comp. Biochem. Physiol. A,* 68, 523, 1981.

511. **Arpadjan, S. and Nakova, D.,** Direct determination of Fe, Zn, Cu, and Mn in milk powder by AA, *Nahrung,* 25, 359, 1981.

512. **Rasmussen, L.,** Determination of trace metals in seawater by CHELEX-100 or solvent extraction technique and atomic absorption spectrometry, *Anal. Chim. Acta,* 125, 117, 1981.

513. **Krivan, V., Geiger, H., and Franz, H. E.,** Determination of Fe, Co, Zn, Se, Rb and Cs in NBS bovine liver blood plasma and erythrocytes by INAA and AAS, *Fresenius' Z. Anal. Chem.,* 305, 399, 1981.

514. **Saleh, A. and Udall, J. N.,** Minimizing contamination of specimens for zinc determination, *Clin. Chem. (Winston-Salem, N.C.),* 27, 338, 1981.

515. **Sturgeon, R. E., Berman, S. S., Willie, S. N., and Desauliners, J. A. H.,** Preconcentration of trace elements from seawater with silica-immobilized 8-hydroxy quinoline, *Anal. Chem.,* 53, 2337, 1981.

516. **Normakhmatov, R. N. and Muradova, S. B.,** Mineral composition of pecan kernels, *Khlebopek. Konditer. Prom.,* 7, 37, 1981.

517. **Farmer, J. G.,** The analytical chemist in studies of metal pollution in sediment cores, *Anal. Proc. (London),* 18, 249, 1981.

518. **Julshamn, K.,** Studies on major and minor elements in molluscs in Western Norway. VII. The contents of 12 elements including copper, zinc, cadmium and lead in common mussel (*Mytilus edulis*) and brown seaweed (*Ascophyllum nodusum*) relative to the distance from the industrial sites in Sorfjorden, inner Hardangerfjord, *Fiskeridir. Skr. Ser. Ernaer,* 1, 267, 1981.

519. **Sanks, R. L.,** Flameless Atomic Absorption for Tracing the Fate of Heavy Metals in Soils and Ground Water, Rep. MWRRC-III, W81-02491, OWRT-A 109-MONT (I); Order No. PB81-189821, 1980; available from the NTIS, Springfield, Va., *Govt. Rep. Announce. Index (U.S.),* 81, 3565, 1981.

520. **Takiyama, K., Ishii, Y., and Yoshimura, I.,** Determination of metals in vegetables by ashing in a teflon crucible, *Mukogawa Joshi Daigaku Kiyo Shokumotsu-hen,* 28, F15, 1981 (in Japanese).

521. **Amini, M. K., Defreese, J. D., and Hathaway, L. R.,** Comparison of three instrumental spectroscopic techniques for elemental analysis of Kansas shales, *Appl. Spectrosc.,* 35, 497, 1981.

522. **Watson, M. E.,** Interlaboratory comparison in the determination of nutrient concentrations of plant tissue, *Commun. Soil Sci. Plant Anal.,* 12, 601, 1981.

523. **Suzuki, K. T. and Maitani, T.,** Metal-dependent properties of metallothionein. Replacement in vitro of zinc in zinc-thionein with copper, *Biochem. J.,* 199, 289, 1981.

524. **Bank, H. L., Robson, J., Bigelow, J. B., Morrison, J., Spell, L. H., and Kantor, R.,** Preparation of fingernails for trace element analysis, *Clin.Chim. Acta,* 116, 179, 1981.

525. **Ogata, K., Murakami, K., Tanabe, S., and Imanari, T.,** Determination of zinc and citric acid in human seminal plasma, *Rinsho Kagaku,* 10, 136, 1981 (in Japanese).

526. **Magnusson, B. and Westerlund, S.,** Solvent extraction procedures combined with back-extraction for trace metal determinations by atomic absorption spectrometry, *Anal. Chim. Acta,* 131, 63, 1981.

527. **Nurbekov, M. K., Favorova, O. O., Dmitrienko, S. G., Bolokina, I. A., and Kieselev, L. L.,** Role of zinc ions in bovine tryptophanyl-tRNA-synthetase activity, *Mol. Biol. (Moscow),* 15, 1000, 1981.

528. **Mitchell, G. E.,** Trace metal levels in Queensland dairy products, *Aust. J. Dairy Technol.,* 36(2), 70, 1981.

529. **Brunner, W., Heckner, H., and Sansoni, B.,** Fully automatic simultaneous operation of several atomic absorption spectrometers in the analytical service laboratory of a reasearch center by use of a process computer, *Spektrometertagung,* 13, 99, 1981 (in German).

530. **Adelman, H., Jennis, S. W., and Katz, S. A.,** Interlaboratory analysis of sewage sludge, *Am. Lab.,* 31, December 1981.

531. **Wesenberg, G. B. R., Fosse, G., and Rasmussen, P.,** The effect of graded doses of cadmium on lead, zinc and copper content of target and indicator organs in rats, *Int. J. Environ. Stud.,* 17 (3-4), 191, 1981.

532. **Shinotesuto, K. K.,** Determination of zinc in serum by affinity chromatography, Jpn. Kokai Tokkyo Koho JP 81, 114, 761 (cl. GOIN 33/84), appl. 80/16, 045, Feb. 14, 1980, September 9, 1981.

533. **Blakemore, W. M. and Billedeau, S. M.,** Analysis of laboratory animal feed for toxic and essential elements by atomic absorption and inductively coupled argon plasma emission spectrometry, *J. Assoc. Off. Anal. Chem.,* 64(6), 1284, 1981.

534. **Kirleis, A. W., Sommers, L. E., and Nelson, D. W.,** Heavy metal content of groats and hulls of oats grown on soil treated with sewage sludge, *Cereal Chem.,* 58(6), 530, 1981.

535. **Shaburova, V. P., Ruchkin, E. D., Yudelevich, I. G., and Kruglov, L. M.,** Atomic-absorption method of analysis for piezoceramic materials of the lead zirconate titanate system, *Izv. Sib. Otd. Akad. Nauk SSSR Ser. Khim. Nauk,* (5), 95, 1981.

536. **Knuuttila, M., Lappalainen, R., and Lammi, S.,** Zinc concentration of human subgingival calculus related to fluoride, magnesium and calcium contents, *Scand. J. Dent. Res.,* 89(5), 412, 1981.

537. **Mattsson, P. and Albanus, L.,** Cadmium and some other elements in liver and kidney from moose *(Alces alces), Var. Foeda,* 33(8-9), 335, 1981.

538. **Favini, G., Brillo, A., and Bosi, E.,** Determination of zinc in antisulfur varnishes, *Riv. Ital. Sostange Grasse,* 58, 343, 1981.

539. **Hosaki, K.,** Measurement of trace metals (such as zinc) in living bodies, *Ghugai Iyaku,* 34, 450, 1981 (in Japanese).

540. **Clegg, M. S., Keen, C. L., Loennerdal, B., and Hurley, L. S.,** Influence of ashing techniques on the analysis of trace elements in biological samples. II. Dry ashing, *Biol. Trace Elem. Res.,* 3, 237, 1981.

541. **Elson, C. M., Bem, E. M., and Ackman, R. G.,** Determination of heavy metals in a menhaden oil after refining and hydrogenation using several analytical methods, *J. Am. Oil. Chem. Soc.,* 58, 1024, 1981.

542. **Kayakirilmaz, K. and Koksal, O.,** Determination of copper, iron and zinc in breast milk by atomic absorption spectrophotometry and effect of lactation stages and cultural, social and economical conditions of mother on the amount of these minerals, *Doga Seri C,* 5, 151, 1981.

543. **Brovko, I. A. and Tursanov, A. T.,** Extraction-atomic absorption determination of cadmium, cobalt, copper, manganese, nickel and zinc in natural waters using diphenylcarbazone as the extraction reagent, *Uzb. Khim. Zh.,* 3, 18, 1981.

544. **Burba, P. and Schaefer, W.,** Atomic absorption spectrometric determination (flame-AAS) of heavy metals in bacterial leach liquors (Jarosite) after their analytical separation on cellulose, *Erzmetall,* 34, 582, 1981.

545. **Viala, A., Gouezo, F., Mallet, B., Fondarai, J., Cano, J. P., Sauve, J. M., Grimaldi, E., and Deturmeny, E.,** Determination of four trace metals (lead, cadmium, chromium and zinc) in atmospheric dust in Marseilles (France) between 1977 and 1979, *Pollut. Atmos.,* 91, 207, 1981.

546. **Pilipenko, A. T. and Samchuk, A. I.,** Extraction-atomic absorption determination of trace elements in natural waters, *Khim. Tekhnol. Vody,* 3, 343, 1981.

547. **Dyakova, N. P., Kulachenko, S. P., Kulachenko, V. P., and Schastlivenko, V. A.,** Determination of trace elements in hair, *Veterinariya (Moscow),* 10, 62, 1981 (in Russian).

548. **Fahim, R. B., Zaki, M. I., El-Roudi, A. M., and Hassan, A. M. A.,** Formation and activity of certain chromite catalysts, *J. Res. Inst. Catal. Hokkaido Univ.,* 29, 25, 1981.

549. **Khalighie, J., Ure, A. M., and West, T. S.,** Atom-trapping atomic absorption spectrometry of arsenic, cadmium, lead, selenium and zinc in air-acetylene and air-propane flame, *Anal. Chim. Acta,* 131, 27, 1981.

550. **Satsmadjis, J. and Voutsino-Taliadouri, F.,** Determination of trace metals at concentrations above the linear calibration range by electrothermal atomic absorption spectrometry, *Anal. Chim. Acta,* 131, 83, 1981.

551. **Berndt, H. and Messerschmidt, J.,** Microanalytical method of flame atomic absorption and emission spectrometry (platinum loop method). II. Determination of As, Bi, Cd, Hg, Pb, Sb, Se, Te, Tl, Zn, Li, Na, K, Cs, Sr and Ba, *Spectrochim. Acta,* 36b, 809, 1981.

552. **Chauhan, O. S., Garg, B. S., Singh, R. P., and Singh, I.,** Rapid spectrophotometric method for determination of zinc in milk, *Talanta,* 28, 399, 1981.

553. **Guevremont, R.,** Organic matrix modifiers for direct determination of zinc in seawater by graphite furnace atomic absorption spectrometry, *Anal. Chem.,* 53, 911, 1981.

554. **Iwata, Y., Matsumoto, K., Haraguchi, H., Fuwa, K., and Okamoto, K.,** Proposed certified reference material for pond sediment, *Anal. Chem.,* 53, 1136, 1981.

555. **Jastrow, J. D., Zimmerman, C. A., Dvorak, A. J., and Hinchman, R. R.,** Plant growth and trace element uptake on acidic coal refuse amended with lime or fly ash, *J. Environ. Qual.,* 10, 154, 1981.

556. **Katz, S. A., Jennis, S. W., Mount, T., Tout, R. E., and Chatt, A.,** Comparison of sample preparation methods for the determination of metals in sewage sludges by flame atomic absorption spectrometry, *Anal. Chem.,* 9, 209, 1981.

557. **Kozma, M., Szerdahelyi, P., and Kosa, P.,** Histochemical detection of zinc and copper in various neurons of the central nervous system, *Acta Histochem.,* 69, 12, 1981.

558. **Levi, S., Fortin, R. C., and Purdy, W. C.,** Electrothermal atomic absorption spectrometric techniques for the determination of zinc and copper in microliter and submicroliter volumes of aqueous and serum matrixes, *Anal. Chim. Acta,* 127, 103, 1981.

559. **Makino, T. and Takahara, K.,** Direct determination of plasma copper and zinc in infants by atomic absorption with discrete nebulization, *Clin. Chem. (Winston-Salem, N.C.),* 27, 1445, 1981.

560. **Mendiola Ambrosio, J. M. and Gonzalez Lopez, A.,** Analysis of cadmium and zinc in PVC composites by atomic absorption spectrophotometry, *Rev. Plast. Mod.,* 41, 413, 1981 (in Spanish).

561. **Sheng-jun, M., Shi-yan, Y. B., Wen-sheng, Z., and Yong-mao, L.,** Atomic absorption determination of high concentrations of calcium, magnesium, lead and zinc in ores, *Fen Hsi Hua Hsueh,* 9, 12, 1981 (in Chinese).

562. **Polo-Diez, L., Hernandez-Mendez, J., and Rodriguez-Gonzalez, J. A.,** Analytical applications of emulsions in atomic absorption spectrophotometry: determination of zinc in undecenoate ointments using aqueous inorganic standards, *Analyst (London),* 106, 737, 1981.

563. **Wolff, E. W., Landy, M. P., and Peel, D. A.,** Preconcentration of cadmium, copper, lead and zinc in water at the 10^{-12} mg/g level by absorption onto tungsten wire followed by flameless atomic absorption spectrometry, *Anal. Chem.*, 53, 1566, 1981.

564. **Young, R. S.,** Analysis of nickel refinery slimes and residues, *Talanta*, 28, 25, 1981.

565. **Crespo, S., Soriano, E., Sampera, C., and Balasch, J.,** Zinc and copper distribution in excretory organs of the dogfish *Scytiorhinus canicula* and chloride cell response following treatment with zinc sulfate, *Mar. Biol.*, 65, 117, 1981.

566. **Kovarskii, N. Y., Kovekovdova, L. T., Pryazhevskaya, I. S., Belenkii, V. S., Shapovalov, E. N., and Popkova, S. M.,** Preconcentration of trace elements from seawater by electrodeposited magnesium hydroxide, *Zh. Anal. Khim.*, 36, 2264, 1981.

567. **Arafat, N. M. and Glooschenko, W. A.,** Method for the simultaneous determination of arsenic, aluminum, iron, zinc, chromium and copper in plant tissue without the use of perchloric acid, *Analyst (London)*, 106, 1174, 1981.

568. **Peters, H. J. and Hoehler, H.,** Direct determination by flame atomic absorption spectrometry of the trace elements magnanese, copper and zinc in biological tissues, *Dtsch. Gesundheitswes.*, 36, 1919, 1981.

569. **Sekiya, T., Tanimura, H., and Hikasa, Y.,** Simplified determination of copper, zinc and manganese in plasma and bile by the flameless atomic absorption spectrometry, *Arch. Jpn. Chir.*, 50, 729, 1981.

570. **Fiala, K. and Studeny, M.,** Use of a dry method for mineralization of plant material to improve chemical analysis for determination of major and trace elements from a common weighed sample, *Ved. Pr. Vysk. Ustavu Podoznalectva Vyz. Rastl. Bratislave*, 10, 167, 1981.

571. **Rao, N. C. and Rao, B. S. N.,** Trace element content of Indian foods and the dietaries, *Indian J. Med. Res.*, 73, 904, 1981.

572. **Beni, R. and Ott, W. R.,** The effect of pH on the durability of lithium-zinc phosphate glasses, *Glass Technol.*, 22, 182, 1981.

573. **West, T. S.,** The atomic spectroscopy of biosignificant trace element in soils in relation to plant and animal nutrition, *Bunseki Kagaku*, 30, 5103, 1981.

574. **Tateuchi, S.,** Determination of copper, zinc, iron and manganese in commercial premix by atomic absorption photometry, *Shiryo Kenkyu Hokoku*, 7, 183, 1981.

575. **Brzozowska, B. and Zwadzka, T.,** Atomic absorption spectrophotometry in determination of lead, cadmium, zinc and copper in meat products, *Rocz. Panstw. Zakl. Hig.*, 32, 323, 1981.

576. **Bogden, J. D., Zadzielski, E., Weiner, B., Oleske, J. A., and Aviv, A.,** Release of zinc from disposable coils during hemodialysis, *Trace Subst. Environ. Health*, 15, 121, 1981.

577. **Li, Q.,** Atomic absorption spectrometric determination of microamounts of copper, zinc, nickel, cobalt, cadmium and lead, *Fenxi Huaxue*, 9, 718, 1981.

578. **Uchino, E., Tsuzuki, T., Inoue, K., and Kishi, R.,** Concentration of iron copper, zinc and lead in three tissues of young (22 days old) and adult (290 days old) rats, *Hokkaidoritsu Eisei Kenkyushoho*, 31, 117, 1981.

579. **Kolasa, F., Krajewski, J., and Draminski, W.,** Blood serum zinc concentrations in healthy pregnant women at term and in prolonged pregnancies, *Ginekol. Pol.*, 52, 973, 1981.

580. **Bourcier, D. R., Sharma, R. P., and Brinkerhoff, C. R.,** Cadmium-copper interaction: tissue accumulation and subcellular distribution of cadmium in mice after simultaneous administration of cadmium and copper, *Trace Subst. Environ. Health*, 15, 190, 1981.

581. **Kiilrich, S., Christensen, M. S., Naesroft, J., and Christensen, C.,** Serum and plasma zinc concentrations with special reference to standardized sampling procedure and protein status, *Clin. Chim. Acta*, 114, 117, 1981.

582. **Balaes, G. E. E. and Robert, R. V. D.,** Determination by atomic absorption spectrophotometry of impurities in manganese dioxide, *Natl. Inst. Metall. Repub. S. Afr. Rep.*, No. 2094, 1981.

583. **Borg, H., Edin, A., Holm, K., and Skoeld, E.,** Determination of metals in fish livers by flameless atomic absorption spectroscopy, *Water Res.*, 15, 1291, 1981.

584. **Boyer, K. W., Jones, J. W., Linscott, D., Wright, S. K., Stroube, W., and Cunningham, W.,** Trace element levels in tissues from cattle fed a sewage sludge amended diet, *J. Toxicol. Environ. Health*, 8, 281, 1981.

585. **Christensen, E. R. and Nan-Kwang, C.,** Fluxes of arsenic, lead, zinc and cadmium to Green Bay and Lake Michigan sediments, *Environ. Sci. Technol.*, 15, 553, 1981.

586. **Costantini, S., Macri, A., and Vernillo, I.,** Atomic absorption spectrophotometric analysis of mineral elements in milk-base feeds for zootechnical use, *Riv. Soc. Ital. Sci. Aliment.*, 10, 231, 1981 (in Italian).

587. **Haralambie, G.,** Serum zinc in athletes in training, *Int. J. Sports Med.*, 2, 135, 1981.

588. **Massee, R. and Maessen, F. J. M. J.,** Losses of silver, arsenic, cadmium, selenium and zinc traces from distilled water and artificial seawater by sorption on various container surfaces, *Anal. Chim. Acta*, 127, 181, 1981.

589. **Mattera, V. D., Jr., Arbige, V. A., Jr., Tomellini, S. A., Erbe, D. A., Doxtader, M. M., and Force, R. K.**, Evaluation of wash solutions as a preliminary step for copper and zinc determinations in hair, *Anal. Chim. Acta*, 124, 409, 1981.

590. **Salmela, S., Vouri, E., and Kilpio, J. O.**, Effect of washing procedures on trace element content of human hair, *Anal. Chim. Acta*, 125, 131, 1981.

591. **Scheuhammer, A. M. and Cherian, M. G.**, Influence of manganese on the distribution of essential trace elements. I. Regional distribution of manganese, sodium, potassium, zinc, iron and copper in rat brain after chronic manganese exposure, *Toxicol. Appl. Pharmacol.*, 61, 227, 1981.

592. **Schou, M. and Holt, E.**, Zinc concentrations in human plasma and in rat plasma and tissues during lithium administration, *Acta Pharmacol. Toxicol.*, 49, 298, 1981.

593. **Taguchi, M., Takagi, H., Iwashima, K., and Yamagata, N.**, Metal content of shark muscle powder biological reference material, *J. Assoc. Off. Anal. Chem.*, 64, 260, 1981.

594. **Tunstall, M., Berndt, H., Sommer, D., and Ohls, K.**, Direct determination of trace elements in aluminum with flame atomic absorption spectrometry and ICP emission spectrometry — a comparison, *Erzmetall*, 34, 588, 1981.

595. **Yang, F. and Ni, Z.**, Determination of zinc in seawater by graphite furnace atomic absorption spectrometry using citric acid as a matrix modifier, *Huanjing Kexue*, 2, 423, 1981.

596. **Karring, M., Pohjanvirta, R., Rahko, T., and Korpela, H.**, The influence of dietary molybdenum and copper supplementation on the contents of serum uric acid and some trace elements in cocks, *Acta Vet. Scand.*, 22, 289, 1981.

597. **Borg, H.**, Trace Metals in Natural Water. An Analytical Intercomparison. Rapp.-Naturvardsverket (Swed.) SNV PM 1463, Forskningssekretariatet, Statens Naturvaardsverk, Solna, Sweden, 1981 (in Swedish).

598. **Lintsen, D.**, Atomic absorption determination of calcium, magnesium, zinc and manganese in plant tissues by the wet ashing method, *Agric. Tec. (Santiago)*, 41, 273, 1981 (in Spanish).

599. **Greenwood, R.**, Distribution of heavy metals (copper, lead and zinc) in rural areas of Rio de Janeiro (Brazil), *Rev. Bras. Geocienc.*, 11, 98, 1981 (in Portuguese).

600. **Tateuchi, S.**, Determination of copper, zinc, iron and manganese in commercial premix by atomic absorption photometry, *Shiryo Kenkyu Hokoku*, 7, 183, 1981.

601. **Senesi, N. and Polemio, M.**, Trace element addition to soil by application of NPK fertilizers, *Fert. Res.*, 2, 289, 1981.

602. **Ichikuni, M. and Gautam, M.**, Chemical composition of prince feather leaves and seeds. Its usefulness as geochemical indicator of heavy metals in soils, *Chikyu Kagaku*, 15, 25, 1981.

603. **Fosse, G., Justesen, N. P. B., and Wesenberg, G. B. R.**, Microstructure and chemical composition of fossil mammalian teeth, *Calcif. Tissue Int.*, 33, 521, 1981.

604. **Burton, L. C., Uppal, P. N., and Dwight, D. W.**, Cross diffusion of cadmium and zinc into copper (I) sulfide formed in zinc cadmium sulfide thin flame, *Conf. Rec. IEEE Photovoltaic Spec. Conf.*, 15, IEEE, Piscataway, N.J., 780, 1981.

605. **Suddendorf, R. F., Gajan, R. J., and Kenneth, W.**, Utilization of atomic spectroscopy for the determination of lead and cadmium in zinc salts used as food additives, *De. At. Plasma Spectrochim. Anal., Proc. Int. Winter Conf.*, Barnes, R. M., Ed., Heyden, London, 1981.

606. **Barrow, G. I., Collins, V. G., Conchie, E. C., and Cooper, R. L.**, Zinc in potable waters by atomic absorption spectrophotometry 1980, Methods Exam. Waters Assoc. Mater., Department of the Environment, London, U.K., 1981.

607. **Kawahara, H., Yamada, T., Nakamura, M., Tomoda, T., Kobayashi, H., Saijo, A., Kawata, Y., and Hikari, S.**, Solubility of metal components into tissue culture medium from dental amalgam, *Shika Rikogaku Zasshi*, 22, 285, 1981.

608. **Ohtake, M., Chiba, R., Mochizuki, K., and Tada, K.**, Zinc and copper concentrations in human milk and in serum from exclusively breast-fed infants during the first 3 months, *Tohoku J. Exp. Med.*, 135, 335, 1981.

609. **Kane, J. S. and Smith, H.**, Analysis of Egyptian Geological Survey and Mining Department samples by rapid rock and atomic absorption procedures, *U.S. Geol. Surv. Open-File Rep.*, 81, 1, 1981.

610. **Xia, L.**, Flame atomic absorption spectrophotometric determination of copper, cadmium, lead and zinc in tea leaves, *Fenxi Huaxue*, 9, 498, 1981 (in Chinese).

611. **Henrion, G. and Gelbrecht, J.**, Determination of trace metals in molybdenum (VI) compounds using extraction/flame AAS, *Z. Chem.*, 21, 453, 1981.

612. **Jones, E. A. and Dixon, K.**, The separation of trace elements in manganese dioxide, *Natl. Inst. Metall. Repub. S. Afr. Rep.*, No. 2131, 1981.

613. **Wei, J. and Geng, T.**, Determination of available cadmium, lead, copper and zinc in calcareous soil, *Fenxi Huaxue*, 9, 565, 1981.

614. **Fugas, M.**, Metals in airborne atmospheric particles, *Zest. Atmos.*, 9, 13, 1981 (in Serbo-Croatian).

615. **Marlier-Geets, O., Heck, J. P., Barideau, L., and Rocher, M.,** Method for Determination of Heavy Metals in Sewage Sludge, Their Distribution According to Their Origin and Their Concentration Variation with Time, Comm. Eur. Communities. (rep) EUR 7076 Charact., Treat. Use Sewage Sludge, 284-290, Serv. Sci. Sol., Fac. Sci. Agrono. Etat, Gembloux, Belgium, 1981 (in French).

616. **Heider, P. J.,** Uranium Occurrence in (the Mining District of) California near Bucaramanga (Colombia), Rep. INIS-mf-6712, 1980 (in German); INIS Atomindex 12, Abstr. No. 641045, 1981; available from the NTIS, Springfield, Va.

617. **Inhat, M.,** Analytical approach to the determination of copper, zinc, cadmium and lead in natural fresh waters, *Int. J. Environ. Anal. Chem.*, 10, 217, 1981.

618. **Hirano, K., Iida, K., Shimada, T., Iguchi, K., and Nagasaki, Y.,** Study on practical application of monitoring method of total amounts of heavy metals in waste water on absorption of chelating resins and ion-exchange resins, *Zenkoku Kogaiken Kaishi*, 6, 9, 1981 (in Japanese).

619. **Foerster, M. and Lieser, K. H.,** Determination of traces of heavy metals in inorganic salts and organic solvents by energy-dispersive X-ray fluorescence analysis or the flameless atomic absorption spectrometry after enrichment on a cellulose exchanger, *Fresenius' Z. Anal. Chem.*, 309, 355, 1981 (in German).

620. **Hayashi, T., Nishizawa, M., and Yamagishi, T.,** Chemical studies on *Paeoniae radix.* IV. Relation between gallotannin contents and discoloration of *Paeoniae radix*, *Hokkaidoritsu Eisei Kenkyushoho*, 31, 23, 1981.

621. **Jin, K., Matsuda, K., and Chiba, Y.,** Concentrations of some metals in whole blood and plasma of normal adult subjects, *Hokkaidoritsu Eisei Kenkyushoho*, 31, 16, 1981.

622. **Okamoto, A., Ohmori, M., and Ishibashi, T.,** Analysis of the trace elements in natural food dyes, *Annu. Rep. Osaka City Inst. Public Health Environ. Sci.*, 43, 98, 1981.

623. **Mishima, M., Hoshiai, T., Watanuki, T., and Sugawara, K.,** Ion-exchange and atomic absorption spectrometric determination of trace metals in human teeth, *Koshu Eiseiin Kenkyu Hokoku*, 30, 40, 1981.

624. **Balaes, G. E., Dixon, K., Russell, G. M., and Wall, D. J.,** Analysis of activated carbon as used in the carbon-in-pulp process, for gold and eight other constituents, *S. Afr. J. Chem.*, 35, 4, 1982.

625. **Briggs, R. W. and Armitage, I. M.,** Evidence for site-selective metal binding in calf liver metallothionein, *J. Biol. Chem.*, 257, 1259, 1982.

626. **Danielsson, L. G., Magnusson, B., and Zhang, K.,** Matrix interference in the determination of trace metals by graphite furnace AAS after CHELEX-100 preconcentration, *At. Spectrosc.*, 3, 39, 1982.

627. **Foote, J. W. and Delves, H. T.,** Rapid determination of albumin-bound zinc in human serum by simple affinity chromatography and atomic absorption spectrophotometry, *Analyst (London)*, 107, 121, 1982.

628. **Gardiner, P. E., Roesick, E., Roesick, U., Braetter, P., and Kynast, G.,** Application of gel filtration, immunonephelometry and electrothermal atomic absorption spectrometry to the study of the distribution of copper, iron and zinc bound constituents in human amniotic fluid, *Clin. Chim. Acta*, 120, 103, 1982.

629. **Jacyszyn, K., Walas, J., Malinowski, A., Latkowski, T., and Cwynar, L.,** Concentration of heavy metals in pregnant women, *Zentralbl. Gynaekol.*, 104, 117, 1982 (in German).

630. **Janota-Bassalik, L. and Glabowna, M.,** Atomic absorption spectrophotometry of milk for prognosis of mastitis, *Milchwissenschaft*, 37, 13, 1982 (in English).

631. **Kauffman, R. E., Saba, C. S., Rhine, W. E., and Eisentraut, K. J.,** Quantitative multielement determination of metallic wear species in lubricating oils and hydraulic fluids, *Anal. Chem.*, 54, 975, 1982.

632. **Lendermann, B. and Hundeshagen, D.,** Use of multielement standards for calibration in water analysis by AAS, *Z. Anal. Chem.*, 310, 415, 1982.

633. **Olsen, K. B., Wilkerson, C. L., Toste, A. P., and Hays, D. J.,** Isolation of metallic complexes in shale oil and shale oil retort waters, *NBS Spec. Publ.*, 618, 105, 1982.

634. **Rawat, N. S., Sinha, J. K., and Sahoo, B.,** Atomic absorption spectrophotometric and X-ray studies of respirable dusts in Indian coal mines, *Arch. Environ. Health*, 37, 32, 1982.

635. **Salmon, S. G. and Holcombe, J. A.,** Alteration of metal release mechanisms in graphite furnace atomizers by chemisorbed oxygen, *Anal. Chem.*, 54, 630, 1982.

636. **Shaw, J. C., Bury, A. J., Barber, A., Mann, L., and Taylor, A.,** Micromethod for the analysis of zinc in plasma or serum by atomic absorption spectrophotometry using graphite furnace, *Clin. Chim. Acta*, 118, 229, 1982.

637. **Szerdahelyi, P., Kozma, M., and Ferke, A.,** Zinc deficiency-induced trace element concentration and localization changes in the central nervous system of albino rat during postnatal development. II. Atomic absorption spectrophotometric examinations, *Acta Histochem.*, 70, 173, 1982.

638. **Whitehouse, R. C., Prasad, A. S., Rabbani, P. I., and Cossack, Z. T.,** Zinc in plasma, neutrophils, lymphocytes and erythrocytes as determined by flameless atomic absorption spectrophotometry, *Clin. Chem. (Winston-Salem, N.C.)*, 28, 475, 1982.

639. **Kumina, D. M. and Karyakin, A. V.,** Effect of the nature of organic compounds on the atomic absorption of certain elements, *Zh. Prikl. Spektrosk.*, 36, 143, 1982.

640. **Sakla, A. B., Badran, A. H., and Shalaby, A. M.,** Determination of elements by atomic absorption spectrometry after destruction of blood in the oxygen flask, *Mikrochim. Acta,* 1, 483, 1982.

641. **Jelinek, P., Illek, J., and Jagos, P.,** Levels of zinc, manganese and copper in blood plasma, liver, hair, gonads and accessory sexual glands of coypu males, *Zivocisna Vyroba,* 27, 223, 1982.

642. **Abdulla, M., Norden, A., Schersten, B., Svensson, S., Thulin, T., and Oeckerman, P. O.,** The intake and urinary excretion of electrolytes and trace elements, in *Trace Element Metab. Man Anim., Proc. 4th Int. Symp.,* Gawthorne, J. M., Howell, J. M., and White, C. L., Eds., Springer Verlag, Berlin, 1982, 81.

643. **Fischer, W. R. and Fechter, H.,** Analytical determination and fractionation of copper, zinc, lead, cadmium, nickel and cobalt in soils and underwater soils, *Z. Pflanzeneraehr. Bodenkd.,* 145, 151, 1982.

644. **Ikebe, K. and Tanaka, R.,** Determination of heavy metals in blood by atomic absorption spectrometry. II, *Osaka Furitsu Koshu Eisei Kenkyusho Kenkyu Hokoku Shokuhin Eisei Hen,* 11, 43, 1980; CA 96, 137220y, 1982.

645. **Harada, Y.,** Determination of trace impurities in high-purity aluminum by flame atomic absorption spectrometry after coprecipitation with nickel hydroxide, *Bunseki Kagaku,* 31, 130, 1982.

646. **Lappalainen, R., Knuuttila, M., Iammi, S., Alhava, E. M., and Olkkonen, H.,** Zinc and copper content in human cancellous bone, *Acta Orthop. Scand.,* 53, 51, 1982.

647. **Landberger, S., Jervis, R. E., Aufreiter, S., and Van Loon, J. C.,** The determination of heavy metals (aluminum, manganese, iron, nickel, copper, zinc, cadmium and lead) in urban snow using an atomic absorption graphite furnace, *Chemosphere,* 11, 237, 1982.

648. **Burton, L. C., Uppal, P. N., and Dwight, D. W.,** Cross diffusion of cadmium and zinc into cuprous sulfide formed on zinc cadmium sulfide ($Zn_x Cd_{1-x} S$) thin films, *J. Appl. Phys.,* 53, 1538, 1982.

649. **Takahashi, T. and Asano, N.,** Thermodynamic studies of solid aluminum-zinc alloys, *Niihama Kogyo Koto Semmon Gakko Kiyo Rikogaku Hen,* 18, 78, 1982.

650. **Graham, P. P., Bittel, R. J., Bovard, K. P., Lopez, A., and Williams, H. L.,** Mineral element composition of bovine spleen and separated spleen components, *J. Food Sci.,* 47, 720, 1982.

651. **Aoki, T. and Yuza, C.,** Changes in serum constituents following endotoxic shock in rabbits, *Jikeikai Med. J.,* 29, 25, 1982.

652. **Brudzynska, K., Ploszajaska, A., Roszczyk, R., and Walter, Z.,** Interaction of zinc ions with DNA-dependent RNA polymerases A, B and C isolated from calf thymus, *Mol. Biol. Rep.,* 8, 77, 1982.

653. **Baldini, M., Grossi, M., Micco, C., and Stacchini, A.,** Presence of metal in cereal. II. Contamination of 1978 Italian rice, *Riv. Soc. Ital. Sci. Aliment.,* 11, 23, 1982.

654. **Sekiya, T., Tanimura, H., and Hikasa, Y.,** Study on the dosage of trace elements in total parenteral nutrition, *Jutsugo Taisha Kenkyu Kaishi,* 14, 114, 1980; CA 96, 18923v, 1982.

655. **Suwirma, S., Surtipanti, S., and Thamzil, L.,** Distribution of heavy metals mercury, lead, cadmium, chromium, copper and zinc in fish, *Majalah BATAN,* 13, 9, 1980 (in Indonesian); CA 96, 15797w, 1982.

656. **Thamzil, L., Suwirma, S., and Surtipanti, S.,** Determination of heavy metals in the stream of Sunter River, *Majalah BATAN,* 13, 41, 1980 (in Indonesian); CA 96, 24538q, 1982.

657. **Popov, G. K. and Shamgunov, A. N.,** Determination of iron, copper and zinc in a small amount of a biological material by atomic absorption spectrophotometric analysis, *Deposited Doc. VINITI,* 4618, VINITI, Moscow, U.S.S.R., 1980 (in Russian); CA 96, 65145c, 1982.

658. **Kasperek, K., Iyenger, G. V., Feinendegen, L. E., Hashish, S., and Mahfouz, M.,** Multielement analysis of fingernail, scalp hair and water samples from Egypt (a preliminary study), *Sci. Total Environ.,* 22, 149, 1982.

659. **Ough, C. S., Crowell, E. A., and Benz, J.,** Metal content of California wines, *J. Food Sci.,* 47, 825, 1982.

660. **Helgeland, K., Haider, T., and Jonsen, J.,** Copper and zinc in human serum in Norway. Relationship to geography, sex and age, *Scand. J. Clin. Lab. Invest.,* 42, 35, 1982.

661. **Cirlin, E. H. and Housley, R. M.,** Distribution and evolution of zinc, cadmium and lead in Apollo 16 regolith samples and the average uranium-lead ages of the parent rocks, *Geochim. Cosmochim. Acta Suppl.,* 16, 529, 1982.

662. **Yamagata, N.,** Interlaboratory comparison study on the reliability of environmental analysis. Soil and sediment 1978-1980, *Bunseki Kagaku,* 31, T1, 1982.

663. **Shamrai, Z. Y., Sergeeva, K. I., Naftal, I. P., and Dymova, M. S.,** Atomic absorption analysis of copper-manganese alloys, *Zavod. Lab.,* 48, 73, 1982.

664. **Takenaka, S.,** Analysis of heavy metals in human gallstone and renal calculus, *Tetsudo Yakugaku Kankyu Nempo,* 28, 82, 1980 (in Japanese); CA 97, 12145j, 1982.

665. **Andoh, K., Saitoh, Y., Takatani, A., Takahashi, F., Tazuya, Y., Tsunajima, K., Motoki, C., Yasuoka, K., Yamaji, Y., and Matsuoka, C.,** The measurement and evaluation of zinc and copper in foods by the methods of microwave oven-based wet digestion technique and atomic absorption spectrophotometry, *Tokushima Bunri Daigaku Kankyu Kiyo,* 25, 113, 1982 (in Japanese).

666. **Korenev, A. P., Garanzha, A. P., Konovalov, G. S., and Ovsyannikova, T. V.,** Manganese, iron, zinc, and lead detection by atomic absorption in extracts from bottom sediments, *Gidrokhim. Mater.*, 80, 83, 1982.

667. **Egorova, K. A., Ermakova, T. I., Shkol'nikov, V. M., and Shirokova, G. B.,** Rapid micromethod for testing of hydraulic oils, *Khim. Tekhnol. Topl. Masel*, 7, 38, 1982 (in Russian).

668. **Hollies, N. R. S., Krejci, V., and Smith, I. T.,** The transfer of chemicals from clothing to skin, *Text. Res. J.*, 52, 370, 1982.

669. **Raja, K. B., Leach, P. M., Smith, G. P., McCarthy, D., and Peters, T. J.,** The concentration and subcellular localization of zinc, magnesium and calcium in human polymorphonuclear leukocytes, *Clin. Chim. Acta*, 123, 19, 1982.

670. **Adam, J.,** Determination of copper, zinc, and lead in mine waters, *Hutn. Listy*, 37, 276, 1982 (in Czech).

671. **Tulley, R. T. and Lehmann, H. P.,** Method for simultaneous determination of cadmium and zinc in whole blood by atomic absorption spectrophotometry and measurement in normotensive and hypertensive humans, *Clin. Chim. Acta*, 122, 189, 1982.

672. **Gregorio, P., Siracusano, C., and Toscano, G.,** Contents of iron, copper, zinc, magnesium and manganese in infant foods. I. Homogenized baby food, *Ig. Mod.*, 77, 348, 1982.

673. **Ines, G. G., Leonardo, B. K., Valdebenito, S., Rita, M., and Luz, V. Q.,** Technique for the preparation of mixed human saliva samples for the determination of copper, zinc, and manganese by flameless atomic absorption spectrometry, *Bol. Soc. Chil. Quim.*, 27, 340, 1982.

674. **Schmidt, L. H., Meissner, D., and Lenski, K. H.,** Results and perspectives for standardization of trace element analysis in biological material, *Zentralbl. Pharm. Pharmakother. Laboratoriumsdiagn.*, 121, 444, 1982 (in German).

675. **Cullinane, J. P. and Whelan, P. M.,** Copper, cadmium and zinc in seaweeds from the south coast of Ireland, *Mar. Pollut. Bull.*, 13, 205, 1982.

676. **Balkas, T., Tugrul, S., and Salihoglu, I.,** Trace metal levels in fish and crustacea from northern Mediterranean coastal waters, *Mar. Environ. Res.*, 6, 281, 1982.

677. **Brovko, I. A.,** Diphenylcarbazone as an extraction reagent in atomic absorption, *Deposited Doc. VINITI*, 2842-81, 1981; CA 97, 65661p, 1982.

678. **Agudelo, M. I., Kustin, K., and Robinson, W. E.,** Blood chemistry of *Boltenia ovifera*, *Comp. Biochem. Physiol. A*, 72a, 161, 1982.

679. **Brzezinska, A.,** Some remarks on the determination of trace metals in marine samples by electrothermal atomization, *Eur. Spectrosc. News*, 40, 19, 1982.

680. **Taliadouri-Voutsinou, F.,** Trace metals in marine organisms from the Saronikos Gulf (Greece), *J. Etud. Pollut. Mar. Mediterr.*, 5, 1981; CA 97, 34323a, 1982.

681. **Ebdon, L., Ellis, A. T., and Ward, R. W.,** Aspects of chloride interference in zinc determination by atomic absorption spectroscopy with electrothermal atomization, *Talanta*, 29, 297, 1982.

682. **Kovalskii, V. V., Letunova, S. V., and Alekseeva, S. A.,** Accumulation of nickel and other elements in soil microflora biomass in the South Urals subregion biosphere, *Spurenelem.-Symp.: Nickel*, 3rd, Anke, M., Schneider, H., and Brueckner, C., Eds., Friedrich-Schiller University, Jena, East Germany, 1980, 163 (in Russian); CA 97, 95913q, 1982.

683. **Fudagawa, N., Nakamura, S., and Kawase, A.,** Comparison of tungsten, molybdenum and tantalum ribbon atomizers in atomic absorption spectrometry, *Bunseki Kagaku*, 31, 324, 1982.

684. **Mahalingam, T. R., Geetha, R., Thiruvengadasamy, A., and Mathews, C. K.,** Analysis of trace metals in sodium by flameless atomic absorption spectrophotometry, *Mater. Behav. Phys. Chem. Liq. Met. Syst. (Proc. Conf.)*, Borgstedt, H. U., Ed., Plenum Press, New York, 1982, 329.

685. **Seeling, W., Feist, H., Gruenert, A., Heinrich, H., and Luckner, L.,** The influence of drawing technique and sample preparation on zinc determination in serum and plasma. Statistical distribution of plasma zinc in male and female blood donors, *Magnesium Bull.*, 4, 18, 1982.

686. **Zhou, Z. and Mao, X.,** A new chelate forming resin bearing xylenol orange group and its application, *Gaodeng Xuexiao Huaxue Xuebao*, 3, 181, 1982.

687. **Bogden, J. D., Zadzielski, E., Weiner, B., Oleske, J. A., and Aviv, A.,** Release of some trace metals from disposable coils during hemodialysis, *Am. J. Clin. Nutr.*, 36, 403, 1982.

688. **Gallego, M., Garcia-Vargas, M., and Valcarcel, M.,** Analytical application of picolinealdehyde salicyloylhydrazone. III. Extraction and determination of zinc by atomic absorption spectrophotometry, *Microchem. J.*, 27, 328, 1982.

AUTHOR INDEX

SUBJECT INDEX (ZINC)

SUBJECT INDEX (BIOLOGY)

CADMIUM, Cd (ATOMIC WEIGHT 112.41)

Cadmium occurs in small quantities associated with zinc ores. It is mostly used in the preparation of alloys and bearings alloys with low coefficients of friction and great resistance to fatigue. It is extensively used in electroplating, soldering, batteries, as a barrier to control atomic fission, and in paint pigments.

Cadmium is toxic and care should be taken when its compounds are handled. Determination of cadmium is required in biological samples, food, water, fertilizers, and soil, etc., due to its toxic nature. Analysis of slab zinc, ores, alloys, zinciferrous materials, paints, pigments, and dental amalgam, etc. is also required for the cadmium content. Samples can be prepared by dissolution in hot hydrochloric acid or aqua regia and followed by sulfuric acid treatment. Fusion is preferred for hard-to-dissolve samples.

No chemical or spectral interference has been observed in air-acetylene flame using the atomic absorption method for cadmium determination. High concentration of silicate interference has been reported. A microsampling system increases the cadmium detection and sensitivity.

Standard Solution

To prepare 1000 mg/ℓ cadmium solution, dissolve 1.0000 g of cadmium metal in 50 mℓ of 1:1 hydrochloric acid. Dilute to 1 ℓ with 1% (v/v) hydrochloric acid.

Instrumental Parameters

Wavelength	228.8 nm
Slit	0.7 nm
Light source	Hollow cathode lamp
Flame	Air-acetylene, oxidizing (lean, blue)
Sensitivity	0.01 mg/ℓ
Detection limit	0.002 mg/ℓ
Optimum range	2 mg/ℓ
Secondary wavelength	326.1 nm

REFERENCES

1. **Elwell, W. T. and Gidley, J. A. F.**, *Atomic Absorption Spectrophotometry*, Pergamon Press, New York, 1961.
2. **Willis, J. B.**, Determination of lead and other heavy metals in urine by atomic absorption spectroscopy, *Anal. Chem.*, 34, 614, 1962.
3. **Poluektov, N. S. and Vitkin, R. A.**, Atomic absorption flame photometric determination of cadmium, *Zh. Anal. Khim.*, 17, 935, 1962.
4. **Slavin, W., Sprague, S., Rieders, P., and Cordova, V.**, The determination of certain toxicological trace metals by atomic absorption spectrophotometry, *At. Absorpt. Newsl.*, No. 17, January 1964.
5. **Stupar, J., Podobnik, B., and Korosin, J.**, Determination of cadmium in uranium compounds by atomic absorption spectrophotometry, *Rep. Nukl. Inst. Josef. Stefan, Ljubljana*, R-447, 1965.
6. **Takeuchi, T., Suzuki, M., and Yanagisawa, M.**, Some observations on the determination of metals by atomic absorption spectroscopy combined with extractions, *Anal. Chim. Acta*, 36, 258, 1966.
7. **Wilson, J.**, The determination of cadmium in stainless steel by atomic absorption spectroscopy, *Anal. Chim. Acta*, 35, 123, 1966.
8. **Bell, G. F.**, The analysis of aluminum alloys by means of atomic absorption spectrophotometry, *At. Absorpt. Newsl.*, 5, 73, 1966.
9. **Dagnall, R. M., West, T. S., and Young, P.**, Determination of cadmium by atomic fluorescence and atomic absorption spectrophotometry, *Talanta*, 13, 803, 1966.
10. **Farrar, B.**, Determination of cadmium in ore and magnesium in rock samples, *At. Absorpt. Newsl.*, 5, 62, 1966.
11. **Pulido, P., Fuwa, K., and Vallee, B. L.**, Determination of cadmium in biological materials by atomic absorption spectrophotometry, *Anal. Biochem.*, 14, 393, 1966.
12. **Sattur, T. W.**, Routine atomic absorption analysis on non-ferrous alloys and plant intermediates, *At. Absorpt. Newsl.*, 5, 37, 1966.
13. **Fixman, M. and Boughton, L.**, Minerals assay for silver, zinc and cadmium, *At. Absorpt. Newsl.*, 5, 33, 1966.
14. **Taskaeva, T. P.**, Atomic absorption determination of cadmium in aqueous solutions of the cadmium chloride-potassium chloride lithium chloride eutectic mixture, *Ser. Khim. Nauk*, 3, 105, 1967.
15. **Berman, E.**, Determination of cadmium, thallium and mercury in biological materials, *At. Absorpt. Newsl.*, 6, 57, 1967.
16. **Miller, W. J., Lampp, B., Powell, G. W., Salloti, C. A., and Blackmon, D. M.**, Influence of a high level of dietary cadmium on cadmium content in milk, excretion and cow performance, *J. Dairy Sci.*, 50, 1404, 1967.
17. **Ramakrishna, T. V., Robinson, J. W., and West, P. W.**, Determination of copper, cadmium and zinc by atomic absorption spectroscopy, *Anal. Chim. Acta*, 37, 20, 1967.
18. **Lehnert, G., Schaller, K. H., and Haas, T.**, The atomic absorption spectrophotometric determination of cadmium in serum and urine, *Z. Klin. Chem. Klin. Biochem.*, 6, 174, 1968.
19. **Torres, F.**, The Quantitative Determination of Lead, Cadmium and Indium in Microgram Amounts by Atomic Absorption Spectrophotometry, SL-TM-68-4, Sandia Laboratories, Albuquerque, 1968.
20. **Wall, H. and Rhodes, C.**, Measurement of salt effect in the determination of cadmium by atomic absorption, *Can. Spectrosc.*, 13, 51, 1968.
21. **Yamada, J., Lida, C., and Yamasaki, K.**, Atomic absorption spectrophotometric determination of trace amounts of cadmium in silicates with the absorption tube technique, *Bunseki Kagaku*, 18, 1088, 1969.
22. **Kaikov, D. A. and L'vov, B. V.**, Atomic absorption determination of trace amounts of cadmium in a graphite cuvet, *Zh. Prikl. Spektrosk.*, 10, 867, 1969 (in Russian).
23. **Lehnert, G., Clavis, G., Schaller, K. H., and Haas, T.**, Cadmium determination in urine by atomic absorption spectrometry as a screening test in industrial medicine, *Br. J. Ind. Med.*, 26, 156, 1969.
24. **L'vov, B. V. and Khartsyzov, A. D.**, Atomic absorption spectral analysis of radioactive preparations of zinc and cadmium, *Zh. Anal. Khim.*, 24, 936, 1969.
25. **McCrackan, J. D., Vecchione, M. C., and Longo, S. L.**, Ion exchange separation of traces of tin, cadmium and zinc from copper and their determination by atomic absorption spectrophotometry, *At. Absorpt. Newsl.*, 8, 102, 1969.
26. **Robinson, J. L., Barnekow, R. G., and Lott, P. F.**, Rapid determination of cadmium and copper in plating wastes and river water by atomic absorption spectroscopy, *At. Absorpt. Newsl.*, 8, 60, 1969.
27. **Curry, A. S. and Knott, A. R.**, Normal levels of cadmium in human liver and kidney in England, *Clin. Chim. Acta*, 30, 115, 1970.
28. **Kono, T. and Kobayashi, S.**, Determination of trace amounts of cadmium by atomic absorption spectrometry combined with extraction of iodide-MIBK, *Bunseki Kagaku*, 19, 1491, 1970.

29. **Mizuno, T., Harada, A., Kudo, Y., and Hasegawa, N.,** Determination of cadmium traces in zirconium and its alloy by atomic absorption spectrophotometry, *Bunseki Kagaku,* 19, 251, 1970.

30. **Silvester, M. D. and McCarthy, W. J.,** The intensity of electrodeless discharge lamps containing cadmium, mercury and manganese, *Spectrochim. Acta,* 25b, 229, 1970.

31. **Westerlund-Helmerson, U.,** Determination of lead and cadmium in blood by a modification of the Hessel method, *At. Absorpt. Newsl.,* 9, 133, 1970.

32. **Bazhov, A. S. and Zherebenko, A. V.,** Determination of the width of the resonance line of cadmium absorption in a flame, *Zh. Prikl. Spektrosk.,* 12, 403, 1970 (in Russian).

33. **Yamada, J., Iida, C., and Yamasaki, K.,** Atomic absorption spectrophotometric determination of cadmium in silicates using the absorption tube technique combined with a ring burner, *Bunseki Kagaku,* 19, 1259, 1970.

34. **Tati, M. and Katagiri, Y.,** Determination of cadmium in organs by atomic absorption spectrophotometry, *Igaku To Seibutsu,* 81, 131, 1970.

35. **Zook, E., Greene, F., and Morris, E.,** Nutrient composition of selected wheats and wheat products. Distribution of manganese, copper, nickel, zinc, magnesium, lead, tin, cadmium, chromium and selenium as determined by atomic absorption spectroscopy and calorimetry, *Cereal Chem.,* 47, 720, 1970.

36. **Yamamoto, Y., Kumamuru, T., and Hayashi, Y.,** Solvent extraction of anions with metal chelate cations. Determination of ppb levels of cadmium by atomic absorption spectrophotometry combined with solvent extraction by the Tris (1,10-phenanthroline) cadmium perchlorate ion pair, *Nippon Kagaku Zasshi,* 92, 345, 1971.

37. **Tonouchi, S., Suzuki, T., Sotobayashi, T., and Koyama, S.,** Determination of cadmium by atomic absorption spectrometry combined with solvent extraction with high molecular weight amines, *Bunseki Kagaku,* 20, 1453, 1971.

38. **Belyaev, Y. I., Pchelintsev, A. M., Zvereva, N. F., and Kostin, B. I.,** Atomic absorption determination of traces of elements in rocks using the pulsed thermal atomization of solid samples. Determination of cadmium, *Zh. Anal. Khim.,* 26, 492, 1971.

39. **Headridge, J. B. and Smith, D. R.,** Induction furnace for the determination of cadmium in solutions and zinc base metals by atomic absorption spectroscopy, *Talanta,* 18, 247, 1971.

40. **Kuroha, T., Tsukahara, I., and Shibuya, S.,** Determination of microamounts of cadmium and zinc in copper, nickel, aluminum and uranium metals by solvent extraction-atomic absorption spectrophotometry, *Bunseki Kagaku,* 20, 1137, 1971.

41. **Kuwata, K., Hisatomi, K., and Hasegawa, T.,** The rapid determination of trace amounts of cadmium and copper in river and sea water by atomic absorption spectroscopy, *At. Absorpt. Newsl.,* 10, 111, 1971.

42. **Lener, J. and Bibr, B.,** Determination of traces of cadmium in biological materials by atomic absorption spectrophotometry, *J. Agric. Food Chem.,* 19, 1011, 1971.

43. **Jaros, J.,** Determination of cadmium, cobalt, chromium, magnesium, manganese and nickel in uranium (and its) compounds by atomic absorption spectrophotometry, *Jad. Energ.,* 17, 307, 1971 (in Czech).

44. **Leucke, W. and Emmermann, R.,** The application of the boat technique for lead, zinc, silver and cadmium in soil samples, *At. Absorpt. Newsl.,* 10, 45, 1971.

45. **Murthy, G. K., Rhea, V., and Peeler, J. T.,** Levels of antimony, cadmium, chromium, cobalt, manganese and zinc in institutional diets, *Environ. Sci. Technol.,* 5, 436, 1971.

46. **Tamamoto, Y., Kumamaru, T., Hayashi, Y., and Kanke, M.,** Determination of a trace amount of cadmium, zinc, lead and copper in water by atomic absorption spectrophotometry, *Bunseki Kagaku,* 20, 347, 1971.

47. **Christian, C. M., II and Robinson, J. W.,** The direct determination of cadmium and mercury in the atmosphere, *Anal. Chim. Acta,* 56, 466, 1971.

48. **Eisler, R.,** Cadmium poisoning in *Fundulus heteroclitus* (pisces: cyprinodontidae) and other marine organisms, *J. Fish Res. Board Canada,* 28, 1225, 1971.

49. **Kanke, M., Hayashi, Y., Kumamaru, T., and Yamamoto, Y.,** Determination of ppb levels of cadmium, zinc, lead and copper extracted as their dithizonates into nitrobenzene by atomic absorption spectrophotometry, *Nippon Kagaku Zasshi,* 92, 983, 1971.

50. **Bergner, K. C. and Lang, B.,** The determination of iron, copper, zinc, manganese and cadmium in grape juice and wine by atomic absorption spectrophotometry, *Dtsch. Lebensm. Rundsch.,* 67, 121, 1971.

51. **Bohnstedt, U.,** Determination of traces of lead, bismuth, zinc, and cadmium in alloys. An example of the use of combined analytical procedures, *DEW Tech. Ber.,* 11, 101, 1971 (in German).

52. **Nordberg, G. F., Piscator, M., and Lind, B.,** Distribution of cadmium among protein fractions of mouse liver, *Acta Pharmacol. Toxicol.,* 29, 456, 1971.

53. **Hauser, T. R., Hinners, T., and Kent, J. L.,** Atomic absorption determination of cadmium and lead in whole blood by a reagent-free method, *Anal. Chem.,* 44, 1819, 1972.

54. **John, M. K., Chuah, H. H., and Van Laerhoven, C. J.,** Cadmium contamination of soil and its uptake by oats, *Environ. Sci. Technol.,* 6, 555, 1972.

55. **John, M. K., Van Laerhoven, C. J., and Chuah, H. H.,** Factors affecting plant uptake and phytotoxicity of cadmium added to soils, *Environ. Sci. Technol.,* 6, 1005, 1972.

56. **Kahn, H. L., Fernandez, F. J., and Slavin, S.,** The determination of lead and cadmium in soils and leaves by atomic absorption spectroscopy, *At. Absorpt. Newsl.,* 11, 42, 1972.

57. **Kubota, N. and Imai, Y.,** Silver plating obtained from cyanide baths. Determination of nickel, zinc and cadmium in silver plating solution by atomic absorption, *Kinzoku Hyomen Gijutsu,* 23, 157, 1972.

58. **Lewis, G. P., Jusko, W. J., Coughlin, L. L., and Hartz, S.,** Contribution of cigarette smoking to cadmium accumulation in man, *Lancet,* 5, 291, 1972.

59. **Aston, S. R., Chester, R., Griffiths, A., and Riley, J. P.,** Distribution of cadmium in north Atlantic deep sea sediments, *Nature (London),* 239, 393, 1972.

60. **Lagerwerff, J. V. and Brower, D. L.,** Exchange adsorption of trace quantities of cadmium in soils treated with chlorides of aluminum, calcium and sodium, *Soil Sci. Soc. Am. Proc.,* 36, 734, 1972.

61. **Marks, G. E., Moore, C. E., Kanabrocki, E. L., Oester, Y. T., and Kaplan, E.,** Determination of trace elements in human tissue. I. Cd, Fe, Zn, Mg, and Ca, *Appl. Spectrosc.,* 26, 523, 1972.

62. **Menden, E. E., Elia, V. J., Michael, L. W., and Petering, H. G.,** Distribution of cadmium and nickel of tobacco during cigarette smoking, *Environ. Sci. Technol.,* 6, 830, 1972.

63. **Moten, L.,** Quantitative determination of cadmium in water soluble color additives by atomic absorption spectroscopy, *J. Assoc. Off. Anal. Chem.,* 55, 1145, 1972.

64. **Owa, T., Hiiro, K., and Tanaka, T.,** Determination of trace amounts of pollutants in sea and river waters by atomic absorption spectrometry. Determination of ppb amounts of cadmium in sea water by coprecipitation-solvent extraction-concentration and atomic adsorption spectrometry, *Bunseki Kagaku,* 21, 878, 1972.

65. **Prudnikov, E. D. and Shapkina, U. S.,** Atomic absorption microdetermination of trace amounts of cadmium in a flame, *Izv. Vyssh. Zaved. Khim. Khim. Technol.,* 15, 944, 1972.

66. **Scharpf, L. G., Jr., Hill, I. D., Wright, P. L., Plank, J. B., Keplinger, M. L., and Calandra, J. C.,** Effect of sodium nitrilotriacetate on toxicity, teratogenicity and tissue distribution of cadmium, *Nature (London),* 239, 231, 1972.

67. **Holroyd, P. M. and Snodin, D. J.,** Determination of lead and cadmium in tap water, rainwater, milk and urine, *J. Assoc. Public Anal.,* 10, 110, 1972.

68. **Shiraishi, N., Masegawa, T., Hisayuki, T., and Takahashi, H.,** Dissolution of methyl isobutyl ketone in various salt solutions. Determination of cadmium and lead by atomic absorption, *Bunseki Kagaku,* 21, 705, 1972.

69. **Takagi, Y., Kiyotani, J., and Satake, M.,** Atomic absorption spectrometric determination of cadmium in unfinished rice by solvent extraction with dithizone, *Nippon Kagaku Kaishi,* 10, 1983, 1972.

70. **Yamamoto, Y., Kumamaru, T., Hayashi, Y., and Kanke, M.,** Effect of solvent extraction on the atomic absorption spectrophotometric determination of ppm levels of cadmium with dithizone, *Talanta,* 19, 953, 1972.

71. **Cheremisinoff, P. N. and Habib, Y. H.,** Cadmium, chromium, lead and mercury: plenary account for water pollution. I. Occurrence, toxicity and detection, *Water Sewage Works,* 119, 73, 1972.

72. **Takagi, Y. and Satake, M.,** Low temperature ashing technique for the determination of cadmium in unpolished rice by atomic absorption, *Nippon Kagaku Kaishi,* 2207, 1972.

73. **Williams, C. H., David, D. J., and Lismaa, O.,** Determination of cadmium in soils, plants and fertilizers by dithizone extraction and atomic absorption spectroscopy, *Commun. Soil Sci. Plant Anal.,* 3, 399, 1972.

74. **Rudnevskii, N. K., Maksimov, D. E., Shabanova, T. M., and Lazareva, L. P.,** Spectroscopic determination of cadmium in solutions by excitation of the spectrum in a hollow cathode discharge with an applied magnetic field, *Zh. Prikl. Spektrosk.,* 16, 356, 1972.

75. **Vens, M. D. and Lauwerys, R.,** Simultaneous determination of lead and cadmium in blood and urine by coupling ion exchange resin chromatography and atomic absorption spectrophotometry, *Arch. Mal. Prof.,* 33, 97, 1972 (in French).

76. **Piscator, M. and Lind, B.,** Cadmium, zinc, copper and lead in human renal cortex, *Arch. Environ. Health,* 24, 426, 1972.

77. **Buchauer, M. J.,** Contamination of soil and vegetation near a zinc smelter by zinc, cadmium, copper and lead, *Environ. Sci. Technol.,* 7, 131, 1973.

78. **Chaube, S., Nishimura, H., and Swinyard, C. A.,** Zinc and cadmium in normal human embryos and fetuses. Analysis by atomic absorption spectrophotometry, *Arch. Environ. Health,* 26, 237, 1973.

79. **Doolan, K. J. and Smythe, L. E.,** Synergistic influence of zinc(II) in the extraction-atomic absorption determination of trace cadmium in waters, *Talanta,* 20, 241, 1973.

80. **Ediger, R. D. and Coleman, R. L.,** Determination of cadmium in blood by a Delves cup technique, *At. Absorpt. Newsl.,* 12, 3, 1973.

81. **Holak, W.,** Determination of traces of lead and cadmium in foods by atomic absorption spectrophotometry using the "sampling boat", *At. Absorpt. Newsl.,* 12, 63, 1973.

82. **Murthy, L., Menden, E. E., Eller, P. M., and Petering, H. G.,** Atomic absorption determination of zinc, copper, cadmium and lead in tissues solubilized by aqueous tetra methyl ammonium hydroxide, *Anal. Biochem.,* 53, 365, 1973.

83. **Nakagawa, R.,** Atomic absorption analysis of zinc, lead and cadmium in hot spring waters, *Nippon Kagaku Kaishi,* 514, 1973.

84. **Okusu, H., Ueda, Y., Ota, K., and Kawano, K.,** Determination of cadmium, zinc, copper and lead in waste water by atomic absorption spectrometry, *Bunseki Kagaku,* 22, 84, 1973.

85. **Sorensen, J. R. J., Levin, L. S., and Petering, H. G.,** Cadmium, copper, lead, mercury and zinc concentrations in the hair of individuals living in the United States, *Interface,* 2, 17, 1973.

86. **Cernik, A. A.,** A preliminary procedure for the determination of cadmium in blood, *At. Absorpt. Newsl.,* 12, 163, 1973.

87. **Doyle, J. J., Pfander, W. H., Grebing, S. E., and Pierce, J. C., II,** Effects of dietary cadmium on growth and tissue levels in sheep, in *Trace Substances in Environmental Health,* Vol. 6, Hemphill, D. D., Ed., University of Missouri, Columbia, 1973, 181.

88. **Gajan, R. J., Gould, J. R., Watts, J. O., and Fiorino, J. A.,** Collaborative study of a method for the atomic absorption spectrophotometric and polarographic determination of cadmium in food, *J. Assoc. Off. Anal. Chem.,* 56, 876, 1973.

89. **Gish, C. D. and Christensen, R. E.,** Cadmium, nickel, lead and zinc in earthworms from roadside soil, *Environ. Sci. Technol.,* 7, 1060, 1973.

90. **Gluskoter, H. J. and Lindahl, P. C.,** Cadmium: mode of occurrence in Illinois coals, *Science,* 181, 264, 1973.

91. **Krinitz, B. and Franco, V.,** Collaborative study of an atomic absorption method for the determination of lead and cadmium extracted from glazed ceramic surfaces, *J. Assoc. Off. Anal. Chem.,* 56, 869, 1973.

92. **Kubasik, N. P. and Volosin, M. T.,** Simplified determination of urinary cadmium, lead and thallium with use of carbon rod atomization and atomic absorption spectrophotometry, *Clin. Chem.,* 19, 954, 1973.

93. **Lagerwerff, J. V., Bower, D. L., and Biersdorf, G. T.,** Accumulation of cadmium, copper, lead and zinc in soils and vegetation in the proximity of a smelter, in *Trace Substances in Environmental Health,* Vol. 6, Hemphill, D. D., Ed., University of Missouri, Columbia, 1973, 71.

94. **Lieberman, K. W.,** Determination of cadmium in biological fluids by the Delves modification of atomic absorption spectrophotometry, *Clin. Chim. Acta,* 46, 217, 1973.

95. **Linnman, L., Andersson, A., Nilsson, K. O., Lind, B., Kjellstroem, T., and Friberg, L.,** Cadmium uptake by wheat from sewage sludge used as a plant nutrient source. Comparative study using flameless atomic absorption and neutron activation analysis, *Arch. Environ. Health,* 27, 45, 1973.

96. **Mikhailova, T. P., Klimashova, B. P., and Babeuva, L. V.,** Determination of zinc, cadmium, copper and iron in aluminum and gallium nitrides using an atomic absorption method, *Izv. Sib. Otd. Akad. Nauk SSSR Ser. Khim. Nauk,* 135, 1973.

97. **Olsen, R. D. and Sommerfeld, M. R.,** A technique for extraction and storage of water samples for Mn, Cd and Pb determination by atomic absorption spectroscopy, *At. Absorpt. Newsl.,* 12, 165, 1973.

98. **Parker, C. R., Rowe, J., and Sandoz, D. P.,** Methods of environmental cadmium analysis by flameless atomization, *Am. Lab.,* 5, 53, 1973.

99. **Perry, H. M., Jr., Pierce, J. A., Hirst, R. N., Purifoy, J. E., Perry, E. F., and Cruz, M.,** Elevated pulmonary cadmium in emphysematous subjects without known cadmium exposure, in *Trace Substances in Environmental Health,* Vol. 6, Hemphill, D. D., Ed., University of Missouri, Columbia, 1973, 207.

100. **Pinkerton, C., Hammer, D. I., Bridbord, K., Creason, J. P., Kent, J. L., and Murthy, G. K.,** Human milk as a dietary source of cadmium and lead, in *Trace Substances in Environmental Health,* Vol. VI, Hemphill, D. D., Ed., University of Missouri, Columbia, 1973, 39.

101. **Robinson, J. W., Wolcott, D. K., Slevin, P. J., and Hindman, G. D.,** Determination of cadmium in atomic absorption in air, water, seawater and urine with a radio frequency carbon atomizer, *Anal. Chim. Acta,* 66, 13, 1973.

102. **Shigematsu, T., Matsui, M., and Fujino, G.,** Determination of cadmium by atomic absorption with a heated carbon tube atomizer. Application to seawater, *Bunseki Kagaku,* 22, 1162, 1973.

103. **Sorensen, J. R. J., Melby, E. G., Nord, P. J., and Petering, H. J.,** Interferences in the determination of metallic elements in human hair. Evaluation of zinc, copper, lead and cadmium using atomic absorption spectrophotometry, *Arch. Environ. Health,* 27, 36, 1973.

104. **Jensen, R. E. and Kaehler, M.,** A simultaneous determination of zinc and cadmium, *Anal. Chim. Acta,* 67, 465, 1973.

105. **Brockhaus, A.,** Quantitative detection of lead, cadmium and zinc in blood and organs, *Staub Reinhalt. Luft,* 33, 437, 1973 (in German).

106. **Snodin, D. J.,** Determination of lead and cadmium in baby foods, *J. Assoc. Public Anal.,* 11, 112, 1973.

107. **Reusmann, G. and Westphalen, J.,** Automated determination of lead, zinc and cadmium in vegetable materials, *Staub Reinhalt. Luft,* 33, 435, 1973.

108. **Hahn, C.,** Determination of lead, cadmium and zinc in fired glazes and engobe decorations using atomic absorption spectroscopy, *Silikat J.,* 12, 38, 1973.
109. **Van Loon, J. C., Lichwa, J., and Ruttan, D.,** Study of the determination and distribution of cadmium in samples collected in a heavily industrialized and urbanized region (Metropolitan Toronto), *Int. J. Environ. Anal. Chem.,* 3, 147, 1973.
110. **Smit, A. and Backe-Hansen, K.,** Determination of contamination of zinc compounds by lead and cadmium, *Acta Pharm. Suec.,* 10, 254, 1973.
111. **Kaplan, P. D., Blackstone, M., and Richdale, N.,** Direct determination of contamination of cadmium, nickel and zinc in rat lungs by atomic absorption spectrophotometry, *Arch. Environ. Health,* 27, 387, 1973.
112. **Langmyhr, F. J., Thomassen, Y., and Massoumi, A.,** Atomic absorption spectrometric determination of lead and cadmium in silicon, ferrosilicon and ferromanganese by direct atomization from the solid state, *Anal. Chim. Acta,* 67, 460, 1973.
113. **Machata, G. and Bindes, R.,** Determination of traces of lead, thallium, zinc and cadmium in biological material by flameless atomic absorption spectrophotometry, *Z. Rechtsmed.,* 73, 29, 1973.
114. **King, W. G., Rodriguez, J. M., and Wai, C. M.,** Losses of trace concentrations of cadmium from aqueous solution during storage in glass containers, *Anal. Chem.,* 46, 771, 1974.
115. **Kubota, J., Mills, E. L., and Oglesby, R. T.,** Lead, cadmium, zinc, copper and cobalt in streams and lake waters of Cayuga Lake Basin, New York, *Environ. Sci. Technol.,* 8, 243, 1974.
116. **Amore, F.,** Determination of cadmium, lead, thallium and nickel in blood by atomic absorption spectrometry, *Anal. Chem.,* 46, 1597, 1974.
117. **Dudas, M. J.,** The quantitative determination of cadmium in soils by solvent extraction and flame and flameless atomic absorption spectroscopy, *At. Absorpt. Newsl.,* 13, 109, 1974.
118. **Falchuk, K. H., Evenson, M., and Vallee, B. L.,** Multichannel atomic absorption instrument. Simultaneous analysis of zinc, copper and cadmium in biologic materials, *Anal. Biochem.,* 62, 225, 1974.
119. **Ganje, T. J. and Page, A. L.,** Rapid acid dissolution of plant tissue for cadmium determination by atomic absorption spectrophotometry, *At. Absorpt. Newsl.,* 13, 131, 1974.
120. **Halpin, M. K. and Carroll, D. M.,** Light sensitivity of tests for cadmium on ceramic tableware, *Nature (London),* 247, 197, 1974.
121. **Hinners, T. A., Bumgarner, J. E., and Simmons, W. S.,** Extraction of cadmium from rice, *At. Absorpt. Newsl.,* 13, 146, 1974.
122. **Jensen, F. O., Dolezal, J., and Langmyhr, F. J.,** Atomic absorption spectrometric determination of cadmium, lead and zinc in salts or salt solutions by hanging mercury drop electrodeposition and atomization in a graphite furnace, *Anal. Chim. Acta,* 72, 245, 1974.
123. **Kaminski, E. E.,** Interference of aluminum in the atomic absorption determination of cadmium using sodium diethyldithiocarbamate as chelating agent, *Anal. Chem.,* 46, 1304, 1974.
124. **Kjellstrom, T., Lind, B., Linnman, L., and Nordberg, G.,** A comparative study on methods for cadmium analysis of grain with an application to pollution evaluation, *Environ. Res.,* 8, 92, 1974.
125. **Krinitz, B.,** Effect of light on the extraction of lead and cadmium from glazed ceramic surface, *J. Assoc. Off. Anal. Chem.,* 57, 966, 1974.
126. **Langmyhr, F. J., Sundli, A., and Jonsen, J.,** Atomic absorption spectrometric determination of cadmium and lead in dental material by atomization directly from the solid state, *Anal. Chim. Acta,* 73, 81, 1974.
127. **Baetz, R. A. and Kenner, C. T.,** Determination of low levels of cadmium in foods using a chelating ion exchange resin, *J. Assoc. Off. Anal. Chem.,* 57, 14, 1974.
128. **Brodie, K. G. and Matousek, J. P.,** Determination of cadmium in air by nonflame atomic absorption spectrometry, *Anal. Chim. Acta,* 69, 200, 1974.
129. **Childs, E. A. and Gaffke, J.,** Organic solvent extraction of lead and cadmium from aqueous solutions for atomic absorption spectrophotometric measurement, *J. Assoc. Off. Anal. Chem.,* 57, 360, 1974.
130. **Childs, E. A. and Gaffke, J.,** Possible interference in the measurement of lead and cadmium from elements found in fish muscle, *J. Assoc. Off. Anal. Chem.,* 57, 365, 1974.
131. **Holmes, C. W., Slade, E. A., and McLerran, C. J.,** Migration and redistribution of zinc and cadmium in marine esturine system, *Environ. Sci. Technol.,* 8, 255, 1974.
132. **Janssens, M. and Dams, R.,** Determination of cadmium in air particulates by flameless atomic absorption spectrometry with a graphite tube, *Anal. Chim. Acta,* 70, 25, 1974.
133. **Jung, P. D. and Clarke, D.,** Atomic absorption spectrometric determination of cadmium in fungicide formulations, *J. Assoc. Off. Anal. Chem.,* 57, 379, 1974.
134. **Langmyhr, F. J., Solberg, R., and Wold, L. T.,** Atomic absorption spectrometric determination of silver, bismuth and cadmium in sulfide ores by direct atomization from the solid state, *Anal. Chim. Acta,* 69, 267, 1974.
135. **Langmyhr, F. J., Thomassen, Y., and Massoumi, A.,** Atomic absorption spectrometric determination of copper, lead, cadmium and manganese in pulp and paper by the direct atomization technique, *Anal. Chim. Acta,* 68, 305, 1974.

136. **Lund, W. and Larsen, B. V.,** The application of electrodeposition techniques to flameless atomic absorption. I. The determination of cadmium with a tungsten filament, *Anal. Chim. Acta,* 70, 299, 1974.

137. **Rattonetti, A.,** Determination of soluble cadmium, lead, silver and indium in rain water and steam water with the use of flameless atomic absorption, *Anal. Chem.,* 46, 739, 1974.

138. **Ross, R. T. and Gonzalez, J. G.,** The direct determination of cadmium in biological samples by selective volatilization and graphite tube reservoir atomic absorption spectrometry, *Anal. Chim. Acta,* 70, 443, 1974.

139. **Langmyhr, F. J., Stubergh, J. R., Thomassen, Y., Hanssen, J. E., and Dolezal, J.,** Atomic absorption spectrometric determination of cadmium, lead, silver, thallium and zinc in silicate rocks by direct atomization from the solid state, *Anal. Chim. Acta,* 71, 35, 1974.

140. **Lund, W. and Larsen, B. V.,** Application of electrodeposition techniques to flameless atomic absorption spectrometry. Determination of cadmium in seawater, *Anal. Chim. Acta,* 72, 57, 1974.

141. **Mikhailova, T. P. and Slyadneva, N. A.,** Atomic absorption determination of zinc and cadmium in industrial solutions, *Zh. Anal. Khim.,* 29, 1912, 1974.

142. **Schumacher, E. and Umland, F.,** Improved fast destruction method for the determination of cadmium in body fluids using a graphite tube atomizer, *Z. Anal. Chem.,* 270, 285, 1974.

143. **Talmi, Y.,** Determination of zinc and cadmium in environmentally based samples by the radiofrequency spectrometric source, *Anal. Chem.,* 46, 1005, 1974.

144. **Ullucci, P. A. and Hwang, J. Y.,** Determination of cadmium in biological materials by atomic absorption, *Talanta,* 21, 745, 1974.

145. **Utsumi, S., Okutani, T., and Ozawa, T.,** Atomic absorption spectrophotometric determination of ultramicro amounts of cadmium by the extraction of cadmium iodide-zephir amine ion pairs, *Nippon Kagaku Kaishi,* 6, 1073, 1974.

146. **Yasuda, S. and Kakiyama, H.,** Determination of trace amounts of cadmium by flameless atomic absorption spectroscopy using a carbon rod atomizer, *Bunseki Kagaku,* 23, 406, 1974.

147. **Brovko, I. A., Nazarov, S. N., and Rish, M. A.,** Atomic absorption determination of zinc, cadmium, cobalt, copper and nickel after their extraction concentration in the di-phenyl carbazone-pyridine-toluene system, *Zh. Anal. Khim.,* 29, 2387, 1974.

148. **Topping, J. J. and MacCrehan, W. A.,** Preconcentration and determination of cadmium in water by reversed-phase column chromatography and atomic absorption, *Talanta,* 21, 1281, 1974.

149. **Woidich, H. and Pfannhauser, W.,** Determination of cadmium in foods, *Z. Lebensm. Unters. Forsch.,* 155, 72, 1974.

150. **Oelschlaeger, W. and Bestenlehner, L.,** Determination of cadmium in biological and other material by atomic absorption spectrophotometry. I. Possibilities of errors and their elimination during ashing and preparation of the analytical solutions, *Landwirtsch. Forsch.,* 27, 62, 1974 (in German).

151. **Oelschlaeger, W. and Buehler, E.,** Determination of cadmium in biological and other material by atomic absorption spectrophotometry. II. Possibilities of errors and their elimination during atomic absorption spectrophotometry, *Landwirtsch. Forsch.,* 27, 70, 1974.

152. **Stoeppler, M., Backhaus, F., Dahl, R., Dumont, M., Hagedorn-Goetz, H., Hilpert, K., Klahre, P., Rutzel, H., Valenta, P., and Nurnberg, H. W.,** Determination of Lead and Cadmium in Biological Matrices, in Proc. Int. Symp. on Recent Advances in the Assessment of the Health Effects of Environmental Pollution, Paris, France, June 24 to 28, 1974.

153. **Belyaev, Y. I., Oreshkin, V. N., and Vnukovskaya, G. L.,** Atomic absorption determination of element traces in rocks by using pulse thermal atomization of solid samples. III. Suppression of the nonresonance absorption in determination of cadmium, silver and thallium, *Zh. Anal. Khim.,* 30, 503, 1975.

154. **Beyer, M. E. and Bond, A. M.,** Simultaneous determination of cadmium, copper, lead and zinc in lead and zinc concentrates by a.c. polarographic methods. Comparison with atomic absorption spectrometry, *Anal. Chim. Acta,* 75, 409, 1975.

155. **Dewit, A., Duwijn, R., Smeyers-Verbeke, J., and Massart, D. L.,** Ion-association extraction combined with low temperature ashing for the determination of cadmium in foodstuffs, *Bull. Soc. Chim. Belg.,* 84, 91, 1975.

156. **Evenson, M. A. and Anderson, C. T., Jr.,** Ultramicro analysis for copper, cadmium and zinc in human liver tissue by use of atomic absorption spectrophotometry and the heated graphite tube atomizer, *Clin. Chem. (Winston-Salem, N.C.),* 21, 537, 1975.

157. **Lauwerys, R., Buchet, J. P., Roels, H., Berlin, A., and Smeats, J.,** Intercomparison program of lead, mercury and cadmium analysis in blood, urine and aqueous solutions, *Clin. Chem. (Winston-Salem, N.C.),* 21, 551, 1975.

158. **Lockwood, T. H. and Limtiaco, L. P.,** Determination of beryllium, cadmium and tellurium in animal tissues using electronically excited oxygen and atomic absorption spectrophotometry, *Am. Ind. Hyg. Assoc. J.,* 36, 57, 1975.

159. **Noguchi, C., Hirayama, H., Shige, T., Jinbo, M., Takahashi, T., Tsuji, K., Nakasato, S., Matsubara, S., Murase, Y., Murui, T., Yamashita, T., and Yishida, J.,** Determination of microamounts of cadmium, copper, nickel and manganese in fats and oils by atomic absorption spectrophotometry, *Yukagaku,* 24, 100, 1975.

160. **Perry, E. F., Koirtyohann, S. R., and Perry, H. M., Jr.,** Determination of cadmium in blood and urine by graphite furnace atomic absorption spectrophotometry, *Clin. Chem. (Winston-Salem, N.C.),* 21, 626, 1975.

161. **Posma, F. D., Balke, J., Herber, R. P. M., and Stuik, E. J.,** Microdetermination of cadmium and lead in whole blood by flameless atomic absorption spectrometry using carbon-tube and carbon-cup as sample cell and comparison with flame studies, *Anal. Chem.,* 47, 834, 1975.

162. **Sperling, K. R.,** Heavy metal determination in seawater and in marine organisms with the aid of flameless AAS. II. Determination of cadmium in biological material, *At. Absorpt. Newsl.,* 14, 60, 1975.

163. **Tominaga, M., Kimura, A., Miyazaki, A., and Umezaki, Y.,** Determination of cadmium by flameless atomic absorption spectrometry using a heated graphite atomizer, *Bunseki Kagaku,* 24, 61, 1975 (in Japanese).

164. **Aihara, M. and Kiboku, M.,** Atomic absorption spectrophotometry of cadmium and copper using solvent extraction with potassium ethyl xanthate-methyl isobutyl ketone, *Bunseki Kagaku,* 24, 447, 1975.

165. **Blood, E. R. and Grant, G. C.,** Determination of cadmium in fish tissue by flamelessc absorption with a tantalum ribbon, *Anal. Chem.,* 47, 1438, 1975.

166. **Briese, L. A. and Giesy, J. P.,** Determination of lead and cadmium associated with naturally occurring organics extracted from surface waters using flameless atomic absorption, *At. Absorpt. Newsl.,* 14, 133, 1975.

167. **Cernik, A. A. and Sayers, M. H.,** Application of blood cadmium determination to industry using a punched disc technique, *Br. J. Ind. Med.,* 32, 155, 1975.

168. **Igarashi, S.,** Determination of cadmium in dust, *Bunseki Kagaku,* 24, 270, 1975 (in Japanese).

169. **Jackson, K. W. and Mitchell, D. G.,** Rapid determination of cadmium in biological tissues by micro-sampling cup atomic absorption spectrometry, *Anal. Chim. Acta,* 80, 39, 1975.

170. **Kirk, M., Perry, E. G., and Arritt, J. M.,** Separation and atomic absorption measurement of trace amounts of lead, silver, zinc, bismuth and cadmium in high nickel alloys, *Anal. Chim. Acta,* 80, 163, 1975.

171. **Kjellstrom, T., Lind, B., Linnmann, L., and Elinder, C. G.,** Variation of cadmium concentration in Swedish wheat and barley, *Arch. Environ. Health,* 31, 321, 1975.

172. **Kono, T. and Nemori, A.,** Extraction of copper, cadmium, lead, silver and bismuth with iodide methyl isobutyl ketone in atomic absorption, *Bunseki Kagaku,* 24, 419, 1975.

173. **Korkisch, J. and Sorio, A.,** Determination of cadmium, copper and lead in natural waters after anion exchange separation, *Anal. Chim. Acta,* 76, 393, 1975.

174. **Krinitz, B.,** Photosensitivity as extraction factor ion test for cadmium in ceramic glazes, *Bull. Am. Ceram. Soc.,* 54, 500, 1975.

175. **Lau, O. W. and Li, K. L.,** Determination of lead and cadmium in paint by atomic absorption spectro-photometry by the Delves microsampling technique, *Analyst (London),* 100, 430, 1975.

176. **Luzar, O. and Sliva, V.,** Determination of cadmium oxide, magnesium oxide, alumina, silica, iron, copper, zinc, lead, cadmium, sodium oxide and potassium oxide in iron ores and agglomerates, *Hutn. Listy,* 30, 55, 1975 (in Czech).

177. **Machata, G.,** The normal level of cadmium, copper and zinc in blood of the Viennese population, *Wien. Klin. Wochensch.,* 87, 484, 1975 (in German).

178. **Musha, S. and Takahashi, T.,** Enrichment of trace amounts of metals utilizing the coagulation of soybean protein. IV. Enrichment of trace cadmium by soybean protein for the atomic absorption analysis, *Bunseki Kagaku,* 24, 540, 1975.

179. **Robbins, W. K. and Walker, H. H.,** Analysis of petroleum for trace metals. Determination of trace quantities of cadmium in petroleum by atomic absorption spectrometry, *Anal. Chem.,* 47, 1269, 1975.

180. **Windemann, H. and Mueller, U.,** Determination of cadmium in tobacco by atomic absorption spectro-photometry, *Mitt. Geb. Lebensmittelunters. Hyg.,* 66, 64, 1975 (in German).

181. **Yamamoto, Y., Kumamaru, T., Kamada, T., Tanaka, T., and Kawabe, M.,** Determination of the ppb level of cadmium, lead and copper in water by a carbon-tube flameless atomic absorption spectropho-tometry combined with dithizone-carbon tetrachloride extraction, *Nippon Kagaku Kaishi,* 5, 836, 1975.

182. **Barnett, W. B. and McLaughlin, E. A.,** The atomic absorption determination of antimony, arsenic, bismuth, cadmium, lead and tin in iron, copper and zinc alloys with the graphite furnace, *Anal. Chim. Acta,* 80, 285, 1975.

183. **Hunt, D. C.,** Screening technique for the presence of lead and cadmium in solid samples by atomic absorption spectrophotometry, *Lab. Pract.,* 24, 411, 1975.

184. **Stoeppler, M., Kuppers, G., Matthes, W., Rutzel, H., and Valenta, P.,** On the accuracy of lead and cadmium determinations in human blood, Int. Conf. on Heavy Metals in the Environment, Toronto, Ont., October 27 to 31, 1975.

185. **Yamamoto, Y., Kumamaru, T., Kamada, T., and Tanaka, T.,** Determination of part-per-10^9 levels of cadmium, lead and copper in water by carbon tube flameless atomic absorption spectrophotometry combined with ammonium pyrrolidine-1-carbodithioate-isobutyl methyl ketone extraction, *Eisei Kagaku*, 21, 71, 1975 (in Japanese).

186. **Vondenhoff, T.,** Determination of lead, cadmium, copper and zinc in plant and animal material by atomic absorption in a flame and in a graphite tube after sample decomposition by the Schoniger technique, *Mitt. GDCh. Fachgruppe Lebensmittelchemie Gerichtl. Chem.*, 29, 341, 1975 (in German).

187. **Ranfft, K.,** Determination of lead and cadmium in feeding stuffs by atomic absorption spectroscopy, *Landwirtsch. Forsch.*, 31, 135, 1975 (in German).

188. **Rudnevskii, N. E., Demarin, V. T., Sklemina, L. V., and Tumanova, A. N.,** Influence of some organic solvents on sensitivity of atomic absorption determination of iron, cadmium, copper, magnesium, sodium, vanadium, selenium and zinc, *Tr. Khim. Khim. Tekhnol.*, 1, 106, 1975.

189. **Ooghe, W. and Kastelijan, H.,** Determination of traces of lead and cadmium in apertifs by flameless atomic absorption spectrophotometry, *Ann. Falsif. Expert. Chim.*, 69, 351, 1976 (in French).

190. **Sighinolfi, G. P. and Santos, A. M.,** Determination of cadmium in silicate materials by flameless atomic absorption spectroscopy, *Mikrochim. Acta*, 1, 447, 1976.

191. **Tsutsumi, S., Koizumi, H., and Yoshikawa, S.,** Atomic absorption spectrophotometric determination of lead, cadmium and copper in foods by simultaneous extraction of the iodides with methyl isobutyl ketone, *Bunseki Kagaku*, 25, 150, 1976.

192. **Zaroogian, G. E. and Cheer, S.,** Accumulation of cadmium by the American oyster, *Crassostrea Virginica*, *Nature (London)*, 261, 408, 1976.

193. **Zook, E. G., Powell, J. J., Hackley, B. M., Emerson, J. A., Brooker, J. R., and Knobl, G. M., Jr.,** National Marine Fisheries Service preliminary survey of selected seafoods for mercury, lead, cadmium, chromium and arsenic content, *J. Agric. Food Chem.*, 24, 47, 1976.

194. **Gong, H. and Suhr, N. H.,** The determination of cadmium in geological materials by flameless atomic absorption spectrometry, *Anal. Chim. Acta*, 81, 297, 1976.

195. **Krinitz, B. and Holak, W.,** Collaborative study of effect of light on cadmium and lead leaching from ceramic glazes, *J. Assoc. Off. Anal. Chem.*, 59, 158, 1976.

196. **Lund, W., Larsen, B. V., and Gundersen, N.,** The application of electrodeposition techniques to flameless atomic absorption spectrometry. III. The determination of cadmium in urine, *Anal. Chim. Acta*, 81, 319, 1976.

197. **Lundren, G.,** Direct determination of cadmium in blood with a temperature controlled heated graphite tube atomizer, *Talanta*, 23, 309, 1976.

198. **Magyar, B. and Wechsler, P.,** Application of flameless atomic absorption to examination of complexes. II. Solubility, stability and distribution of cadmium oxinates, *Talanta*, 23, 95, 1976.

199. **Morris, N. M., Clarke, M. A., Tripp, V. W., and Carpenter, F. G.,** Determination of lead, cadmium and zinc in sugar, *J. Agric. Food Chem.*, 24, 45, 1976.

200. **Kovats, A. and Bohm, B.,** Urinary cadmium determination by flameless atomic absorption spectrophotometry with atomization in a graphite furnace, *Stud. Cercet. Biochim.*, 19, 125, 1976.

201. **Oleru, U. G.,** Kidney, liver, hair and lungs as indicators of cadmium absorption, *Am. Ind. Hyg. Assoc. J.*, 37, 617, 1976.

202. **Ohzeki, K., Abe, M., and Kambara, T.,** Extraction-atomic absorption method for the determination of trace amounts of cadmium with tetrabutyl ammonium iodide, *Bunseki Kagaku*, 25, 333, 1976.

203. **Roberts, A. H. C., Turner, M. A., and Syers, J. K.,** Simultaneous extraction and concentration of cadmium and zinc from soil extracts, *Analyst (London)*, 101, 574, 1976.

204. **Satake, M., Asano, T., Takagi, Y., and Yonekubo, T.,** Determination of copper, zinc, lead, cadmium and manganese in brackish and coastal waters by the combination of chelating ion exchange separation and atomic absorption spectrophotometry, *Nippon Kagaku Kaishi*, 5, 762, 1976.

205. **Solomon, R. L. and Hartford, J. W.,** Lead and cadmium in dusts and soils in a small urban community, *Environ. Sci. Technol.*, 10, 773, 1976.

206. **Beaufays, J. M. and Nangniot, P.,** Comparative study of the determination of cadmium in water, fertilizers and plants using differential pulse polarography and atomic absorption spectrometry, *Analusis*, 4, 193, 1976.

207. **Boline, D. R. and Schrenk, W. G.,** A method for the determination of cadmium in plant material by atomic absorption spectroscopy, *Appl. Spectrosc.*, 30, 607, 1976.

208. **Chambers, J. C. and McClellan, B. E.,** Enhancement of atomic absorption sensitivity for copper, cadmium, antimony, arsenic and selenium by means of solvent extraction, *Anal. Chem.*, 48, 2061, 1976.

209. **Joensson, H.,** Determination of lead and cadmium in milk with modern analytical methods, *Lebensm. Unters. Forsch.,* 160, 1, 1976.

210. **Kerfoot, W. B. and Jacobs, S. A.,** Cadmium occurred in combined wastewater treatment-aquaculture system, *Environ. Sci. Technol.,* 10, 662, 1976.

211. **Koizumi, H. and Yasuda, K.,** Determination of lead, cadmium and zinc using the Zeeman effect in atomic absorption spectrometry, *Anal. Chem.,* 48, 1178, 1976.

212. **Schuller, P. L. and Egan, H.,** Cadmium, Lead, Mercury and Methylmercury Compounds: Review of Methods of Trace Analysis and Sampling with Special Reference to Food, Rep. No. 92-5-I-0094-M-84, Food and Agriculture Organization, Rome, 1976.

213. **Steinnes, E.,** Determination of cadmium in sulfide ores by atomic absorption spectrophotometry after anion exchange separation, *At. Absorpt. Newsl.,* 15, 102, 1976.

214. **Delage, C., Oudart, N., and Guichard, C.,** Determination of cadmium by atomic absorption spectrophotometry with regard to toxicology: determination in food and tobacco, *Ann. Pharm. Fr.,* 34, 315, 1976.

215. **Hirata, S. and Takimura, O.,** Determination of trace amounts of cadmium in suspended substances in sea water by atomic absorption spectrophotometry using a carbon tube atomizer, *Bunseki Kagaku,* 25, 760, 1976.

216. **Ure, A. M. and Mitchell, M. C.,** The determination of cadmium in plant material and soil extracts by solvent extraction and atomic absorption with a carbon rod atomizer, *Anal. Chim. Acta,* 87, 283, 1976.

217. **Horak, O.,** Determination by flameless atomic absorption spectroscopy of lead and cadmium in grain and grass samples taken at different distances from motorways, *Landwirtsch. Forsch.,* 29, 289, 1976.

218. **O'Laughlin, J. W., Hemphill, D. D., and Pierce, J. O.,** *Analytical Methodology for Cadmium in Biological Matter. A Critical Review,* International Lead Zinc Research Organization, New York, 1976.

219. **Wunderlich, E. and Hadeler, W.,** Determination of traces of zinc, cadmium and bismuth in pure copper by AAS, *Z. Anal. Chem.,* 281, 300, 1976.

220. **Bea-Barredo, F., Polo-Polo, C., and Polo-Diez, L.,** The simultaneous determination of gold, silver and cadmium at ppb levels in silicate rocks by atomic absorption spectrometry with electrothermal atomization, *Anal. Chim. Acta,* 94, 283, 1976.

221. **Brodie, K. G. and Stevens, B.,** Measurement of whole blood lead and cadmium at low levels using an automatic sample dispenser and furnace atomic absorption, *J. Anal. Toxicol.,* 1, 282, 1977.

222. **Cohen, E. S. and Kurchatova, G.,** Determination of lead and cadmium traces in atmospheric air by flame and flameless atomic absorption, *Dokl. Bolg. Akad. Nauk,* 30, 1439, 1977.

223. **Del-Castino, P. and Herber, R. F.,** The rapid determination of cadmium, lead, copper and zinc in whole blood by atomic absorption spectrometry with electrothermal atomization. Improvements in precision with a peak-shape monitoring device, *Anal. Chim. Acta,* 94, 269, 1977.

224. **Armannsson, H.,** The use of dithizone extraction and atomic absorption spectrometry for the determination of cadmium, zinc, copper, nickel and cobalt in rocks and sediments, *Anal. Chim. Acta,* 88, 89, 1977.

225. **Bower, N. W. and Ingle, J. D.,** Precision of flame atomic absorption measurements of arsenic, cadmium, calcium, copper, iron, magnesium, molybdenum, sodium and zinc, *Anal. Chem.,* 49, 574, 1977.

226. **Delves, H. T.,** A simple matrix modification procedure to allow the direct determination of cadmium in blood by flame microsampling atomic absorption spectrophotometry, *Analyst (London),* 102, 403, 1977.

227. **Eller, P. M. and Haartz, J. G.,** A study of methods for the determination of lead and cadmium, *J. Am. Ind. Hyg. Assoc.,* 38, 116, 1977.

228. **Florence, T. M. and Batley, G. E.,** Determination of the chemical forms of trace metals in natural waters with special reference to copper, lead, cadmium and zinc, *Talanta,* 24, 151, 1977.

229. **Hinesly, T. D., Jones, R. L., Ziegler, E. L., and Tyler, J. J.,** Effects of annual and accumulative applications of sewage sludge on assimilation of zinc and cadmium by corn (*Zea mays* L.), *Environ. Sci. Technol.,* 11, 182, 1977.

230. **Lekehal, N., Hanocq, M., and Helson-Cambier, M.,** Determination of cadmium traces by atomic absorption after extraction with dipivaloyl methane. Possible application to the study of cadmium in urine, *J. Pharm. Belg.,* 32, 76, 1977.

231. **Mitchell, D. G., Mills, W. N., Ward, A. F., and Aldous, K. M.,** Determination of cadmium, copper and lead in sludges by microsampling cup atomic absorption spectrometry in a nitrous oxide-acetylene flame, *Anal. Chim. Acta,* 90, 275, 1977.

232. **Pakalns, P. and Farrar, Y. J.,** The effect of surfactants on the extraction-atomic absorption spectrophotometric determination of copper, iron, manganese, lead, nickel, zinc, cadmium and cobalt, *Water Res.,* 11, 145, 1977.

233. **Sperling, K. R., Bahr, B., and Kremling, K.,** Heavy metal determination in sea water and in marine organisms with the aid of flameless AAS. IV. Description of a routine method for the determination of cadmium in small samples of biological material, *Z. Lebensm. Unters. Forsch.,* 163, 87, 1977.

234. **Thompson, K. C., Wagstaff, K., and Wheatstone, K. C.,** Method for the minimization of matrix interferences in the determination of lead and cadmium in non-saline waters by atomic absorption spectrophotometry using electrothermal atomization, *Analyst (London)*, 102, 310, 1977.

235. **Vesterberg, O. and Bergstrom, T.,** Determination of cadmium in blood by use of atomic absorption spectroscopy with crucibles — and a rational procedure for dry ashing, *Clin. Chem. (Winston-Salem, N.C.)*, 23, 555, 1977.

236. **Wegscheider, W., Knapp, G., and Spitzy, H.,** Statistical investigations of interferences in graphite furnace atomic absorption spectroscopy. II. Cadmium, *Z. Anal. Chem.*, 283, 97, 1977.

237. **Woodis, T. C., Jr., Hunter, G. B., and Johnson, F. J.,** Statistical studies of matrix effects on the determination of cadmium and lead in fertilizer and plant tissue by flameless atomic absorption spectrometry, *Anal. Chim. Acta*, 90, 127, 1977.

238. **Yasuda, S. and Kakiyama, H.,** Vaporization of atoms and molecules during heating of cadmium, lead and zinc salts in a carbon tube atomizer, *Anal. Chim. Acta*, 89, 369, 1977.

239. **Boyle, E. A. and Edmond, J. M.,** Determination of copper, nickel, and cadmium in sea water by APDC chelate coprecipitation and flameless atomic absorption spectrometry, *Anal. Chim. Acta*, 91, 189, 1977.

240. **Campbell, W. C. and Ottaway, J. M.,** Direct determination of cadmium and zinc in seawater by carbon furnace atomic absorption spectrometry, *Analyst (London)*, 102, 495, 1977.

241. **Capar, S. G.,** Atomic absorption spectrophotometric determination of lead, cadmium, zinc and copper in clams and oysters: collaborative study, *J. Assoc. Off. Anal. Chem.*, 60, 1400, 1977.

242. **Inoue, S., Yotsuyanagi, T., Sasaki, M., and Aomura, K.,** Atomic absorption method for the determination of trace amounts of cadmium, copper, lead and zinc by ion-pair extraction of dimercaptomaleonitrile complexes, *Bunseki Kagaku*, 26, 550, 1977.

243. **Lester, J. N., Harrison, R. M., and Perry, R.,** Rapid flameless atomic absorption analysis of the metallic content of sewage sludges. I. Lead, cadmium and copper, *Sci. Total. Environ.*, 8, 153, 1977.

244. **Manoliu, C. and Tomi, B.,** The determination of lithium, cadmium and copper in aluminum alloys by atomic absorption spectrophotometry, *Rev. Chim.*, 28, 370, 1977.

245. **Meisch, H. U. and Reinle, W.,** Direct determination of cadmium in unicellular green algae by flameless atomic absorption, *Mikrochim. Acta*, 1, 505, 1977.

246. **Menden, E. E., Brockman, D., Choudhury, H., and Petering, H. G.,** Dry ashing of animal tissues for atomic absorption spectrometric determination of zinc, copper, cadmium, lead, iron, manganese, magnesium and calcium, *Anal. Chem.*, 49, 1644, 1977.

247. **Sperling, K. R.,** Determination of heavy metals in seawater and marine organisms by flameless atomic absorption spectrophotometry. VI. Cadmium determination in culture waters from toxicological experiments with marine organisms, *Z. Anal. Chem.*, 287, 23, 1977.

248. **Zawadzke, H., Baralkiewiez, D., and Elbanowska, H.,** Flame atomic absorption spectrometric determination of cobalt, cadmium, lead, nickel, copper and zinc in natural waters, *Chem. Anal.*, 22, 913, 1977.

249. **Pellerin, F. and Goule, J. P.,** Detection and rapid determination of cadmium, copper, lead and zinc in dyes and antioxidants authorized for use in drugs and foodstuffs, *Ann. Pharm. Fr.*, 35, 189, 1977.

250. **Barlow, P. J.,** Microdetermination of lead and cadmium in pasteurized market milks by flameless atomic absorption spectroscopy using a base digest, *J. Dairy Res.*, 44, 377, 1977.

251. **Nekehal, N., Hanocq, M., and Helson-Cambier, H.,** Determination of traces of cadmium by atomic absorption after extraction with dipivaloylmethane. Determination of cadmium in urine, *J. Pharm. Belg.*, 32, 76, 1977.

252. **Parolari, G. and Pezzani, G.,** Determination of cadmium in foods, *Ind. Conserve*, 52, 130, 1977 (in Italian).

253. **Geladi, P. and Adams, F.,** The determination of cadmium, copper, iron, lead and zinc in aerosols by atomic absorption spectrometry, *Anal. Chim. Acta*, 96, 229, 1978.

254. **Glaeser, E.,** Atomic absorption spectroscopic determination of iron, copper, zinc, cadmium, manganese and cobalt in waters by the "injection method" after an extractive concentration, *Acta Hydrochim. Hydrobiol.*, 6, 83, 1978.

255. **Hudnik, V., Gomiscek, S., and Gorenc, B.,** The determination of trace metals in mineral waters. I. Atomic absorption spectrometric determination of cadmium, cobalt, chromium, copper, nickel and lead by a electrothermal atomization after concentration by coprecipitation, *Anal. Chim. Acta*, 98, 39, 1978.

256. **Ishino, F., Matsumae, H., Shibata, K., Ariga, N., and Goshima, F.,** Determination of copper, cadmium and zinc in bone by atomic absorption spectrophotometry, *Bunseki Kagaku*, 27, 232, 1978.

257. **Ryabinin, A. I. and Lazareva, E. A.,** Solvent extraction-atomic absorption determination of copper, silver and cadmium in Black Sea water, *Zh. Anal. Khim.*, 33, 298, 1978.

258. **Vesterberg, O. and Wrangskogh, K.,** Determination of cadmium in urine by graphite furnace atomic absorption spectroscopy, *Clin. Chem. (Winston-Salem, N.C.)*, 24, 681, 1978.

259. **Watling, R. J.,** The use of a slotted tube for the determination of lead, zinc, cadmium, bismuth, manganese and silver by atomic absorption spectrometry, *Anal. Chim. Acta*, 97, 395, 1978.

260. **Yamaguchi, K., Okumura, I., and Deguchi, M.,** Atomic absorption spectrophotometric determination of cadmium in tin metal using solvent extraction with thiothenoyl trifluoroacetone and *o*-phenanthroline, *Bunseki Kagaku,* 27, 125, 1978.

261. **Arpadjan, S. and Kachov, I.,** Atomic absorption determination of cadmium, chromium and zinc in blood serum, *Zentralbl. Pharm. Pharmakother. Laboratoriumsdiagn.,* 117, 237, 1978 (in German).

262. **Boiteau, H. L. and Metayer, C.,** Microdetermination of lead, cadmium, zinc and tin in biological materials by atomic absorption spectrometry after mineralization and extraction, *Analusis,* 6, 350, 1978.

263. **Elson, C. M., Dostal, J., Hynes, D. L., and de Albuquerque, C. A. R.,** Silver, cadmium and lead contents of some rock reference samples, *Geostandards Newsl.,* 11, 121, 1978.

264. **Evans, W. H., Read, J. I., and Lucas, B. E.,** Evaluation of a method for the determination of total cadmium, lead and nickel in foodstuffs using measurement by flame atomic absorption spectrophotometry, *Analyst (London),* 103, 580, 1978.

265. **Hirata, S., Marushita, K., and Takimura, O.,** Determination of cadmium in sediments by atomic absorption spectrophotometry with a carbon tube atomizer, *Bunseki Kagaku,* 27, 543, 1978.

266. **Hoenig, M. and Vanderstappen, R.,** Determination of cadmium, copper, lead, zinc and manganese in plants by flame atomic absorption spectroscopy. Mineralization effects, *Analusis,* 6, 312, 1978.

267. **Kitagawa, K., Shigeyasu, T., and Takeuchi, T.,** Application of the Faraday effect to the trace determination of cadmium by atomic spectroscopy with an electrothermal atomizer, *Analyst (London),* 103, 1021, 1978.

268. **Krinitz, B.,** Rapid screening field test for detecting cadmium and lead extracted from glazed ceramic dinnerware: collaborative study, *J. Assoc. Off. Anal. Chem.,* 61, 1124, 1978.

269. **Langmyhr, F. J. and Kjuns, I.,** Direct atomic absorption spectrometric determination of cadmium, lead and manganese in bone and of lead in ivory, *Anal. Chim. Acta,* 100, 139, 1978.

270. **Page, A. G., Godbole, S. V., Deshkar, S. B., and Joshi, B. D.,** Determination of cadmium and lithium in uranium oxide (U_3O_8) powder by AAS with carbon cup atomization, *Anal. Lett.,* A11, 619, 1978.

271. **Raptis, S. and Muller, K.,** A new method for determination of cadmium in plasma, *Clin. Chim. Acta,* 88, 393, 1978 (in German).

272. **Robbins, W. B. and Caruso, J. A.,** Determination of lead and cadmium in normal human lung tissue by flameless atomic absorption spectroscopy, *Spectrosc. Lett.,* 11, 333, 1978.

273. **Robinson, J. W. and Weiss, S.,** Direct determination of cadmium in whole blood using an RF-heated carbon-bed atomizer for atomic absorption spectroscopy, *Spectrosc. Lett.,* 11, 715, 1978.

274. **Sperling, K. R.,** Determination of heavy metals in seawater and in marine organisms by flameless atomic absorption spectrophotometry. VII. Simple extraction method for determination of extremely low cadmium levels in small seawater samples, *Anal. Chem.,* 292, 113, 1978.

275. **Tanaka, T., Hayashi, Y., and Ishizawa, M.,** Simultaneous determination of cadmium and copper in water by a graphite furnace dual diannel atomic absorption spectrophotometry, *Bunseki Kagaku,* 27, 499, 1978.

276. **Viets, J. G.,** Determination of silver, bismuth, cadmium, copper, lead and zinc in geologic materials by atomic absorption spectrometry with tricaprylyl methyl ammonium chloride, *Anal. Chem.,* 50, 1097, 1978.

277. **Feinberg, M. and Ducauze, C.,** Mineralization of lead and cadmium in animal tissue with regard to their determination by atomic absorption spectrometry, *Bull. Soc. Chim. Fr.,* 11—12, 419, 1978.

278. **Kantor, T., Fodor, P., and Pungor, E.,** Determination of traces of lead, cadmium and zinc in copper by an arc-nebulization and flame atomic absorption technique, *Anal. Chim. Acta,* 102, 15, 1978.

279. **Lewandoska, I.,** Determination of cadmium in plastic utensils, *Rocz. Panstw. Zakl. Hig.,* 29, 295, 1978 (in Polish).

280. **Marushita, K. and Takimura, O.,** Determination of cadmium in marine sediments by atomic absorption spectrometry with a carbon tube atomizer, *Bunseki Kagaku,* 27, 543, 1978.

281. **Viala, A., Cano, J. P., Gouezo, F., Sauve, J. M., Bourbon, P., and Mallet, B.,** A methodical approach for determining lead, cadmium and chromium in atmospheric dusts by flameless atomic absorption spectrometry, *Pollut. Atm.,* 20, 115, 1978 (in French).

282. **Nagahiro, T., Fujino, O., Matsui, M., and Shigematsu, T.,** Determination of cadmium in individual organs and divided shells of sea water clam by atomic absorption spectrometry with a carbon tube atomizer, *Bull. Inst. Chem. Res. Kyoto Univ.,* 56, 274, 1978.

283. **Pilipyuk, V. S., Pyatnitskii, I. V., and Kovalenko, A. A.,** Atomic absorption determination of zinc and cadmium in chloroform extracts of their amine complexes with cinnamic acid, *Ukr. Khim. Zh.,* 44, 1206, 1978 (in Russian).

284. **Schuetze, I. and Mueller, W.,** Determination of trace metals in dietary fats and emulsifiers by flameless atomic absorption spectrometry. I. Determination of bound heavy metals, copper, iron, nickel, zinc, lead and cadmium in dietary fats, *Nahrung,* 22, 777, 1978.

285. **Holm, J.,** Simplified digestion method and measuring technique for determining lead, cadmium and arsenic in animal tissues by atomic absorption spectrometry, *Fleischwirtschaft,* 58, 745, 1978.

286. **Koops, J. and Westerbeck, D.,** Determination of lead and cadmium in pasturized liquid milk by flameless atomic absorption spectrophotometry, *Neth. Milk Dairy J.,* 32, 149, 1978.

287. **Allain, P. and Mauras, Y.,** Microdetermination of lead and cadmium in blood and urine by graphite furnace atomic absorption, *Clin. Chim. Acta,* 91, 41, 1978.

288. **Bruland, K. W., Franks, R. P., Knauer, G. A., and Martin, J. H.,** Sampling and analytical methods for the determination of copper, cadmium, zinc and nickel at the nanogram per liter level in sea water, *Anal. Chim. Acta,* 105, 233, 1979.

289. **Heinrichs, H.,** Determination of bismuth, cadmium and thallium in 33 international standard reference rocks by fractional distillation combined with flameless atomic absorption spectrometry, *Z. Anal. Chem.,* 294, 345, 1979.

290. **Hoenig, M., Vanderstappen, R., and Van Hoeyweghen, P.,** Contamination with cadmium from micropipette tips, *Analusis,* 7, 17, 1979.

291. **Salmela, S. and Vuori, E.,** Contamination with cadmium from micropipette tips, *Talanta,* 26, 81, 1979.

292. **Iu, K. L., Pulford, I. D., and Duncan, H. J.,** Determination of cadmium, cobalt, nickel and lead in soil extracts by dithizone extraction and atomic absorption spectrometry with electrothermal atomization, *Anal. Chim. Acta,* 106, 319, 1979.

293. **Meranger, J. C., Subramanian, K. C., and Chalifoux, C.,** A natural survey for cadmium, chromium, copper, lead, zinc, calcium and magnesium in Canadian drinking water supplies, *Environ. Sci. Technol.,* 13, 707, 1979.

294. **Narasaki, H.,** Determination of cadmium in polished rice by low-temperature ashing and atomic absorption spectrometry, *Anal. Chim. Acta,* 104, 393, 1979.

295. **Patel, B. M., Bhatt, P. M., Gupta, N., Pawar, M. M., and Joshi, B. D.,** Electrothermal atomic absorption spectrometric determination of cadmium, chromium and cobalt in uranium without preliminary separation, *Anal. Chim. Acta,* 104, 113, 1979.

296. **Schmidt, W. and Dietl, F.,** Determination of cadmium in digested soils and sediments and extracts using flameless atomic absorption with zirconium coated graphite tubes, *Z. Anal. Chem.,* 295, 110, 1979.

297. **Weisel, C. P., Fasching, J. L., Piotrowicz, S. R., and Duce, R. A.,** A modified standard addition method for determining cadmium, lead, copper and iron in sea water derived samples by atomic absorption spectroscopy, *Adv. Chem. Ser.,* 172, 134, 1979.

298. **Armannsson, J.,** Dithizone and flame atomic absorption spectrometry for the determination of cadmium, zinc, lead, copper, nickel, cobalt and silver in sea water and biological tissues, *Anal. Chim. Acta,* 110, 21, 1979.

299. **Baechmann, K., Zikos, C., and Spachidis, C.,** Determination of cadmium in high-purity aluminum by flameless atomic absorption spectrometry after volatilization, *Z. Anal. Chem.,* 296, 374, 1979.

300. **Bengtasson, M., Danielsson, L. G., and Magnusson, B.,** Determination of cadmium and lead in sea water after extraction using electrothermal atomization. Minimization of interferences from coextracted sea salts, *Anal. Lett.,* 12, 1367, 1979.

301. **Berndt, H. and Messerschmidt, J.,** A microanalytical method using flame atomic absorption and emission spectrome (platinum loop method). I. Basic method and application to the determination of lead and cadmium in drinking water, *Spectrochim. Acta,* 34b, 241, 1979.

302. **Bozsai, G. and Csanady, M.,** Systematic investigations on the heavy metal pollution (cadmium, lead, copper, zinc, chromium and barium) of drinking water using atomic absorption spectrophotometric methods, *Z. Anal. Chem.,* 297, 370, 1979.

303. **Carmack, G. D. and Evenson, M. A.,** Determination of cadmium in urine by electrothermal atomic absorption spectrometry, *Anal. Chem.,* 51, 907, 1979.

304. **Dabeka, R. W.,** Graphite furnace atomic absorption spectrometric determination of lead and cadmium in foods after solvent extraction and stripping, *Anal. Chem.,* 51, 902, 1979.

305. **Forrester, J. E., Lehecka, V., Johnston, J. R., and Ott, W. L.,** Direct determination of trace quantities of antimony, arsenic, bismuth, cadmium, lead, selenium, silver, tellurium and thallium in high-purity nickel by electrothermal atomic absorption spectrometry, *At. Absorpt. Newsl.,* 18, 73, 1979.

306. **Gardiner, P. E., Ottaway, J. M., and Fell, G. S.,** Accuracy of the direct determination of cadmium in urine by carbon furnace atomic absorption spectrometry, *Talanta,* 26, 841, 1979.

307. **Grainger, F. and Gale, I. G.,** Direct analysis of solid cadmium mercury telluride by flameless atomic absorption using interactive computer processing, *J. Mater. Sci.,* 14, 1370, 1979.

308. **Kitagawa, K., Koyama, T., and Takeuchi, T.,** Correction system for spectroscopic determination of trace amounts of cadmium using the atomic faraday effect with electrothermal atomization, *Analyst (London),* 104, 822, 1979.

309. **Langmyhr, F. J., Eyde, B., and Jansen, J.,** Determination of the total content and distribution of cadmium, copper and zinc in human parotid saliva, *Anal. Chim. Acta,* 107, 211, 1979.

310. **Oreshkin, V. N., Belyaev, Y. I., Vnukovskaya, G. L., and Tatzii, Y. G.,** Determination of cadmium in soils by the method of a direct nonflame atomic absorption analysis, *Pochvovendenie,* 5, 109, 1979.

311. **Sefzik, E.,** Determination of arsenic, lead, cadmium, chromium and selenium with respect to the require-ments of drinking water regulations by means of flameless atomic absorption spectroscopy, *Vom Wasser,* 50, 285, 1979.

312. **Fonds, A. W., Kempf, T., Minderhoud, A., and Sonneborn, M.,** Heavy metal content of various kinds of water, *Reinhalt Wassers,* 78—88, 1979.

313. **Anonymous,** Determination of Trace Metals in Liquid Coke-Oven Effluents by Atomic Absorption Spec-trophotometry, Carbonization Res. Rep. 76, British Carbonization Research Association, Chesterfield, Derbyshire, England, 1979, 17.

314. **Ihida, M., Ishii, T., and Ohnishi, R.,** Precision on the determination of trace elements in coal, *Am. Chem. Soc. Div. Fuel. Chem.,* 24, 262, 1979.

315. **Boyer, K. W., Capar, S. G., Jones, J. W., Suddendorf, R. F., and Forwalter, J.,** Multielement/trace analysis identifies 48 elements in foods. Simultaneous element analysis and data reduction reaches parts per billion, *Food Process.,* 40, 72, 1979.

316. **Carmichael, N. G., Squibb, K. S., and Fowler, B. A.,** Metals in the molluscan kidney: a comparison of two closely related bivalve species (argopecten) using X-ray microanalysis and atomic absorption spec-troscopy, *J. Fish. Res. Board Can.,* 36, 1149, 1979.

317. **Skorka-Trybula, Z. and Kozinska, E.,** Electrolytic concentration of trace amounts of cadmium prior to determination by atomic absorption spectroscopy, *Chem. Anal.,* 24, 657, 1979.

318. **Timinaga, M. and Umezaki, Y.,** Suppression of interferences in the determination of lead and cadmium by graphite furnace atomic absorption spectrometry, *Bunseki Kagaku,* 28, 347, 1979.

319. **Tsalev, D. and Petrova, V.,** Hexamethylene ammonium hexamethylene dithicarbamate-*n*-butyl acetate as an extractant of bismuth, cadmium, cobalt, mercury, indium, lead and palladium from acidic media, *Dokl. Bolg. Akad. Nauk,* 32, 911, 1979.

320. **Tsushida, T. and Takeo, T.,** Direct determination of copper, lead and cadmium in tea infusions by flameless atomic absorption spectrometry, *Agric. Biol. Chem.,* 43, 1347, 1979.

321. **Flanjak, J. and Lee, H. Y.,** Trace metal content of livers and kidneys of cattle, *J. Sci. Food Agric.,* 30, 503, 1979.

322. **Ikebe, K., Tanaka, Y., and Tanaka, R.,** Determination of cadmium in human blood by flameless atomic absorption spectrometry using an automatic sample dispenser, Osaka-Furitsu Koshu Eisei Kenkyusho Kenkyu Hokoku, *Shokunin Eisei Hen,* 9, 41, 1979.

323. **Kitagawa, K., Shigeyasu, T., and Takeuchi, T.,** Application of the atomic farady effect to the trace determination of elements (cadmium, silver and copper): effect of the hyperfine structure on the Zeeman splitting and line-crossing, *Spectrochim. Acta,* 34b, 389, 1979.

324. **Knezevic, G.,** Heavy metals in food. I. Content of cadmium in raw cocoa beans and in semifinished and finished chocolate products, *Dtsch. Lebensm. Rundsch.,* 75, 305, 1979.

325. **Komarek, J., Havel, J., and Sommer, L.,** The use of chelates of copper, nickel, cobalt, cadmium and zinc with heterocyclic azo dyes in the AAS determination of these elements, *Collect. Czech. Chem. Commun.,* 44, 3241, 1979.

326. **Kucharev, E. A. and Zavarzina, E. F.,** Extraction-flame photometric study of solvation of cadmium complexes with methyl isobutyl ketone in extraction from an iodide system, *Zh. Anal. Khim.,* 34, 1914, 1979.

327. **Lagesson, V. and Andrasko, L.,** Direct determination of lead and cadmium in blood and urine by flameless atomic absorption spectrophotometry, *Clin. Chem. (Winston-Salem, N.C.),* 25, 1948, 1979.

328. **Mamontova, S. A. and Pclelintseva, N. F.,** Extraction-atomic absorption determination of lead and cadmium in natural waters, suspensions and sediments, *Zh. Anal. Khim.,* 34, 2231, 1979.

329. **Meranger, J. C. and Subramanian, K. S.,** Direct determination of cadmium in drinking water supplies by graphite furnace atomic absorption spectrometry, *Can. J. Spectrosc.,* 24, 132, 1979.

330. **Pleban, P. A. and Pearson, K. H.,** Determination of cadmium in whole blood and urine by Zeeman atomic absorption spectroscopy, *Clin. Chem. Acta,* 99, 267, 1979.

331. **Ruttner, O. and Jarc, H.,** Residue studies in Austria. I. Investigations on the lead, cadmium and chromium content of meat from upper Austrian cattle, *Wien Tieraerztl. Monatsschr.,* 66, 259, 1979.

332. **Baucells, M., Lacort, G., and Roura, M.,** Determination of major and trace elements in marine sediments of Costa Brava by emission and atomic absorption spectroscopy, *Invest. Pesq.,* 43, 551, 1979.

333. **Sperling, K. R.,** Determination of heavy metals in sea water and marine organisms by flameless atomic absorption spectrophotometry. IX. Determination of cadmium traces in biological materials by simple extraction method, *Z. Anal. Chem.,* 299, 103, 1979.

334. **Strickland, B. C., Chaney, W. R., and Lamoreaux, R. J.,** Cadmium uptake by *pinus resinosa* AIT. Pollen and the effect on cation release and membrane permeability, *Plant Physiol.,* 64, 366, 1979.

335. **Tsujii, K., Kuga, K., Murayama, S., and Yasuda, M.,** Evaluation of high-frequency discharge lamp for atomic absorption and atomic fluorescence spectrometry of cadmium, lead and zinc, *Anal. Chim. Acta,* 111, 103, 1979.

336. **Berta, E. and Olah, J.,** Survey of heavy metal content of waste water sludges in Hungary, *Hidrol. Kozl.,* 59, 469, 1979.

337. **Burdin, K. S., Kurpina, M., and Savelev, T. B.,** Mollusks of the Genus *Mytilus* as possible indicators of the concentration of heavy and transition metals in the marine environment, *Okeanologiya,* 19, 1038, 1979.

338. **Colovos, G., Eaton, W. S., Ricci, G. R., Shepard, L. S., and Wang, H.,** Collaborative Testing of NIOSH Atomic Absorption Method, Publ. 79-144, U.S. Department of Health, Education and Welfare (NIOSH), Washington, D.C., 1979.

339. **El-Enany, F. F., Mahmoud, K. F., and Varma, M. M.,** Single organic extraction for determination of ten heavy metals in sea water, *J. Water Pollut. Control Fed.,* 51, 2545, 1979.

340. **Fujino, O., Matsui, M., and Shigematsu, T.,** Determination of trace metals in natural water by atomic absorption spectrometry. Preconcentration of metals by diethyldithiocarbamate / di-isobutyl ketone solvent extraction, *Mizu Shori Gijutsu,* 20, 201, 1979 (in Japanese).

341. **Korenev, A. P. and Konovalov, G. S.,** Optimization of conditions for determination of some metals (zinc and cadmium) in natural waters and in solutions by an atomic absorption method, *Gidrokhim. Mater.,* 73, 40, 1979.

342. **Palacios Remondo, J. and Ramirez Diaz, S.,** Lead and cadmium levels in liver of monogastric and polygastric animals determined by atomic absorption spectrophotometry, *Rev. Agrochim. Technol. Aliment.,* 19, 279, 1979.

343. **Fricke, F. L., Robbins, W. B., and Caruso, J. A.,** Trace element analysis of food and beverages by atomic absorption spectrometry, *Prog. Anal. At. Spectrosc.,* 2, 185, 1979.

344. **Chowdhury, A. N. and Das, A. K.,** Determination of traces of lead and cadmium in iron ores by non-aqueous atomic absorption spectrophotometry, *Indian J. Chem. Sect. A,* 17, 536, 1979.

345. **Price, W. J.,** New techniques of atomic absorption in clinical and biochemical analysis, *Sci. Ind.,* 12, 22, 1979.

346. **Wesenberg, G. B. R., Fosse, G., Justesen, N. P. B., and Rasmusson, P.,** Lead and cadmium in teeth, bone and kidneys of rats with a standard lead-cadmium supply, *Int. J. Environ. Stud.,* 14, 223, 1979.

347. **Sumino, K., Yamamoto, R., Hatayama, F., Kitamura, S., and Itoh, H.,** Laser atomic absorption spectrometry for histochemistry, *Anal. Chem.,* 52, 1064, 1980.

348. **Thompson, K. C. and Wagstaff, K.,** Simplified method for the determination of cadmium, chromium, copper, nickel, lead and zinc in sewage sludge using atomic absorption spectrophotometry, *Analyst (London),* 105, 883, 1980.

349. **Weigert, P.,** Heavy metal content of chicken eggs, *Z. Lebensm. Unters. Forsch.,* 171, 18, 1980.

350. **Waughrman, G. J. and Bretti, T.,** Interference due to major elements during estimation of trace heavy metals in natural materials by atomic absorption spectrophotometry, *Environ. Res.,* 21, 385, 1980.

351. **Ledl, G., Horak, O., and Janauer, G. A.,** Manganese, iron, copper and other heavy metals in Austrian rivers, *Oesterr. Abwasser Rundsch. OAR Int.,* 25, 28, 1980.

352. **Magnusson, B. and Westerland, S.,** Determination of cadmium, copper, iron, nickel, lead and zinc in Baltic Sea water, *Mar. Chem.,* 8, 231, 1980.

353. **Morse, R. A. and Lisk, D. J.,** Elemental analysis of honeys from several nations, *Am. Bee J.,* 120, 592, 1980.

354. **Nadkarni, R. A.,** Multitechnique multielement analysis of coal and fly ash, *Anal. Chem.,* 52, 929, 1980.

355. **Pande, S. P. and Pendharkar, A. V.,** Some analytical aspects of lead and cadmium determination in drinking water by atomic absorption spectrophotometry, *J. Inst. Chem.,* 52, 141, 1980.

356. **Poldoski, J. E.,** Determination of lead and cadmium in fish and clam tissue by atomic absorption spectrometry with a molybdenum and lanthanum treated prolytic atomizer, *Anal. Chem.,* 52, 1147, 1980.

357. **Rantala, R. T. T. and Loring, D. H.,** Direct determination of cadmium in silicates from a fluoboric-bone and matrix by graphite furnace atomic absorption spectrometry, *At. Spectrosc.,* 1, 163, 1980.

358. **Danielsson, L. G.,** Cadmium, cobalt, copper, iron, lead, nickel and zinc in Indian Ocean water, *Mar. Chem.,* 8, 199, 1980.

359. **Gorshkov, V. V., Orlova, L. P., and Voronkova, M. A.,** Preconcentration and atomic absorption determination of cadmium and lead in natural materials, *Zh. Anal. Khim.,* 35, 1277, 1980.

360. **Guevremont, R.,** Organic matrix modifiers for direct graphite furnace atomic absorption determination of cadmium in sea water, *Anal. Chem.,* 52, 1574, 1980.

361. **Holak, W.,** Analysis of foods for lead, cadmium, copper, zinc, arsenic and selenium using closed system sample digestion. Collaborative study, *J. Assoc. Off. Anal. Chem.,* 63, 485, 1980.

362. **Horler, D. N. H., Barber, J., and Barringer, A. R.,** Multielemental study of plant surface particles in relation to geochemistry and biogeochemistry, *J. Geochem. Explor.,* 13, 41, 1980.

363. **Hubert, J., Candelaria, R. M., and Applegate, H. G.,** Determination of lead, zinc, cadmium and arsenic in environmental samples, *At. Spectrosc.,* 1, 90, 1980.

364. **Amiard, J. C., Amiard-Triquet, C., Metayer, G., Marchand, J., and Ferre, R.,** Study of transfer of cadmium, lead, copper and zinc in esturine food chains. I. Study in the Loire Estuary during summer, 1978, *Water Res.,* 14, 665, 1980.

365. **Ash, C. P. J. and Lee, D. L.,** Lead, cadmium, copper and iron in earthworms from roadside sites, *Ser. A Ecol. Biol.,* 22, 59, 1980.

366. **Barudi, W. and Bielig, H. J.,** Heavy metal content of vegetables which grow above ground and fruits, *Z. Lebensm. Unters. Forsch.,* 170, 254, 1980.

367. **Carter, G. F. and Yeoman, W. B.,** Determination of cadmium in blood after destruction of organic material by low temperature ashing, *Analyst (London),* 105, 295, 1980.

368. **Chakrabarti, C. L., Wan, C. C., and Li, W. C.,** Atomic absorption spectrometric determination of cadmium, lead, zinc, copper, cobalt and iron in oyster tissue by direct atomization from the solid state using the graphite furnace platform technique, *Spectrochim. Acta,* 35b, 547, 1980.

369. **Collela, M. B., Siggia, S., and Barnes, R. M.,** Poly (acrylamidoxime) resin for determination of trace metals in natural waters, *Anal. Chem.,* 52, 2347, 1980.

370. **Sengupta, P.,** Studies on the presence and determination of cadmium in foods by atomic absorption spectrophotometry, *J. Inst. Chem.,* 52, 15, 1980.

371. **Smith, R. G. and Windom, H. L.,** A solvent extraction technique for determining nanogram per liter concentrations of cadmium, copper, nickel and zinc in seawater, *Anal. Chim. Acta,* 113, 39, 1980.

372. **Sperling, K. and Bahr, B.,** Determination of extremely low concentrations of cadmium in blood and urine by flameless AAS. I. Testing of a micromethod, *Z. Anal. Chem.,* 301, 29, 1980.

373. **Sperling, K. R. and Bahr, B.,** Determination of extremely low concentrations of cadmium in blood and urine by flameless AAS. II. Extraction behaviour of cadmium, *Z. Anal. Chem.,* 301, 31, 1980.

374. **Trofimov, N. V., Petrov, B. I., Nekhaev, N. N., and Busev, A. I.,** Atomic absorption determination with the extraction of a cadmium trace impurity in very pure cobalt and alloys of iron and nickel, *Zavod. Lab.,* 46, 15, 1980.

375. **Adams, T. G., Atchison, G. J., and Vetter, R. J.,** Impact of an industrially contaminated lake on heavy metal levels in its effluent stream, *Hydrobiologia,* 69, 187, 1980.

376. **Feinberg, M. and Ducauze, C.,** High temperature ashing of foods for atomic absorption spectrometric determination of lead, cadmium and copper, *Anal. Chem.,* 52, 207, 1980.

377. **Guevremont, R., Sturgeon, R. E., and Berman, S. S.,** Application of EDTA to direct graphite furnace atomic absorption analysis for cadmium in seawater, *Anal. Chim. Acta,* 115, 163, 1980.

378. **Legotte, P. A., Rosa, W. C., and Sutton, D. C.,** Determination of cadmium and lead in urine and other biological samples by graphite furnace atomic absorption spectrometry, *Talanta,* 27, 39, 1980.

379. **Pedersen, B., Willems, M., and Storgaard-Joergensen, S.,** Determination of copper, lead, cadmium, nickel and cobalt in EDTA extracts of soil by solvent extraction and graphite furnace atomic absorption spectrophotometry, *Analyst (London),* 105, 119, 1980.

380. **Petrov, I. I., Tsalev, D. L., and Barsev, A. I.,** Atomic absorption spectrometric determination of cadmium, cobalt, copper, manganese, nickel, lead and zinc in acetate soil extracts, *At. Spectrosc.,* 1, 47, 1980.

381. **Knight, M. J.,** Comparison of Four Digestion Procedures not Requiring Perchloric Acid for the Trace Element Analysis of Plant Material, Rep. ANL/LRP-TM-18, Argonne National Laboratory, Argonne, Ill., 1980; available from the NTIS, Springfield, Va.

382. **Tursunov, A. T., Brovko, I. A., and Nazarov, S. N.,** Use of a "furnace flame" system for the atomic absorption determination of some trace elements in soil extracts, *Uzb. Khim. Zh.,* 5, 5, 1980.

383. **Fink, L. K., Jr., Harriss, A. B., and Schick, L. L.,** Trace Metals in Suspended Perchlorates, Biota and Sediments of the St. Croix, Narraguagus and Union Estuaries and the Goose Cove Region of Penobscot Bay, Rep. W 80-04809, OWRT-A-041-ME (I), 1980; available from the NTIS, Springfield, Va.; *Gov. Rep. Announce. Index (U.S.),* 80(17), 3372, 1980.

384. **Orpwood, B.,** Use of Chelating Ion-Exchange Resins for the Determination of Trace Metals in Drinking waters, Tech. Rep. TR-Water Res. Cent. TR 153, Medmenham Laboratories, Marlow, Bucks., 35pp, 1980.

385. **Inhat, M., Gordon, A. D., Gaynor, J. D., Berman, S. S., Desauliners, A., Stoeppler, M., and Valenta, P.,** Interlaboratory analysis of natural fresh waters for Cu, Zn, Cd and Pb., *Int. J. Environ. Anal. Chem.,* 8, 259, 1980.

386. **Saito, S., Kamoda, M., and Gijutsu, S.,** Sample preparation by dry ashing prior to the determination of metals in sugars, *Kenkyu Kaishi,* 29, 36, 1980.

387. **Takiyama, K., Ishii, Y., and Yoshimura, I.,** Determination of metals in vegetables by ashing in a teflon crucible, *Mukogawa Joshi Daigaku Kiyo, Shokumotsu-hen,* 28, F15, 1980.

388. **Bel'skii, N. K. and Ochertyanova, L. I.,** Atomic absorption determination of the deviation of cadmium chromium selenide ($CdCr_2Se_4$) crystals from the stoichiometric composition, *Zh. Anal. Khim.,* 35, 604, 1980.

389. **Chakrabarti, C. L., Wan, C. C., and Li, W. C.,** Direct determination of traces of copper, zinc, lead, cobalt, iron and cadmium in bovine liver by graphite furnace atomic absorption spectrometry using the solid sampling and the platform techniques, *Spectrochim. Acta,* 35b, 93, 1980.

390. **Schramel, P., Wolf, A., Seif, R., and Klose, B. J.,** New device for ashing of biological materials under pressure, *Z. Anal. Chem.*, 302, 62, 1980.

391. **Sinez, S. A., Cantillo, A., and Helz, G. R.,** Accuracy of acid extraction methods for trace metals in sediments, *Anal. Chem.*, 52, 2342, 1980.

392. **Sperling, K. R.,** Determination of heavy metals in seawater and in marine organisms by flameless atomic absorption spectrophotometry. XII. A method for the determination of "total cadmium" in natural water samples. Limits of mere extraction methods, *Z. Anal. Chem.*, 301, 294, 1980.

393. **Stein, V. B. and McClellan, B. E.,** Enhancement of atomic absorption sensitivity for cadmium, manganese, nickel and silver and determination in submicrogram quantities of cadmium and nickel in environmental samples, *Environ. Sci. Technol.*, 14, 872, 1980.

394. **Wesenberg, G. B. R., Fosse, G., Rasmussen, P., and Justesen, N. P. B.,** Effects of lead and cadmium uptake on zinc and copper levels in hard and soft tissues of rats, *Int. J. Environ. Stud.*, 15, 41, 1980.

395. **White, T. A. and Rolte, G. L.,** Differing effects of cadmium on two varieties of cottonwood *Populus deloides bartr, Environ. Pollut. Ser. A Ecol. Biol.*, 22, 29, 1980.

396. **Zmudzki, J.,** Determination of cadmium in biological material by atomic absorption spectrometry, *Bromatol. Chem. Toksykol.*, 13, 77, 1980.

397. **Abo-Rady, M. D. K.,** Aquatic macrophytes as indicator for heavy metal pollution in the River Seine, *Arch. Hydrobiol.*, 89, 387, 1980.

398. **Brodowsky, H., Oei, Y. S., and Schaller, H. J.,** Thermodynamic properties of palladium-cadmium alloys, *Metallkd.*, 71, 593, 1980.

399. **Brovko, I. A.,** Diphenyl carbazone as a reagent for extraction-atomic absorption determination of cadmium, copper, manganese, nickel and zinc, *Zh. Anal. Khim.*, 35, 2095, 1980.

400. **Budnick, A.,** Determination of cadmium and copper in galvanic coatings by means of atomic absorption, *Microchem. J.*, 25, 531, 1980.

401. **Cool, M., Marcoux, F., Paulin, A., and Mehra, M. C.,** Metallic contaminants in street soils of Moncton, New Brunswick, Canada, *Bull. Environ. Contam. Toxicol.*, 25, 409, 1980.

402. **Davison, W.,** Ultratrace analysis of soluble zinc, cadmium, copper and lead in Windermere Lake water using anode stripping voltammetry and atomic absorption spectroscopy, *Freshwater Biol.*, 10, 223, 1980.

403. **Hoshino, Y., Utsunomiya, T., and Fukui, K.,** Graphite furnace atomic absorption spectrometry utilizing selective concentration onto tungsten wire, *Rep. Res. Lab. Eng. Mater. Tokyo Inst. Technol.*, 5, 109, 1980.

404. **Ishii, T., Hirano, S., Matsuba, M., and Koyanagi, T.,** Determination of trace elements in shellfish, *Nippon Suisan Gakkaishi*, 46, 1375, 1980.

405. **Itokawa, H., Watanabe, K., Tazaki, T., Hayashi, T., and Hayashi, Y.,** Quantitative analysis of metals in crude drugs, *Shoyakugaku Zasshi*, 34, 155, 1980.

406. **Jenniss, S. W., Katz, S. A., and Mount, T.,** Comparison of sample preparation methods for the determination of cadmium and lead in sewage sludge by AAS, *Am. Lab.*, 12, 18, 1980.

407. **Kruse, R.,** Multiple determination of lead and cadmium in fish by electrothermal AAS after wet ashing in commercially available teflon beakers, *Z. Lebensm. Unters. Forsch.*, 171, 261, 1980.

408. **Lindh, U., Prune, D., Nordberg, G., and Wester, P. O.,** Levels of antimony, arsenic, cadmium, copper, lead, mercury, selenium, silver, tin and zinc in bone tissue of industrially exposed workers, *Sci. Total Environ.*, 16, 109, 1980 (in English).

409. **Ogihara, K., Seki, H., and Nagase, K.,** Comparison of the method of bottom sediment analysis and the method of the agricultural soil pollution prevention law, *Nagano-Ken Eisei Kogai Kenkyusho Kenkyu Hokoku*, 2, 88, 1980 (in Japanese).

410. **Oreshkin, V. N., Belyaev, Y. S., Tatzii, Y. G., and Vnukovskaya, G. L.,** Direct simultaneous determination of cadmium, lead and silver in sea, river and colian suspended matter by flameless atomic absorption, *Okeanologiya*, 20, 736, 1980 (in Russian).

411. **Posta, J.,** Determination of the optimum parameters of a combined atomic absorption spectrophotometric plus arc flame method for the analysis of floating dust samples, *Hung. Sci. Instrum.*, 47, 33, 1980.

412. **Riandey, C., Gavinelli, R., and Pinta, M.,** Effect of heating rate on electrothermal atomization in atomic absorption spectrometry: application to the volatile elements cadmium and lead, *Spectrochim. Acta*, 35b, 765, 1980.

413. **Sourova, J. and Capkova, A.,** Determination of trace elements in water with a high iron content by AAS, *Vodni Hospod.*, 30b, 133, 1980 (in Czech).

414. **Stein, V. B., Canelli, E., and Richards, A. H.,** Simplified determination of cadmium, lead and chromium in estuarine waters by flameless atomic absorption, *Int. J. Environ. Anal. Chem.*, 8, 99, 1980.

415. **Tsushida, T. and Takeo, T.,** Determination of copper, lead and cadmium in tea by graphite furnace atomic absorption spectrophotometry, *Nippon Shokuhin Kogyo Gakkaishi*, 27, 585, 1980 (in Japanese).

416. **Tursunov, A. T., Brovko, I. A., and Nazarov, S. N.,** Use of a furnace flame system for the atomic absorption determination of some trace elements in soil extracts, *Uzb. Khim. Zh.*, 5, 5, 1980 (in Russian).

417. **Van Willis, W., El-Ahraf, A., Vinjamoori, D. V., and Aref, K.,** Analysis of animal feed ingredients and soil amendment products produced from beef cattle manure for selected trace metals using atomic absorption spectrophotometry, *J. Food Prot.,* 43, 834, 1980.

418. **Wetzel, L. T. and Bell, J. U.,** Electrothermal atomic absorption spectrophotometry of cadmium in semen, *Clin. Chem. (Winston-Salem, N.C.),* 26, 1795, 1980.

419. **Fernandez, F. J., Beaty, M. M., and Barnett, W. B.,** Use of the L'vov platform for furnace atomic absorption applications, *At. Spectrosc.,* 2, 16, 1981.

420. **Hinderberger, E. J., Kaiser, M. L., and Koirtyohann, S. R.,** Furnace atomic absorption analysis of biological samples using the L'vov platform and matrix modification, *At. Spectrosc.,* 2, 1, 1981.

421. **Laxen, D. P. and Harrison, R. M.,** Cleaning methods for polythene containers prior to the determination of trace metals in fresh water samples, *Anal. Chem.,* 53, 345, 1981.

422. **Martynyuk, T. G., Sevastyanova, N. I., Nikolaeva, V. A., and Mukhtarov, E. I.,** Flameless atomic absorption method for determining the lead and cadmium content of poultry products, *Vopr. Pitan.,* 2, 16, 1981.

423. **Meranger, J. C., Hollebone, B. R., and Blanchette, G. A.,** Effect of storage times, temperatures and containers types on the accuracy of atomic absorption determination of Cd, Cu, Hg, Pb and Zn in whole heparinized blood, *J. Anal. Toxicol.,* 5, 33, 1981.

424. **Meranger, J. C., Subramanian, K. S., and Chalifoux, C.,** Survey of Cd, Co, Cr, Cu, Ni, Pb, Zn, Ca and Mn in Canadian drinking water supplies, *J. Assoc. Off. Anal. Chem.,* 64, 44, 1981.

425. **Alt, F.,** Comparative determination of cadmium in blood by four different methods. Application of furnace atomic absorption spectrometry (background measurement by Zeeman effect and deuterium emitter) and inverse voltametry, *Fresenius' Z. Anal. Chem.,* 308, 137, 1981.

426. **Carpenter, R. C.,** Determination of cadmium, copper, lead and thallium in human liver and kidney tissue by flame atomic absorption spectrometry after enzymic digestion, *Anal. Chim. Acta,* 125, 209, 1981.

427. **Clark, J. R. and Viets, J. G.,** Multielement extraction system for the determination of 18 trace elements in geochemical samples, *Anal. Chem.,* 53, 61, 1981.

428. **Clark, J. R. and Viets, J. G.,** Back-extraction of trace elements from organometallic-halide extracts for determination by flameless atomic absorption spectrometry, *Anal. Chem.,* 65, 1981.

429. **Delves, H. T. and Woodward, J.,** Determination of low levels of cadmium in blood by electrothermal atomization and atomic absorption spectrophotometry, *At. Spectrosc.,* 2, 65, 1981.

430. **Bewers, J. M., Dalziel, J., Yeats, B. A., and Barron, J. L.,** An intercalibration for trace metals in seawater, *Mar. Chem.,* 10, 173, 1981.

431. **Clanet, F., Deloncle, R., and Popoff, G.,** Chelating resin catcher for capture, preconcentration and determination of toxic metal traces (Zn, Cd, Hg and Pb) in waters, *Water Res.,* 15, 591, 1981.

432. **Carleer, R., Francois, J. P., and Van Poucke, L. C.,** Determination of the main, minor and trace elements in lead/tin based solder by atomic absorption spectrophotometry, *Bull. Soc. Chim. Belg.,* 90, 357, 1981.

433. **Capel, I. D., Pinnock, M. H., Dorrell, H. M., Williams, D. C., and Grant, E. C. G.,** Comparison of concentrations of some trace, bulk and toxic metals in the hair of normal and dyslexic children, *Clin. Chem. (Winston-Salem, N.C.),* 27, 879, 1981.

434. **Kimura, M. and Kewanami, K.,** Separation and preconcentration of trace amounts of several metals in sodium perchlorate using activated carbon as a collector, *Nippon Kagaku Kaishi,* 1, 1, 1981 (in Japanese).

435. **Petrov, I., Tsalev, D., and Vasileva, E.,** Pulse nebulization atomic absorption spectrometry after preconcentration from acidic media, *Dokl. Bolg. Akad. Nauk,* 34(5), 679, 1981; CA 95, 146378n, 1981.

436. **Sanks, R. L.,** Flameless Atomic Absorption for Tracing the Fate of Heavy Metals in Soils and Ground Water, Rep. MWR RC-III, W 81-02491, OWRT-A 109-MONT (I), Order No. PB-81-189821, 1980; available from the NTIS, Springfield, Va; *Gov. Rep. Announce Index (U.S.),* 81(7), 3565, 1981.

437. **Berndt, H. and Messerschmidt, J.,** *o,o*-Diethyl dithiophosphate for trace enrichment on activated carbon. I. Analysis of high purity gallium-aluminum-determination of element traces by flame AAS (injection method loop AAS), *Fresenius' Z. Anal. Chem.,* 308(2), 104, 1981 (in German).

438. **Sturgeon, R. R., Berman, S. S., Willie, S. N., and Desauliners, J. A. H.,** Preconcentration of trace elements from seawater with silica-immobilized 8-hydroxyquinoline, *Anal. Chem.,* 53(14), 2337, 1981.

439. **Pleban, P. A., Kerkay, J., and Pearson, K. H.,** Polarized Zeeman effect flameless atomic absorption spectrometry of cadmium, copper, lead and manganese in human kidney cortex, *Clin. Chem. (Winston-Salem, N.C.),* 27, 68, 1981.

440. **Farmer, J. G. and Gibson, M. J.,** Direct determination of cadmium, chromium, copper and lead in siliceous standard reference materials from a fluoboric acid matrix by graphite furnace atomic absorption spectrometry, *At. Spectrosc.,* 2(6), 176, 1981.

441. **Adelman, R., Jenniss, S. W., and Katz, S. A.,** Interlaboratory analysis of sewage sludge, *Am. Lab.,* 31, 1981.

442. **Alt, F., Berndt, H., Messerschmidt, J., and Sommer, D.,** Determination of cadmium, lead and thallium in mineral raw materials after chemical preconcentration using different spectrometric methods, *Spektrometertagung (Votr.),* 13, 331, 1981 (in German).

443. **Shaburova, V. P., Ruchkin, E. D., Yudelevich, I. G., and Kruglov, L. M.,** Atomic-absorption method of analysis for piezoceramic materials of the lead zirconate titanate system, *Izv. Sib. Otd. Akad. Nauk SSSR Ser. Khim. Nauk,* 5, 95, 1981 (in Russian).

444. **Mattson, P., Albanus, L., and Frank, A.,** Cadmium and some other elements in liver and kidney from mouse (*Alces alces*), *Var Foeda,* 33, 335, 1981 (in Swedish).

445. **Wesenberg, G. B. R., Fosse, G., and Rasmussen, P.,** The effect of graded doses of cadmium on lead, zinc and copper content of target and indicator organs in rats, *Int. J. Environ. Stud.,* 17, 191, 1981.

446. **Mueller, H. and Siene, V.,** Quantitative determination of arsenic, lead, cadmium, mercury and selenium in foods by flameless atomic absorption spectrophotometry, *Dtsch. Lebensm. Rundsch.,* 77, 392, 1981.

447. **Blakemore, W. M. and Billedeau, S. M.,** Analysis of laboratory animal feed for toxic and essential elements by atomic absorption and inductively coupled organ plasma emission spectrometry, *J. Assoc. Off. Anal. Chem.,* 64, 1284, 1981.

448. **Kirleis, A. S., Sommers, L. E., and Sommers, D. W.,** Heavy metal content of groats and hulls of oats grown on soil treated with sewage sludges, *Cereal Chem.,* 58(6), 530, 1981.

449. **Brunner, W., Heckner, H., and Sansoni, B.,** Fully automatic simultaneous operation of several atomic absorption spectrometers in the analytical service laboratory of a research center by use of a process computer, *Spektrometeragung (Votr.),* 13, 99, 1981.

450. **Ranchet, J., Menissier, F., Lamathe, J., and Voinovitch, I.,** Interlaboratory comparison of the determinations of cadmium, chromium, copper and lead by flameless atomic absorption spectrometry, *Bull. Liaison Lab. Ponts Chaussees,* 114, 81, 1981.

451. **Prudnikov, E. D., Bradaczek, H., and Labischinski, H.,** Calculation of standard deviation in atomic absorption spectrometry, *Fresenius' Z. Anal. Chem.,* 308(4), 342, 1981 (in German).

452. **Mitchell, G. E.,** Trace metal levels in Queensland dairy products, *Aust. J. Dairy Technol.,* 36(2), 70, 1981.

453. **Van Hattum, B. and De Voogt, P.,** An analytical procedure for the determination of cadmium in human placentas, *Int. J. Environ. Anal. Chem.,* 10(2), 121, 1981.

454. **Halls, D. J.,** Applications of graphite furnace atomic absorption spectrometry in chemical analysis, *Anal. Proc. (London),* 18(8), 344, 1981.

455. **Fukushima, I.,** A Study on Monitoring Methodology for Trace Elements in the Environment Through Chemical Analysis of Hair. Part of a Coordinated Program on Nuclear Methods in Health-Related Monitoring of Trace Elements Pollutants, Rep. IAEA-R-2480/F, International Atomic Energy Agency, Vienna, 1981; INIS Atomindex, 12 (19), Abstr. No. 626438, 1981.

456. **Magnusson, B. and Westerlund, S.,** Solvent extraction procedures combined with back extraction for trace metal determinations by atomic absorption spectrometry, *Anal. Chim. Acta,* 131(1), 63, 1981.

457. **Brovko, I. A. and Tursunov, A. T.,** Extraction-atomic absorption determination of cadmium, cobalt, copper, manganese, nickel and zinc in natural waters using diphenylcarbazone as the extraction reagent, *Uzb. Khim. Zh.,* 3, 19, 1981 (in Russian).

458. **Satsmadjis, J. and Voutsinou-Taliadouri, F.,** Determination of trace metals at concentrations above the linear calibration range by electrothermal atomic absorption spectrometry, *Anal. Chim. Acta,* 131, 83, 1981.

459. **Burba, P. and Schaefer, W.,** Atomic-absorption spectrometric determination (flame-AAS) of heavy metals in bacterial leach liquors (Jarosite) after their analytical separation on cellulose, *Erzmetall,* 34, 582, 1981 (in German).

460. **Berndt, H. and Messerschmidt, J.,** Microanalytical method of flame atomic absorption and emission spectrometry (platinum loop method) II. Determination of As, Bi, Cd, Hg, Pb, Sb, Se, Te, Tl, Zn, Li, Na, K, Ce, Sr and Ba, *Spectrochim. Acta,* 36b, 809, 1981.

461. **Blanusa, M. and Breski, D.,** Comparison of dry and wet ashing procedures for cadmium and iron determination in biological material by atomic absorption spectrophotometry, *Talanta,* 28, 681, 1981.

462. **Bruhn, F. C. and Navarrete, A. G.,** Matrix modification for the direct determination of cadmium in urine by electrothermal atomic absorption spectrometry, *Anal. Chim. Acta,* 130, 209, 1981.

463. **Bye, R.,** Improvement of the sensitivity of the "Sampling Boat" technique in atomic absorption spectrometry. Determination of silver, lead and cadmium, *Z. Anal. Chem.,* 306, 30, 1981.

464. **Jastrow, J. D., Zimmerman, C. A., Dvorak, A. J., and Hinchman, R. R.,** Plant growth and trace element uptake on acidic coal refuse amended with lime or fly ash, *J. Environ. Qual.,* 10, 154, 1981.

465. **Katz, S. A., Jenniss, S. W., Mount, T., Tout, R. E., and Chatt, A.,** Comparison of sample preparation methods for the determination of metals in sewage sludges by flame atomic absorption spectrometry, *Int. J. Environ. Anal. Chem.,* 9, 209, 1981.

466. **Kurfuerst, U. and Rues, B.,** Determination of heavy metals (lead, cadmium, mercury) in settling sludge using Zeeman AAS without chemical decomposition, *Z. Anal. Chem.,* 308, 1, 1981.

467. **Mendiola Ambrosio, J. M. and Gonzalez Lopez, A.,** Analysis of cadmium and zinc in PVC composites by atomic absorption spectrophotometry, *Rev. Plast. Mod.,* 41, 413, 1981 (in Spanish).

468. **Orlova, L. P., Gorshkov, V. V., Voronkova, M. A., and Voskresenskaya, V. S.,** Atomic absorption determination of cadmium in natural objects having a high content of iron and aluminum, *Pochvovdenie,* 4, 104, 1981.

469. **Pleban, P. A., Kerkay, J., and Pearson, K. H.,** Cadmium, copper, lead, manganese and selenium levels and glutathione peroxidase activity in human kidney cortex, *Anal. Lett.,* 14b, 1089, 1981.

470. **Sperling, K. R. and Bahr, B.,** Determination of heavy metals in seawater and marine organisms by flameless atomic absorption spectrophotometry. XIII. Correspondence and some possible sources of error in an intercalibration of cadmium, *Z. Anal. Chem.,* 306, 7, 1981.

471. **Subramanian, K. S. and Meranger, J. C.,** Rapid electrothermal atomic absorption spectrophotometric method for cadmium and lead in human whole blood, *Clin. Chem. (Winston-Salem, N.C.),* 27, 1866, 1981.

472. **Taddia, M.,** Atomic absorption spectrometry of cadmium after solvent extraction with zinc dibenzyl dithiocarbamate, *Microchem. J.,* 26, 340, 1981.

473. **Wolffe, E. W., Landy, M. P., and Peel, D. A.,** Preconcentration of cadmium, copper, lead and zinc in water at the 10^{-12} g/g level by absorption onto tungsten wire followed by flameless atomic absorption spectrometry, *Anal. Chem.,* 53, 1566, 1981.

474. **Wunderlich, E.,** Determination of lead and cadmium in biological material, *Erzmetall ,* 34, 577, 1981 (in German).

475. **Jackson, K. F., Benedik, J. E., and Birholz, F. A.,** Metal distribution in shale oil fractions, *Oil Shale Symp. Proc.,* 14, 75, 1981.

476. **Khalighie, J., Ure, A. M., and West, T. S.,** Atomic-tapping atomic absorption spectrometry of arsenic, cadmium, lead, selenium and zinc in air-acetylene and air-propane flames, *Anal. Chim. Acta,* 131, 27, 1981.

477. **Kovarskii, N. Y., Kovekovdova, L. T., Pryazhevskaya, I. S., Belenkii, V. S., Shapovalov, E. N., and Popkova, S. M.,** Preconcentration of trace elements from seawater by electrodeposited magnesium hydroxide, *Zh. Anal. Khim.,* 36, 2264, 1981.

478. **Pilipenko, A. T. and Samchuk, A. I.,** Extraction-atomic absorption determination of trace elements in natural waters, *Khim. Tekhnol-Vody,* 3, 343, 1981.

479. **Elson, C. M., Bem, E. M., and Ackman, R. G.,** Determination of heavy metals in a menhaden oil after refining and hydrogenation using several analytical methods, *J. Am. Oil Chem. Soc.,* 58, 1024, 1981.

480. **Subramanian, K. S. and Meranger, J. S.,** A rapid electrothermal atomic absorption spectrophotometric method for cadmium and lead in human blood, *Clin. Chem. (Winston-Salem, N.C.),* 27, 1866, 1981.

481. **Nechiporenko, N. I., Protserova, T. K., and Klauchek, V. V.,** Atomic absorption determination of cadmium in the atmosphere at a work zone, *Gig. Tr. Prof. Zabol.,* 10, 55, 1981 (in Russian).

482. **Ueta, T., Mori, K., Nishida, S., and Yoshihara, T.,** Hygienic chemical studies on dental treatment materials. I. Content of cadmium, lead and arsenic in dental casting alloys and acrylic denture resins, *Tokyo Toritsu Eisei Kenkyusho Kenkyu Nempo,* 32, 110, 1981.

483. **Borg, H.,** Trace metals in natural waters. An analytical intercomparison, Rapp-Naturvardsverket (Sweden), SNV PM 1463, Forkingssekretarietet, Statens Naturvaardsverk, Solna, Sweden, 1981 (in Swedish).

484. **Suwirma, S., Surtipanti, S., and Yatim, S.,** Determination of mercury, lead, cadmium and chromium in several sea fishes, *Majalah BATAN,* 14, 2, 1981 (in Indonesian).

485. Analysis of the trace elements in natural food dyes, *Annu. Rep. Osaka City Inst. Public Health Environ. Sci.,* 43, 98, 1981.

486. **Valenta, P.,** Use of voltammetric method for determination of cadmium and mercury in water and wastewater, *Gewaesserschutz Wasser Abwasser,* 52, 248, 1981.

487. **Vackova, M., Kuchar, E., and Zemberyova, M.,** Cadmium in the environment and its determination, *Acta Fac. Rerum Nat. Univ. Comenianae Form. Prot. Nat.,* 7, 321, 1981.

488. **Petrov, I. and L'ochev, I.,** Soil pollution with cadmium and analytical methods for its determination, *Khig. Zdraveopaz.,* 24, 589, 1981.

489. **Mishima, M., Hoshiai, T., Watanuki, T., and Sugawara, K.,** Ion-exchange and atomic absorption spectrometric determination of trace metals in human teeth, *Koshu Eiseiin Kenkyu Hokoku,* 30, 40, 1981.

490. **Senesi, N. and Polemio, M.,** Trace element addition to soil by application of NPK fertilizers, *Fert. Res.,* 2, 289, 1981.

491. **Simon, J. and Liese, T.,** Trace determination of lead and cadmium in bones. Comparison of atomic absorption spectrometry and inverse polarography, *Fresenius' Z. Anal. Chem.,* 309, 383, 1981.

492. **Fosse, G., Justesen, N. P. B., and Wesenberg, G. B. R.,** Microstructure and chemical composition of fossil mammalian teeth, *Calcif. Tissue Int.,* 33, 521, 1981.

493. **Burton, L. C., Uppal, P. N., and Dwight, D. W.,** Cross diffusion of cadmium and zinc into copper (I) sulfide formed on zinc cadmium sulfide thin films, in *Conf. Rec. 15th IEEE Photovoltaic Spec. Conf.,* IEEE, Piscataway, N.J., 1981, 780.

494. **Kaiser, M. L., Koirtyohann, S. R., Hinderberger, F. J., and Taylor, H. E.,** Reduction of matrix interferences in furnace atomic absorption with the L'vov platform, *Spectrochim. Acta,* 36b, 773, 1981.

495. **Kol'tsova, N. G.,** Determination of cesium, iron, cadmium and calcium in silver catalysts by atomic absorption spectroscopy, *Khim. Promst. Ser. Khlornaya Promst.,* 4, 11, 1981.

496. **Zawadzki, H., Szymendera, J., Nowakowski, K., and Karczynaski, F.,** Analytical application of coordination compounds compounds of *N,N'*-diphenylguanidine with zinc (II), cadmium (II), mercury (II) and cobalt (II), *Chem. Anal. (Warsaw),* 26, 237, 1981.

497. **Kane, J. S. and Smith, H.,** Analysis of Egyptian Geological Survey and Mining Department samples by rapid rock and atomic absorption procedures, *U.S. Geol. Survey Open File Rep.,* 81, 59, 1981.

498. **Rohbock, E.,** The effect of airborne heavy metals on automobile passengers in Germany, *Environ. Int.,* 5, 133, 1981.

499. **Xia, L.,** Flame atomic absorption spectrophotometric determination of copper, cadmium, lead and zinc in tea leaves, *Fenxi Huaxue,* 9, 498, 1981.

500. **Jones, E. A. and Dixon, K.,** The separation of trace elements in manganese dioxide, *Natl. Inst. Metall. Repub. S. Afr. Rep.,* No. 2131, 1981.

501. **Borg, H., Edin, A., Holm, K., and Skoeld, E.,** Determination of metals in fish livers by flameless atomic absorption spectroscopy, *Water Res.,* 15, 1291, 1981.

502. **Constantini, S.,** Application of anodic voltammetry to direct determination of lead and cadmium in powdered milk for zootechnical use, *Riv. Soc. Ital. Sci. Aliment.,* 10, 239, 1981.

503. **Wen, J. and Geng, T.,** Determination of available cadmium, lead, copper and zinc in calcareous soil, *Fenxi Huaxue,* 9, 565, 1981.

504. **Fugas, M.,** Metals in airborne atmospheric particles, *Zast. Atoms.,* 9, 13, 1981 (in Serbo-Croatian).

505. **Marlier-Geets, O., Heck, J. P., Barideau, L., and Rocher, M.,** Method for Determination of Heavy Metals in Sewage Sludge, Their Distribution According to Their Origin and Their Concentration Variation with Time, Comm. Eur. Commodities, Rep. EUR 7076, Charact., Treat. Use Sewage Sludge, Serv. Sci. Sol., Fac. Sci. Agrono. Etat, Gembloux, Belgium, 1981, 284 (in French).

506. **Hirano, K., Iida, K., Shimada, T., Iguchi, K., and Nagasaki, Y.,** Study on practical application of monitoring method of total amounts of heavy metals in waste water on adsorption of chelating resins and ion-exchange resins, *Zenkoku Kogaiken Kaishi,* 6, 9, 1981 (in Japanese),

507. **Inhat, M.,** Analytical approach to the determination of copper, zinc, cadmium and lead in natural fresh waters, *Int. J. Environ. Anal. Chem.,* 10, 217, 1981.

508. **Jin, K., Matsuda, K., and Chiba, Y.,** Concentration of some metals in whole blood and plasma of normal adult subjects, *Hokkaidoritsu Eisei Kenkyushoho,* 31, 16, 1981.

509. **Hwang, H. L., Ho, J. S., Ou, H. J., Lee, Y. K., Sun, C. Y., Chen, C. J., and Loferski, J. J.,** Copper (I) sulfide/cadmium sulfide solar cells prepared by organometallic chemical vapor deposition: preliminary stage, in *Conf. Rec. 15th IEEE Photovoltaic Spec. Conf.,*IEEE, Piscataway, N.J., 1981, 1035.

510. **Qiying, L.,** Atomic absorption spectrometric determination of microamounts of copper, zinc, nickel, cobalt, cadmium and lead, *Fenxi Huaxue,* 9, 718, 1981 (in Chinese).

511. **Taliadouri-Voutsinou, F.,** Trace metals in marine organisms from the Saronikos Gulf (Greece), *J. Etud. Mar. Mediterrr.,* 5, 275, 1981.

512. **Brzozowska, B. and Zawadzka, T.,** Atomic absorption spectrophotometry in determination of lead, cadmium, zinc and copper in meat products, *Rocz. Panstw. Zakl. Hig.,* 32, 323, 1981 (in Polish).

513. **Djujic, I., Djordjevic, V., and Radovic, N.,** Lead, cadmium, copper, arsenic and mercury in some additives (for meat), *Technol. Mesa,* 22, 355, 1981.

514. **Brovko, I. A.,** Diphenylcarbazone as an extraction reagent in atomic absorption, *Deposited Doc. VINITI 1981,* 2842-81, VINITI, Moscow, U.S.S.R., 1981 (in Russian).

515. **Delves, H. T. and Woodward, J.,** The determination of low levels of cadmium in blood by electrothermal atomization and atomic absorption spectrophotometry, in *3rd Int. Conf. Heavy Met. Environ.,* CEP Consult. Ltd., Edinburgh, U.K., 1981, 622.

516. **Sutton, D. C., Rosa, W. C., and Legotte, P. A.,** Analytical measurements of selected metals in samples from a human metabolic study, *Trace Subst. Environ. Health,* 15, 270, 1981.

517. **Reboredo, F., Carrondo, M. J. T., Ganho, R. M. B., and Oliveira, J. F. S.,** Use of a rapid flameless atomic absorption method for the determination of the metallic content of sediments in the Tejo estuary, Portugal, in *3rd Int. Conf. Heavy Met. Environ.,* CEP Consult. Ltd., Edinburgh, U.K., 1981, 587.

518. **Lecomte, J., Mericam, P., and Astruc, M.,** A new scheme for the speciation of lead and cadmium in polluted river water, in *3rd Int. Conf. Heavy Met. Environ.,*CEP Consult. Ltd., Edinburgh, U.K., 1981, 678.

519. **Ashworth, M. J. and Farthing, R. H.,** Analysis of cadmium in marine samples, *Int. J. Environ. Anal. Chem.,* 10, 35, 1981.

520. **Balaes, G. E. E. and Robert, R. V. D.,** Determination by atomic absorption spectrophotometry of impurities in manganese dioxide, *Natl. Inst. Metall. Repub. S. Afr. Rep.,* No. 2094, 1981.

521. **Batley, G. E.,** *In situ* electrodeposition for the determination of lead and cadmium in sea water, *Anal. Chim. Acta,* 124, 121, 1981.

522. **Borg, H., Edin, A., Holm, K., and Skoeld, E.,** Determination of metals in fish livers by flameless atomic absorption spectroscopy, *Water Res.,* 15, 1291, 1981.

523. **Boyer, K. W., Jones, J. W., Linscott, D., Wright, S. K., Stroube, W., and Cunningham, W.,** Trace element levels in tissues from cattle fed a sewage sludge-amended diet, *J. Toxicol. Environ. Health,* 8, 281, 1981.

524. **Brzozowska, B. and Zawadzka, T.,** Determination of lead, cadmium, zinc and copper in vegetable products by atomic absorption spectrometry, *Rocz. Panstw. Zakl. Hig.,* 32, 9, 1981.

525. **Christensen, E. R. and Nan-Kwang, C.,** Fluxes of arsenic, lead, zinc and cadmium in Green Bay and Lake Michigan sediments, *Environ. Sci. Technol.,* 15, 553, 1981.

526. **Costantini, S., Macri, A., and Vernillo, I.,** Atomic absorption spectrophotometric analysis of mineral elements in milk-base feeds for zootechnical use, *Riv. Soc. Ital. Sci. Aliment.,* 10, 231, 1981 (in Portuguese).

527. **Debus, H., Hanle, W., Scharmann, A., and Wirz, P.,** Trace element detection by forward scattering using a cintinuum light source, *Spectrochim. Acta,* 36b, 1015, 1981.

528. **Foulkes, F. R., Smith, J. W., Kalia, R., and Kirk, D. W.,** Effect of cadmium impurities on the electrowinning of zinc, *J. Electrochem. Soc.,* 128, 2307, 1981.

529. **Massee, R. and Maessen, F. J. M. J.,** Losses of silver, arsenic, cadmium, selenium and zinc traces from distilled water and artificial seawater by sorption on various container surfaces, *Anal. Chim. Acta,* 127, 181, 1981.

530. **Salmela, S., Vouri, E., and Kilpio, J. O.,** Effect of washing procedures on trace element content of human hair, *Anal. Chim. Acta,* 125, 131, 1981.

531. **Shimizu, T., Shijo, Y., and Sakai, K.,** Determination of cadmium in human urine by graphite furnace atomic absorption spectrometry, *Bunseki Kagaku,* 30, 85e, 1981.

532. **Tunstall, M., Berndt, H., Sommer, D., and Ohls, K.,** Direct determination of trace elements in aluminum with flame atomic absorption spectrometry and ICP emission spectrometry. A comparison, *Erzmetall,* 34, 588, 1981.

533. **Viala, A., Gouezo, F., Mallet, B., Fondarai, J., Cano, J. P., Sauve, J. M., Grimaldi, F., and Deturmeny, E.,** Determination of four trace metals (lead, cadmium, chromium and zinc) in atmospheric dust in Marseille between 1977 and 1979, *Pollut. Atmos.,* 91, 107, 1981.

534. **Wunderlich, E.,** Determination of lead and cadmium in biological material, *Erzmetall,* 34, 577, 1981 (in German).

535. **Sharma, R. P., McKenzie, J. M., and Kjellstrom, T.,** Analysis of submicrogram levels of cadmium in whole blood, urine and hair by graphite furnace atomic absorption spectroscopy, *J. Anal. Toxicol.,* 6, 135, 1982.

536. **Wittig, H. and Faulhaber, E.,** Quantitative determination of toxic elements in microbiologically produced technical protein feeds. II. Determination of traces of lead and cadmium by atomic absorption spectrometry following extraction, *Nahrung,* 26, 277, 1982.

537. **Ueno, G.,** Atomic absorption spectrophotometric determination of cadmium, lead and copper in sugars using iodide-methyl isobutyl ketone extraction, *Seito Gijutsu Kenkyu Kaishi,* 30, 11, 1982.

538. **Viala, A., Gouezo, F., Mallet, B., Fondarai, J., Grimaldi, F., and Cano, J. P.,** Air pollution by cadmium in Marseille (France), *Toxicol. Eur. Res.,* 4, 25, 1982.

539. **Mojo, L., Martella, S., and Martino, G.,** Seasonal concentrations of heavy metals (mercury, cadmium and lead) in some marine organisms in the central Mediterranean Sea, *Mem. Biol. Mar. Oceanogr.,* 10, 27, 1980 (in Italian); CA 97, 97982k, 1982.

540. **Uratani, F., Yoshinaka, T., and Miyagi, M.,** The determination of trace amounts of elements by atomic absorption spectrochemical analysis. IV. Graphite furnace atomizer equipped with temperature feedback controller, *Osaka-furitsu Kogyo Gijutsu Kenkyusho Hokoku,* 77, 22, 1980; CA 97, 103428c, 1982.

541. **Mahalingam, T. R., Geetha, R., Thiruvengadasamy, A., and Mathews, C. K.,** Analysis of trace metals in sodium by flameless atomic absorption spectrophotometry, *Mater. Behav. Phys. Chem. Liq. Met. Syst. Proc. Conf.,* 1981, 329, 1982.

542. **Schaller, K. H. and Zober, A.,** Renal excretion of toxicologically relevant metals in occupationally non-exposed individuals, *Aerztl. Lab.,* 28, 209, 1982.

543. **Petrov, I. and Tsalev, D.,** Determination of cadmium in soils by electrothermal absorption spectrometry, *Dokl. Bolg. Akad. Nauk,* 35, 467, 1982.

544. **Hoenig, M. and Wollast, R.,** The possibilities and limitations of electrothermal atomization in atomic absorption spectrometry for the direct determination of trace metals in seawater, *Spectrochim. Acta,* 37b, 399, 1982.

545. **Yanagisawa, M., Kitagawa, K., and Tauge, S.,** Separative column atomizer for the direct determination of trace amounts of volatile elements by atomic absorption spectrometry, *Spectrochim. Acta,* 37b, 493, 1982.

546. **Balkas, T., Tugrul, S., and Salihoglu, I.,** Trace metal levels in fish and crustacea from Northeastern Mediterranean coastal waters, *Mar. Environ. Res.,* 6, 281, 1982.

547. **Ohta, K., Smith, B., Suzuki, M., and Winefordner, J. D.,** The determination of atom vapor diffusion coefficients by high temperature gas chromatography with atomic absorption detection, *Spectrochim. Acta,* 37b, 343, 1982.

548. **Pinta, M., De Kersabiec, A. M., and Richard, M. L.,** Possible uses of the Zeeman effect for correcting nonspecific absorptions in atomic absorption, *Analusis,* 10, 207, 1982.

549. **Akaiwa, H., Kawamoto, H., Koizumi, T., and Schindler, P. W.,** Preconcentration of cadmium (II) using synergistic extraction with dithizone and trioctylphosphine oxide, *Bunseki Kagaku,* 31, E151, 1982.

550. **Senften, H. and Pfenninger, H.,** Testing of beer for undesirable trace elements, *Brau. Rundsch.,* 93, 89, 1982.

551. **Zhang, X.,** Determination of chromium, cadmium and lead in biological samples by flame atomic absorption spectroscopy, *Zhonghua Yufangyixue Zazhi,* 16, 177, 1982.

552. **Brzezinska, A.,** Some remarks on the determination of trace metals in marine samples by electrothermal atomization, *Eur. Spectrosc. News,* 40, 19, 1982.

553. **Walker, T. I., Glover, J. W., and Powell, D. G. M.,** Effects of length, locality and tissue type on mercury and cadmium content of the commercial scallop, *Pacten alba* Tate, from Port Phillip Bay, Victoria, *Aust. J. Mar. Freshwater Res.,* 33, 547, 1982.

554. **Prokof'ev, A. K., Oradovskii, S. G., and Georgiesvskii, V. V.,** Flameless atomic absorption method for the determination of copper, lead and cadmium in marine bottom sediments, *Tr. Gos. Okeanogr. In-ta,* 162, 51, 1981; CA 97, 119764v, 1982.

555. **Alt, F., Berndt, H., and Messerschmidt, J.,** Determination of the toxic elements lead, cadmium and thallium in liver and human blood by differential analytical methods, *Aerztl. Lab.,* 28, 243, 1982.

556. **Bourcier, D. R., Sharma, R. P., and Brinkerhoff, C. R.,** Cadmium-copper interaction: tissue accumulation and subcellular distribution of cadmium in mice after simultaneous administration of cadmium and copper, *Trace Subst. Environ. Health,* 15, 190, 1981; CA 97, 139751s, 1982.

557. **Mannino, S.,** Determination of lead, copper and cadmium in wines using anode dissolution potentiometry, *Riv. Vitic. Enol.,* 35, 297, 1982.

558. **Burylev, B. P., Kulish, N. G., and Zabenko, L. U.,** Atomic absorption determination of cadmium in soils of the Krasnodar region, *Ref. Zh. Khim.,* Abstr. No. 14 G 223, 1982.

559. **Ritter, C. J.,** The dry-ashing method of preparing sewage sludge for cadmium and lead determinations by AAS, *Am. Lab.,* 14, 72, 1982.

560. **Zhou, Z. and Mao, X.,** A new chelate-forming resin bearing xylenol orange group and its applications, *Gaodeng Xuexiao Xuebao,* 3, 181, 1982.

561. **Hasan, M. Z., Kumar, A., and Pande, S. P.,** Determination of cadmium in water by flameless atomic absorption spectrophotometry, *Res. Ind.,* 27, 8, 1982.

562. **Sturgeon, R. E., Desauliners, J. A. H., Berman, S. S., and Russell, D. S.,** Determination of trace metals in esturine sediments by graphite furnace atomic absorption spectrometry, *Anal. Chim. Acta,* 134, 283, 1982.

563. **Olney, C. E., Schauer, P. S., McLean, S., Lu, Y., and Simpson, K. L.,** International study on Artemia. VIII. Comparison of the chlorinated hydrocarbons and heavy metals in five different strains of newly hatched artemia and a laboratory-reared marine fish, *Proc. Int. Symp. Brine Shrimp Artemia,* 3, Universa Press, Wetteren, Belgium, 1980, 343; CA 96, 81005n, 1982.

564. **Sperling, K. R.,** Determination of heavy metals in seawater and in marine organisms by flameless atomic absorption spectrophotometry. XIV. Comments on the usefulness of organohalides as solvents for the extraction of heavy metal (cadmium) complexes, *Fresenius' Z. Anal. Chem.,* 310, 254, 1982.

565. **Khalighie, J., Ure, A. M., and West, T. S.,** Atom-trapping absorption spectrometry with water-cooled metals collector tube, *Anal. Chim. Acta,* 134, 271, 1982.

566. **Rawat, N. S., Sinha, J. K., and Sahoo, B.,** Atomic absorption spectrophotometric and X-ray studies of respirable dusts in Indian coal mines, *Arch. Environ. Health,* 37, 32, 1982.

567. **Monteil, A. and Welte, B.,** Comparison of the different methods of attack during the determination of micropollutants in the sediments, C20-I/C20-13, *Acta 2nd, Cong. Mediterr. Ing. Quim.,* F.O.I.M., Barcelona, Spain, 1981; CA 96, 154622y, 1982.

568. **Ikebe, K. and Tanaka, R.,** Determination of heavy metals in blood by atomic absorption spectrometry. II, *Osaka-Puritsu Koshu Eisei Kenkyusho Kenkyu Hokoku Shokuhin Eisei Hen,* 11, 43, 1980 (in Japanese); CA 96, 137220y, 1982.

569. **Ranchet, J., Menissier, F., Lamathe, J., and Voinovitch, I.,** Interlaboratory comparison: the determination of cadmium, chromium, copper and lead in standard solutions by flameless atomic absorption spectrometry, *Analusis,* 10, 71, 1982.

570. **Kaufmann, R. E., Saba, C. S., Rhine, W. E., and Eisentraut, K. J.,** Quantitative multielement determination of metallic wear species in lubricating oils and hydraulic fluids, *Anal. Chem.,* 54, 975, 1982.

571. **Heinrichs, H. and Keltsch, H.,** Determination of arsenic, bismuth, cadmium, selenium and thallium by atomic absorption spectrometry with a volatilization technique, *Anal. Chem.,* 54, 1211, 1982.

572. **Peters, H. J. and Koehler, H.,** Remarks on direct flame atomic absorption analysis in determining the trace elements cadmium, cobalt and chromium in biological materials, *Zentralbl. Pharmakother. Laboratoriumsdiagn.,* 121, 129, 1982.

573. **Harada, Y.,** Determination of trace impurities in high-purity aluminum by flame atomic absorption spectrometry after coprecipitation with nickel hydroxide, *Bunseki Kagaku,* 31, 130, 1982.

574. **Hoenig, M., Lima, C., and Dupire, S.,** Validity of atomic absorption spectrometric determination of cadmium, cobalt, chromium, nickel and lead in animal tissues, *Analusis,* 10, 132, 1982 (in French).

575. **Landsberger, S., Jervis, R. E., Aufreiter, S., and Van Loon, J. C.,** The determination of heavy metals (aluminum, manganese, iron, nickel, copper, zinc, cadmium and lead) in urban snow using an atomic absorption graphite furnace, *Chemosphere,* 11, 237, 1982.

576. **Burton, L. C., Uppal, P. N., and Dwight, D. W.,** Cross diffusion of cadmium and zinc into cuprous sulfide formed on zinc cadmium sulfide ($Zn_xCd_{1-x}S$) thin films, *J. Appl. Phys.,* 53, 1538,1982.

577. **Bangia, T. R., Kartha, K. N. K., Varghese, M., Dhawala, B. A., and Joshi, B. D.,** Chemical separation and electrothermal atomic absorption spectrophotometric determination of cadmium, cobalt, copper and nickel in high-purity uranium, *Fresenius' Z. Anal. Chem.,* 310, 410, 1982.

578. **Bragin, G. Y. and Sadagov, Y. M.,** Spatial distribution of temperature in graphite tubular furnace of electrothermal atomizers, *Zh. Prikl. Spektrosk.,* 36, 185, 1982.

579. **Severin, G., Schumacher, E., and Umland, F.,** Determination of cadmium by flameless atomic absorption spectroscopy. II. Modification of graphite furnace tubes by aluminum oxide, *Fresenius' Z. Anal. Chem.,* 311, 205, 1982 (in German).

580. **Schilcher, H.,** Residues and impurities in medicinal plants and drug preparations. I. Report: on the determination of value and quality control of drugs, *Planta Med.,* 44, 65, 1982.

581. **Thamzil, L., Suwirma, S., and Surtipanti, S.,** Distribution of heavy metals in the stream of Sunter River, *Majalah BATAN,* 13, 41, 1982 (in Indonesian).

582. **Surwirma, S., Surtipanti, S., and Thamzil, L.,** Distribution of heavy metals mercury, lead, cadmium, chromium, copper and zinc in fish, *Majalah BATAN,* 13, 9, 1980 (in Indonesian); CA 96, 15797w, 1982.

583. **Muhlbaier, J., Stevens, C., Graczyk, D., and Tisue, T.,** Determination of cadmium in Lake Michigan by mass spectrometric isotope dilution analysis or atomic absorption spectrometry following electrodeposition, *Anal. Chem.,* 54, 496, 1982.

584. **Danielsson, L. G., Magnusson, B., and Zhang, K.,** Matrix interference in the determination of trace metals by graphite furnace AAS after Chelex-100 preconcentration, *At. Spectrosc.,* 3, 39, 1982.

585. **Kasperek, K., Iyengar, G. V., Feinendegen, L. E., Hashish, S., and Mahfouz, M.,** Multielement analysis of fingernail, scalp hair and water samples from Egypt (a preliminary study), *Sci. Total. Environ.,* 22, 149, 1982.

586. **Baldini, M., Grossi, M., Micco, C., and Stacchini, A.,** Presence of metal in cereal. II. Contamination of 1978 Italian rice, *Riv. Soc. Ital. Sci. Aliment.,* 11, 23, 1982.

587. **Cirlin, E. H. and Housley, R. M.,** Distribution and evaluation of zinc, cadmium and lead in Apollo 16 regolith samples and the average uranium-led ages of the parent rocks, *Geochim. Cosmochim. Acta Suppl.,* 16, 529, 1982.

588. **Yamagata, N.,** Interlaboratory comparison study on the reliability of environmental analysis of soil and sediment 1978—80, *Bunseki Kagaku,* 31, T1, 1982 (in Japanese).

589. **Shen, W., Zhang, Y., Teng, G., and Mao, J.,** Determination of cadmium in foods, *Zhonghua Yufangyixue Zazhi,* 16, 45, 1982.

590. **Fuwa, K.,** Metallothionein: historical aspects of trace spectrochemical analysis, *Dev. Toxicol. Environ. Sci.,* 9, 1, 1982.

591. **Fischer, W. R. and Fechter, H.,** Analytical determination and fractionation of copper, zinc, lead, cadmium, nickel and cobalt in soils and underwater soils, *Z. Pflanzenernaehr. Bodenkd.,* 145, 151, 1982 (in German).

592. **Wesenberg, G. B.,** Effect of cadmium on calcium, magnesium and phosphorus content of rat molars, *Scand. J. Dent. Res.,* 90, 95, 1982.

593. **Briggs, R. W. and Armitage, I. M.,** Evidence for site-selective metal binding in calf liver metallothionein, *J. Biol. Chem.,* 257, 1259, 1982.

594. **Cammann, K. and Andersson, J. T.,** Increased sensitivity and reproducibility through signal averaging in ranges near and instrumental limit of detection. Thallium and cadmium trace determinations in rock samples, *Z. Anal. Chem.,* 310, 45, 1982.

595. **Lendermann, B. and Hundeshagen, D.,** Use of multielement standards for calibration in water analysis by AAS, *Z. Anal. Chem.,* 310, 415, 1982.

596. **Maitani, T. and Suzuki, K.,** Changes of essential metal levels in selected tissues and splenomegaly induced by the injection of suspending cadmium salt into mice, *Toxicol. Appl. Pharmacol.,* 62, 219, 1982.

597. **Salmon, S. G. and Holcombe, J. A.,** Alteration of metal release mechanisms in graphite furnace atomizers by chemisorbed oxygen, *Anal. Chem.,* 54, 630, 1982.

598. **Severin, G., Schumacher, E., and Umland, F.,** Determination of cadmium by flameless atomic absorption spectroscopy. I. Surface treatment of graphite furnace tubes by carbide-forming metals, *Z. Anal. Chem.,* 311, 201, 1982 (in German).

599. **Wachter, C. and Weisweller, W.,** Extraction of thallium and simultaneous extraction of lead, cadmium and thallium for their quantitative analysis using flame atomic absorption spectroscopy, *Mikrochim. Acta,* 1, 307, 1982.

AUTHOR INDEX

SUBJECT INDEX (CADMIUM)

MERCURY, Hg (ATOMIC WEIGHT 200.59)

Mercury occurs freely in nature, but not abundantly. It was known to ancient Hindus and Chinese. It is used in making thermometers, barometers, diffusion pumps, mercury vapor lamps, advertising signs, pesticides, dental preparations, antifouling paints and pigments, batteries, catalysts, explosives, and also in medicines. Mercury is used in the extraction of gold from ores by forming the amalgam. Organic mercurials are used as fungicides, slimicides, and antiseptics, etc.

Mercury vapors and its soluble salts are highly toxic. Care must be taken when mercury and its compounds are handled. Determination of mercury is required in ores, minerals, biological samples, fish, food, air soil, water, waste, and other environmental samples. Its trace analysis is important due to its toxicity.

Preparation of sample solution is very important due to volatilization loss of mercury. Decomposition with acids or fusion for organic samples should be performed under controlled and closed conditions.

In an air-acetylene flame, mercury I and mercury II show different sensitivities and cause interference. This can be corrected by the addition of 1 mℓ of freshly prepared 20% (w/v) stannous chloride to each solution before analysis. Reducing agent will also produce same effect. Large concentrations of cobalt will absorb at the 253.7-nm resonance line. A 1000 mg/ℓ cobalt solution produces approximately 10% absorption. The flame method is usually not recommended unless it is absolutely necessary, as the elemental mercury formed is readily lost due to evaporation. For improved sensitivity mercury cold vapor should be used.

Standard Solution

To prepare 1000 mg/ℓ solution, dissolve 1.080 g of mercury (II) oxide (HgO) in 1:1 hydrochloric acid and dilute to 1 ℓ with deionized water.

Instrumental Parameters

Wavelength	253.7 nm
Slit	0.7 nm
Light source	Hollow cathode lamp
Flame	Air-acetylene, oxidizing (lean, blue)
Sensitivity	2.5—4.2 mg/ℓ
Detection limit	0.28 mg/ℓ
Optimum range	200 mg/ℓ

REFERENCES

1. **Woodson, T. T.,** A new mercury vapor detector, *Rev. Sci. Instrum.,* 10, 308, 1939.
2. **Zuehlke, C. W. and Ballard, A. E.,** Photometric method for estimation of minute amounts of mercury, *Anal. Chem.,* 22, 953, 1950.
3. **Lindstrom, O.,** Rapid microdetermination of mercury by spectrophotometric flame combustion, *Anal. Chem.,* 31, 461, 1959.
4. **Willis, J. B.,** Determination of lead and other heavy metals in urine by atomic absorption spectroscopy, *Anal. Chem.,* 34, 614, 1962.
5. **Fledman, C. and Dhumwad, R.,** An atomic absorption tube for use with an atomizer burner: application to the determination of mercury, in Proc. 6th Conf. Anal. Chem. in Nuclear Reactor Technology, AEC TID 7655, Gatlinburg, 1962, 1963; available from the NTIS, Springfield, Va.
6. **Poluektov, N. S. and Vitkun, R. A.,** Atomic absorption determination of mercury by means of a flame method, *Zh. Anal. Khim.,* 18, 37, 1963.
7. **Allan, J. E.,** Absorption flame photometry below 2000 Å, 4th Australian Spectroscopy Conf., August 1963.
8. **Poluektov, N. S., Vitkun, R. A., and Zelyukova, Y. V.,** Determination of milligram amounts of mercury by atomic absorption in the gaseous phase, *Zh. Anal. Khim.,* 19, 937, 1964.
9. **Vaughn, W. W. and McCarthy, J. H., Jr.,** An instrumental technique for the determination of submicrogram concentrations of mercury in soils, rocks and gas, *U.S. Geol. Surv. Prof. Pap.,* 501-D, D123, 1964.
10. **Kuznetsov, Yu. N. and Chabovskii, L. P.,** Rapid atomic absorption determination of mercury in powdered samples, *Ulch. Zap. Tsentr. Nauchno-Issled. Inst. Olvyan. Prom.,* 75, 1964.
11. **Pappas, E. G. and Rosenberg, L. A.,** Determination of submicrogram quantities of mercury by cold vapor atomic absorption photometry, *J. Assoc. Off. Anal. Chem.,* 49, 792, 1966.
12. **Schachter, M. M.,** Apparatus for cold vapor atomic absorption of mercury, *J. Assoc. Off. Anal. Chem.,* 49, 778, 1966.
13. **Barringer, A. R.,** Interference-free spectrometer for high sensitivity mercury analysis of soils, rocks and air, *Trans. Inst. Min. Metall.,* B75, 120, 1966.
14. **Mulford, C. E.,** Low-temperature ashing for determination of volatile metals by atomic absorption spectroscopy, *At. Absorpt. Newsl.,* 5, 135, 1966.
15. **Hingle, D. N., Kirkbright, G. F., and West, T. S.,** Some observations on the determination of mercury by atomic absorption spectroscopy in an air-acetylene flame, *Analyst (London),* 92, 752, 1967.
16. **Berman, E.,** Determination of cadmium, thallium and mercury in biological materials, *At. Absorpt. Newsl.,* 6, 57, 1967.
17. **Brandenberger, H. and Bader, H.,** The determination of nanogram levels of mercury in solution by a flameless atomic absorption technique, *At. Absorpt. Newsl.,* 6, 101, 1967.
18. **Tindall, F. M.,** Mercury analysis by atomic absorption spectrophotometry, *At. Absorpt. Newsl.,* 6, 104, 1967.
19. **Yamamoto, Y., Kumamaru, T., Hayashi, Y., and Otani, Y.,** Indirect determination of mercury by atomic absorption spectrophotometry with a zinc hollow cathode lamp, *Anal. Lett.,* 1, 955, 1968.
20. **Brandenberger, H. and Bader, H.,** The determination of mercury by flameless atomic absorption. II. A static vapor method, *At. Absorpt. Newsl.,* 7, 53, 1968.
21. **Hatch, W. R. and Ott, W. L.,** Determination of sub-microgram quantities of mercury by atomic absorption spectrophotometry, *Anal. Chem.,* 40, 2085, 1968.
22. **Ling, C.,** Portable atomic absorption photometer for determining nanogram quantities of mercury in the presence of interfering substances, *Anal. Chem.,* 40, 1876, 1968.
23. **Manning, D. C. and Fernandez, F.,** Cobalt spectral interference in the determination of mercury, *At. Absorpt. Newsl.,* 7, 24, 1968.
24. **Browner, R. F., Dagnall, R. M., and West, T. S.,** Atomic fluorescence spectroscopy. Atomic fluorescence and atomic absorption of thallium and mercury with electrodeless discharge tubes as sources, *Talanta,* 16, 75, 1968.
25. **Thilliez, G.,** Rapid and accurate determination of mercury in air and biological samples by atomic absorption spectrophotometry, *Chim. Anal.,* 50, 1968 (in French).
26. **Igoshin, A. M. and Bogusevich, L. N.,** Atomic absorption determination of mercury in water without use of a flame, *Gidrokhim. Mater.,* 47, 150, 1968 (in Russian).
27. **L'vov, B. V. and Khartsyzov, A. D.,** Atomic absorption determination of sulfur, phosphorus, iodine and mercury from resonance lines in the vacuum ultraviolet spectral region, *Zh. Prikl. Spektrosk.,* 11, 413, 1969.
28. **Poluektov, N. S. and Zolyukova, Yu. V.,** Atomic absorption determination of traces of mercury in alkali-metal hydroxides, *Zavod. Lab.,* 35, 186, 1969 (in Russian).

29. **Chau, Y. K. and Saitoh, H.,** Determination of submicrogram quantities of mercury in lake waters, *Environ. Sci. Technol.,* 4, 839, 1969.
30. **Delaughter, B.,** Mercury determination in industrial plant atmospheres by atomic absorption spectrophotometry, *At. Absorpt. Newsl.,* 9, 1969.
31. **Fishman, M. J.,** Determination of mercury in water, *Anal. Chem.,* 42, 1462, 1970.
32. **Goulden, P. D. and Afghan, B. K.,** An Automated Method for Determining Mercury in Water, No. 27, Inland Waters Branch Tech. Bull. Canadian Department of Energy, Mines and Resources, Ottawa, Canada, 1970.
33. **Jeffus, M. T., Elkins, J. S., and Kenner, C. T.,** Determination of mercury in biological materials, *J. Assoc. Off. Anal. Chem.,* 53, 1172, 1970.
34. **Kalb, G. W.,** The determination of mercury in water and sediment samples by flameless atomic absorption, *At. Absorpt. Newsl.,* 9, 84, 1970.
35. **Lindstedt, G.,** A rapid method for the determination of mercury in urine, *Analyst (London),* 95, 264, 1970.
36. **Manning, D. C.,** Nonflame methods for mercury determination by atomic absorption. A review, *At. Absorpt. Newsl.,* 9, 97, 1970.
37. **Manning, D. C.,** Compensation for broad-band absorption interference in the flameless atomic absorption determination of mercury, *At. Absorpt. Newsl.,* 9, 109, 1970.
38. **Mesman, B. B. and Smith, B. S.,** Determination of mercury in urine by atomic absorption utilizing the APDC/MIBK extraction and boat technique, *At. Absorpt. Newsl.,* 9, 81, 1970.
39. **Moffitt, A. E., Jr. and Kupel, R. E.,** A rapid method employing impregnated charcoal and atomic absorption spectrophotometry for the determination of mercury in atmospheric, biological and aquatic samples, *At. Absorpt. Newsl.,* 9, 113, 1970.
40. **Mittelhauser, H. M.,** Mercury analysis by flameless atomic absorption, *At. Absorpt. Newsl.,* 9, 134, .
41. **Shimomura, S., Fukumoto, Y., Hashimoto, M., and Tanase, Y.,** Effect of reducing agents on determination of mercury by atomic absorption method, *Bunseki Kagaku,* 19, 1296, 1970.
42. **Silvester, M. D. and McCarthy, W. J.,** The intensity of electrodeless discharge lamps containing cadmium, mercury and manganese, *Spectrochim. Acta,* 25b, 229, 1970.
43. **Hwang, J. Y. and Ullucci, P. A.,** Analysis of mercury by flameless AAS, *Am. Lab.,* 2, 50, 1970.
44. **Harper, W. H.,** Determination of parts per million of mercury in eye-drops by atomic absorption, *Proc. Soc. Anal. Chem.,* 7, 104, 1970.
45. **Osland, R.,** Atomic absorption determination of mercury using a vapor technique, *Spectrovision,* 24, 11, 1970.
46. **Mesman, B. B., Smith, B. S., and Pierce, J. O., II,** Determination of mercury in urine by atomic absorption, *Am. Ind. Hyg. Assoc. J.,* 31, 701, 1970.
47. **April, R. W. and Hume, D. N.,** Environmental mercury. Rapid determination in water at nanogram levels, *Science,* 170, 849, 1970.
48. **Cumont, G.,** Determination of mercury by atomic absorption spectrophotometry, *Ann. Falsif. Expert. Chim.,* 64, 115, 1971 (in French).
49. **Duffer, J. K.,** Determination of mercury at the parts-per-billion level, *J. Paint Technol.,* 43, 67, 1971.
50. **Krause, L. A., Henderson, H., Shotwell, H. P., and Culp, D. A.,** Analysis for mercury in urine, blood, water and air, *Am. Ind. Hyg. Assoc. J.,* 32, 331, 1971.
51. **Magos, L.,** Selective atomic absorption determination of inorganic mercury and methyl mercury in undigested biological samples, *Analyst (London),* 96, 847, 1971.
52. **Ribeiro, N. T.,** Determination of mercury (as the preservative thiomersal) in vaccines by atomic absorption spectrophotometry, *Rev. Farm. Bioquim. Univ. Paulo,* 9, 357, 1971 (in Portuguese).
53. **Olivier, M.,** Method for the determination of traces of mercury by flameless atomic absorption, *Z. Anal. Chem.,* 257, 187, 1971 (in German).
54. **Moffitt, A. E., Jr. and Kupel, R. E.,** Rapid method employing impregnated charcoal and atomic absorption spectrophotometry for the determination of mercury (in air, water and biological materials), *Am. Ind. Hyg. Assoc. J.,* 32, 614, 1971.
55. **Cavallaro, A. and Elli, G.,** Determination of mercury in foods, industrial waste, liquors and potable water by flameless atomic absorption, *Boll. Lab. Chim. Prov.,* 22, 168, 1971 (in Italian).
56. **Schlesinger, M. D. and Schultz, H.,** Analysis for mercury in coal, *U.S. Bur. Mines Tech. Prog. Rep.,* 43, 1971.
57. **Christian, C. M., II and Robinson, J. W.,** The direct determination of cadmium and mercury in the atmosphere, *Anal. Chim. Acta,* 56, 466, 1971.
58. **Braun, R. and Husbands, A. P.,** Determination of low levels of mercury by cold-vapor atomic absorption, *Spectrovision,* 26, 2, 1971.
59. **Thorpe, V. A.,** Determination of mercury in food products and biological fluids by aeration and flameless atomic absorption spectrophotometry, *J. Assoc. Off. Anal. Chem.,* 54, 206, 1971.

60. **Lindstedt, G. and Skare, I.,** Microdetermination of mercury in biological samples. II. An apparatus for rapid automatic determination of mercury in digested samples, *Analyst (London),* 96, 223, 1971.

61. **Weissberg, B. G.,** Determination of mercury in soils by flameless atomic absorption spectrometry, *Econ. Geol.,* 66, 1042, 1971.

62. **Stainton, M. P.,** Syringe procedure for transfer of nanogram quantities of mercury vapor for flameless atomic absorption spectrophotometry, *Anal. Chem.,* 43, 625, 1971.

63. **Brooks, J. D. and Wolfram, W. E.,** Trace mercury determination, *Am. Lab.,* 3, 54, 1971.

64. **Uthe, J. F., Armstrong, F. A. J., and Tam, K. C.,** Determination of trace amounts of mercury in fish tissues: results of a North American check sample study, *J. Assoc. Off. Anal. Chem.,* 54, 866, 1971.

65. **Hadeishi, T. and McLaughlin, R. D.,** Hyperfine Zeeman effect. A.A. Spectrometer for mercury, *Science,* 174, 404, 1971.

66. **Goleb, J. A.,** Use of a permanent mercury source for spectroscopic studies of varying amounts of mercury, *Appl. Spectrosc.,* 25, 94, 1971.

67. **Smith, J. D., Nicholson, R. A., and Moore, P. J.,** Mercury in water of the tidal Thames, *Nature (London),* 232, 303, 1971.

68. **Omang, S. H.,** Determination of mercury in natural waters and effluents by flameless atomic absorption spectrophotometry, *Anal. Chim. Acta,* 53, 415, 1971.

69. **Melton, J. R., Hoover, W. L., and Howard, P. A.,** Determination of mercury in soils by flameless atomic absorption, *Soil Sci. Soc. Am. Proc.,* 35, 850, 1971.

70. **Hey, H.,** Atomic absorption spectrometry as a mercury specific detecting system for gas chromatography, *Z. Anal. Chem.,* 256, 361, 1971.

71. **Munns, R. K. and Holland, D. C.,** Determination of mercury in fish by flameless atomic absorption. Collaborative study, *J. Assoc. Off. Anal. Chem.,* 54, 202, 1971.

72. **Fabbrini, A., Modi, G., Signorelli, L., and Simiani, G.,** Determination of total mercury and methyl mercury in canned fish products: results and considerations, *Boll. Lab. Chim. Prov.,* 22, 339, 1971 (in Italian).

73. **Law, S. L.,** Absorption enhancement by an organometallic in aqueous solution-methyl mercury in dilute acid, *At. Absorpt. Newsl.,* 10, 75, 1971.

74. **Omang, S. H. and Paus, P. E.,** Trace determination of mercury in geological materials by flameless atomic absorption spectroscopy, *Anal. Chim. Acta,* 56, 393, 1971.

75. **Schaller, K. H., Strasser, P., Woitowitz, R., and Szadkowski, D.,** Quantitative determination of traces of mercury in urine after electrolytic concentration, *Z. Anal. Chem.,* 256, 123, 1971 (in German).

76. **Umezaki, Y. and Iwamoto, K.,** Determination of submicrogram amounts of mercury in water by flameless atomic absorption spectrophotometry. Differentiation of inorganic and organic mercury, *Bunseki Kagaku,* 20, 173, 1971.

77. **Carisano, A., Riva, M., and Daghetta, A.,** Determination of mercury in tuna fish, *Ind. Conserve,* 46, 194, 1971 (in Italian).

78. **Anderson, D. H., Evans, J. H., Murphy, J. J., and White, W. W.,** Determination of mercury by a combustion technique using gold as a collector, *Anal. Chem.,* 43, 1511, 1971.

79. **Doherty, P. E. and Dorsett, R. S.,** Determination of trace concentrations of mercury in environmental water samples, *Anal. Chem.,* 43, 1887, 1971.

80. **Bailey, B. W. and Lo, F. C.,** Automated method for determination of mercury, *Anal. Chem.,* 43, 1525, 1971.

81. **Lee, D. C. and Laufmann, C. W.,** Determination of submicrogram quantities of mercury in pulp and paperboard by flameless atomic absorption spectrometry, *Anal. Chem.,* 43, 1127, 1971.

82. **Armstrong, F. A. J. and Uthe, J. F.,** Semi-automated determination of mercury in animal tissue, *At. Absorpt. Newsl.,* 10, 101, 1971.

83. **Bailey, B. W. and Lo, F. C.,** Cold water atomic absorption determination of mercury in coal, *J. Assoc. Off. Anal. Chem.,* 54, 1447, 1971.

84. **Cumont, G.,** Flameless atomic absorption spectrophotometry for the determination of mercury (I), *Chim. Anal.,* 53, 634, 1971.

85. **Hadeishi, T. and Klingebiel, U. I.,** Hyperfine Zeeman effect atomic absorption spectrometer for mercury, *Science,* 174, 404, 1971.

86. **Hoover, W. L., Melton, J. R., and Howard, P. A.,** Determination of trace amounts of mercury in foods by flameless atomic absorption, *J. Assoc. Off. Anal. Chem.,* 54, 860, 1971.

87. **Joensuu, O. I.,** Mercury-vapor detector, *Appl. Spectrosc.,* 25, 526, 1971.

88. **Hwang, J. Y., Ullucci, P. A., and Malenfant, A. L.,** Determination of mercury by a flameless atomic absorption technique, *Can. Spectrosc.,* 16, 100, 1971.

89. **Kahn, H. L.,** A mercury analysis system, *At. Absorpt. Newsl.,* 10, 58, 1971.

90. **Casiewicz, T. A. and Dinan, F. J.,** Concentration of mercury in the manufacture of fish protein concentrate by isopropyl alcohol extraction of sheepshead and carp, *Environ. Sci. Technol.,* 6, 726, 1972.

91. **Kunkel, E.,** Determination of traces of mercury by wickhold combustion and flameless atomic absorption, *Z. Anal. Chem.,* 258, 337, 1972 (in German).

92. **Gutenmann, W. H., Lisk, D. J., and Grier, N.,** Flameless atomic absorption response of mercury without added reducing agent, *Bull. Environ. Contam. Toxicol.,* 8, 138, 1972.

93. **Iskandar, I. K., Syers, J. K., Jacobs, L. W., Keeney, D. R., and Gilmour, J. T.,** Determination of total mercury in sediments and soils, *Analyst (London),* 97, 388, 1972.

94. **Fukamachi, K., Morimoto, M., and Tokunaga, T.,** Indirect determination of microamounts of mercury by atomic absorption spectrophotometry using the replacement reaction, *Bunseki Kagaku,* 21, 1173, 1972.

95. **Uthe, J. F., Solomon, J., and Grift, B.,** Rapid semimicro method for the determination of methyl mercury in fish tissue, *J. Assoc. Off. Anal. Chem.,* 55, 583, 1972.

96. **Okuno, I., Wilson, R. A., and White, R. E.,** Determination of mercury in biological samples by flameless atomic absorption after combustion and mercury-silver amalgamation, *J. Assoc. Off. Anal. Chem.,* 55, 96, 1972.

97. **Evans, R. J., Bails, J. D., and D'Itri, F. M.,** Mercury levels in muscle tissues of reserved museum fish, *Environ. Sci. Technol.,* 6, 901, 1972.

98. **Gonzalez, J. G. and Ross, R. T.,** Interfacing of an atomic absorption spectrophotometer with a gas-liquid chromatograph for the determination of trace quantities of alkyl mercury compounds in fish tissue, *Anal. Lett.,* 5, 683, 1972.

99. **O'Gorman, J. V., Suhr, N. H., and Walker, P. L., Jr.,** The determination of mercury in some American coals, *Appl. Spectrosc.,* 26, 44, 1972.

100. **Kressin, I.,** Rapid separation and determination of mercury, *Talanta,* 19, 197, 1972.

101. **Wolber, D. R. and Bosshart, R. E.,** Inexpensive absorption cell for use in the determination of mercury by the flameless atomic absorption technique, *Anal. Chem.,* 44, 1546, 1972.

102. **Windham, R. L.,** Simple device for compensation of broad-band absorption interference in flameless atomic absorption determination of mercury, *Anal. Chem.,* 44, 1334, 1972.

103. **Saha, J. C. and Lee, Y. W.,** Interference of fats in the determination of mercury residues in fish by atomic absorption spectrometry, *Bull. Environ. Contam. Toxicol.,* 7, 301, 1972.

104. **Monteil, A.,** Determination of mercury in water by flameless atomic absorption, *Analusis,* 1, 66, 1972 (in French).

105. **Rains, T. C. and Menis, O.,** Determination of submicrogram amounts of mercury in standard reference materials by flameless atomic absorption spectrometry, *J. Assoc. Off. Anal. Chem.,* 55, 1339, 1972.

106. **Kubasik, N. P., Sine, H. E., and Volosin, M. T.,** Rapid analysis for total mercury in urine and plasma by flameless atomic absorption analysis, *Clin. Chem. (Winston-Salem, N.C.),* 18, 1326, 1972.

107. **Klemm, W. A. and Fetter, N. R.,** An improved apparatus for the flameless atomic absorption determination of mercury, *At. Absorpt. Newsl.,* 11, 108, 1972.

108. **Kato, K., Ando, A., and Kishimoto, T.,** Determination of trace amount of mercury in rocks and soils by a flameless atomic absorption spectrophotometry, *Bunseki Kagaku,* 21, 1057, 1972.

109. **Robinson, J. W., Slevin, P. J., Hindman, G. D., and Wolcott, D. K.,** Non-flame atomic absorption in the vacuum ultraviolet region. Direct determination of mercury in air at the 184.9 nm resonance line, *Anal. Chim. Acta,* 61, 431, 1972.

110. **Jones, A. M., Jones, Y., and Stewart, W. D. P.,** Mercury in marine organisms of the Tay region, *Nature (London),* 238, 164, 1972.

111. **Holak, W., Krintz, B., and Williams, J. C.,** Simple, rapid digestion technique for the determination of mercury in fish by flameless atomic absorption, *J. Assoc. Off. Anal. Chem.,* 55, 741, 1972.

112. **Ciusa, W., Giaccio, M., and Ottombrini, G.,** Proportioning methods in analysis of mercury traces by atomic absorption, *Ann. Chim.,* 62, 426, 1972.

113. **Applequist, M. D., Katz, A., and Turekian, K. K.,** Distribution of mercury in the sediments of New Haven (Conn.) harbor, *Environ. Sci. Technol.,* 6, 1123, 1972.

114. **Lidums, V.,** Determination of mercury in small quantities by direct combustion with cold vapor atomic absorption, *Chem. Ser.,* 2, 159, 1972.

115. **Jones, R. L. and Hinesly, T. D.,** Total mercury content in morrow plot soils over a period of 63 years, *Soil. Sci. Soc. Am. Proc.,* 36, 921, 1972.

116. **Nicholson, R. A. and Smith, J. D.,** Source unit for flameless atomic absorption determination of mercury, *Lab. Pract.,* 21, 638, 1972.

117. **Cheremisinoff, P. N. and Habib, Y. H.,** Cadmium, chromium, lead and mercury: plenary account for water pollution. I. Occurrence, toxicity and detection, *Water Sewage Works,* 119(7), 73, 1972.

118. **Magos, L. and Clarkson, T.,** Atomic absorption determination of total, inorganic and organic mercury in blood, *J. Assoc. Off. Anal. Chem.,* 55, 966, 1972.

119. **Aston, S. R. and Riley, J. P.,** The determination of mercury in rocks and sediments, *Anal. Chim. Acta,* 59, 349, 1972.

120. **Masri, M. S. and Friedman, M.,** Mercury uptake by polyamine-carbohydrates, *Environ. Sci. Technol.,* 6, 745, 1972.

121. **Topping, G. and Pirie, J. M.,** Determination of inorganic mercury in natural waters, *Anal. Chim. Acta,* 62, 560, 1972.

122. **Klein, D. H.,** Mercury and other metals in urban soils, *Environ. Soil. Technol.,* 6, 560, 1972.

123. **Workman, E. J.,** Determination of mercury, indium and tellurium in mercury indium tellurium, *Analyst (London),* 97, 703, 1972.

124. **Skare, I.,** Microdetermination of mercury in biological samples. III. Automated determination of mercury in urine, fish and blood samples, *Analyst (London),* 97, 148, 1972.

125. **Clark, D., Dagnall, R. M., and West, T. S.,** The determination of mercury by atomic absorption spectrophotometry with the "Delves sampling cup" technique, *Anal. Chim. Acta,* 58, 339, 1972.

126. **Henry, H. G., Stever, K. R., Barry, W. L., and Heady, H. H.,** Determination of mercury in low-grade ores, *Appl. Spectrosc.,* 26, 288, 1972.

127. **Cranston, R. E. and Buckley, D. E.,** Mercury pathways in a river estuary, *Environ. Sci. Technol.,* 6, 274, 1972.

128. **Dagnall, R. M., Manfield, J. M., Silvester, M. D., and West, T. S.,** Relative sensitivities of the 184.9 and 253.7 nm mercury lines in atomic absorption spectrometry, *Nature (London) Phys. Sci.,* 235, 156, 1972.

129. **Marinenko, J., May, I., and Dinnin, J. I.,** Determination of mercury in geological materials by flameless atomic absorption spectrometry, *U.S. Geol. Surv. Prof. Pap.,* 800b, B151, 1972.

130. **Huffman, C., Jr., Rahill, R. L., Shaw, V. E., and Norton, D. R.,** Determination of mercury in geological materials by flameless atomic absorption spectroscopy, *U.S. Geol. Surv. Prof. Pap.,* 800-C, 203, 1972.

131. **Stephens, R.,** The application of a fast pulse atom reservoir to the determination of mercury, *Anal. Lett.,* 5, 851, 1972.

132. **Band, R. B. and Wilkinson, N. M.,** Interference in the determination of mercury in mineralized samples by the wet reduction-flameless atomic absorption method, *J. Geochem. Explor.,* 1, 195, 1972.

133. **Malaiyandi, M. and Barrett, J. P.,** Wet oxidation method for the determination of submicrogram quantities of mercury in cereal grains, *J. Assoc. Off. Anal. Chem.,* 55, 951, 1972.

134. **Coyne, R. V. and Collins, J. A.,** Loss of mercury from water during storage, *Anal. Chem.,* 44, 1093, 1972.

135. **Kopp, J. F., Longbottom, M. C., and Lobring, L. B.,** "Cold vapor" method for determining mercury in water, *J. Am. Water Works Assoc.,* 64, 20, 1972.

136. **Eichner, K.,** Examination of eggs for mercury by atomic absorption spectrophotometry, *Lebensmittelchemie Gerichtl. Chem.,* 26, 240, 1972 (in German).

137. **Dowd, G. and Thomas, R. S.,** Electronic design parameters for optimizing photometric mercury detectors, *Chem. Instrum.,* 5, 231, 1973.

138. **Bouchard, A.,** Determination of mercury after room temperature digestion by flameless atomic absorption, *At. Absorpt. Newsl.,* 12, 115, 1973.

139. **Gilbert, T. R. and Hume, D. N.,** Improved apparatus for determination of mercury by flameless atomic absorption, *Anal. Chim. Acta,* 65, 461, 1973.

140. **Lech, J. F., Siemer, D. D., and Woodriff, R.,** Determination of mercury in air samples by flameless atomic absorption, *Spectrochim. Acta,* 28b, 435, 1973.

141. **Hahne, H. C. H. and Kroontje, W.,** The simultaneous effect of pH and chloride concentrations upon mercury (II) as a pollutant, *Soil Sci. Soc. Am. Proc.,* 37, 838, 1973.

142. **Harsanyi, E., Polos, L., Bezur, L., and Pungor, E.,** Determination of mercury in water by flameless atomic absorption, *Magy. Kem. Foly.,* 79, 471, 1973.

143. **Aaroe, B. and Salvesen, B.,** Determination of mercury in fish homogenates by cold vapor atomic absorption spectroscopy, *Medd. Nor. Farm. Selsk.,* 35, 49, 1973.

144. **Shimomura, S., Hayashi, Y., Morita, H., and Shinohara, T.,** Behaviour of mercury vapor upon irradiation by a mercury lamp in an inert gas, *Anal. Lett.,* 6, 769, 1973.

145. **Rosain, R. M. and Wai, C. H.,** Rate of loss of mercury from aqueous solution when stored in various containers, *Anal. Chim. Acta,* 65, 279, 1973.

146. **Nord, P. J., Kadaba, M. P., and Sorensen, J. R. J.,** Mercury in human hair. Study of residents of Los Alamos, N.M. and Pasadena, Calif. by cold vapor atomic absorption spectrophotometry, *Arch. Environ. Health,* 27, 40, 1973.

147. **Lockhart, W. L., Uthe, J. F., Kenney, A. R., and Mehrie, P. M.,** Studies on methyl mercury in Northern Pike, in *Trace Substances in Environmental Health,* Vol. 6, Hemphill, D. D., Ed., University of Missouri, Columbia, 1973, 115.

148. **Long, S. J., Scott, D. R., and Thompson, R. J.,** Atomic absorption determination of elemental mercury collected from ambient air on silver wool, *Anal. Chem.,* 45, 2227, 1973.

149. **Gardner, D. and Riley, J. P.,** Distribution of dissolved mercury in the Irish Sea, *Nature (London),* 241, 526, 1973.

150. **Carr, R. A. and Wilkniss, P. E.,** Mercury short-term storage of natural waters, *Environ. Sci. Technol.,* 7, 62, 1973.

151. **Fitzgerald, W. F. and Lyons, W. B.,** Organic mercury compounds in coastal waters, *Nature (London),* 242, 452, 1973.

152. **Graf, E., Polos, L., Bezur, L., and Pungor, E.,** Determination of mercury in water by a flameless atomic absorption method, *Magy. Kem. Foly.,* 79, 471, 1973 (in Hungarian).

153. **Tsujino, R., Ueda, S., and Morita, T.,** Study of ashing for the rapid determination of mercury in oily biological materials, *Bunseki Kagaku,* 22, 591, 1973.

154. **Kamada, T., Hayadhi, Y., Kumamaru, T., and Yamamoto, Y.,** Flameless atomic absorption spectro-photometry for determination of inorganic and organic mercury at the parts per billion level in water using the vapor phase equilibrated with the solution, *Bunseki Kagaku,* 22, 1481, 1973.

155. **Harsanyi, E., Polos, L., and Pungor, E.,** Enhancement of sensitivity for the determination of mercury in waters, *Anal. Chim. Acta,* 67, 229, 1973.

156. **Holden, A. V.,** Mercury and organochlorine residue analysis of fish and aquatic mammals, *Pestic. Sci.,* 4, 399, 1973.

157. **Archer, M. A., Stillings, B. R., Tannenbaum, S. R., and Wang, D. I. C.,** Reduction in mercury content of fish protein concentration by enzymatic digestion, *J. Agric. Food Chem.,* 21, 1116, 1973.

158. **Wright, F. C., Palmer, J. S., and Riner, J. C.,** Accumulation of mercury in tissues of cattle, sheep and chickens given the mercurial fungicide, Penagen 15, orally, *J. Agric. Food Chem.,* 21, 414, 1973.

159. **Sorensen, J. R. J., Levin, L. S., and Petering, H. G.,** Cadmium, copper, lead, mercury and zinc concentrations in the hair of individuals living in the United States, *Interface,* 2, 17, 1973.

160. **Omang, S. H.,** Trace determination of mercury in biological materials by flameless atomic absorption spectrometry, *Anal. Chim. Acta,* 63, 247, 1973.

161. **Head, P. C. and Nicholson, R. A.,** Cold vapor technique for the determination of mercury in geological materials involving its reduction with tin (II) chloride and collection on gold wire, *Analyst (London),* 98, 53, 1973.

162. **Noel, W. A.,** Some notes on the use of the Perkin-Elmer flameless mercury system, *At. Absorpt. Newsl.,* 12, 62, 1973.

163. **Deitz, F. J., Sell, J. L., and Bristol, D.,** Rapid, sensitive method for determination of mercury in a variety of biological samples, *J. Assoc. Off. Anal. Chem.,* 56, 378, 1973.

164. **Lopez-Escobar, L. and Hume, D. N.,** Ozone as releasing agent. Determination of trace mercury in organic matrices by cold vapor atomic absorption, *Anal. Lett.,* 6, 343, 1973.

165. **Dagnall, R. M., Manfield, J. M., Silvester, M. D., and West, T. S.,** Atomic absorption and emission spectrometry of mercury at 184.9 nm, *Spectrosc. Lett.,* 6, 183, 1973.

166. **Aston, S. R., Bruty, D., Chester, R., and Padgham, R. C.,** Mercury in lake sediments: a possible indicator of technological growth, *Nature (London),* 241, 450, 1973.

167. **Hoggins, P. E. and Brooks, R. R.,** Instrumental parameters for determination of mercury by flameless atomic absorption spectrophotometry, *J. Assoc. Off. Anal. Chem.,* 56, 1306, 1973.

168. **Lo, F. C. and Bush, B.,** Modified procedure for determining mercury in coal by cold vapor atomic absorption spectrophotometry, *J. Assoc. Off. Anal. Chem.,* 56, 1509, 1973.

169. **Hoellerer, G. and Hoffman, J.,** Determination of traces of mercury in biological materials, *Z. Lebensmitt. Unters. Forsch.,* 150, 277, 1973 (in German).

170. **LaFleur, P. D.,** Retention of mercury when freeze-drying biological materials, *Anal. Chem.,* 45, 1534, 1973.

171. **Heppleston, P. B. and French, M. C.,** Mercury and other metals in British seals, *Nature (London),* 243, 302, 1973.

172. **Corte, G. L., Dowd, G., and Monkman, J. L.,** Wide range continuous mercury vapor monitor, *Chem. Instrum.,* 5, 221, 1973—74.

173. **Jambor, J.,** Contribution to the determination of mercury by atomic absorption spectroscopy, *Scr. Chem.,* 3, 55, 1973.

174. **Thistlewaite, P. J. and Trease, M.,** Determination of mercury by a simple atomic absorption method, *J. Chem. Educ.,* 51, 687, 1974.

175. **Kirkbright, G. F. and Wilson, P. J.,** Application of a demountable hollow cathode lamp as a source for the direct determination of sulfur, iodine, arsenic, selenium and mercury by atomic absorption flame spectrometry, *Anal. Chem.,* 46, 1414, 1974.

176. **Vitkun, R. A., Zelyukova, Y. V., and Poluektov, N. S.,** Flameless atomic absorption determination of mercury in selenium and tellurium preparations using formaldehyde as a reducing agent, *Ukr. Khim. Zh.,* 40, 1304, 1974 (in Russian).

177. **Krinitz, B. and Holak, W.,** Simple, rapid digestion techniques for the determination of mercury in sea food by flameless atomic absorption spectroscopy, *J. Assoc. Off. Anal. Chem.,* 57, 568, 1974.

178. **Voyce, D. and Zeitlin, H.,** Separation of mercury from seawater by adsorption colloid flotation and analysis by flameless atomic absorption, *Anal. Chim. Acta,* 69, 27, 1974.

179. **Ure, A. M.,** The determination of mercury by non-flame atomic absorption and fluorescence spectrometry. A review, *Anal. Chim. Acta,* 76, 1, 1974.

180. **Vitkun, R. A., Poluektov, N. S., and Zelyukova, Y. V.,** Ascorbic acid as reducing agent in the flameless atomic absorption determination of mercury, *Zh. Anal. Khim.,* 29, 691, 1974.

181. **Ure, A. M. and Shand, C. A.,** Determination of mercury in soils and related materials by cold vapor atomic absorption spectrometry, *Anal. Chim. Acta,* 72, 63, 1974.

182. **Nakamachi, H., Okamoto, K., and Kusumi, I.,** Determination of mercury by flameless atomic absorption using charcoal as an absorbant, *Bunseki Kagaku,* 23, 10, 1974.

183. **Siemer, D. and Woodriff, R.,** Application of the carbon rod atomizer to the determination of mercury in the gaseous products of oxygen combustion of solid samples, *Anal. Chem.,* 46, 597, 1974.

184. **Lieu, V. T., Cannon, A., and Huddleston, W. E.,** Nonflame atomic absorption attachment for trace mercury determination, *J. Chem. Educ.,* 51, 752, 1974.

185. **Least, C. J., Jr., Rejent, T. A., and Lees, H.,** Modification of a cold vapor technique for the determination of mercury in urine, *At. Absorpt. Newsl.,* 13, 4, 1974.

186. **Feldman, C.,** Preservation of dilute mercury solutions, *Anal. Chem.,* 46, 99, 1974.

187. **Issaq, H. J. and Zielinski, W. L.,** Hot atomic absorption spectrometry method for the determination of mercury at the nanogram and subnanogram level, *Anal. Chem.,* 46, 1436, 1974.

188. **Ankersmit, R., Barjhoux, J., Cappellina, F., Carter, W. T., Deetman, A. A., Durr, W., Killens, C., Lutz, J., Melard, P., Norberg, S. A., Nouyrigat, F., Olivier, M., Pouillot, M., Reiners, P., and Romeis, H.,** Standardization of methods for the determination of traces of mercury. I. Determination of total inorganic mercury in inorganic samples, *Anal. Chim. Acta,* 72, 37, 1974.

189. **Kivalo, P., Visapa, A., and Backman, R.,** Atomic absorption determination of mercury in fish using the Coleman MAS-50 mercury analizer, *Anal. Chem.,* 46, 1814, 1974.

190. **Huckabee, J. W., Feldman, C., and Talmi, Y.,** Mercury concentrations in fish from the Great Smoky Mountains National Park, *Anal. Chim. Acta,* 70, 41, 1974.

191. **Dujmovic, M. and Winkler, H. A.,** Determination of mercury in water and effluent by flameless atomic absorption, *Chem. Ztg.,* 98, 233, 1974.

192. **Brandvold, L. A. and Marson, S. J.,** Acid-pressure technique for analysis of mercury in ores and concentrates, *At. Absorpt. Newsl.,* 13, 125, 1974.

193. **Bisogni, J. J., Jr. and Lawrence, A. W.,** Determination of submicrogram quantities of monomethyl mercury in aquatic samples, *Environ. Sci. Technol.,* 8, 850, 1974.

194. **Fitzgerald, W. F., Lyons, W. B., and Hunt, C. D.,** Cold-trap preconcentration method for the determination of mercury in seawater and in other natural materials, *Anal. Chem.,* 46, 1882, 1974.

195. **Vitkun, R. A., Zelyukova, Y. V., and Poluektov, N. S.,** Atomic absorption determination of mercury, *Zavod. Lab.,* 40, 949, 1974.

196. **Giovanoli-Jakubczak, T. and Berg, G. G.,** Measurement of mercury in human hair, *Arch. Environ. Health,* 28, 139, 1974.

197. **Giovanoli-Jakubczak, T., Greenwood, M. R., Smith, J. C., and Clarkson, T. W.,** Determination of total and inorganic mercury in hair by flameless atomic absorption and of methyl mercury by gas chromatography, *Clin. Chem. (Winston-Salem, N.C.),* 20, 222, 1974.

198. **Gerasimova, L. I. and Ponomareva, V. A.,** Atomic absorption determination of nanogram concentrations of mercury in soils, *Zavod. Lab.,* 40, 360, 1974.

199. **Olafsson, J.,** Determination of nanogram quantities of mercury in sea water, *Anal. Chim. Acta,* 68, 207, 1974.

200. **Anderson, D. H., Murphy, J. J., and White, W. W.,** Gelatin as a matrix for a mercury reference material, *Anal. Chem.,* 44, 2099, 1974.

201. **Bretthauer, E. W., Moghissi, A. A., Snyder, S. S., and Mathews, N. W.,** Determination of submicrogram amounts of mercury by the oxygen bomb combustion, *Anal. Chem.,* 46, 445, 1974.

202. **Alberts, J. J., Schindler, J. E., Miller, R. W., and Carr, P. W.,** Mercury determinations in natural waters by persulfate oxidation, *Anal. Chem.,* 46, 434, 1974.

203. **Siemer, D., Lech, J., and Woodriff, R.,** Application of carbon rod atomizer for the analysis of mercury in air, *Appl. Spectrosc.,* 28, 68, 1974.

204. **Galeno, N.,** Research on mercury contained in fish consumed in the city of Turin, possible ingestion rate of the metal by the population, *Vet. Ital.,* 25, 645, 1974 (in Italian).

205. **Uthe, J. F. and Armstrong, F. A. J.,** Microdetermination of mercury and organomercury compounds in environmental materials, *Toxicol. Environ. Chem. Rev.,* 2, 45, 1974.

206. **Nakamura, Y. and Moriki, H.,** Flameless atomic absorption spectrophotometry for the determination of inorganic mercury. I. Determination by reduction aeration single path method. *Eisei Kagaku,* 20, 300, 1974 (in Japanese).

207. **Tanaka, K., Fukaya, K., Fukui, S., and Kanno, S.,** Environmental studies on mercury. II. Conditions for determination of mercury in fish and sediments by silica tube combustion-gold amalgamation method, *Eisei Kagaku,* 20, 344, 1974 (in Japanese).

208. **Tanaka, K., Fukaya, K., Fukui, S., and Kanno, S.,** Environmental studies on mercury. III. Rapid flameless atomic absorption determination of organic mercury by silica tube combustion-gold amalgamation method, *Eisei Kagaku,* 20, 349, 1974.

209. **Van Loon, J. C.,** Mercury input to the environment resulting from products and effluents from municipal sewage treatment plants, *Environ. Pollut.,* 7, 141, 1974.

210. **Stein, P. C., Campbell, E. E., and Moss, W. D.,** Mercury in man, *Arch. Environ. Health,* 29, 125, 1974.

211. **Jenne, E. A.,** Leads to the mercury literature, *At. Absorpt. Newsl.,* 13, 106, 1974.

212. **Hall, E. T.,** Mercury in commercial canned food, *J. Assoc. Offic. Anal. Chem.,* 57, 1068, 1974.

213. **Woidich, H. and Pfannhauser, W.,** Quantitative analysis of mercury in biological material. Digestion and atomic absorption determination in the vapor phase, *Z. Lebensm. Unters. Forsch.,* 155, 271, 1974.

214. **Antonacopoulos, N.,** Determination of mercury in fish and fish products. I. On the methods of sample preparation and measurement by flameless AA (cold vapor procedure), *Chem. Mikrobiol. Technol. Lebensm.,* 3, 8, 1974.

215. **Baltisberger, R. J. and Knudson, C. L.,** The differentiation of submicrogram amounts of inorganic and organomercury in water by flameless atomic absorption spectrometry, *Anal. Chim. Acta,* 73, 265, 1974.

216. **Lauwerys, R., Buchet, J. P., Roels, H., Berlin, A., and Smeets, J.,** Intercomparison program of lead, mercury and cadmium analysis in blood, urine and aqueous solutions, *Clin. Chem. (Winston-Salem, N.C.),* 21, 551, 1975.

217. **Agemian, H. and Chau, A. S. Y.,** A method for the determination of mercury in sediments by the automated cold vapor atomic absorption technique after digestion, *Anal. Chim. Acta,* 75, 297, 1975.

218. **Heinrichs, H.,** Determination of mercury in water, rocks, coal and petroleum with flameless atomic absorption spectrophotometry, *Z. Anal. Chem.,* 273, 197, 1975.

219. **Carlsen, J. B.,** Thiol group determination in proteins by flameless atomic absorption spectrometry of mercury, *Anal. Biochem.,* 64, 53, 1975.

220. **Hawley, J. E. and Ingle, J. D., Jr.,** Improvements in cold vapor atomic absorption determination of mercury, *Anal. Chem.,* 47, 719, 1975.

221. **Bothner, M. H. and Robertson, D. E.,** Mercury contamination of sea water samples stored in polyethylene containers, *Anal. Chem.,* 47, 592, 1975.

222. **Chilov, S.,** Determination of small amounts of mercury, *Talanta,* 22, 205, 1975.

223. **Dassani, S. D., McClellan, B. E., and Gordon, M.,** Submicrogram level determination of mercury in seeds, grains and food products by cold vapor atomic absorption spectrophotometry, *J. Agric. Food Chem.,* 23, 671, 1975.

224. **Corte, G. L. and Dubois, L.,** Determination of trace amounts of mercury in rock sample, *Mikrochim. Acta,* 1, 69, 1975.

225. **Dogan, S. and Haerdi, W.,** Preconcentration and separation of mercury traces by reduction on metallic copper and mercury determination by nonflame atomic absorption spectrometry, *Anal. Chim. Acta,* 76, 345, 1975 (in French).

226. **Ebbestad, U., Gunderson, N., and Torgrimsen, T.,** A simple method for the determination of inorganic mercury and methyl mercury in biological samples by flameless atomic absorption, *At. Absorpt. Newsl.,* 14, 142, 1975.

227. **Floyd, M. and Sommers, L. E.,** Determination of alkylmercury compounds in lake sediments by steam distillation flameless atomic absorption, *Anal. Lett.,* 8, 525, 1975.

228. **Knauer, E. and Milliman, G. E.,** Analysis of petroleum for trace metals. Determination of mercury in petroleum and petroleum products, *Anal. Chem.,* 47, 1263, 1975.

229. **Litman, R., Finston, H. L., and Williams, E. T.,** Evaluation of sample treatments for mercury determination, *Anal. Chem.,* 47, 2364, 1975.

230. **Murphy, J. J.,** Determination of mercury in coals by peroxide digestion and cold vapor atomic absorption spectrophotometry, *At. Absorpt. Newsl.,* 14, 151, 1975.

231. **Musha, S. and Takahashi, Y.,** Enrichment of trace amounts of metals utilizing the coagulation of soybean protein. III. Enrichment of trace mercury by soybean protein for the flameless atomic absorption analysis, *Bunseki Kagaku,* 24, 535, 1975.

232. **Nishigaki, N.,** Methylmercury and selenium in umbilical cords of inhabitants of Minimata area, *Nature (London),* 258, 324, 1975.

233. **Teeny, F. M.,** Rapid method for the determination of mercury in fish tissue by atomic absorption spectrometry with an argon-hydrogen flame, *J. Agric. Food Chem.,* 23, 668, 1975.

234. **Thompson, R. D. and Hoffman, T. J.,** Determination of mercury-containing pharmaceuticals by vapor phase atomic absorption spectroscopy, *J. Pharm. Sci.,* 64, 1863, 1975.

235. **Toffaletti, J. and Savory, J.,** Use of sodium borohydride for determination of total mercury in urine by atomic absorption spectrometry, *Anal. Chem.*, 47, 2091, 1975.

236. **Vitkun, R. A., Kravchenko, T. B., Zelyukova, Y. V., and Poluektov, N. S.,** Atomic absorption determination of mercury in waters using concentration, *Zavod. Lab.*, 41, 663, 1975.

237. **Crecelius, E. A., Bothner, M. H., and Carpenter, R.,** Geochemistries of arsenic, antimony, mercury and related elements in sediments of Puget Sound, *Environ. Sci. Technol.*, 1, 325, 1975.

238. **Itsuki, K., Abukawa, J., and Tofuku, Y.,** Determination of trace mercury in metallic selenium by flameless atomic absorption spectroscopy, *Bunseki Kagaku*, 24, 782, 1975.

239. **Newton, M. P. and Davis, D. G.,** Determination of mercury using a copper wire atomizer for flameless atomic absorption spectroscopy, *Anal. Lett.*, 8, 729, 1975.

240. **Potter, L., Kidd, D., and Standiford, D.,** Mercury in Lake Powell. Bioamplification of mercury in man-made desert reservoir, *Environ. Sci. Technol.*, 9, 41, 1975.

241. **Fitzgerald, W. F.,** Mercury analysis in seawater using cold-trap preconcentrations and gas phase detection, *Adv. Chem. Ser.*, 147, 1975.

242. **Takeda, M., Inamasu, Y., and Tomida, T. et al.,** On mercury and selenium contained in tuna fish tissues, *J. Shimonaseki Univ. Fish.*, 23, 145, 1975 (in Japanese).

243. **Trujillo, P. E. and Campbell, E. E.,** Development of a multistage air sampler for mercury, *Anal. Chem.*, 47, 1629, 1975.

244. **Agemian, H. and Chau, A. S. Y.,** An improved digestion method for the extraction of mercury from environmental samples, *Analyst (London)*, 101, 91, 1976.

245. **Deldime, P. and Tran-Trieu, V.,** Contribution to the determination of mercury traces by atomic absorption photometry, *Anal. Lett.*, 9, 169, 1976.

246. **Gardner, D.,** A rapid method for the determination of mercury in air by flameless atomic absorption spectrometry, *Anal. Chim. Acta*, 82, 321, 1976.

247. **Helsby, C. A.,** Determination of mercury in fingernails and body hair, *Anal. Chim. Acta*, 82, 427, 1976.

248. **Handzel, M. R. and Jamieson, D. M.,** Determination of mercury in fish, *Anal. Chem.*, 48, 926, 1976.

249. **Kacprzak, J. L. and Chvojka, R.,** Determination of methyl mercury in fish by flameless atomic absorption spectroscopy and comparison with the acid digestion method for total mercury, *J. Assoc. Off. Anal. Chem.*, 59, 153, 1976.

250. **Koirtyohann, S. R. and Khalid, M.,** Variables in the determination of mercury by cold vapor atomic absorption, *Anal. Chem.*, 48, 136, 1976.

251. **Lyashenko, S. D. and Stepanov, A. S.,** Flameless atomic absorption determination of mercury using sodium ultra hydroborate as the reducing agent, *Zh. Anal. Khim.*, 31, 279, 1976.

252. **Owens, J. W., Gladney, E. S., Issaq, H. J., and Zielinski, W. L., Jr.,** Exchange of comments: loss of mercury during flameless atomic absorption spectrometry, *Anal. Chem.*, 48, 787, 1976.

253. **Ramelow, G. J. and Balkas, T. I.,** A simple mercury analysis system for use with commercial atomic absorption spectrophotometers, *At. Absorpt. Newsl.*, 15, 55, 1976.

254. **Seeger, R.,** A simple digestion method for the determination of mercury in mushrooms by flameless atomic absorption spectroscopy, *At. Absorpt. Newsl.*, 15, 45, 1976.

255. **Svistov, P. F. and Turkin, Y. I.,** Atomic absorption determination of mercury in waste water, *Zavod. Lab.*, 42, 155, 1976.

256. **Vitkun, R. A., Zelyukova, Y. V., Kravchenko, T. B., and Poluektov, N. S.,** Flameless atomic absorption determination of mercury in the presence of gold, *Zh. Anal. Khim.*, 31, 388, 1976.

257. **Zelyukova, Y. V., Vitkun, R. A., Kravchenko, T. B., and Poluektov, N. S.,** Atomic determination of mercury in air, *Gig. Sanit.*, 1, 66, 1976.

258. **Zook, E. G., Powell, J. J., Hackley, B. M., Emerson, J. A., Brooker, J. R., and Knobl, G. M., Jr.,** National marine fisheries service preliminary survey of selected seafoods for mercury, lead, cadmium, chromium and arsenic content, *J. Agric. Food Chem.*, 24, 47, 1976.

259. **Dogan, S. and Haerdi, W.,** Some applications of the rapid separation of mercury on metallic copper to environmental samples with determination by flameless atomic absorption spectrometry, *Anal. Chim. Acta*, 84, 89, 1976.

260. **Dusci, L. J. and Hackett, L. P.,** Rapid digestion and flameless atomic absorption spectroscopy of mercury in fish, *J. Assoc. Off. Anal. Chem.*, 59, 1183, 1976.

261. **Kirkwood, D.,** Determination of mercury: modification to the static vapor technique for flameless atomic absorption spectroscopy, *Lab. Pract.*, 25, 233, 1976.

262. **Kothandarman, P. and Dallmeyer, J. F.,** Improved desicator for mercury cold vapor technique, *At. Absorpt. Newsl.*, 15, 120, 1976.

263. **Kuwae, Y., Hasegawa, T., and Shono, T.,** Rapid determination of mercury in solid samples by high-frequency induction heating and atomic absorption spectrometry, *Anal. Chim. Acta*, 84, 185, 1976.

264. **Littlejohn, D., Fell, G. S., and Ottaway, J. M.,** Modified determination of total inorganic mercury in urine by cold vapor atomic absorption spectrometry, *Clin. Chem. (Winston-Salem, N.C.)*, 22, 1719, 1976.

265. **Matsunaga, K., Ishida, T., and Oda, T.,** Extraction of mercury from fish for atomic absorption spectrometric determination, *Anal. Chem.,* 48, 1421, 1976.

266. **Pearce, I. D., Brooks, R. R., and Reeves, R. D.,** Digestion of fish samples for mercury determination by flameless atomic absorption spectrophotometry, *J. Assoc. Off. Anal. Chem.,* 59, 655, 1976.

267. **Rooney, R. C.,** Use of sodium borohydride for cold vapor atomic absorption determination of trace amounts of inorganic mercury, *Analyst (London),* 101, 678, 1976.

268. **Calder, I. T. and Miller, J. H.,** Application of an atomic absorption spectrophotometric method for the determination of organomercurial preservatives in eye drops, *J. Pharm. Pharmacol. Suppl.,* 28, 25p, 1976.

269. **Matsunaga, K. and Takahashi, S.,** Cold vapor atomic absorption spectrometric determination of nanogram amounts of organic mercury in sediments or aquatic organisms, *Anal. Chim. Acta,* 87, 487, 1976.

270. **Helsby, C. A.,** Development and application of cold-vapor techniques for determination of mercury in biological materials, *Mikrochim. Acta,* 1, 307, 1976.

271. **Brun, S. and Cayrol, M.,** Determination of mercury in food products and viological materials, *Ann. Falsif. Expert. Chim.,* 69, 107, 1976 (in French).

272. **Schuler, P. L. and Egan, H.,** Cadmium, Lead, Mercury and Methylmercury Compounds: Review of Methods of Trace Analysis and Sampling with Special Reference to Food, FAO Rep. No. 92-5-I-00094-M-84, Food and Agriculture Organization, Rome, 1976.

273. **Adler, J. F. and Hickman, D. A.,** Determination of mercury by atomic absorption spectrometry with graphite tube atomization, *Anal. Chem.,* 49, 336, 1977.

274. **Gladney, E. S. and Owens, J. W.,** Determination of mercury by carrier free combustion separation and flameless atomic absorption spectrometry, *Anal. Chim. Acta,* 90, 271, 1977.

275. **Nicholson, R. A.,** Rapid thermal decomposition technique for the atomic absorption determination of mercury in rocks, soils and sediments, *Analyst (London),* 102, 399, 1977.

276. **Simpson, W. R. and Nickless, G.,** Rapid versatile method for determining mercury at subnanogram levels by cold vapor atomic absorption spectroscopy, *Analyst (London),* 102, 86, 1977.

277. **Tindall, F. M.,** Hints on chemical analysis. I. Tungsten determination by atomic absorption spectrophotometry. II. Revised notes on gold and silver determination by atomic adsorption spectrophotometry. III. Mercury in copper concentrates by atomic absorption spectrophotometry, *At. Absorpt. Newsl.,* 16, 37, 1977.

278. **Velghe, N., Campe, A., and Claeys, A.,** Use of an absorption cell without windows for the cold vapor determination of mercury using the HGA-72, *At. Absorpt. Newsl.,* 16, 28, 1977.

279. **White, W. W. and Murphy, P. J.,** Determination of mercury in silver or silver nitrate by atomic absorption spectrometry, *Anal. Chem.,* 49, 255, 1977.

280. **Zhusipbekov, A. Z., Gladyshev, V. P., Vilyams, G. F., Fidel, I. N., and Khisamutdinova, R. I.,** Study of air samples for the mercury vapor content by a chemical method and the atomic absorption method, *Gig. Sanit.,* 4, 84, 1977.

281. **Gaffin, S. and Hornung, H.,** Rapid determination of mercury in urine by flameless atomic absorption spectrometry, *Clin. Toxicol.,* 10, 345, 1977.

282. **Imaeda, K., Ohsawa, K., and Wako, M.,** Determination of atmospheric mercury by flameless atomic absorption spectrometry, *Bunseki Kagaku,* 26, 651, 1977.

283. **Janssen, J. H., Vanden Enk, J. E., Bult, R., and deGroot, D. C.,** Determination of total mercury in air by atomic absorption spectrometry after collection on manganese dioxide, *Anal. Chim. Acta,* 92, 71, 1977.

284. **Munns, R. K. and Holland, D. C.,** Rapid digestion and flameless atomic absorption spectroscopy of mercury in fish: collaborative study, *J. Assoc. Off. Anal. Chem.,* 60, 833, 1977.

285. **Narasaki, H., Down, J. L., and Ballah, R.,** Enhancement of the sensitivity of the cold vapor atomic absorption spectrophotometric method towards mercury by using a sintered glass bubbler and a magnetic stirrer, *Analyst (London),* 102, 537, 1977.

286. **Taguchi, M., Yasuda, K., Dokiya, Y., Shimizu, M., and Toda, S.,** Mercury determination in fish samples by flameless atomic absorption spectrometry: sampling and wet digestion, *Bunseki Kagaku,* 26, 438, 1977.

287. **Taguchi, M., Yasuda, K., Dokiya, Y., Shimizu, M., and Toda, S.,** Mercury determination in fish samples by flameless atomic absorption spectrometry: comparison between the combustion gold trap method and the wet digestion — reduction method, *Bunseki Kagaku,* 26, 496, 1977.

288. **Toth, J. R. and Ingle, J. D.,** Determination of mercury in manganese nodules and crusts by cold vapor atomic absorption spectrometry, *Anal. Chim. Acta,* 92, 409, 1977.

289. **Tsai, W. and Shiau, L.,** Determination of mercury in edible oils by combustion and atomic absorption spectrophotometry, *Anal. Chem.,* 49, 1641, 1977.

290. **Yamazaki, S., Dokiya, Y., Hayashi, T., Toda, S., and Fuwa K.,** Determination of trace amounts of mercury by long absorption cell-cold vapor atomic absorption spectrometry, *Nippon Kagaku Kaishi,* 8, 1148, 1977.

291. **Fujiwara, K., Sato, K., and Fuwa, K.,** Atomic absorption spectrometry of mercury using a graphite furnace atomizer, *Bunseki Kagaku,* 26, 772, 1977.

292. **Greenwood, M. R., Dhahir, P., Clarkson, T. W., Farant, P., Chartrand, A., and Khayat, A.,** Epidemiological experience with the Mago's reagents in the determination of different forms of mercury in biological samples by flameless atomic absorption, *J. Anal. Toxicol.,* 1, 265, 1977.

293. **Szprengier, T.,** Modified method for determining mercury in biological samples by atomic absorption spectrophotometry, *Med. Water.,* 33, 182, 1977 (in Polish).

294. **Yoshida, Y. and Murozumi, M.,** Determination of chemical components at the ppb level in polar snow strata. VII. Determination of trace mercury by flameless atomic absorption spectrophotometry, *Bunseki Kagaku,* 26, 789, 1977.

295. **Dujmovic, M.,** Determination of Bi, Hg, Sb, Sn, Te and Pb in aqueous solutions by flameless atomic absorption spectroscopy. *GIT Fachz. Lab.,* 21, 861, 1977.

296. **Dittrich, K. and Mueller, H.,** Studies on mercury trace determination in solutions and solids by flameless atomic absorption spectroscopy, *Chem. Anal.,* 23, 81, 1977.

297. **Dittrich, K., Wenrich, R., and Werner, B.,** Studies on the determination of mercury traces and gases by flameless atomic absorption and atomic fluorescence, *Chem. Anal.,* 23, 71, 1978.

298. **Egaas, E. and Julshamn, K.,** A method for the determination of selenium and mercury in fish products using the same digestion procedure, *At. Absorpt. Newsl.,* 17, 135, 1978.

299. **Mitsuhashi, T., Morita, H., and Shimomura, S.,** Atomic absorption determination of total mercury by a combined iron (III)-sodium borohydride reduction reagent, *Bunseki Kagaku,* 27, 666, 1978.

300. **Rabenstein, D. L.,** The chemistry of methylmercury toxicology, *J. Chem. Educ.,* 55, 292, 1978.

301. **Stuart, D. C.,** Radiotracer investigation of the cold-vapor atomic absorption method of analysis for trace mercury, *Anal. Chim. Acta,* 101, 429, 1978.

302. **Agemian, H. and Cheam, V.,** Simultaneous extraction of mercury and arsenic from fish tissues and an automated determination of arsenic by atomic absorption spectrometry, *Anal. Chim. Acta,* 101, 193, 1978.

303. **Ahmed, S., Dil, W., Choudhri, S. A., and Ejaz, M.,** Extraction of mercury (II) from aqueous thiocyanate solutions with 5-(n-pyridyl) nonane in benzene and its subsequent atomic absorption spectrometric determination, *Talanta,* 25, 563, 1978.

304. **Burns, D. T., Glockling, F., Mahale, V. B., and Swindall, W. J.,** Techniques for the determination of mercury in silicon-containing organomercurials using atomic absorption spectrophotometry, *Analyst (London),* 103, 985, 1978.

305. **Campe, A., Velghe, N., and Claeys, A.,** Determination of inorganic, phenyl and total mercury in urine, *At. Absorpt. Newsl.,* 17, 100, 1978.

306. **Chapman, J. F. and Dale, L. S.,** A simple apparatus for the spectrometric determination of mercury by the cold vapor-atomic absorption technique, *Anal. Chim. Acta,* 101, 203, 1978.

307. **Agemian, H. and Chau, A. S.,** Automated method for the determination of total dissolved mercury in fresh and saline waters by ultraviolet digestion and cold vapor atomic absorption spectrometry, *Anal. Chem.,* 50, 13, 1978.

308. **Bishay, T. Z.,** A comparative study of radioactivation, atomic absorption and spectrophotometric techniques applied to the determination of mercury in lead, *J. Radioanal. Chem.,* 43, 147, 1978.

309. **Jirka, A. M. and Carter, M. J.,** Automated determination of mercury in sediments, *Anal. Chem.,* 50, 91, 1978.

310. **Ramelow, G. and Hornung, H.,** An investigation into possible mercury losses during lyophilization of marine biological samples, *At. Absorpt. Newsl.,* 17, 59, 1978.

311. **Teeny, F.,** Simple combustion-vapor trapping technique for determination of mercury in fish, *J. Assoc. Off. Anal. Chem.,* 61, 43, 1978.

312. **Tong, S.,** Stationary cold vapor atomic absorption spectrometric method for mercury determination, *Anal. Chem.,* 50, 412, 1978.

313. **Velghe, N., Campe, A., and Claeys, A.,** Semi-automated determination of mercury in fish, *At. Absorpt. Newsl.,* 17, 37, 1978.

314. **Velghe, N., Campe, A., and Claeys, A.,** Semi-automated cold vapor determination of inorganic and methyl mercury in fish by direct injection of tissue in the aeration cell, *At. Absorpt. Newsl.,* 17, 139, 1978.

315. **Yamamoto, Y., Kumamaru, T., and Shiraki, A.,** Comparative study of sodium borohydride tablet and tin (II) chloride reducing systems in the determination of mercury by atomic absorption spectrophotometry, *Z. Anal. Chem.,* 292, 273, 1978.

316. **Davies, I. M.,** Determination of methyl mercury in the muscle of marine fish by cold vapor atomic absorption spectrometry, *Anal. Chim. Acta,* 102, 184, 1978.

317. **Hocquellet, P.,** Application of electrothermal atomization to the determination of arsenic, antimony, selenium and mercury by atomic absorption spectrometry, *Analusis,* 6, 426, 1978.

318. **Knutti, R. and Balsiger, C.,** Routine determination of urinary mercury concentration in workers exposed to mercury, *GIT Labor-Med.,* 3, 201, 1978 (in German).

319. **Malissa, H., Maly, K., and Till, T.,** Atomic absorption spectrophotometric determination of mercury in roots of teeth and in jawbones, *Z. Anal. Chem.,* 293, 141, 1978.

320. **Millward, C. E. and Le Bihan, A.,** Flameless atomic absorption analysis of mercury in model aquatic system, *Water Res.,* 12, 979, 1978.

321. **Purushottam, A. and Naidu, P. P.,** Taking nonatomic absorption into account in an atomic absorption determination of mercury by the cold vapor method, *Curr. Aci.,* 47, 88, 1978.

322. **Slemr, F., Seiler, W., and Schuster, G.,** Mercury in the troposphere, *Ber. Bunsenges. Phys. Chem.,* 82, 1142, 1978 (in German).

323. **Takahashi, J., Tanabe, K., Haraguchi, H., and Fuwa, K.,** A conventional absorption cell system for vacuum UV atomic absorption spectrometry of mercury, *Bunseki Kagaku,* 27, 738, 1978.

324. **Castello, G. and Kanitz, S.,** Determination of organomercury compounds in food and man using gas chromatography and flameless atomic absorption spectrophotometry, *Boll. Chim. Lab. Prov.,* 4, 57, 1978 (in Italian).

325. **Dogan, S. and Haerdi, W.,** Preconcentration of silver wool of volatile organo-mercury compounds in neutral waters and air and the determination of mercury by flameless atomic absorption spectrometry, *Int. J. Environ. Anal. Chem.,* 5, 157, 1978.

326. **Kurchatova, G. and Koen, E.,** Flameless atomic absorption determination of traces of mercury in air, *Khig. Zdraveopaz.,* 21, 490, 1978 (in Bulgarian).

327. **Rakuns, S. A. and Smythe, L. E.,** Mercury in cereals and cereal products on sale in NSW, *Food Technol. Aust.,* 30, 271, 1978.

328. **Knechtel, J. R. and Fraser, J. L.,** Wet digestion method for the determination of mercury in biological and environmental samples, *Anal. Chem.,* 51, 315, 1979.

329. **Kunert, I., Komarek, J., and Sommer, L.,** Determination of mercury by atomic absorption spectrometry with cold vapor and electrothermal technique, *Anal. Chim. Acta,* 106, 285, 1979.

330. **Sadin, Y. and Deldime, P.,** Determination of mercury by flameless atomic absorption spectroscopy, *Anal. Lett.,* 12, 563, 1979.

331. **Sharma, D. C. and Davis, P. S.,** Direct determination of mercury in blood by use of sodium borohydride reduction and atomic absorption spectrophotometry, *Clin. Chem. (Winston-Salem, N.C.),* 25, 769, 1979.

332. **Shum, G. T., Freeman, H. C., and Uthe, J. F.,** Determination of organic (methyl) mercury in fish by graphite furnace atomic absorption spectrophotometry, *Anal. Chem.,* 51, 414, 1979.

333. **Stuart, D. C.,** Factors affecting peak shape in cold-vapor atomic absorption spectrometry for mercury, *Anal. Chim. Acta,* 106, 411, 1979.

334. **Yoshida, Z. and Motojima, K.,** Rapid determination of mercury in air with gold-coated quartz wool as collector, *Anal. Chim. Acta,* 106, 405, 1979.

335. **Fonds, A. W., Kempt, T., Minderhoud, A., and Sonneborn, M.,** Heavy metal content of various kinds of water, *Reinhalt Wassers,* 78, 1979.

336. **Agemian, H. and Dasilva, J. A.,** Automatic method for the determination of total mercury in fresh and saline waters and sediments, *Anal. Chim. Acta,* 104, 285, 1979.

337. **Bouzanne, M., Sire, J., and Voinovitch, I. A.,** Determination of mercury in water by atomic absorption spectrometry after fixation on silver wool or activated carbon, *Analusis,* 7, 62, 1979.

338. **Casas, A. and Vaquer, R.,** Preliminary results of mercury in CRPG and ANRT rock and mineral standards, *Geostandards Newsl.,* 3, 53, 1979.

339. **Jackwerth, E., Willmer, P. G., Hohn, R., and Berndt, H.,** A simple accessory for the determination of mercury and the hydride-forming elements (As, Bi, Sb, Se and Te) using flameless atomic absorption, *At. Absorpt. Newsl.,* 18, 66, 1979.

340. **Karalis, V. N. and Tarasov, V. V.,** Source of the mercury 253.7 nm resonance line for atomic absorption measurements, *Zh. Prikl. Spektrosk.,* 30, 732, 1979.

341. **Bye, R. and Paus, P. E.,** Determination of alkyl mercury compounds in fish tissue with an atomic absorption spectrometer used as a specific gas chromatographic detector, *Anal. Chim. Acta,* 107, 169, 1979.

342. **Chvojka, R. and Kacprzak, J. L.,** Accuracy of the selective reduction method for determining methyl mercury in fish, *J. Assoc. Off. Anal. Chem.,* 62, 1179, 1979.

343. **Grainger, F. and Gale, I. G.,** Direct analysis of solid cadmium mercury telluride by flameless atomic absorption using interactive computer processing, *J. Mater. Sci.,* 14, 1370, 1979.

344. **Hoffman, E., Ludke, C., and Tilch, J.,** Determination of mercury by flameless AAS at the 184.9 nm resonance line, *Spectrochim. Acta,* 34b, 301, 1979 (in German).

345. **Kuldvere, A. and Andreassen, B. Th.,** Determination of mercury in seaweed by atomic absorption spectrophotometry using the Perkin-Elmer MHS. I, *At. Absorpt. Newsl.,* 18, 106, 1979.

346. **Lutze, R. L.,** Implementation of a sensitive method for determining mercury in surface waters and sediments by cold vapor atomic absorption spectrophotometry, *Analyst (London),* 104, 979, 1979.

347. **Naganuma, A., Satoh, H., Yamamoto, R., Suzuki, T., and Imura, N.,** Effect of selenium on determination of mercury in animal tissues, *Anal. Biochem.,* 98, 287, 1979.

348. **Nakano, K., Takada, T., and Fujita, K.,** Determination of mercury by direct heating of mercury-adsorbed ion-exchange resin in electrochemical atomic absorption spectrometry, *Chem. Lett.*, 7, 869, 1979.

349. **Nelson, L. A.,** Brominating solution for the preconcentration of mercury from natural waters, *Anal. Chem.*, 51, 2289, 1979.

350. **Slemr, F., Seiler, W., Eberting, C., and Roggendorf, P.,** The determination of total gaseous mercury in air at background levels, *Anal. Chim. Acta*, 110, 35, 1979.

351. **Taguchi, M., Yasuda, K., Hashimoto, M., and Toda, S.,** Some improvements in mercury determination in marine organisms by atomic absorption spectrometry, *Bunseki Kagaku*, 28, 33 T, 1979.

352. **Tsalev, D. and Petrova, V.,** Hexamethylene ammonium hexamethylene dithiocarbamate-n-butyl acetate as an extractant of bismuth, cadmium, cobalt, mercury, indium, lead and palladium from acidic media, *Dokl. Bolg. Akad. Nauk*, 32, 911, 1979.

353. **Zvonarev, B. A., Zyrin, N. G., and Obukhov, A. I.,** Atomic absorption determination of mercury in soils by a cold vapor method, *Pochvovedenie*, 6, 153, 1979.

354. **Abo-Rady, M. D.,** Determination of mercury in water, fish, plant and sediment samples by atomic absorption spectroscopy, *Z. Anal. Chem.*, 299, 187, 1979.

355. **Aliseda, J. A., Ankersmit, R., Ashley, G. W., Barjhoux, J., Bult, R., Carter, W. T., Durr, W., Garbayo, J., and Garcia, M.,** Standardization of methods for the determination of traces of mercury. V. Determination of total mercury in water, *Anal. Chim. Acta*, 109, 209, 1979.

356. **Baltisberger, R. J., Hildebrand, D. A., Grieble, D., and Ballantine, T. A.,** A study of the disproportionation of mercury (I) induced by gas sparging in acidic aqueous solutions for cold vapor atomic absorption spectrometry, *Anal. Chim. Acta*, 111, 111, 1979.

357. **Capelli, R., Fezia, C., Franchi, A., and Zanicchi, G.,** Extraction of methyl mercury from fish and its determination by atomic absorption spectroscopy, *Analyst (London)*, 104, 1197, 1979.

358. **Dabrowski, J. and Debska, W.,** Determination of mercury in poppy straw, *Herba Pol.*, 25, 155, 1979.

359. **Flanjak, J. and Lee, H. Y.,** Trace metal content of livers and kidneys of cattle, *J. Sci. Food Agric.*, 30, 503, 1979.

360. **Hoffman, E., Luedke, C., and Tilch, J.,** Determination of mercury by flameless atomic absorption of the 184.9 nm resonance line, *Spectrochim. Acta*, 34b, 301, 1979.

361. **Mizunama, H., Morita, H., Sakurai, H., and Shimomura, S.,** Selective atomic absorption determination of inorganic and organic mercury by combined use of iron (III) and sodium tetrahydroborate, *Bunseki Kagaku*, 28, 695, 1979.

362. **Rubenstein, D. A. and Soares, J. H., Jr.,** The effect of selenium on the biliary excretion and tissue deposition of two forms of mercury in the broiler clock, *Poult. Sci.*, 58, 1289, 1979.

363. **Shan, X. and Ni, Z.,** Matrix modification for the determination of mercury using electrothermal atomic absorption spectrometry, *Acta Chim. Sinica*, 37, 261, 1979.

364. **Baluja, G. and Gonzalez, M. J.,** Contribution to the simultaneous residual analysis of inorganic and organic mercury compounds by chromatographic-spectroscopic techniques, *Rev. Agroquim. Technol. Aliment.*, 19, 270, 1979.

365. **Berta, E. and Olah, J.,** Survey of heavy metal content of wastewater sludges in Hungary, *Hidrol. Kozl.*, 59, 469, 1979.

366. **Bogden, J. D., Kemp, F. W., Troiana, R. A., Jortner, B. S., Timpone, C., and Giulani, D.,** Elevated renal copper from mercury exposure, *Trace Subst. Environ. Health*, 13, 353, 1979.

367. **Fitzgerald, W. F. and Gill, G. A.,** Subnanogram determination of mercury and gas phase detection applied to atmospheric analysis, *Anal. Chem.*, 51, 1714, 1979.

368. **Garcia Frades, J. P., Sanz Diaz, F., and Lacamara Bescos, R.,** Rapid prolytic method for determining total mercury in soils and rocks, *Quim. Ind.*, 25, 505, 1979.

369. **Luca, C., Danet, A. F., and Tarabic, M.,** Instrument for the determination of mercury in residues, *Rev. Chim.*, 30, 260, 1979.

370. **Schuetze, I. and Mueller, W.,** Determination of trace elements of edible fats and emulsifiers by flameless atomic absorption spectrometry. II. Determination of mercury in edible fats and emulsifiers, *Nahrung*, 23, 867, 1979.

371. **Dogan, S. and Haerdi, W.,** Rapid separation on copper powder of total mercury in blood and determination of mercury by flameless atomic absorption spectrometry, *Int. J. Environ. Anal. Chem.*, 6, 327, 1979.

372. **Sighinolfi, G. P., Gorgoni, C., and Santos, A. M.,** Atomic absorption determination of ultratrace elements in geological materials by vapor, hydride-forming techniques. I. Mercury, *Geostandards Newsl.*, 4, 223, 1980.

373. **Terada, K., Morimoto, K., and Kiba, T.,** Chromatographic concentration of mercury in seawater with 2-mercaptobenzothiazole supported on silica gel, *Bull. Chem. Soc. Jpn.*, 53, 1605, 1980.

374. **Tuncel, G. and Yavuz Ataman, O.,** Design and evaluation of a new absorption cell for cold vapor mercury determination by atomic absorption spectrometry, *At. Spectrosc.*, 1, 126, 1980.

375. **Weigert, P.,** Heavy metal content of chicken eggs, *Z. Lebensm. Unters. Forsch.*, 171, 18, 1980.

376. **Yamagami, E., Tateishi, S., and Hashimoto, A.,** Application of a chelating resin to the determination of trace amounts of mercury in natural waters, *Analyst (London),* 105, 491, 1980.

377. **Abo-Rady, M. D. K.,** Aquatic macrophytes as indicator for heavy metal pollution in the river Siene, *Arch. Hydrobiol.,* 89, 387, 1980 (in German).

378. **Kuklin, Y. S. and Korobeinikova, L. G.,** Atomic absorption method for the determination of mercury in the soil, *Khim. Selsk. Khoz.,* 18, 49, 1980.

379. **Lawrence, K. E., White, M., Potts, R. A., and Bertrand, R. D.,** Cold-vapor determination of mercury, *Anal. Chem.,* 52, 1391, 1980.

380. **Nagase, H., Ose, Y., Sato, T., Ishikawa, T., and Mitani, K.,** Differential determination of alkylmercury and inorganic mercury in river sediment, *Int. J. Environ. Anal. Chem.,* 7, 261, 1980.

381. **Osawa, K., Kaoru, K., and Imaeda, K.,** Determination of mercury compounds with an aluminum foil thin layer plate, *Bunseki Kagaku,* 29, 431, 1980.

382. **Seritti, A., Petrosino, A., Ferrara, R., and Barghigiani, C.,** A contribution to the determination of "reactive" and "total" mercury in seawater, *Environ. Technol. Lett.,* 1, 50, 1980.

383. **Shiraishi, N. and Kuroda, T.,** Determination of mercury in coastal seawater, *Bunseki Kagaku,* 29, T1, 1980.

384. **Dumarey, R., Heindryckx, R., and Dams, R.,** Determination of mercury in environmental standard reference materials, *Anal. Chim. Acta,* 118, 381, 1980.

385. **Gardner, D.,** Use of magnesium perchlorate as dessicant in the syringe injection technique for determination of mercury by cold vapor atomic absorption spectrometry, *Anal. Chim. Acta,* 119, 167, 1980.

386. **Idzikowski, A. and Michalewska, M.,** Determination of mercury in copper ores by flameless atomic absorption spectroscopy, *Chem. Anal.,* 25, 35, 1980.

387. **Kirkbright, G. F., Shan, H. C., and Snook, R. D.,** Evaluation of some matrix modification procedures for use in the determination of mercury and selenium by atomic absorption spectroscopy with a graphite tube electrothermal atomizer, *At. Spectrosc.,* 1, 85, 1980.

388. **Korunova, V. and Dedina, J.,** Determination of trace concentrations of mercury in biological materials after digestion under pressure in nitric acid catalyzed by vanadium pentoxide, *Analyst (London),* 105, 48, 1980.

389. **Jonsen, J., Helgeland, K., and Steinnes, E.,** Trace elements in human serum. Regional distribution in Norway, in *Geomedical Aspects of Present and Future Research 1978,* Laag, J., Ed., Global Book Resources, Ltd., London, England, 1980.

390. **Barudi, W. and Bielig, H. J.,** Heavy metal content of vegetables which grow above ground and fruits, *Z. Lebensm. Unters. Forsch.,* 170, 254, 1980.

391. **Bogden, J. D., Kemp, F. W., Troiano, R. A., Jortner, B. S., Timpone, C., and Guiliani, D.,** Effect of mercuric chloride and methylmercury chloride exposure on tissue concentrations of six essential minerals, *Environ. Res.,* 21, 350, 1980.

392. **Colella, M. B., Siggia, S., and Barnes, R. M.,** Poly (acrylamidoxime) resin for determination of trace metals in natural waters, *Anal. Chem.,* 52, 2347, 1980.

393. **Colett, D. L., Fleming, D. E., and Taylor, G. A.,** Determination of alkyl mercury in fish by steam distillation and cold vapor atomic absorption spectrophotometry, *Analyst (London),* 105, 897, 1980.

394. **Palliers, M. and Gernez, G.,** New technique for the determination of mercury and elements easily giving volatile compounds by atomic absorption spectrometry, *Analusis,* 8, 23, 1980.

395. **Pollock, E. N.,** The determination of mercury in air, *At. Spectrosc.,* 1, 78, 1980.

396. **Seimer, D. D. and Hageman, L.,** Determination of mercury in water by furnace atomic absorption spectrometry after reduction and aeration, *Anal. Chem.,* 52, 105, 1980.

397. **Slovak, Z. and Docekalova, H.,** Electrothermal atomization of mercury in the presence of thiols, *Anal. Chim. Acta,* 115, 111, 1980.

398. **Tanabe, K., Takahashi, J., Haraguchi, H., and Fuwa, K.,** Vacuum-ultraviolet atomic absorption spectrometry of mercury with cold vapor generation, *Anal. Chem.,* 52, 453, 1980.

399. **Tong, S. L. and Leow, W. K.,** Stationary cold vapor atomic absorption spectrometric attachment for determination of total mercury in undigested fish samples, *Anal. Chem.,* 52, 581, 1980.

400. **Ambe, M. and Niikura, N.,** Determination of mercury in biological and environmental samples by cold vapor atomic absorption spectrometry: comparison of sample treatments, *Bunseki Kagaku,* 29, 5T, 1980.

401. **Haraguchi, H., Takahashi, J., Tanabe, K., Akai, Y., Homma, A., and Fuwa, K.,** A conventional system for nondispersive vacuum-ultraviolet atomic absorption spectrometry of mercury, *Bunseki Kagaku,* 29, 348, 1980.

402. **Matsueda, T.,** Determination of traces of mercury in water by Zeeman effect atomic absorption spectrometry after preconcentration using activated carbon, *Bunseki Kagaku,* 29, 110, 1980.

403. **Minagawa, K., Takizawa, Y., and Kifune, I.,** Determination of very low levels of inorganic and organic mercury in natural waters by cold vapor atomic absorption spectrometry after preconcentration on a chelating resin, *Anal. Chim. Acta,* 115, 103, 1980.

404. **Muzykov, G. G. and Prostetsov, G. P.,** Atomizer for determining mercury by the atomic absorption method of coldstream, *Zavod. Lab.,* 46, 234, 1980.

405. **Rombach, N., Apel, R., and Tschochner, F.,** Trace determination of arsenic, barium and mercury in plastics with flameless atomic absorption spectroscopy and ICP emission spectral analysis, *GIT Fachz. Lab.,* 24, 1165, 1980.

406. **Sakamoto, H. and Kamada, M.,** Determination of trace amounts of mercury in solid samples using permanganate solution as a collector by cold vapor atomic absorption spectrophotometry, *Butsurigaku Kagaku,* 13, 63, 1980.

407. **Itsuki, K. and Ikeda, T.,** Determination of trace mercury in sulfur and pyrite by flameless AA spectrometry, *Bunseki Kagaku,* 29, 861, 1980.

408. **Baroccio, A. and Moauro, A.,** Determination of some elements in vegetal samples by instrumental neutron activation. Comparison with the data obtained from atomic absorption. CNEN-RT/CHI (80), Comitato Nazionale Energia Nucleare, Casaccio Nuclear Studies Center, Rome, Italy, 1980, 15 (in Italian).

409. **Fink, L. K., Jr., Harris, A. B., and Schick, L. L.,** Trace Metals in Suspended Particulates, Biota and Sediments of the St. Croix, Narraguagus and Union Estuaries and the Goose Cove Region of Penobscot Bay, Rep. W 80-04809, OWRT A-041-ME(I), 1980 available from the NTIS, Springfield, Va.; *Govt. Announce. Index (U.S.),* 80, 3372, 1980.

410. **Goulden, P. D. and Anthony, D. H. J.,** Chemical speciation of mercury in natural waters, *Anal. Chim. Acta,* 120, 129, 1980.

411. **Karmanova, N. G. and Pogrebnyak, Y. F.,** Determination of mercury in powdered samples of sulfide ores by an atomic absorption spectral method with a "graphite-capsule-adaptor" atomizer, *Zh. Prikl. Spektrosk.,* 33, 813, 1980.

412. **Koshima, H. and Onishi, H.,** Collection of mercury from artificial seawater with activated carbon, *Talanta,* 27, 795, 1980.

413. **Lindh, U., Brune, D., Nordberg, G., and Wester, P. O.,** Levels of Sb, As, Cd, Cu, Pb, Hg, Se, Ag, Sn and Zn in bone tissue of industrially exposed workers, *Sci. Total Environ.,* 16, 109, 1980 (in English).

414. **Posta, J.,** Determination of the optimum parameters of a combined atomic absorption spectrophotometric plus arc flame method for the analysis of floating dust samples, *Hung. Sci. Instrum.,* 47, 33, 1980.

415. **Van Willis, W., El-Ahraf, A., Vinjamoori, D. V., and Aref, K.,** Analysis of animal feed ingredients and soil amendment products produced from beef cattle manure for selected trace metals using atomic absorption spectrophotometry, *J. Food Prot.,* 43, 834, 1980.

416. **Szakacs, O., Lasztity, A., and Horvath, Z.,** Breakdown of organic mercury compounds by hydrochloric acid-permanganate or bromine monochloride solution for the determination of mercury by cold vapor atomic absorption spectrometry, *Anal. Chim. Acta,* 121, 219, 1980.

417. **Harms, U.,** Trace analysis of mercury at the ng/g level, *Z. Lebensm. Unters. Forsch.,* 172, 118, 1981 (in German).

418. **Meranger, J. C., Hollebone, B. R., and Blanchette, G. A.,** Effects of storage times, temperature and container types on the accuracy of atomic absorption determination of Cd, Cu, Hg, Pb and Zn in whole heparinized blood, *J. Anal. Toxicol.,* 5, 33, 1981.

419. **Narasaki, H.,** Determination of trace mercury in milk products and plastics by combustion in an oxygen bomb and cold vapor atomic absorption spectrometry, *Anal. Chim. Acta,* 125, 187, 1981.

420. **Schierling, P. and Schaller, K. H.,** Quantitative determination of metallic mercury in air, *At. Spectrosc.,* 2, 92, 1981.

421. **Wigfield, D. C., Croteau, S. M., and Perkins, S. L.,** Elimination of the matrix effect in the cold vapor atomic absorption analysis of mercury in human hair samples, *J. Anal. Toxicol.,* 5, 52, 1981.

422. **Zelyukova, Y. V., Vitkun, R. A., Didorenko, T. O., and Poluekyov, N. S.,** Flameless atomic absorption determination of mercury in environmental samples, *Zh. Anal. Khim.,* 36, 454, 1981 (in Russian).

423. **Coyle, P. and Hartley, T.,** Automated determination of mercury in urine and blood by the Magos reagent and cold vapor atomic absorption spectrometry, *Anal. Chem.,* 53, 354, 1981.

424. **Dumarey, R., Heindryckx, R., and Dams, R.,** Modified Perkin-Elmer MAS-50 Spectrometer for mercury determinations, *At. Spectrosc.,* 2, 51, 1981.

425. **El-Ahraf, A., Van Willis, W., and Vinjamoori, D. V.,** Sodium hypophosphite as reducing agent for determination of submicrogram quantities of mercury in animal feeds and manures, *J. Assoc. Off. Anal. Chem.,* 64, 9, 1981.

426. **Farant, J. P., Brisette, D., Moncion, L., Bigras, L., and Chartrand, A.,** Improved cold-vapor atomic absorption technique for the microdetermination of total and inorganic mercury in biological samples, *J. Anal. Toxicol.,* 5, 47, 1981.

427. **Freiser, H.,** Study on the exchange reaction of nickel (II)-dithizone with mercury (II) by atomic absorption spectrophotometry, *Bunseki Kagaku,* 30, 139, 1981.

428. **Sanemasa, I., Takagi, E., Deguchi, T., and Nagai, H.,** Preconcentration of inorganic mercury with an anion-exchange resin and direct reduction-aeration measurements by cold-vapor atomic absorption spectrometry, *Anal. Chim. Acta,* 130(1), 149, 1981.

429. **Julshamn, K.,** Studies on major and minor elements in molluscs in Western Norway. VII. The contents of 12 elements including copper, zinc, cadmium and lead in common muscel (*Mytilus edulis*) and brown seaweed (*Ascophyllum nodosum*) relative to the distance from the industrial sites in Sorfjorden, inner Hardangerfjord, *Fiskerider. Skr. Ser. Ernaer.*, 1(5), 267, 1981.

430. **Aldrighetti, V.,** Evaluation of mercury vapor amalgamation efficiency on gold sponge at the submilligram level by flameless atomic absorption spectrophotometry, *At. Spectrosc.*, 2, 13, 1981.

431. **Bourcier, D. R. and Sharma, R. P.,** Stationary cold vapor technique for the determination of submicrogram amounts of mercury in biological tissues by flameless atomic absorption, *J. Anal. Toxicol.*, 5, 65, 1981.

432. **Oda, C. E. and Ingle, J. D., Jr.,** Continuous flow cold vapor atomic absorption determination of mercury, *Anal. Chem.*, 53, 2030, 1981.

433. **Clanet, F., Deloncle, R., and Popoff, G.,** Chelating resin for capture, preconcentration and determination of toxic metal traces (Zn, Cd, Hg and Pb) in waters, *Water Res.*, 15(5), 591, 1981.

434. **Capel, I. D., Pinnock, M. H., Dorrell, H. M., Williams, D. C., and Grant, E. C. G.,** Comparison of concentrations of some trace, bulk and toxic metals in the hair of normal and dyslexic children, *Clin. Chem. (Winston-Salem, N.C.)*, 27, 879, 1981.

435. **Kimura, M. and Kawanami, K.,** Separation and preconcentration of trace amounts of several metals in sodium perchlorate using activated carbon as a collector, *Nippon Kagaku Kaishi*, 1, 1, 1981 (in Japanese).

436. **Cowgill, U. M.,** The chemical composition of bananas. Market basket values, *Biol. Trace Elem. Res.*, 3(1), 33, 1981.

437. **Ohsawa, K., Suzuki, K., and Imaeda, K.,** Simultaneous determination of mercury and oxygen in mercury compounds by atomic absorption spectrometry and coulometric titrimetry, *Bunseki Kagaku*, 30(5), 300, 1981.

438. **Nadkarni, R. A.,** Determination of volatile elements in coal and other organic materials by oxygen bomb combustion, *Am. Lab.*, 13(8), 22, 24, 27, 1981.

439. **Manyashin, Y. A. and Zusman, B. L.,** Flameless atomic absorption method for the determination of mercury in biological samples, *Gig. Tr. Prof. Zabol.*, 2, 48, 1981 (in Russian).

440. **Marigo, A., Pasquetto, A., Nova, A., and Zanetti, R.,** Trace metal analysis by atomic absorption, *Inquinamento*, 23(1), 31, 1981.

441. **Fleishmann, D. and Schwab, H.,** Enrichment of mercury from town gas, Ger.(East) 146,503 (Cl. GOIN 31/6), 11 Feb, 1981; CA 95, 153646b, 1981.

442. **Oda, C. E. and Ingle, J. D., Jr.,** Continuous flow cold vapor atomic absorption determination of mercury, *Anal. Chem.*, 53(13), 2030, 1981.

443. **Amini, M. K., Defreese, J. D., and Hathaway, L. R.,** Comparison of three instrumental spectroscopic techniques for elemental analysis of Kansas shales, *Appl. Spectrosc.*, 35(5), 497, 1981.

444. **Oda, C. E. and Ingle, J. D., Jr.,** Speciation of mercury by cold vapor atomic absorption spectrometry with selective reduction, *Anal. Chem.*, 53, 2304, 1981.

445. **Suddendorf, R. F.,** Interference of selenium in tellurium in the determination of mercury by cold vapor generation atomic absorption spectrometry, *Anal. Chem.*, 53, 2234, 1981.

446. **Aldeman, H., Jenniss, S. W., and Katz, S. A.,** Interlaboratory analysis of sewage sludge, *Am. Lab.*, 31, 1981.

447. **Kahn, H. L., Chritiano, L. C., Dulude, G. F., and Sotera, J. J.,** Automated hydride analysis, *Am. Lab.*, 13(11), 136, 138, 141, 1981.

448. **Didorenko, T. O., Vitkun, R. A., and Zelyukova, Y. V.,** Mercury determination by flameless atomic absorption method, *Lab. Delo*, 9, 536, 1981 (in Russian).

449. **Gadea Carrera, E., Marti Veciana, A., and Sanchez Perez, M. V.,** Development of a rapid and sensitive method for the environmental determination of organic mercury, phenyl mercury acetate by atomic absorption spectrophotometry, Libro Actas-Congr. Nac. Med. Hig. Segur. Trab., 9th, 2, 781, Madrid, Spain, 1981 (in Spanish).

450. **Szakacs, O., Lasztity, A., and Horvath, Z.,** Determination of total mercury in various waters by cold-vapor atomic absorption spectrometry following the breakdown of organomercury compounds by hydrochloric acid permanganate or bromine monochloride, *Magy. Kem. Foly.*, 87(10), 458, 1981 (in Hungarian).

451. **Mueller, B., Wenzel, B., and Schroeder, L.,** Quantitative determination of phenylmercury acetate in the presence of inorganic mercury salts using flameless atomic absorption, *Z. Chem.*, 21(10), 367, 1981 (in German).

452. **Nadkarni, R. A.,** Determination of volatile elements in coal and other organic materials by oxygen bomb combustion, *Int. Lab.*, September 1981.

453. **Perry, J. A., Farrell, R. F., and Mackie, A. J.,** Modification of a commercial atomic absorption spectrophotometer for cold-vapor determination of mercury, *Rep. Invest. U.S. Bur. Mines*, RI 8573, 1, 1981.

454. **Sugiyama, G. R. K. K.,** Determination of Mercury in Waters, Jpn. Kokai Tokkyo Koho JP 81,100,360 (Cl.GOIN 31/00); appl. 80/2 879, January 14, 1980; August 12, 1981.

455. **Mueller, H. and Sieps, V.,** Quantitative determination of arsenic, lead, cadmium, mercury and selenium in foods by flameless atomic absorption spectrophotometry, *Dtsch. Lebensm. Rundsch.,* 77(11), 392, 1981 (in German).

456. **Denton, G. R. W. and Burdon-Jones, C.,** Influence of temperature and salinity on the uptake distribution and depuration of mercury, cadmium and lead by the black-lip oyster *Saccostrea eschinata, Mar. Biol. (Berlin),* 64(3), 317, 1981.

457. **Blakemore, W. M. and Billedeau, S. M.,** Analysis of laboratory animal feed for toxic and essential elements by atomic absorption and inductively coupled argon plasma emission spectrometry, *J. Assoc. Off. Anal. Chem.,* 64, 1284, 1981.

458. **Fukushima, I.,** A Study on Monitoring Methodology for Trace Elements in the Environment Through Chemical Analysis of Hair. Part of a Coordinated Program on Nuclear Methods in Health-Related Monitoring of Trace Element Pollutants, IAEA-R-2480/F, International Atomic Energy Agency, Vienna, 1981; INIS Atomindex 12, Abstr. No. 626438, 1981.

459. **Yao, T., Akino, M., and Musha, S.,** Atomic absorption spectrometric determination of mercury in water at the picogram-per millimeter levels after preconcentration onto dithiocarbamate bonded silica gel, *Bunseki Kagaku,* 30, 740, 1981.

460. **Miyagawa, H.,** Microdetermination of mercury by carbon tube flameless atomic absorption spectroscopy, *Nenpo-Fukui-Ken Kogyo Shikenjo,* 69, 1981.

461. **Berndt, H. and Messerschmidt, J.,** Microanalytical method of flame atomic absorption and emission spectrometry (platinum loop method). II. Determination of As, Bi, Cd, Hg, Sb, Se, Te, Tl, Zn, Li, Na, K, Cs, Sr and Ba, *Spectrochim. Acta,* 36b, 809, 1981.

462. **Clanet, F., Deloncle, R., and Popoff, G.,** Chelating resin catcher for capture, preconcentration and determination of toxic metal traces (Zn, Cd, Hg and Pb) in waters, *Water Res.,* 15, 591, 1981.

463. **Halasz, A., Polyak, K., and Gegus, E.,** Processes taking place in the graphite tube: anion and matrix effects in atomic absorption spectrometry. II. Determination of mercury, *Mikrochim. Acta,* 2, 229, 1981.

464. **Haraguchi, H., Takahashi, J., Tanabe, K., and Fuwa, K.,** Novel instrumentation of a nondispersive vacuum ultraviolet atomic absorption spectrophotometer for mercury, *Spectrochim. Acta,* 36b, 719, 1981.

465. **Jastrow, J. D., Zimmerman, C. A., Dvorak, A. J., and Hinchman, R. R.,** Plant growth and trace element uptake on acidic coal refuse amended with lime or fly ash, *J. Environ. Qual.,* 10, 154, 1981.

466. **Kurfuerst, U. and Rues, B.,** Determination of heavy metals (lead, cadmium, mercury) in setting sludge using Zeeman AAS without chemical decomposition, *Z. Anal. Chem.,* 308, 1, 1981.

467. **Margler, L. W. and Mah, R. A.,** Thin layer chromatographic and atomic absorption spectrophotometric determination of methyl mercury, *J. Assoc. Off. Anal. Chem.,* 64, 1017, 1981.

468. **Oda, C. E. and Ingle, J. D., Jr.,** Continuous flow cold vapor atomic absorption determination of mercury, *Anal. Chem.,* 53, 2030, 1981.

469. **Okouchi, S. and Sasaki, S.,** Measurement of the solubility of metallic mercury in hydrocarbons by the cold vapor atomic absorption method, *Bull. Chem. Soc. Jpn.,* 54, 2513, 1981.

470. **Oster, O.,** Direct determination of mercury in urine by cold vapor atomic absorption spectroscopy, *J. Clin. Chem. Clin. Biochem.,* 19, 471, 1981.

471. **Scott, J. E. and Ottaway, J. M.,** Determination of mercury vapor in air using a passive gold wire sampler, *Analyst (London),* 106, 1076, 1981.

472. **Suddendorf, R. F., Watts, J. D., and Boyer, K.,** Simplified apparatus for determination of mercury by atomic absorption and inductively coupled plasma emission spectroscopy, *J. Assoc. Off. Anal. Chem.,* 64, 1105, 1981.

473. **Tarui, T. and Tokairin, H.,** Determination of mercury in petroleum by flameless atomic absorption spectrometry after acid decomposition, *Bunseki Kagaku,* 30, 776, 1981.

474. **Wittmann, Z.,** Determination of mercury by atomic absorption spectrophotometry, *Talanta,* 28, 271, 1981.

475. **Robinson, J. W. and Skelly, E. M.,** The direct determination of mercury in hair by atomic absorption spectroscopy at the 184.9 nm resonance line, *Spectrosc. Lett.,* 14, 519, 1981.

476. **Liu, G. and Huang, C.,** A rapid method for determination of mercury in urine, *Xingixue,* 12, 50, 1981.

477. **Romanova, I. B. and Kashparova, E. V.,** Determination of mercury content in cultural fluids, *Deposited Doc. VINITI,* 227, VINITI, Moscow, U.S.S.R., 1981 (in Russian).

478. **Savichev, E. I., Dudnik, A. L., Shugurov, E. V., Kosnyrev, V. S., and Koifman, M. D.,** Device for Atomic Absorption Determination of Mercury in Powder-Like Materials, USSR, SU 868, 492 (Cl. GOIN 21/31), September 30, 1981, appl. 2874, 936, January 28, 1980; from *Otkrytiya, Izobret., Prom. Obraztsy, Tovarnye Znaki,* 36, 179, 1981.

479. **Koshima, H. and Onishi, H.,** Desorption of mercury after collection from solution with activated carbon, *Bunseki Kagaku,* 30, 672, 1981.

480. **Baucells, M., Lacort, G., and Roura, M.,** Atomic absorption spectrophotometric determination of mercury in marine sediments, *An. Quim. Ser. A,* 77, 59, 1981.

481. **Ludwicki, J.,** Determination of mercury level in wheat bran, *Rocz. Panstw. Zakl. Hig.,* 32, 309, 1981.

482. **Kvietkus, K. and Spauskas, S.,** Use of a potassium permanganate absorption solution to determine mercury vapor concentrations in atmospheric air, *Fiz. Atmos.,* 7, 147, 1981.

483. **Takahashi, Y., Shiozawa, Y., Kawai, T., Shimizu, K., Kato, H., and Hasegawa, J.,** Elements dissolved from three types of amalgams in synthetic saliva, *Aichi Gakuin Daigaku Shigakkaishi,* 19, 107, 1981.

484. **Senesi, N. and Polemio, M.,** Trace element addition to soil by application of NPK fertilizers, *Fert. Res.,* 2, 289, 1981.

485. **Kawahara, H., Yamada, T., Nakamura, M., Tomoda, T., Kobayashi, H., Saijo, A., Kawata, Y., and Hikari, S.,** Solubility of metal components into tissue culture medium from dental amalgams, *Shika Rikogaku Zasshi,* 22, 285, 1981.

486. **Zawadzki, H., Szymendera, J., Kowakowski, K., and Karczynski, F.,** Analytical application of co-ordination compounds of *N,N'*diphenylguanidine with Zn(II), Cd(II), Hg(II) and cobalt(II), *Chem. Anal. (Warsaw),* 26, 237, 1981.

487. **Musznska-Zimma, E.,** Occupational exposure to metallic mercury in workers of dental surgeries, *Pol. Tyg. Lek.,* 36, 1195, 1981.

488. **Valenta, P.,** Use of voltammetric method for determination of cadmium and mercury in water and waste-water, *Gewaesserschutz Wasser Abwasser,* 52, 248, 1981.

489. **Kruse, R.,** Experiences in the trace analysis of mercury, lead and cadmium in fish using modern methods of determination, *Fisch Umwelt,* 10, 19, 1981 (in German).

490. **Djujic, I., Djordjevic, V., and Radovic, N.,** Lead, cadmium, copper, arsenic and mercury in some additives (for meat), *Technol. Mesa,* 22, 355, 1981.

491. **Kermoshchuk, J. O. and Warner, P. O.,** A modified method for determination of ambient particulate mercury at trace levels in suspended dust, in *3rd Int. Conf. Heavy Met. Environment,* CEP Consult. Ltd., Edinburgh, U.K., 1981, 603.

492. **Balaes, G. R. E. and Robert, R. V. D.,** Determination by atomic absorption spectrophotometry of impurities in manganese dioxide, *Natl. Inst. Metall. Repub. S. Afr. Rep.,* No. 2094, 1981.

493. **Costantini, S., Macri, A., and Vernillo, I.,** Atomic absorption spectrophotometric analysis of mineral elements in milk-base feeds for zootechnical use, *Riv. Soc. Ital. Sci. Aliment.,* 10, 231, 1981 (in Italian).

494. **Didorenko, T. O., Vitkun, R. A., and Zelyukova, Y. V.,** Mercury determination by non-flame atomic absorption method, *Lab. Delo,* 9, 536, 1981.

495. **Mitchell, G. E.,** Trace metal levels in Queensland dairy products, *Aust. J. Dairy Technol.,* 36, 70, 1981.

496. **Shu-wei, P., Qang-kui, Q., and Jing-fang, S.,** Sequential chemical extraction for speciation of mercury in river sediments, *Huanjing Kexue Xuebao,* 1, 234, 1981 (in Chinese).

497. **Ralston, G. B. and Crisp, E. A.,** Action of organic mercurials on the erythrocyte membrane, *Biochim. Biophys. Acta,* 649, 98, 1981.

498. **Taguchi, M., Tagaki, H., Iwashima, K., and Yamagata, N.,** Metal content of shark muscle powder biological reference material, *J. Assoc. Off. Anal. Chem.,* 64, 260, 1981.

499. **Chapman, J. F. and Dale, L. S.,** Use of alkaline permanganate in the preparation of biological materials for the determination of mercury by atomic absorption spectrometry, *Anal. Chim. Acta,* 134, 1981.

500. **Mandal, S. and Das, A. K.,** Application of an anion exchange resin to the determination of trace mercury in natural waters by cold vapor atomic absorption spectrometry, *At. Spectrosc.,* 3, 56, 1982.

501. **Flanagan, F. J., Moore, R., and Aruscavage, P. J.,** Mercury in geologic reference samples, *Geostandards Newsl.,* 6, 25, 1982.

502. **Kuldvere, A.,** Apparent and real reducing ability of polypropylene in cold vapor atomic absorption spec-trophotometric determinations of mercury, *Analyst (London),* 107, 179, 1982.

503. **Torse, G., Desimoni, E., and Palmisano, F.,** Determination of mercury vapor in air using electrothermal atomic absorption spectrometry with an electrostatic accumulation furnace, *Analyst (London),* 107, 96, 1982.

504. **Zelentsova, L. V., Yudelevich, I. G., and Chanyshova, T. A.,** Flameless atomic absorption determination of mercury in highly pure metals, *Izv. Sib. Otd. Akad. Nauk SSSR Ser. Khim. Nauk,* 1, 130, 1982.

505. **Lendero, L. and Krivan, V.,** Detection or preatomization losses of mercury in the graphite tube with the tracer technique, *Anal. Chem.,* 54, 579, 1982.

506. **Metil, N. I., Taushan, M. D., Chagir, T. S., and Shevchuk, I. A.,** Flameless atomic absorption method for the determination of blood mercury content, *Lab. Delo,* 1, 25, 1982.

507. **Frick, D. A. and Tallman, D. E.,** Flow cell for the determination of mercury in water by electrodeposition followed by atomic absorption spectrometry, *Anal. Chem.,* 54, 1217, 1982.

508. Apparatus for Determining Mercury in Environmental Samples, K. K. Chikyu Kagaku, Jpn. Kokai Tokkyu Koho JP 82,17,857 (Cl. GOIN 31/12) January 29, 1982; appl. 80/92,035, July 5, 1980.

509. **Egawa, H., Kuroda, T., and Shiraishi, N.,** Studies of selective adsorption resins. XV. Determination of a trace amount of ''methylmercury'' in sea water using chelating resins, *Nippon Kagaku Kaishi,* 4, 685, 1982.

510. **Suwirma, S., Surtipanti, S., and Yatim, S.,** Determination of mercury, lead, cadmium and chromium in several sea fishes, *Majalah BATAN,* 14, 2, 1981; CA 96, 211955k, 1982.

511. **Halasz, A., Polyak, K., and Gegus, E.,** Determination of mercury by electrothermal atomization following enrichment by extraction, *Magy. Kem. Foly.,* 88, 139, 1982.

512. **Katalevskii, N. I., Anikanov, A. M., and Semenov, A. D.,** Determination of mercury in natural waters by a flameless atomic absorption method, *Metod. Analiza Mor. Vod. Tr. Sov.-Bulg. Sotrudnichestva, L.,* 78, 1981, from *Ref. Zn. Khim.,* Abstr. No. 7G164, 1982.

513. **Suwirma, S., Surtipanti, S., and Thamzil, L.,** Distribution of heavy metal mercury, lead, cadmium, chromium, copper and zinc in fish, *Majalah BATAN,* 13, 9, 1982 (in Indonesian); CA 96, 15797w, 1982.

514. **Thamzil, L., Suwirma, S., and Surtipanti, S.,** Determination of heavy metals in the stream of Sunter River, *Majalah BATAN,* 13, 41, 1980 (in Indonesian); CA 96, 24538q, 1982.

515. **Tauferova, J. and Zdenek, S.,** Study of the mercury level in potable waters, *Sb. Ved. Pr. Vys. Sk. Chemickotechnol. Pardubice,* 43, 87, 1980 (in Czech); CA 96, 11424z, 1982.

516. **Lawrence, J. and Lawrence, B.,** Mysterious ring appears at miniscus of bottled mercury, *Ind. Res. Dev.,* 1982.

517. **Arai, F., Uziie, A., Iizuka, T., and Saito, T.,** Mercury concentration in Northern Kanto loam, *Nippon Kagaku Kaishi,* 1, 151, 1982.

518. **Senften, H. and Pfenninger, H.,** Testing of beer for undesirable trace elements, *Brau. Rundsch.,* 93, 89, 1982.

519. **Iyengar, G. V., Kasperek, K., Feinedegen, L. E., Wang, Y. X., and Weese, H.,** Determination of cobalt, copper, iron, mercury, manganese, antimony, selenium and zinc in milk samples, *Sci. Total Environ.,* 24, 267, 1982.

520. **Doolan, K. J.,** The determination of traces of mercury in solid fuels by high temperature combustion and cold-vapor atomic absorption spectrometry, *Anal. Chim. Acta,* 140, 187, 1982.

521. **Bourcier, D. R., Sharma, R. P., and Brown, D. B.,** A stationary cold vapor method for atomic absorption measurement of mercury in blood and urine for exposure screening, *Am. Ind. Hyg. Assoc. J.,* 43, 329, 1982.

522. **Cavallaro, A.,** Mercury pollution: flameless atomic absorption analysis, *Rischi Tossic. Inquin. Met.: Cromo Mercurio, (Conv. Naz.)* 145-70 , 1980 (in Italian); CA 97, 50655r, 1982.

523. **Balkas, T., Tugrul, S., and Salihoglu, I.,** Trace metal levels in fish and crustaces from Northeastern Mediterranean coastal waters, *Mar. Environ. Res.,* 6, 281, 1982.

524. **Lugowska, M. and Rubel, S.,** A comparison of potentiometric, spectrophotometric and atomic absorption spectrometric methods for the determination of mercury in wastes, *Anal. Chim. Acta,* 138, 397, 1982.

525. **Prokof'ev, A. K. and Stepanchenko, T. V.,** Determination of labile mercury in sea water and total mercury in bottom sediments by atomic absorption in cold vapors, *Tr. Gos. Okeanogr. In-Ta,* 162, 34, 1981 (in Russian); from *Ref. Zh. Khim.,* Abstr. No. 9G165, 1982.

526. **Sullivan, J. R. and Delfino, J. J.,** The determination of mercury in fish, *J. Environ. Sci. Health Part A,* A17, 265, 1982.

527. **Kovar, K. A., Jarre, G., Lautenschlaeger, W., and Maassen, J.,** Determination of mercury in homeopathic preparations and medical substances and drugs, *Arch. Pharm.,* 315, 662, 1982.

528. **Vitkun, R. A., Didorenko, T. O., Zelyukova, Y. V., and Poluektov, N. S.,** Determination of mercury using flameless atomic absorption with dihydroxymaleic acid as the reducing agent, *Zh. Anal. Khim.,* 37, 833, 1982.

529. **Srivastava, A. K. and Tandon, S. G.,** Studies on mercury pollution: microdetermination of mercury in biological materials by cold vapor atomic absorption spectrometry, *Int. J. Environ. Anal. Chem.,* 11, 221, 1982.

530. **Ni, Z. and Yang, F.,** Determination of mercury in soil by graphite furnace atomic absorption spectrometry using citric acid as matrix modifier, *Huanjing Huaxue,* 1, 83, 1982.

531. **Mojo, L., Martella, S., and Martino, G.,** Seasonal concentrations of heavy metals (mercury, cadmium, lead) in some marine organisms in the central Mediterranean Sea, *Mem. Biol. Mar. Oceanogr.,* 10, 27, 1980 (in Italian); CA 97, 97982k, 1982.

532. **Walker, T. I., Glover, J. W., and Powell, D. G. M.,** Effect of length, locality and tissue type on mercury and cadmium content of the commercial scallop, Pecten alba Tate, from Port Phillip Bay, Victoria, *Aust. J. Mar. Freshwater Res.,* 33, 547, 1982.

533. **Takla, G. and Valijianian, V.,** Determination of mercury in pharmaceutical products by atomic absorption spectrophotometry using a carbon rod atomizer, *Analyst (London),* 107, 378, 1982.

534. **Bartha, A. and Ikrenyi, K.,** Interfering effects on the determination of low concentrations of mercury in geological materials by cold-vapor atomic absorption spectrometry, *Anal. Chim. Acta,* 139, 329, 1982.

535. **Campe, A., Velghe, N., and Claeys, A.,** Determination of inorganic, phenyl and alkyl mercury in hair, *At. Spectrosc.,* 3, 122, 1982.

536. **Yanagisawa, M., Kitagawa, K., and Tsuge, S.,** Separative column atomizer for the direct determination of trace amounts of volatile elements by atomic absorption spectrometry, *Spectrochim. Acta,* 37b, 493, 1982.

AUTHOR INDEX

SUBJECT INDEX (MERCURY)

Group IIIA Elements — B, Al, Ga, In, and Tl

BORON, B (ATOMIC WEIGHT 10.811)

Boron is found in nature as boric acid and in other combination forms in siliceous rocks. It is used in the preparation of special alloys, refractory compounds, atomic energy materials, high-energy fuels, ceramic materials, enamel, paint pigments, food preservatives, and heat-resistant glass. Its varied industrial use requires quantitative determination in these materials.

Sample preparation requires extra care to avoid loss of boron during the decomposition process. Digestion should be carried out in a closed vessel. Boron in crystalline form is not attacked by acids or alkalis. The amorphous boron is soluble in concentrated nitric acid and sulfuric acid. Alkali fusion is preferred for crystalline and amorphous boron. Boric acid is soluble in water and tartaric acid and is slightly soluble in other acids.

For the determination of boron using the atomic absorption spectrophotometric method, not much literature is available. Titration methods for a large amount of boron and the carminic acid method using UV spectrophotometry for trace level determination are predominantly in practice.

In the determination of boron using the atomic absorption method, sodium causes interference when present in excessively large concentration than that of boron. This effect can be minimized by reducing the red feather height, but with a decrease in sensitivity. Background correction is required for samples containing high solids, high acid concentrations, organic solvents, or any other complex matrix which may cause light scatter or molecular absorption. All analytical ranges may be extended by the rotation of the burner head.

Standard Solution

To prepare 500 mg/ℓ solution, dissolve 28.60 g of boric acid (H_3BO_3) in 100 mℓ of deionized water and dilute to 1 ℓ.

Instrumental Parameters

Wavelength	249.7 nm
Slit width	0.7 nm
Light source	Hollow cathode lamp
Flame	Nitrous oxide-acetylene, fuel rich, reddish white cone
Sensitivity	13 mg/ℓ
Detection limit	0.70 mg/ℓ
Optimum range	400 mg/ℓ

Secondary wavelength and its sensitivity

208.9 nm	27 mg/ℓ

REFERENCES

1. **Slavin, W., Sprague, S., and Manning, D. C.,** Detection limits in analytical atomic absorption spectrophotometry, *At. Absorpt. Newsl.,* No. 18, February 1964.
2. **Goleb, J. A.,** An attempt to determine the boron natural abundance ratio B^{11}/B^{10} by atomic absorption spectrophotometry, *Anal. Chim. Acta,* 36, 130, 1966.
3. **Slavin, W., Venghiattis, A., and Manning, D. C.,** Some recent experience with the nitrous oxide-acetylene flame, *At. Absorpt. Newsl.,* 5, 84, 1966.
4. **Manning, D. C.,** The determination of boron, beryllium, germanium and niobium using the nitrous oxide-acetylene flame, *At. Absorpt. Newsl.,* 6, 35, 1967.
5. **Bader, H. and Brandenberger, H.,** Boron determination in biological materials by atomic absorption spectrophotometry, *At. Absorpt. Newsl.,* 7, 1, 1968.
6. **Slavin, W.,** *Atomic Absorption Spectroscopy,* John Wiley & Sons, New York, 1968.
7. **Harris, R.,** Determination of small quantities of boron by atomic absorption spectrophotometry, *At. Absorpt. Newsl.,* 8, 42, 1969.
8. **Melton, J. R., Hoover, W. L., and Howard, P. A.,** Atomic absorption spectrophotometric determination of water-soluble boron in fertilizers, *J. Assoc. Off. Anal. Chem.,* 52, 950, 1969.
9. **Sanders, J.,** Notes on the effect of lamp-gas filling on the atomic absorption of boron, *Resonance Lines,* 1, 7, 1969.
10. **Weger, S. J., Hossner, L. R., and Ferrara, L. W.,** Determination of boron in fertilizers by atomic absorption spectrophotometry, *J. Agric. Food Chem.,* 17, 1276, 1969.
11. **Melton, J. R., Hoover, W. L., Howard, P. A., and Ayers, J. L.,** Atomic absorption spectrophotometric determination of boron in plants, *J. Assoc. Off. Anal. Chem.,* 53, 682, 1970.
12. **Weger, S. J., Hossner, L. R., and Ferrara, L. W.,** Determination of boron in fertilizers by atomic absorption spectrophotometry, *At. Absorpt. Newsl.,* 9, 58, 1970.
13. **Holak, W.,** Atomic absorption determination of boron in foods, *J. Assoc. Off. Anal. Chem.,* 54, 1138, 1971.
14. **Holak, W.,** Collaborative study of the determination of boric acid in foods by atomic absorption spectrophotometry, *J. Assoc. Off. Anal. Chem.,* 55, 890, 1972.
15. **Hayashi, Y., Matsushita, S., Kumamaru, T., and Yamamoto, Y.,** Indirect atomic absorption determination of boron by solvent extraction as tris (1,10-phenanthroline) cadmium tetrafluoroborate, *Talanta,* 20, 414, 1973.
16. **Spielholtz, G. I., Toralballa, G. C., and Willsen, J. J.,** Determination of total boron in sea water by atomic absorption spectroscopy, *Mikrochim. Acta,* 4, 649, 1974.
17. **Clinton, O. E.,** Curcumin method for boron compatible with an atomic absorption system of plant analysis, *N.Z. J. Sci.,* 17, 445, 1974.
18. **Akama, Y., Nakai, T., and Kawamura, F.,** Extraction and atomic absorption spectrophotometric determination of lead with 4-benzoyl-3-methyl-1-phenyl-5-pyrazolone, *Bunseki Kagaku,* 25, 496, 1976.
19. **Elton-Bolt, R. R.,** The determination of boron in plant tissue byatomic absorption spectrometry, *Anal. Chim. Acta,* 86, 281, 1976.
20. **Chapman, J. F. and Dale, L. S.,** Improved sensitivity for boron and silicon in flame spectrometry by a fluoride evolution technique, *Anal. Chim. Acta,* 89, 363, 1977.
21. **Korovin, V. A. and Kalinina, N. M.,** Flame photometric determination of boron in atomic power plant water, *Teploenergetika,* 5, 25, 1977.
22. **Hannaford, P. and Lowe, R. M.,** Determination of boron isotope ratios by atomic absorption spectrometry, *Anal. Chem.,* 49, 1852, 1977.
23. **Pickett, E. E. and Franklin, M. L.,** Filter flame photometer for determination of boron in plants, *J. Assoc. Off. Anal. Chem.,* 60, 1164, 1977.
24. **DeVries, L. E. and Gubner, E.,** Determination of main components and impurities in lithium-boron alloys, *Anal. Chem.,* 50, 694, 1978.
25. **Horta, A. M. and Curtius, A. J.,** Rapid extraction-atomic absorption determination of boron in sea water, *Anal. Chim. Acta,* 96, 207, 1978.
26. **Glasser, E.,** Burnup determination of boron-containing neutron absorbers by atomic absorption spectroscopy, *Kernenenergie,* 21, 235, 1978.
27. **Thevenot, F. and Goeuriot, P.,** Analysis of nonmetallic borides. Interstitial compounds of boron, *Analysis,* 6, 359, 1978 (in French).
28. **Aznarez-Alduan, J. and Bonilla Palo, A.,** Determination of boron by atomic absorption spectroscopy after extraction, *An. Quim.,* 74, 756, 1978 (in Spanish).
29. **Szydlowski, F. J.,** Boron in natural waters by atomic absorption spectrometry with electrothermal atomization, *Anal. Chim. Acta,* 106, 121, 1979.

30. **Semenenko, K. A., Zuikova, N. V., Slepnev, S. N., and Kuzyokov, Y. Y.,** Flame photometric determination of boron using an air acetylene flame, *Vestn. Mosk. Univ. Ser. 2 Khim.,* 20, 369, 1979.

31. **Boyer, K. W., Capar, S. G., Jones, J. W., Suddendorf, R. F., and Forwalter, J.,** Multielement trace analysis identifies 48 elements in foods. Simultaneous element analysis and data reduction reaches parts per billion, *Food Process.,* 40, 72, 1979.

32. **Fricke, F. L., Robbins, W. B., and Caruso, J. A.,** Trace element analysis of food and beverages by atomic absorption spectrometry, *Prog. Anal. At. Spectrosc.,* 2, 185, 1979.

33. **Han, S. S., and Hyon, C. G.,** Determination of boron with metal-cationic complex of 1,10-phenanthroline. II. Indirect extraction and atomic absorption spectrophotometry of boron with the cationic iron complex of 1,10-phenanthroline, *Punsok Hwahak,* 17, 10, 1979 (in Korean).

34. **Grallath, E., Tschoepel, P., Koelblin, G., Stix, U., and Toelg, G.,** Determination of traces of boron in metals, silicon and quartz by spectrophotometry and emission spectrometry with plasma excitation (CMP, ICP) after distillation and extraction of tetrafluoroborate (−) ion associates, *Z. Anal. Chem.,* 302, 40, 1980.

35. **Manoliu, C., Popescu, O., Balasaa, T., Tamas, V., and Georgescu, I.,** Study on determination of metal traces in solvents, *Rev. Chim.,* 31, 291, 1980.

36. **Morse, R. A. and Lisk, D. J.,** Elemental analysis of honeys from several nations, *Am. Bee J.,* 120, 592, 1980.

37. **Nadkarni, R. A.,** Multitechnique and multielemental analysis of coal and fly ash, *Anal. Chem.,* 52, 929, 1980.

38. **Drake, L.,** Analysis of wool glass by atomic absorption spectroscopy, *Chem. N.Z.,* 44, 98, 1980.

39. **Cowgill, U. M.,** The chemical composition of bananas. Market basket value, *Biol. Trace Elem. Res.,* 3, 33, 1981.

40. **Watson, M. E.,** Interlaboratory comparison in the determination of nutrient concentrations of plant tissue, *Commun. Soil. Sci. Plant Anal.,* 12, 60, 1981.

41. **Baliza, S. V. and Soledade, L. E. S.,** Applications of atomic absorption in molecular analysis (spectrophotometry), *Congr. Anu. ABM,* 36th Congress of the Brazilian Metals Association, Sao Paulo, Brazil, 1981, 333 (in Portuguese).

42. **Maio, A.,** Narrow double Zeeman (DZ) transitions of oriented Boron-12 nuclei, *Hyperfine Interact,* 10, 1175, 1981.

43. **Van der Geugten, R. P.,** Determination of boron in river water with flameless atomic absorption spectrometry (graphite furnace technique), *Z. Anal. Chem.,* 306, 13, 1981.

44. **Young, R. S.,** Analysis of nickel refinery slimes and residues, *Talanta,* 28, 25, 1981.

45. **Kagami, N., Miyazaki, H., and Hatcho, N.,** Determination of boron by atomic absorption spectroscopy, *Kanzei Chuo Bunsekisho Ho,* 22, 97, 1981.

46. **DiGiuseppe, T. G., Estes, R., and Davidovits, P.,** Boron atom reactions. III. Rate constants with water, hydrogen peroxide, alcohols and ethers, *J. Phys. Chem.,* 86, 260, 1982.

47. **Olsen, K. B., Wilkerson, C. L., Toste, A. P., and Hayes, D. J.,** Isolation of metallic complexes in shale oil and shale oil retort waters, *NBS Spec. Publ.,* 618, 105, 1982.

AUTHOR INDEX

SUBJECT INDEX (BORON)

ALUMINUM, Al (ATOMIC WEIGHT 26.9815)

Aluminum occurs widely in nature, usually in combination with other elements. It has extensive uses in building materials, manufacturing, aircraft, railroad, automotive, and marine equipment, in the food industry in handling, processing, and packing, and in the pharmaceutical and chemical industries.

Considering the uses of aluminum, there are not many analytical techniques available for determination. Its direct determination is the most difficult one due to other accompanying elements. Most of the analytical methods are applied after a successive separation using precipitation methods. Precipitation of other metal ions along with aluminum is a common source of error in its quantitative determination. The atomic absorption method is a valuable method for aluminum and in most cases is followed after the separation from other elements.

The chemical, metallurgical, and commercial aspects of the metal and its alloys should be taken into consideration to prepare the sample solution. Dissolution can be alkaline or acidic depending on the single metal or mixture analyses. Halogen acids dissolve aluminum readily. High-purity metal is resistant to acid attack, but dissolves rapidly in aqua regia or sodium hydroxide.

Partial ionization of aluminum in nitrous oxide-acetylene flame may be suppressed by the addition of potassium nitrate or potassium chloride (1000 to 2000 mg/ℓ final potassium concentration) to blank, standard, and sample solutions. Presence of nickel, manganese, cobalt, chromium, and high silicate concentrations decrease the sensitivity. Iron and acid concentrations greater than 0.2% also decrease the aluminum sensitivity. The aluminum signal is reported to be enhanced in the presence of iron, titanium, fluoroborate, and acetic acid. Background correction should be used for samples containing high solids, high acid concentrations, and organic solvents.

Standard Solution

To prepare 1000 mg/ℓ solution, dissolve 1.000 g of aluminum wire in 50 mℓ concentrated hydrochloric acid with gentle heating. Cool and dilute to 1 ℓ with deionized water. A small drop of mercury can be used as a catalyst for complete dissolution.

Instrumental Parameters

Wavelength	309.3 nm
Slit width	0.7 nm
Light source	Hollow cathode lamp/electrodeless discharge lamp
Flame	Nitrous oxide-acetylene, reducing (rich and red)
Sensitivity	1.0 mg/ℓ
Detection limit	0.03 mg/ℓ
Optimum range	10—100 mg/ℓ

Secondary wavelengths and their sensitivities

396.2 nm	1.0 mg/ℓ
308.2 nm	1.0 mg/ℓ
394.4 nm	1.0 mg/ℓ
237.3 nm	2.0 mg/ℓ
237.6 nm	4.0 mg/ℓ
257.5 nm	4.0—6.0 mg/ℓ
256.8 nm	6.0—7.0 mg/ℓ

REFERENCES

1. **Dowling, F. B., Chakrabarti, C. L., and Lyles, G. R.,** Atomic absorption spectroscopy of aluminum, *Anal. Chim. Acta,* 28, 392, 1963.
2. **Manning, D. C.,** The determination of aluminum by atomic absorption spectroscopy, *At. Absorpt. Newsl.,* 24, 1964.
3. **Chakrabarti, C. L., Lyles, G. R. and Dowling, F. B.,** The determination of aluminum by atomic absorption spectroscopy, *Anal. Chim. Acta,* 29, 489, 1963.
4. **Nikolaev, G. I. and Aleskovsky, V. B.,** Atomic absorption micromethod for aluminum determinations in pure metals and alloys, *Zh. Anal. Khim.,* 18, 816, 1963 (in Russian).
5. **Novoselov, V. A. and Aidarov, T. K.,** Spectrographic determination of the trace elements Ag, Cu, Pb, Bi, Cd and Al in solutions by using a hollow cathode source, *Tr. Khim. Khim. Tekhnol.,* 108, 1964.
6. **Wilson, L.,** A Complete Scheme for the Analysis of Aluminum Alloys by Atomic Absorption Spectroscopy, Aeron. Res. Lab. Rept. Met. 52, Aeronautical Research Laboratories, Commonwealth Department of Supply, Melbourne, Australia, 1964.
7. **Amos, M. D. and Thomas, P. E.,** The determination of aluminum in aqueous solution by atomic absorption spectroscopy, *Anal. Chim. Acta,* 32, 139, 1965.
8. **Nikolaev, G. I.,** Atomic absorption method for the determination of aluminum and zinc admixtures in solid samples of refractory metals, *Zh. Anal. Khim.,* 20, 445, 1965.
9. **Bell, G. F.,** The analysis of aluminum alloys by means of atomic absorption spectrophotometry, *At. Absorpt. Newsl.,* 5, 73, 1966.
10. **Deily, J. B.,** Determination of aluminum in trialkylaluminums by atomic absorption spectroscopy, *At. Absorpt. Newsl.,* 5, 119, 1966.
11. **Mansell, R. E., Emmel, H. W., and McLaughlin, E. L.,** Analysis of magnesium and aluminum alloys by atomic absorption spectroscopy, *Appl. Spectrosc.,* 20, 231, 1966.
12. **Friend, K. B. and Diefenderfer, A. J.,** Determination of refractory oxide elements by atomic absorption spectrometry with the plasma jet, *Anal. Chem.,* 38, 1763, 1966.
13. **Slavin, S. and Slavin, W.,** Fully automatic analysis of used aircraft oils, *At. Absorpt. Newsl.,* 5, 106, 1966.
14. **Scott, T. C., Roberts, E. D., and Cain, D. A.,** Determination of minor constituents in ferrous materials by atomic absorption spectrophotometry, *At. Absorpt. Newsl.,* 6, 1, 1967.
15. **Cepacho-Delgado, L. and Manning, D. C.,** The determination by atomic absorption spectroscopy of several elements, including silicon, aluminum and titanium in cement, *Analyst (London),* 92, 553, 1967.
16. **Crow, R. F., Hime, W. G., and Connolly, J. D.,** Analysis of Portland cement by atomic absorption, *J. Res. Dev. Lab. Portland Cem. Assoc.,* 9(2), 60, 1967.
17. **Druckman, D.,** Titanium, aluminum and iron analysis in polypropylene by atomic absorption spectrophotometry, *At. Absorpt. Newsl.,* 6, 113, 1967.
18. **Hartley F. R. and Ingles, A. S.,** The determination of aluminum in wool by atomic absorption spectrophotometry, *Analyst (London),* 92, 622, 1967.
19. **Jursik, M. L.,** Application of atomic absorption to the determination of Al, Fe, and Ni in the same uranium-base sample, *At. Absorpt. Newsl.,* 6, 21, 1967.
20. **LaFlamme, Y.,** Determination of aluminum in soils by atomic absorption spectroscopy, *At. Absorpt. Newsl.,* 6, 70, 1967.
21. **Myers, D.,** The determination of aluminum, vanadium and iron in titanium alloys and iron in zirconium alloys, *At. Absorpt. Newsl.,* 6, 89, 1967.
22. **Ramakrishna, T. V., West, P. W., and Robinson, I. W.,** The determination of aluminum and beryllium by atomic absorption spectroscopy, *Anal. Chim. Acta,* 39, 81, 1967.
23. **Katz, S. A.,** The direct and rapid determination of alumina and silica in silicate rocks and minerals by atomic absorption spectroscopy, *Am. Mineral.,* 53, 283, 1968.
24. **Slavin, W.,** *Atomic Absorption Spectroscopy,* John Wiley & Sons, New York, 1968.
25. **Van Loon, J. C.,** Determination of aluminum in high silica materials, *At. Absorpt. Newsl.,* 7, 3, 1968.
26. **Gomez Coedo, A. and Jiminez Seco, J. L.,** Determination of iron, manganese, calcium, magnesium, aluminum and silicon in iron ores by atomic absorption, *Rev. Metal.,* 4, 58, 1968 (in Spanish).
27. **Gomez Coedo, A. and Jiminez Seco, J. L.,** Determination of Fe, Mn, Ca, Mg, Al, Y, S, in minerals by atomic absorption, *Rev. Metal. Ceram.,* 4, 58, 1968.
28. **Endo, Y., Hata, T., and Nakahara, Y.,** Atomic absorption determination of calcium, magnesium, manganese, copper, zinc and aluminum in iron ores, *Jpn. Anal.,* 17, 679, 1968.
29. **Yuan, T. L. and Breland, H. L.,** Evaluation of atomic absorption methods for determinations of aluminum, iron and silicon in clay and soil extracts, *Soil Sci. Soc. Am. Proc.,* 33, 868, 1969.
30. **Tumanov, A. A. and Petukhova, V. G.,** Determination of traces of aluminum and silicon in their oxides and of aluminum in metallic chromium, *Zavod. Lab.,* 35, 654, 1969 (in Russian).

31. **Hansen, R. K., Bachman, R. Z., O'Laughlin, J. W., and Banks, C. V.,** Determination of aluminum in vanadium metal by atomic absorption spectrophotometry, *Anal. Chim. Acta,* 46, 217, 1969.
32. **Koenig, P., Schmitz, K. H., and Thiemann, E.,** Determination of aluminum in low and high alloy steels, in ores and slags by atomic absorption, *Z. Anal. Chem.,* 244, 232, 1969.
33. **Lundy, R. G. and Watje, W. F.,** Determination of aluminum, iron, chromium, nickel and zinc in inhibited red fuming nitric acid by atomic absorption spectrophotometry, *At. Absorpt. Newsl.,* 8, 124, 1969.
34. **Raad, A. T., Protz, R., and Thomas, R. L.,** Determination of sodium dithionate and ammonium oxalate extractable iron, aluminum and manganese in soils by atomic absorption spectroscopy, *Can. J. Soil Sci.,* 49, 89, 1969.
35. **Spencer, D. W. and Sachs, P. L.,** A study of potential interferences on the determination of particulate aluminum in sea water using atomic absorption spectrometry, *At. Absorpt. Newsl.,* 8, 65, 1969.
36. **Stirling, A. J. and Westwood, W. D.,** Investigation of the sputtering of aluminum using atomic absorption spectroscopy, *J. Appl. Phys.,* 41, 742, 1970.
37. **Koenig, P., Schmitz, K. H., and Thiemann, E.,** Determination of aluminum in low and high alloy steels in ores and slags by atomic absorption, *At. Absorpt. Newsl.,* 9, 103, 1970.
38. **Holak, W.,** Atomic absorption spectrophotometry of iron and aluminum in baking powders, *J. Assoc. Off. Anal. Chem.,* 53, 887, 1970.
39. **Goto, H., Kakita, Y., and Namiki, M.,** Determination of aluminum in iron and steel by flame photometry, *Bunseki Kagaku,* 19, 1211, 1970.
40. **Ferris, A. P., Jepson, W. B., and Shapland, R. C.,** Evaluation and correction of interference between aluminum silicon and iron in atomic absorption spectrophotometry, *Analyst (London),* 95, 574, 1970.
41. **Khera, A. K., Steinnes, E., and Oeien, A.,** Comparative study of aluminum determination in ammonium acetate extracts of soil by spectrophotometry, activation analysis and atomic absorption, *Acta Agric. Scand.,* 20, 33, 1970.
42. **Urbain, H. and Varlot, M.,** Study of physico-chemical interferences in aluminum determination by atomic absorption (nitrous oxide acetylene flame), *Rev. GAMS,* 6, 373, 1970 (in French).
43. **Thulborne, C. and Scholes, P. H.,** Determination of Aluminum in Iron and Steel Making Slags by an Atomic Absorption Method, B.I.S.R.A. Open Rep. MG/D/676/70, British Iron and Steel Research Association, Scarborough, England, 1970.
44. **Paus, P. E.,** Interference study of alkali metals on aluminum determinations by atomic absorption spectrophotometry, *Anal. Chim. Acta,* 54, 164, 1971.
45. **Kudd, Y., Hasegawa, N., and Yamashita, T.,** Determination of aluminum, cobalt, chromium, copper, iron, manganese, molybdenum, nickel, tin and vanadium in titanium alloys by atomic absorption spectrometry, *Bunseki Kagaku,* 20, 1319, 1971.
46. **Hofer, A.,** Use of ammonium fluoride in the determination of aluminum by atomic absorption spectrometry in the air-acetylene flame, *Z. Anal. Chem.,* 253, 206, 1971 (in German).
47. **Fassel, V. A., Rasmusson, J. D., and Kniseley, R. N.,** Flame emission spectrometric determination of aluminum, cobalt, chromium, copper, manganese, niobium and vanadium in low and high alloy steels, *Anal. Chem.,* 42, 186, 1971.
48. **Cobb, W. D. and Harrison, T. S.,** Determination of alumina in iron ores, slags and refractory materials by atomic absorption spectroscopy, *Analyst (London),* 96, 764, 1971.
49. **Roos, J. T. H.,** Determination of traces of aluminum in lithium fluoride, *Anal. Chim. Acta,* 61, 136, 1972.
50. **Begak, O. U. and Nikolaev, G. I.,** Determination of aluminum in master alloys by an atomic absorption method in an acetylene-air flame, *Zavod. Lab.,* 38, 1461, 1972.
51. **Fishman, M. J.,** Determination of aluminum in water, *At. Absorpt. Newsl.,* 11, 46, 1972.
52. **Headridge, J. B. and Sowerbutts, A.,** Atomic absorption spectrophotometric determination of total aluminum in steel after its dissolution in a pressure bomb, *Analyst (London),* 98, 57, 1973.
53. **Karkhanis, P. P. and Anfinsen, J. R.,** Atomic absorption spectrophotometric determination of aluminum in pharmaceuticals, *J. Assoc. Off. Anal. Chem.,* 56, 358, 1973.
54. **Legrand, G., Louvrier, J., Musikas, N., and Voinovitch, I. A.,** Determination of silica, alumina, ferric oxide, lime and magnesia in cements by X-ray fluorescence and atomic absorption spectrometry, *Analysis,* 2, 9, 1973.
55. **West, A. C., Kniseley, R. N., and Fassel, V. A.,** Interelement effects in the flame: spectroscopic determination of aluminum, molybdenum and vanadium in a nitrous oxide-acetylene flame formed on a circular slot burner, *Anal. Chem.,* 45, 815, 1973.
56. **Brivot, F., Cohort, I., Legrand, G., Louvrier, J., and Voinovitch, I.,** Comparative determinations by atomic absorption spectrometry and by spectrophotocalorimetry of nickel, copper, chromium, manganese, aluminum, and silicon in certain steels, *Analysis,* 2, 570, 1973.
57. **Furuta, H. and Nakamura, K.,** Effect of ligand on the atomic absorption intensity of aluminum and its application to the determination of aluminum in aluminum alloys, *Nippon Kagaku Kaishi,* 1122, 1973.
58. **Smith, B. H.,** Determination of aluminum by atomic absorption spectrophotometry after chelation with 8-hydroxyquinoline and extraction with isobutyl methyl ketone, *Lab. Pract.,* 22, 100, 1973.

59. **Smith, D. and McLain, M. E.**, Comparison of instrumental neutron activation analysis and atomic absorption spectrophotometry for aluminum determination in phosphate minerals, *Radiochem. Radioanal. Lett.*, 16, 89, 1974.

60. **Stahlavska, A., Prokopova, H., and Tuzar, M.**, Use of spectral-analytical methods in drug analysis. Determination of bismuth, aluminum, magnesium, calcium, titanium and silicon in antacids by atomic absorption spectrophotometry, *Pharmazie*, 29, 140a, 1974.

61. **Atsuya, I. and Sugiura, N.**, Determination of microamounts of aluminum by atomic absorption spectrometry using a heated graphite atomizer. Application to analysis of steel, *Bunseki Kagaku*, 23, 1170, 1974.

62. **Lerner, L. A. and Tikhomirova, E. I.**, Atomic absorption determination of the total amount of iron and aluminum in soils, *Pochvovedenie*, 6, 114, 1974.

63. **Musil, J.**, Interferences in the determination of aluminum by means of atomic absorption spectrophotometry, *Z. Anal. Chem.*, 271, 352, 1974.

64. **Roelandts, I., Bologne, G., Dupain-Klerky, L., and Czichosz, R.**, Separation of microgram quantities of aluminum from silver matrices prior to its determination by atomic absorption spectrophotometry, *Sep. Sci.*, 9, 445, 1974.

65. **Fuchs, C., Brasche, M., Paschen, K., Nordbeck, H., Quellhorst, E., and Peek, U.**, Aluminum determination in serum by flameless atomic absorption, *Clin. Chim. Acta*, 52, 71, 1974 (in German).

66. **Manoliu, C., Tomi, B., and Petruc, F.**, Determination of silicon, aluminum and sodium in a molecular sieve by emission and atomic absorption flame photometry, *Rev. Chim.*, 24, 991, 1974.

67. **Keliher, P. N. and Wohlers, C. C.**, Spectral line profile measurements from calcium, silver and aluminum hollow cathode lamps, *Appl. Spectrosc.*, 29, 198, 1975.

68. **Condylis, A. and Mejean, B.**, Recent applications of atomic absorption to metallurgical analysis. Determination of low concentrations of the elements aluminum, barium, titanium, tin, antimony and tantalum, *Analusis*, 3, 94, 1975.

69. **Cobb, W. D., Foster, W. W., and Harrison, T. S.**, Determination of aluminum at low levels in steel by atomic absorption spectrophotometry, *Lab. Pract.*, 24, 143, 1975.

70. **Dassani, S. D., McClellan, B. E., and Gordon, M.**, Determination of aluminum, calcium, manganese and titanium in ferrosilicon alloys by atomic absorption spectrometry, *J. Agric. Food Chem.*, 23, 671, 1975.

71. **Guest, R. J. and MacPherson, D. R.**, Use of flame procedures in metallurgical analysis. II. Determination of aluminum in sulfide and silicate minerals and in ores and slags, *Anal. Chim. Acta*, 78, 299, 1975.

72. **Kometani, T. Y. and Wiegmann, W.**, Measurement of gallium and aluminum in a molecular beam epitaxy chamber by atomic absorption spectrometry, *J. Vac. Sci. Technol.*, 12, 933, 1975.

73. **Luzar, O. and Sliva, V.**, Determination of calcium oxide, magnesium oxide, alumina, silica, iron, copper, zinc, lead, cadmium, sodium oxide and potassium oxide in iron ores and agglomerates, *Hutn. Listy*, 30, 55, 1975 (in Czech).

74. **Shaw, F. and Ottaway, J. M.**, Determination of trace amount of aluminum and other elements in iron and steel by atomic absorption spectrometry with carbon furnace atomization, *Analyst (London)*, 100, 217, 1975.

75. **Dolinsek, F., Stupar, J., and Spenko, M.**, Determination of aluminum in dental enamel by the carbon cup atomic absorption method, *Analyst (London)*, 100, 884, 1975.

76. **Alfrey, A. C., LeGendre, G. R., and Kachny, W. D.**, The dialysis encephalopathy syndrome. Possible aluminum intoxication, *N. Engl. J. Med.*, 294, 184, 1976.

77. **Krishnan, S. S., Quittkat, S., and Crapper, D. R.**, Atomic absorption analysis for traces of aluminum and vanadium in biological tissue. A critical evaluation of the graphite furnace atomizer, *Can. J. Spectrosc.*, 21, 25, 1976.

78. **LeGendre, G. R. and Alfrey, A. C.**, Measuring picogram amounts of aluminum in biological tissue by flameless atomic absorption analysis of a chelate, *Clin. Chem. (Winston-Salem, N.C.)*, 22, 53, 1976.

79. **Maruta. T., Minegishi, K., and Sudoh, G.**, The flameless atomic absorption spectrometric determination of aluminum with a carbon atomization system, *Anal. Chim. Acta*, 81, 313, 1976.

80. **Thompson, K. C. and Godden, R. G.**, The application of a wide slot nitrous oxide-nitrogen-acetylene burner for the atomic absorption spectrophotometric determination of aluminum, arsenic and tin in steels by the single-pulse nebulization technique, *Analyst (London)*, 101, 96, 1976.

81. **Brashnarov, A. and Popova, S.**, Characteristics of atomic absorption determination of aluminum, *Dokl. Bolg. Akad. Nauk*, 29, 995, 1976.

82. **McDermott, J. R. and Whitehill, I.**, Determination of aluminum in biological tissue by flameless atomic absorption spectrometry, *Anal. Chim. Acta*, 85, 195, 1976.

83. **Pszonicki, L. and Krupinski, M.**, Application of aluminum to the elimination of interferences in vanadium determination by atomic absorption, *Chem. Anal.*, 21, 743, 1976.

84. **Kaehny, W. D., Hegg, A. P., and Alfrey, A. C.**, Gastrointestinal absorption of aluminum from aluminum-containing antacids, *N. Engl. J. Med.*, 296, 1389, 1977.

85. **Persson, J., Frech, W., and Cedergren, A.,** Determination of aluminum in low alloy and stainless steels by flameless atomic absorption spectrometry, *Anal. Chim. Acta*, 89, 119, 1977.

86. **Tekula-Buxbaum, P.,** Determination of small quantities of aluminum and silicon in tungsten metal and tungsten oxides by atomic absorption spectrophotometry, *Mikrochim. Acta*, 1, 145, 1977.

87. **Wise, W. M. and Solksy, S. D.,** Determination of aluminum in glass by atomic absorption spectroscopy, *Anal. Lett.*, 10, 273, 1977.

88. **Langmyhr, F. J. and Tsalev, D. L.,** Atomic absorption spectrometric determination of aluminum in whole blood, *Anal. Chim. Acta*, 92, 79, 1977.

89. **Persson, J. A., Frech, W., and Cedargren, A.,** Investigations of reactions involved in flameless atomic absorption procedures. IV. A theoretical study of factors influencing the determination of aluminum, *Anal. Chim. Acta*, 92, 85, 1977.

90. **Celis, J. P., Helsen, J. A., Hermons, P., and Roos, J. H.,** The determination of alumina in a copper matrix by atomic absorption spectrometry, *Anal. Chim. Acta*, 92, 413, 1977.

91. **Pilate, A., Geladi, P., and Adams, F.,** Determination of aluminum in aerosols by flameless atomic absorption spectrometry, *Talanta*, 24, 512, 1977.

92. **Germestani, K., Blotcky, A. J., and Rack, E. P.,** Comparison between neutron activation analysis and graphite furnace atomic absorption spectrometry for trace aluminum determination in biological materials, *Anal. Chem.*, 50, 144, 1978.

93. **Kono, T.,** Determination of aluminum by atomic absorption spectrophotometry using oxygen sandwiched air-acetylene flame, *Bunseki Kagaku*, 27, 591, 1978.

94. **Liu, J. H. and Huber, C. O.,** Aluminum determination by atomic emission spectrometry with calcium atomization inhibition titration, *Anal. Chem.*, 50, 1253, 1978.

95. **Moriyama, K., Goda, A., and Harimaya, S.,** Determination of trace amounts of aluminum in iron and steel by flameless atomic absorption spectrometry, *Tetsu To Hagane*, 64, 1424, 1978.

96. **Pegon, Y.,** Determination of aluminum in biological liquids by flameless atomic absorption, *Anal. Chim. Acta*, 101, 385, 1978 (in French).

97. **LaBrecque, J. J., Mendelovici, E., Villalba, R. E., and Bellorin, C. C.,** The determination of total iron in Venezualan laterites: the investigation of interferences of aluminum and silicon on the determination of iron in the fluoboric-boric acid matrix by atomic absorption, *Appl. Spectrosc.*, 32, 57, 1978.

98. **Purushottam, A. and Naidu, P. P.,** Rapid atomic absorption spectrophotometric method for estimation of reactive and total silica, iron, aluminum and titanium in bauxite, *Curr. Sci.*, 47, 88, 1978.

99. **Gorsky, J. E. and Dietz, A. A.,** Determination of aluminum in biological samples by atomic absorption spectrophotometry with a graphite furnace, *Clin. Chem. (Winston-Salem, N.C.)*, 24, 1485, 1978.

100. **Julshamn, K., Anderson, K. J., Willassen, Y., and Braekkan, O. R.,** A routine method for the determination of aluminum in human tissue samples using standard addition and graphite furnace atomic absorption spectrophotometry, *Anal. Biochem.*, 88, 552, 1978.

101. **Terashima, S.,** Atomic absorption determination of manganese, iron, copper, nickel, cobalt, lead, zinc, silicon, aluminum, calcium, magnesium, sodium, potassium, titanium and strontium in manganese nodules, *Chishitsu Chosasho Geppo*, 29, 401, 1978 (in Japanese).

102. **McKown, M. M., Tschrin, C. R., and Lee, P. P. F.,** Investigation of Matrix Interferences for Atomic Absorption Spectroscopy of Sediments, Rep. No. EPA-600/7-78/085, U.S. Environmental Protection Agency, Washington, D.C., 1978.

103. **Marinescu, I. and Tamas, M.,** Phytochemical study on poplar buds as possible propolis source. Some aspects of trace elements contained by propolis and poplar buds, *Apic. Rom.*, 54, 14, 1979.

104. **Carrondo, M. J., Lester, J. N., and Perry, R.,** Determination of aluminum, calcium, iron and magnesium in sewages and sewage effluent by a rapid electrothermal atomic absorption spectroscopic method, *Analyst (London)*, 104, 831, 1979.

105. **Carrondo, M. J., Lester, J. N., and Perry, R.,** An investigation of a flameless atomic absorption method for determination of aluminum, calcium, iron and magnesium in sewage sludge, *Talanta*, 26, 929, 1979.

106. **La Brecque, J. J.,** Decomposition and determination of aluminum and silicon in Venezuelan laterites by atomic absorption spectroscopy, *Chem. Geol.*, 26, 321, 1979.

107. **Matsusaki, K., Yoshino, T., and Yamamoto, Y.,** A method for the removal of chloride interference in determination of aluminum by atomic absorption spectrometry with a graphite furnace, *Talanta*, 26, 377, 1979.

108. **Bower, N. W. and Ingle, J. D.,** Precision of flame atomic absorption spectrometric measurements of aluminum, chromium, cobalt, europium, lead, manganese, nickel, potassium, selenium, silicon, titanium and vanadium, *Anal. Chem.*, 51, 72, 1979.

109. **Malhotra, P. D., Naidu, P. P., and Rao, T. J.,** Method for decomposition of chromite for analysis by atomic absorption spectrophotometry, *Curr. Sci.*, 48, 584, 1979.

110. **Boyer, K. W., Capar, S. G., Jones, J. W., Suddendorf, R. F., and Forwalter, J.,** Multielement/trace analysis identifies 48 elements in foods. Simultaneous element analysis and data reduction reaches parts per billion, *Food Process.*, 40, 72, 1979.

111. **Capar, S. G. and Gould, J. H.,** Lead, fluoride and other elements in bonemeal supplements, *J. Assoc. Off. Anal. Chem.*, 62, 1054, 1979.

112. **Carrondo, M. J., Lester, J. N., and Perry, R.,** Electrothermal atomic absorption determination of total aluminum (including zeolite type A) in waters and waste waters, *Anal. Chim. Acta,* 111, 291, 1979.

113. **Dittrich, K.,** Molecular absorption spectrometry with electrothermal volatilization in a graphite tube. III. A study of the determination of fluoride traces by aluminum fluoride, gallium fluoride, indium fluoride and thallium fluoride molecular absorption, *Anal. Chim. Acta,* 111, 123, 1979.

114. **King, S. W., Wills, M. R., and Savory, J.,** Serum binding of aluminum, *Res. Commun. Chem. Pathol. Pharmacol.*, 26, 161, 1979.

115. **Lambert, J. B., Szpunar, C. B., and Buikstra, J. E.,** Chemical analysis of excavated human bone from middle and late woodland sites, *Archaeometry,* 21, 115, 1979.

116. **Shevchuk, I. A. and Alemasova, A. S.,** Atomic absorption determination of aluminum in bronzes, *Zavod. Lab.*, 45, 1101, 1979.

117. **Smrhova, A., Janacek, J., Fialova, E., Smrha, L., and Hruby, Z.,** Automatic evaluation of data obtained by atomic absorption spectrophotometric analysis of oxide inclusions in steel, *Hutn. Listy,* 34, 728, 1979 (in Czech).

118. **Ibarra, J. V., Romero, C., and Gavilan, J. M.,** Analysis of coal ash by atomic absorption spectroscopy, *Afinidad,* 36, 117, 1979 (in Spanish).

119. **Baucells, M., Lacort, G., and Roura, M.,** Determination of major and trace elements in marine sediments of Costa Brava by emission and atomic absorption spectroscopy, *Invest. Pesq.,* 43, 551, 1979.

120. **Ballantyne, A. K., Anderson, D. W., and Stonehouse, H. B.,** Problems associated with extracting Fe and Al from Saskatchewan soils by pyrophosphate and low speed centrifugation, *Can. J. Soil. Sci.,* 60, 141, 1980.

121. **Brenner, I. B., Watson, A. E., Russell, G. M., and Goncalves, M.,** New approach to determination of major and minor constituents in silicate and phosophate rocks, *Chem. Geol.,* 28, 321, 1980.

122. **Brown, J. R., Saba, C. S., Rhine, W. E., and Eisentraut, K. J.,** Particle size independent spectrometric determination of wear metals in aircraft lubricating oils, *Anal. Chem.,* 52, 2365, 1980.

123. **David, D. J.,** Determination of aluminum in plant materials by atomic absorption spectrophotometry, *Commun. Soil Sci. Plant Anal.,* 11, 189, 1980.

124. **De la Guardia-Cirugeda, M., Berenguyer-Navarro, V., and Guinon-Segura, J. L.,** Use of emulsions in flame emission spectrometry. I. Aluminum determination after extraction as oxinate in methyl isobutyl ketone, *Analysis,* 8, 166, 1980.

125. **Horler, D. N. H., Barber, J., and Barringer, A. R.,** Multielemental study of plant surface particles in relation to Geochemistry and biogeochemistry, *J. Geochem. Explor.,* 13, 41, 1980.

126. **Nadkarni, R. A.,** Multitechnique multielemental analysis of coal and fly ash, *Anal. Chem.,* 52, 929, 1980.

127. **Kozusnikova, J.,** Determination of low aluminum contents in silicon steels, *Hutn. Listy,* 35, 440, 1980.

128. **Kupchella, L. and Syty, A.,** Determination of nickel, manganese, copper and aluminum in chewing gum by nonflame atomic absorption spectrometry, *J. Agric. Food Chem.,* 28, 1035, 1980.

129. **Morse, R. A., and Lisk, D. J.,** Elemental analysis of honeys from several nations, *Am. Bee J.,* 120, 592, 1980.

130. **Stolzenburg, T. R. and Andren, A. W.,** Simple acid digestion method for the determination of ten elements in ambient aerosols by flame atomic absorption spectrometry, *Anal. Chim. Acta,* 118, 377, 1980.

131. **Tornheim, K., Gilbert, T. R., and Lowenstein, J. M.,** Metal contaminants in commercial preparation of nucleotides, *Anal. Biochem.,* 103, 87, 1980.

132. **Win, H. and Morin, T. J.,** Characterization of a mixed acid aluminum etchant, *Plat. Surf. Finish.,* 67, 54, 1980.

133. **Yoshimura, C. and Hata, Y.,** Effect of nitrogen compounds on atomic absorption spectrophotometry of aluminum using air acetylene flame, *Bunseki Kagaku,* 29, 119, 1980.

134. **Alderman, F. R. and Gitelman, H. J.,** Improved electrothermal determination of aluminum in serum by atomic absorption spectroscopy, *Clin. Chem. (Winston-Salem, N.C.),* 26, 258, 1980.

135. **Lieu, V. T. and Woo, D. H.,** Application of flame and graphite furnace atomic absorption spectroscopy to the determination of metals in oil-field injection water and its suspended solids, *At. Spectrosc.,* 1, 149, 1980.

136. **Caroli, S. and Senofonte, O.,** Comparative studies of the hollow cathode and glow discharge radiation sources for aluminum and graphite, *Can. J. Spectrosc.,* 25, 73, 1980.

137. **Couri, D., Liss, L., and Ebner, K.,** Determination of aluminum in biological samples, *Neurotoxicology,* 1, 17, 1980.

138. **Katskov, D. A. and Grinshtein, L. L.,** Study of the evaporation of beryllium, magnesium, calcium, strontium, barium and aluminum from a graphite surface by an atomic absorption method, *Zh. Prikl. Spektrosk.,* 33, 1004, 1980.

139. **L'vov, B. V. and Ryabchuk, G. N.,** Determination of aluminum in a graphite furnace, *Zh. Prikl. Spektrosk.,* 33, 1013, 1980.

140. **Maeda, T., Nakagawa, M., Kawakatsu, M., and Tanimoto, Y.,** Determination of metallic elements in serum and urine by flame and graphite furnace atomic absorption spectrophotometry, *Shimadzu Hyoron,* 37, 1980 (in Japanese).

141. **Matsubara, K., Ota, N., Taniguchi, M., and Narita, K.,** Atomic absorption spectrophotometric determination of small amounts of arsenic, tin and aluminum in steel by the hydride generation method and the injection method, *Trans. Iron Steel Inst. Jpn.,* 20, B406, 1980.

142. **Smeyers-Verbeke, J., Verbeden, D., and Massart, D. L.,** Determination of aluminum in biological fluids by means of graphite furnace atomic absorption spectrometry, *Clin. Chim. Acta,* 108, 67, 1980.

143. **Tardon, S.,** Rapid determination of small quantities of inorganic constituents of suspended dust and dust fallout, *Chem. Prum.,* 30, 195, 1980 (in Czech).

144. **Toda, W., Lux, J., and Van Loon, J. C.,** Determination of aluminum in solutions from gel filtration chromatography of human serum by electrothermal atomic absorption spectrometry, *Anal. Lett.,* 13, 1105, 1980.

145. **Tsunoda, K., Haraguchi, H., and Fuwa, K.,** Studies on the occurrence of atoms and molecules of aluminum, gallium, indium and their monohalides in an electrothermal carbon furnace, *Spectrochim. Acta,* 35b, 715, 1980.

146. **Pletneva, T. I.,** Atomic absorption determination of trace contaminants in potassium and sodium nitrates, *Khim. Promst. Ser. Reakt. Osobo Chist. Veshchestva,* 1, 36, 1981; CA 95, 72644X, 1981.

147. **Cowgill, U. M.,** The chemical composition of bananas. Market basket values, *Biol. Trace Elem. Res.,* 3(1), 33, 1981.

148. **Aksyk, A. F., Merzlyakova, N. M., and Tarkhova, L. P.,** Determination of trace elements in desalinated seawater by an atomic absorption method, *Gig. Sanit.,* 5, 72, 1981 (in Russian), CA 95, 8593US, 1981.

149. **Capel, I. D., Pinnock, M. H., Dorrell, H. M., Williams, D. C., and Grant, E. C. G.,** Comparison of concentrations of some trace, bulk and toxic metals in the hair of normal and dyslexic children, *Clin. Chem. (Winston-Salem, N.C.),* 27, 879, 1981.

150. **Bertram, H. P.,** Aluminum determination in body fluids, *Nieren-Hochdruckkr,* 10, 188, 1981.

151. Atomic absorption analysis of silicon nitride, Tokyo Shibura Co. Ltd. Jpn. Kokai Tokkyo Koho 81, 57, 936, (Cl. GOIN 21/31); May 20, 1981.

152. **Yudelevich, I. G., Buyanova, L. M., Beisel, N. F., and Kozhanova, L. A.,** Atomic absorption determination of concentration profiles of major components of $A_{111}B_V$ type semiconductor films, *At. Spectrosc.,* 2, 165, 1981.

153. **Chang, J. G. and Graff, R. L.,** Spectrochemical determination of impurities in uranium following solvent extraction using tributyl phosphate in methyl isobutyl ketone, and metal-bearing ores, *ASTM STP,* 747, 106, 1981.

154. **Klein, A. A.,** Analysis of low-alloy steel using a sequential atomic absorption spectrophotometer equipped with a autosampler, *ASTM STP,* 747, 29, 1981.

155. **Rains, T. C.,** Determination of Al, Ba, Ca, Pb, Mg and Ag in ferrous alloys by atomic emission and atomic absorption spectrometry, *ASTM STP,* 747, 93, 1981.

156. **Adelman, H., Jenniss, S. W., and Katz, S. A.,** Interlaboratory analysis of sewage sludge, *Am. Lab.,* 31, 1981.

157. **Coutinho, C. A., Azevedo, J. C., and Arruda, E. C.,** Rapid determination of soluble aluminum for process control in the fabrication of killed steels, *Congr. Anu. ABM,* 36th, Congress of the Brazilian Metals Association, Sao Paulo, Brazil, 1981, 319 (in Portuguese).

158. **Halls, D. J.,** Applications of graphite furnace atomic absorption spectrometry in clinical analysis, *Anal. Proc. (London),* 18, 344, 1981.

159. **Wu, T. and Liu, J.,** Rapid direct determination of aluminum in steel and iron by atomic absorption spectrometry using a nitrous oxide-acetylene flame, *Fen Hsi Hua Hsueh,* 9, 282, 1981 (in Chinese).

160. **Jones, A. and Dixon, K.,** The separation of trace elements in manganese dioxide, *Natl. Inst. Metall. Repub. S. Afr. Rep.,* No. 2131, 1981.

161. **Kaiser, M. L., Koirtyohann, S. R., Hinderberger, E. J., and Taylor, H. E.,** Reduction of matrix interferences in furnace atomic absorption with the L'vov platform, *Spectrochim. Acta,* 36b, 773, 1981.

162. **Donaldson, E. M.,** Determination of aluminum in iron, steel and ferrous and nonferrous alloys by atomic absorption spectrophotometry after a mercury-cathode separation and extraction of the aluminum-acetyl acetone complex, *Talanta,* 28, 461, 1981.

163. **Gardiner, P. E., Ottaway, J. M., Fell, G. S., and Halls, D. J.,** Determination of aluminum in blood plasma or serum by electrothermal atomic absorption spectrometry, *Anal. Chim. Acta,* 128, 57, 1981.

164. **Gorsky, J. E. and Dietz, A. A.,** Aluminum concentrations in serum of hemodialysis patients, *Clin. Chem. (Winston-Salem, N.C.),* 27, 932, 1981.

165. **Jastrow, J. D., Zimmerman, C. A., Dvorak, A. J., and Hinchman, R. R.,** Plant growth and trace element uptake on acidic coal refuse amended with lime or fly ash, *J. Environ. Qual.,* 10, 154, 1981.
166. **King, S. W., Wills, M. R., and Savory, J.,** Electrothermal atomic absorption spectrometric determination of aluminum in blood serum, *Anal. Chim. Acta,* 128, 221, 1981.
167. **Oster, O.,** Aluminum content of human serum determined by atomic absorption spectroscopy with a graphite furnace, *Clin. Chim. Acta,* 114, 53, 1981.
168. **Young, R. S.,** Analysis of nickel refinery slimes and residues, *Talanta,* 28, 25, 1981.
169. **Arafat, N. M. and Glooschenko, W. A.,** Method for simultaneous determination of arsenic, aluminum, iron, zinc, chromium and copper in plant tissue without use of perchloric acid, *Analyst (London),* 106, 1174, 1981.
170. **Balaes, G. E. E. and Robert, R. V. D.,** Determination by atomic absorption spectrophotometry of impurities in manganese dioxide, *Natl. Inst. Metall. Repub. S. Afr. Rep.,* No. 2094, 1981.
171. **Domingues, A.,** Technology for pressure decomposition in silicate analysis by atomic absorption, *Ceramica,* 27, 1, 1981 (in Portuguese).
172. **Kauffman, R. E., Saba, C. S., Rhine, W. E., and Eisentraut, K. J.,** Quantitative multielement determination of metallic wear species in lubricating oils and hydraulic fluids, *Anal. Chem.,* 54, 975, 1982.
173. **Playle, R., Gleed, J., Jonassen, R., and Kramer, J. R.,** Comparison of atomic absorption spectrometric, spectrophotometric and fluorimetric methods for determination of aluminum in water, *Anal. Chim. Acta,* 134, 369, 1982.
174. **Slavin, W., Carnrick, G. R., and Manning, D. C.,** Magnesium nitrate as matrix modifier in the stabilized temperature platform furnace, *Anal. Chem.,* 54, 621, 1982.
175. **Verbeek, A. A., Mitchell, M. C., and Ure, A. M.,** Analysis of small samples of rock and soil by atomic absorption and emission spectrometry after lithium metabotate fusion/nitric acid dissolution procedure, *Anal. Chim. Acta,* 135, 215, 1982.
176. **Wawschinek, O., Petek, W., Lang, J., and Holzer, H.,** Determination of aluminum in human plasma, *Mikrochim. Acta,* 1, 335, 1982 (in English).
177. **Yokel, R. A.,** Hair as an indicator of excessive aluminum exposure, *Clin. Chem. (Winston-Salem, N.C.),* 28, 662, 1982.
178. **Slavin, W., Carnrick, G. R., and Manning, D. C.,** Graphite-tube effects on perchloric acid interferences on aluminum and thallium in the stabilized-temperature platform furnace, *Anal. Chim. Acta,* 138, 103, 1982.
179. **Guillard, O., Piriou, A., Mura, P., and Reiss, D.,** Precautions necessary when assaying aluminum of chronic hemodialyzed patients, *Clin. Chem. (Winston-Salem, N.C.),* 28, 1714, 1982.
180. **Tarui, T. and Tsuchida, Y.,** Determination of silica and alumina in fuel oils by atomic absorption spectrometry, *Sekiyu Gakkaishi,* 25, 158, 1982 (in Japanese).
181. **Jones, E. A.,** The separation and determination of trace elements in chromic oxide, *Rep. MINTEK,* MI5, Council for Mineral Technology, Randburg, South Africa, 1982.
182. **Jenke, D. R. and Diebold, F. E.,** Characterization of phosphite ores, *Anal. Chem.,* 54, 1008, 1982.
183. **Shen, D. H. and Bennion, D. N.,** Lithium-Aluminum Alloy Electrodes in Lithium Chlorate, Rept. TR-2, Order No. AD-AIO5902, 1981; available from the NTIS, Springfield, Va.; *Gov. Rep. Announce. Index (U.S.),* 82, 713, 1982.
184. **Adhemar, J. P., Laederich, J., Jaudon, M. C., Masselot, J. P., Buisson, C., Galli, A., and Kleinknecht, D.,** Dialysis encephalopathy. Diagnostic and prognostic value of clinical and EEG signs and aluminum levels in serum and cerebrospinal fluid, in *Proc. 17th Eur. Dial. Transplant Assoc. 1980,* 1980, 234; 1980, CA 96, 82213r, 1982.
185. **Landberger, S., Jervis, R. E., Aufreiter, S., and Van Loon, J. C.,** The determination of heavy metals (aluminum, manganese, iron, nickel, copper, zinc, cadmium and lead) in urban snow using an atomic absorption graphite furnace, *Chemosphere,* 11, 237, 1982.
186. **Kikkawa, S. and Koizumi, M.,** Ferrocyanide anion bearing magnesium aluminum hydroxide, *Mater. Res. Bull.,* 17, 191, 1982.
187. **Clavel, J. P., Jaudon, M. C., and Galli, A.,** Determination of serum aluminum: new estimation of normal values, *Ann. Biol. Clin.,* 40, 51, 1982 (in French).
188. **Karpukhin, A. I., Platonov, I. G., and Shestakov, E. I.,** Organomineral compounds in podzolic soils on calcareous sandy loams, *Pochvovedenie,* 3, 37, 1982 (in Russian).
189. **Bragin, G. Y. and Sadagov, Y. M.,** Spatial distribution of temperature in graphite tubular furnaces of electrothermal atomizers, *Zh. Prikl. Spektrosk.,* 36, 185, 1982.
190. **Brodie, K. and Routh, R. W.,** Trace metal analysis of biological samples with the GTA-95 graphite tube atomizer, *Varian Instrum. Appl.,* 16, 18, 1982.
191. **Li, C. and Yan, W.,** Graphite furnace atomic absorption determination of aluminum in serum and urine samples, *Beijing Yixueyuan Xuebo,* 14, 38, 1982 (in Chinese).
192. **Toshiba Corp. Jpn.,** Kokai Tokkyo Koho JP, Determination of Aluminum in Gallium Arsenide, 82 22, 539 (Cl.GOIN 21/31) February 1982; appl. 80/96, 232, July 16, 1980.

AUTHOR INDEX

SUBJECT INDEX (ALUMINUM)

GALLIUM, Ga (ATOMIC WEIGHT 69.72)

Gallium is a very widely distributed element, but occurs in trace amounts in aluminum minerals, iron and zinc ores, and in coal ashes. It is most often used in doping semiconductors to produce devices for the solid state industry, in high-temperature thermometers, in making low-melting alloys, and in forming mirrors on glass and porcelain. Gallium arsenide is capable of converting electricity directly into coherent light.

Gallium samples can be easily dissolved in hydrochloric, nitric, aqua regia, or sulfuric acid and the residue is fused with pyrosulfate or alkali carbonate for dissolution. Gallium and its compounds are slightly toxic, precautions should be taken when handling.

In air-acetylene flame, no interferences have been reported. Partial ionization of gallium in nitrous oxide-acetylene flame can be suppressed by the addition of potassium nitrate or potassium chloride (200 mg/ℓ final potassium concentration) to all the solutions, including blanks.

Standard Solution

To prepare 1000 mg/ℓ of solution dissolve 1.000 g of metallic gallium in minimum amount of aqua regia with gentle heating or in 20 mℓ of 1:1 nitric acid. Dilute to 1 ℓ with 1% (v/v) hydrochloric acid.

Instrumental Parameters

Wavelength	287.4 nm
Slit width	0.7 nm
Light source	Hollow cathode lamp
Flame	Nitrous oxide-acetylene, reducing (rich, red) or air-acetylene
Sensitivity	0.4—1.3 mg/ℓ
Detection limit	0.1 mg/ℓ
Optimum range	60 mg/ℓ

Secondary wavelengths and their sensitivities

294.4 nm	1.1—1.3 mg/ℓ
417.2 nm	1.5—1.8 mg/ℓ
403.3 nm	2.8 mg/ℓ
250.0 nm	9.7—12.0 mg/ℓ
245.0 nm	12.0—13.0 mg/ℓ
272.0 nm	23—25.0 mg/ℓ

REFERENCES

1. **Gatehouse, B. M. and Willis, J. B.**, Performance of a simple atomic absorption spectrophotometry, *Spectrochim. Acta,* 17, 710, 1961.
2. **Slavin, W., Sprague, S., and Manning, D. C.**, Detection limit in analytical atomic absorption spectrophotometry, *At. Absorpt. Newsl.,* 18, 1964.
3. **Gupta, H. K. L., Amore, F. J., and Beltz, D. F.**, The determination of gallium by atomic absorption spectrometry, *At. Absorpt. Newsl.,* 7, 107, 1968.
4. **Popham, R. E. and Schrenk, W. G.**, Atomic absorption characteristics of gallium and indium, *Spectrochim. Acta,* 24b, 223, 1969.
5. **Pollock, E. N.**, Gallium and germanium in limonite by atomic absorption, *At. Absorpt. Newsl.,* 10, 77, 1971.
6. **Lypka, G. N. and Chow, A.**, A comparison of methods for the determination of gallium in ores, *Anal. Chim. Acta,* 60, 65, 1972.
7. **Langmyhr, F. G. and Rasmussen, S.**, Atomic absorption spectrometric determination of gallium and indium in inorganic materials by direct atomization from the solid state in a graphite furnace, *Anal. Chim. Acta,* 72, 79, 1974.
8. **Popova, S. A., Bezur, L., Polos, L., and Pungor, E.**, Determination of gallium by extraction and atomic absorption measurement, *Z. Anal. Chem.,* 270, 180, 1974.
9. **Chow, A. and Lipinsky, W.**, Determination of gallium by atomic absorption spectrometry, *Anal. Chim. Acta,* 76, 87, 1975.
10. **Nakahara, T. and Musha, S.**, The determination of gallium by atomic absorption spectrometry in premixed inert gas (entrained air-hydrogen) flames, *Anal. Chim. Acta,* 75, 305, 1975.
11. **Kometani, T. Y. and Wiegmann, W.**, Measurement of gallium and aluminum in a molecular beam epitaxy chamber by atomic absorption spectrometry, *J. Vac. Sci. Technol.,* 12, 933, 1975.
12. **Sauer, K. H. and Nitsche, M.**, Photometric and atomic absorption spectrometric determination of low gallium levels in steels, *Arch. Eisenhuettenwes.,* 47, 153, 1976.
13. **Ohta, K. and Suzuki, M.**, Flameless atomic absorption spectrometry of gallium with a metal atomizer, *Anal. Chim. Acta,* 85, 83, 1976.
14. **Pelosi, C. and Attolini, G.**, Determination of gallium by atomic absorption spectrometry with a graphite furnace atomizer, *Anal. Chim. Acta,* 84, 179, 1976.
15. **De, D. K., Das, A. K., and Banerjee, S.**, Determination of gallium in bauxite and silicate rock samples by solvent extraction. Atomic absorption spectrophotometry, *Indian J. Chem.,* 15a, 666, 1977.
16. **Dittrich, K.**, Atomic spectroscopic trace analysis in $A^{III}B^V$ semiconductor microsamples. II. Determination of trace gallium in indium and indium arsenide and phosphide by atomic absorption spectrometry with electrothermal atomization, *Talanta,* 24, 735, 1977.
17. **Dittrich, K.**, Molecular absorption spectrometer by electrothermal volatilization in a graphic furnace. I. Principles of the method and studies of the molecular absorption of gallium and indium halides, *Anal. Chim. Acta,* 97, 59, 1978.
18. **Newman, R. A.**, Flameless atomic absorption spectrometry determination of gallium in biological materials, *Clin. Chem. Acta,* 86, 195, 1978.
19. **Sukhoveeva, L. N., Spivakov, B. Y., Karyakin, A. V., and Zolotov, Y. A.**, Atomic absorption determination of gallium in flame and graphite furnaces, *Zh. Anal. Khim.,* 34, 693, 1979.
20. **Korkisch, J., Steffan, I., Nonaka, J., and Arrhenius, G.**, Chemical analysis of manganese nodules. V. Determination of gallium after anion-exchange separation, *Anal. Chim. Acta,* 109, 181, 1979.
21. **Popova, S., Minchev, L., and Gagov, V.**, Atomic absorption determination of gallium in mineral raw materials and industrial wastes. I, *Rudodobiv,* 34, 41, 1979.
22. **Dittrich, K., Schneider, S., Spivakov, B. Y., Suchowejewa, L. N., and Zolotov, Y. A.**, Evaporation and plasma processes in graphite cuvettes for atomic absorption spectrometry. II. Determination of gallium and indium inaqueous solutions and organic extracts by molecular and atomic absorption spectrometry, *Spectrochim. Acta,* 34b, 257, 1979.
23. **Hadeishi, T. and Kimura, H.**, Direct measurements of concentration of trace elements in gallium arsenide crystals by Zeeman atomic absorption spectrometry, *J. Electrochem. Soc.,* 126, 1988, 1979.
24. **Dittrich, K.**, Molecular absorption spectrometry with electrothermal volatilization in a graphite tube. III. A study of the determination of fluoride traces by aluminum fluoride, gallium fluoride, indium fluoride and thallium fluoride molecular absorption, *Anal. Chim. Acta,* 111, 123, 1979.
25. **Montaser, A. and Mehrabzadeh, A. A.**, Determination of calcium, erbium, europium, gallium, indium, potassium, sodium, molybdenum and tungsten by atomic absorption spectrometry with an electrothermal graphite braid atomizer, *Anal. Chim. Acta,* 111, 297, 1979.

26. **Shushunova, A., Demarin, V. T., Makin, G. I., Sklemina, L. V., Rudnevskii, N. K., and Aleksandrov, Y. A.,** Combination of gas chromatography with atomic absorption spectrophotometry for analysis of gallium and indium organometallic compounds, *Zh. Anal. Khim.,* 35, 349, 1980.

27. **Sukhoveeva, L. N., Butrimenko, G. G., and Spivakov, B. Y.,** Atomic absorption determination of gallium and indium in a graphite furnace by evaporation from a substrate, *Zh. Anal. Khim.,* 35, 649, 1980.

28. **Zelentsova, L. V., Yudelevich, I. G., and Chanysheva, I. A.,** Application of atomic absorption and emission spectrometry for analyzing high purity substances. Analysis of high-purity gallium, *Zh. Anal. Khim.,* 35, 515, 1980.

29. **Caroli, S., Alimonti, A., and Violante, N.,** Determination of gallium in biological samples by means of the hollow cathode discharge, *Spectrosc. Lett.,* 13, 313, 1980.

30. **Daidoji, H.,** Molecular absorption spectra of gallium salts in flame, *Bunseki Kagaku,* 29, 389, 1980.

31. **Tsunoda, K., Haraguchi, H., and Fuwa, K.,** Studies on the occurrence of atoms and molecules of aluminum, gallium, indium and their monohalides in an electrothermal carbon furnace, *Spectrochim. Acta,* 35b, 715, 1980.

32. **Katskov, D. A., Grinshtein, I. L., and Kruglikov, L. P.,** Study of the evaporation of the metals indium, gallium, thallium, germanium, tin, lead, antimony, bismuth, selenium and tellurium from a graphite surface by the atomic absorption method, *Zh. Prikl. Spektrosk.,* 33, 804, 1980.

33. **Roza, M.,** The nature of the magma type of the Chamoli metavolcanics using minor and trace elements, *Indian J. Earth Sci.,* 7, 119, 1980.

34. **Kuga, K.,** Determination of gallium by graphite furnace atomic absorption spectrometry using a zirconium inpregnated graphite tube, *Bunseki Kagaku,* 30, 529, 1980.

35. **Clark, J. R. and Viets, J. G.,** Multielement extraction system for the determination of 18 trace elements in geochemical samples, *Anal. Chem.,* 53, 61, 1981.

36. **Clark, J. R. and Viets, J. G.,** Back-extraction of trace elements from organometallic halide extracts for determination by flameless atomic absorption spectrometry, *Anal. Chem.,* 53, 65, 1981.

37. **Yudelevich, I. G., Buyanova, L. M., Beisel, N. F., and Kozhanova, L. A.,** Atomic absorption determination of concentration profiles of major components of $As_{III}B_V$ type semiconductor films, *At. Spectrosc.,* 2, 165, 1981.

38. **Costantini, S., Macri, A., and Vernille, I.,** Atomic absorption spectrophotometric analysis of mineral elements in milk-base feeds for zootechnical use, *Riv. Soc. Ital. Sci. Aliment.,* 10, 231, 1981.

39. **Yudelevich, I. G., Gilbert, E. N., and Shelpakova, I. R.,** Comparison of multielement methods of analysis of highly pure gallium, *Zh. Anal. Khim.,* 36, 2393, 1981.

40. **Kogan, A., Talalaev, B. M., Karovaev, M. M., and Nazarova, T. I.,** Atomic absorption determination of gallium in nonplatinum catalysts of ammonia oxidation, *Pr-vo Azot. Udobrenii,* M.85, 1981 (in Russian), from *Zh. Khim.,* Abstr. No. 13G211, 1982.

41. **Nakamura, K., Fujimori, M., Tsuchiya, H., and Orri, H.,** Determination of gallium in biological materials by electrothermal atomic absorption spectrometry, *Anal. Chim. Acta,* 138, 129, 1982.

42. **Salmon, S. G. and Holcombe, J. A.,** Alteration of metal release mechanisms in graphite furnace atomizers by chemisorbed oxygen, *Anal. Chem.,* 54, 630, 1982.

43. **Zelentsova, L. V., Yudelevich, I. G., and Chanysheva, T. A.,** Use of atomic absorption spectrometry for the analysis of highly pure substances. Analysis of highly pure indium and gallium, *Izb. Sib. Otd. Akad. Nauk SSSR Ser. Khim. Nauk,* 1, 132, 1982.

44. **Ondov, J. M. and Biermann, A. H.,** Effects of Particle-Control Devices on Atmospheric Emissions of Minor and Trace Elements from Coal Combustion, EPA-600/9-80-039d, Symp. Transfer Util. Part. Control. Technol. 2nd., PB 81-122228, U.S. Environmental Protection Agency, Office of Research and Development, Corvalis, Ore., 1980 454-85, CA 96, 90897g, 1982.

45. **Caroli, S., Alimonti, A., Femmine, P. D., and Shukla, S. K.,** Determination of gallium in tumor-affected tissues by means of spectroscopic techniques. A comparative study, *Anal. Chim. Acta,* 136, 225, 1982.

AUTHOR INDEX

SUBJECT INDEX (GALLIUM)

INDIUM, In (ATOMIC WEIGHT 114.82)

Indium is a rare element and occurs in igneous rocks, minerals, sulfides and sulfo salts, and iron, lead and copper ores. It is most frequently associated with zinc materials. Indium is used in making low-melting alloys, bearing alloys, germanium transistors, rectifiers, thermistors and photoconductors, etc. It can be plated on metal and evaporated onto a glass to form a mirror which is resistant to atmospheric corrosion.

Determination of indium is required in ores, alloys, rocks, and minerals, etc. Indium is reported as toxic to humans and proper hygienic precautions should be taken to handle it. Sample solution can be prepared by the dissolution in hydrochloric acid; if necessary, a small amount of nitric acid can be added.

Aluminum, iron, silicon, magnesium, copper, zinc, tin, and phosphate in the 100-fold excess concentration than that of indium produce interferences in the atomic absorption method. Partial ionization of indium can be suppressed with the addition of 0.1% or more potassium nitrate or potassium chloride solution to all the solutions, including blanks.

Standard Solution

To prepare 1000 mg/ℓ solution, dissolve 1.000 g of indium metal in a minimum volume of 1:1 hydrochloric acid. Add few drops of nitric acid and heat gently. Cool and dilute to 1 ℓ with deionized water.

Instrumental Parameters

Wavelength	303.9 nm
Slit width	0.7 nm
Light source	Hollow cathode lamp
Flame	Air-acetylene, oxidizing (lean, blue)
Sensitivity	0.18—0.76 mg/ℓ
Detection limit	0.5 mg/ℓ
Optimum range	15 mg/ℓ

Secondary wavelengths and their sensitivities

325.6 nm	0.2—0.8 mg/ℓ
410.5 nm	0.5—2.5 mg/ℓ
451.1 nm	0.6—2.6 mg/ℓ
256.0 nm	2.0—9.1 mg/ℓ
271.0 nm	4.0—12.0 mg/ℓ
275.4 nm	5.0—21.0 mg/ℓ

REFERENCES

1. **Slavin, W., Sprague, S., and Manning, D. C.,** Detection limits in analytical atomic absorption spectrophotometry, *At. Absorpt., Newsl.,* 18, 1964.
2. **Mulford, C. E.,** Solvent extraction technique for atomic absorption spectroscopy, *At. Absorpt. Newsl.,* 5, 88, 1966.
3. **Sattur, T. W.,** Routine atomic absorption analysis on non-ferrous alloys and plant intermediates, *At. Absorpt. Newsl.,* 5, 37, 1966.
4. **Ivanov, N. P., Minervina, L. V., Baranov, S. V., Pofralidi, L. G., and Olikov, N. P.,** Tubes without electrodes with high frequency excitation of the spectrum of In, Ga, Bi, Sb, Tl, Pb, Mg, Ca and Cu as a radiation source in atomic absorption analysis, *Zh. Anal. Khim.,* 21, 1129, 1966.
5. **Spitzer, H. and Teski, G.,** Determination of indium in pyrites, pyrite cinders and intermediate products of chlorinating roasting by atomic absorption spectrometry, *Z. Anal. Chem.,* 241, 218, 1968 (in German).
6. **Torres, F.,** The Quantitative Determination of Lead, Cadmium and Indium in Microgram Amounts by Atomic Absorption Spectrophotometry, SL-TM-68-4, Sandia Laboratories, Albuquerque, N.M., 1968.
7. **Kartasheva, L. I., Kraulinya, E. K., and Liepa, S. Y.,** Electrodeless high frequency sources of thallium and indium resonance radiation, *Zh. Prikl. Spektrosk.,* 8, 1206, 1968 (in Russian).
8. **Popham, R. E. and Schrenk, W. G.,** Atomic absorption characteristic of gallium and indium, *Spectrochim. Acta,* 24b, 223, 1969.
9. **Tarasevich, N. I., Kozyreva, G. V., and Ivanov, N. P.,** Microdetermination of indium by atomic absorption, *Vestn. Mosk. Univ. Khim.,* 12, 461, 1971.
10. **Browner, R. F. and Winefordner, J. D.,** Temperature effects in pressure broadening of indium spectral lines in flames, *Spectrochim. Acta,* 27b, 257, 1972.
11. **Workman, E. J.,** Determination of mercury, indium and tellurium in mercury indium tellurium, *Analyst (London),* 97, 703, 1972.
12. **Haraguchi, H., Shiraishi, M., and Fuwa, K.,** Recombination reaction between indium and chlorine in air-acetylene flame observed by molecular flame absorption spectroscopy, *Chem. Lett.,* 251, 1973.
13. **Masson, D. B. and Pradhan, S. S.,** Measurement of vapor pressure of indium over alpha silver-indium using atomic absorption, *Met. Trans.,* 4, 991, 1973.
14. **Nikolaev, G. I., Podgornaya, V. I., Kalinin, S. K., and Zakharenko, V. M.,** Determination of indium submicroamounts by the atomic absorption method in a graphite vessel, *Zh. Anal. Khim.,* 29, 155, 1974.
15. **Rattonetti, A.,** Determination of soluble cadmium, lead, silver and indium in rain water and steam water with the use of flameless atomic absorption, *Anal. Chem.,* 46, 739, 1974.
16. **Langmyhr, F. J. and Rasmussen, S.,** Atomic absorption spectrometric determination of gallium and indium in inorganic materials by direct atomization from the solid state in a graphite furnace, *Anal. Chim. Acta,* 72, 79, 1974.
17. **Miksovsky, M. and Rubeska, I.,** Determination of antimony, indium and thallium by atomic absorption spectrometry, *Chem. Listy,* 68, 299, 1974.
18. **Fujiwara, K., Haraguchi, H., and Fuwa, K.,** Response surface and atomization mechanism in air-acetylene flames. Acid interference in atomic absorption of copper and indium, *Anal. Chem.,* 47, 1670, 1975.
19. **Nakahara, T. and Musha, S.,** Atomic absorption spectrometric determination of indium in premixed inert gas entrained air-acetylene flame, *Anal. Chim. Acta,* 80, 47, 1975.
20. **Tarasevich, N. I., Kozyreva, G. V., and Portugal'skaya, Z. P.,** Extraction-atomic absorption determination of indium, bismuth and lead trace impurities in rocks and soils, *Vestn. Mosk. Univ. Khim.,* 16, 241, 1975.
21. **Haraguchi, H. and Fuwa, K.,** Application of molecular flame absorption spectroscopy in the elucidation of chemical interference in indium atomic absorption spectrometry, *Bull. Chem. Soc. Jpn.,* 3056, 1975.
22. **Haraguchi, H. and Fuwa, K.,** Atomic and molecular absorption spectra of indium in air-acetylene flame, *Spectrochim. Acta,* 30b, 535, 1975.
23. **Terenteva, L. A. and Yudelevich, I. G.,** Atomic absorption method for the layer-by-layer determination of the constitution of oxidized layers of indium antimonide, *Izv. Sib. Otd. Akad. Nauk SSSR Ser. Khim. Nauk,* 6, 94, 1978.
24. **Yudelevich, I. G. and Beizel, N. F.,** Determination of indium and thallium in semiconductor silicon by flameless atomic absorption spectrophotometry, *Izv. Sib. Otd. Akad. Nauk SSSR Ser. Khim. Nauk,* 6, 94, 1978.
25. **Dittrich, K., Scheneider, S., Spivakov, B. Y., Suchowejewa, L. N., and Zolotov, J. A.,** Evaporation and plasma processes in graphite cuvettes for atomic absorption spectrometry. II. Determinations of gallium and indium in aqueous solutions and organic extracts by molecular and atomic absorption spectrometry, *Spectrochim. Acta,* 34b, 257, 1979.

26. **Tsalev, D. and Petrova, V.,** Hexamethylene ammonium hexamethylene dithiocarbamate-n-butyl acetate as an extractant of bismuth, cadmium, cobalt, mercury, indium, lead and palladium from acidic media, *Dokl. Bolg. Akad. Nauk,* 32, 911, 1979.

27. **Dittrich, K.,** Molecular absorption spectrometry with electrothermal volatilization in a graphite tube. III. A study of the determination of fluoride traces by aluminum fluoride, gallium fluoride, indium fluoride and thallium fluoride molecular absorption, *Anal. Chim. Acta,* 111, 123, 1979.

28. **Montaser, A. and Mehrabzadeh, A. A.,** Determination of calcium, erbium, europium, gallium, indium, potassium, sodium, molybdenum and tungsten by atomic absorption spectrometry with an electrothermal graphite braid atomizer, *Anal. Chim. Acta,* 111, 297, 1979.

29. **Pilipenko, A. T. and Samchuk, A. I.,** Extraction-atomic absorption determination of indium in minerals and rocks, *Zh. Anal. Khim.,* 34, 2128, 1979.

30. **Spivakov, B. Y., Sukhoreeva, L. N., Dittrich, K., Karyakin, A. V., and Zolotov, Y. A.,** Atomic absorption determination of indium in extracts and aqueous solutions using a graphite furnace and flame, *Zh. Anal. Khim.,* 34, 1947, 1979.

31. **Gomez Coedo, A., Lopez, M. T. D., and Sistiaga, J. M.,** Application of the techniques of atomic absorption and plasma spectrometry (ICP) to the determination of indium and thallium in zinc and its alloys, *Rev. Metal.,* 15, 379, 1979 (in Spanish).

32. **Sukhoreeva, L. N., Butrimenko, G. G., and Spivakov, B. Y.,** Atomic absorption determination of gallium and indium in a graphite furnace by evaporation from a substrate, *Zh. Anal. Khim.,* 35, 649, 1980.

33. **Tsunoda, K., Haraguchi, H., and Fuwa, K.,** Studies in the occurrence of atoms and molecules of aluminum, gallium, indium and their monohalides in an electrothermal carbon furnace, *Spectrochim. Acta,* 35b, 715, 1980.

34. **Popova, S. and Minchev, L.,** Atomic absorption determination of indium in mineral matter and industrial tailings, *Rudodobiv,* 35, 25, 1980 (in Bulgarian).

35. **Shushunova, A. F., Demarin, V. T., Makin, G. I., Sklemina, L. V., Rudnevskii, N. K., and Aleksandrov, Y. A.,** Combination of gas chromatography with atomic absorption spectrophotometry for analysis of gallium and indium organometallic compounds, *Zh. Anal. Khim.,* 35, 349, 1980.

36. **Katskov, D. A., Grinshtein, I. L., and Kruglikova, L. P.,** Study of the evaporation of the metals, indium, gallium, thallium, germanium, tin, lead, mercury, bismuth, selenium and tellurium from a graphite surface by the atomic absorption method, *Zh. Prikl. Spektrosk.,* 33, 804, 1980.

37. **Daidoji, H.,** Spectroscopic study of the behaviour of indium compounds in a flame, *Nippon Kagaku Kaishi,* 11, 1718, 1980.

38. **Kimura, M. and Kawanami, K.,** Separation of preconcentration of trace amounts of several metals in sodium perchlorate using activated carbon as a collector, *Nippon Kagaku Kaishi,* 1, 1, 1980.

39. **Volynskii, A. B., Subochev, A. I., Spivakov, B. Y., Slavnyi, V. A., and Zolotov, Y. A.,** Atomic absorption determination of elements on a multi-channel spectrometer. Analysis of metallic indium, *Zh. Anal. Khim.,* 36, 98, 1981.

40. **Mashireva, L. G., Ryabtsev, N. G., Legasova, M. M., Kuznetsova, N. V., Orlov, A. S., and Bir, T. Z.,** Atomic absorption determination of indium in solutions of indium acetyl acetonates, *Zavod. Lab.,* 47, 39, 1981.

41. **Clark, J. R. and Viets, J. G.,** Multielement extraction system for the determination of 18 trace elements in geochemical samples, *Anal. Chem.,* 53, 61, 1981.

42. **Clark, J. R. and Viets, J. G.,** Back-extraction of trace elements from organometallic halide extracts for determination by flameless atomic absorption spectrometry, *Anal. Chem.,* 53, 65, 1981.

43. **Berndt, H. and Messerschmidt, J.,** *o,o*-Diethyl dithiophosphate for trace enrichment on activated carbon. I. Analysis of high purity gallium and aluminum-determination of element traces by flame AAS (injection method and loop AAS), *Z. Anal. Chem.,* 308, 104, 1981 (in German).

44. **Yudelevich, I. G., Buyanova, L. M., Beisel, N. F., and Kozhanova, L. A.,** Atomic absorption determination of concentration profiles of major components of $A_{III}B_V$ type semiconductor films, *At. Spectrosc.,* 2, 165, 1981.

45. **Aihara, M. and Kiboku, M.,** extraction and atomic absorption spectrophotometric determination of indium and tellurium by using potassium xanthates, *Bunseki Kagaku,* 30, 295, 1981.

46. **Talapova, O. M. and Levin, I. S.,** Extraction of indium with bis (2-ethyl hexyl) hydrogen phosphate and bis (2-ethyl hexyl) hydrogen dithiophosphate and its atomic absorption determination, *Izv. Sib. Otd. Akad. Nauk SSSR Ser. Khim. Nauk,* 2, 60, 1981.

47. **Tunstall, M., Berndt, H., Sommer, D., and Ohls, K.,** Direct determination of trace elements in aluminum with flame atomic absorption spectrometry and ICP emission spectrometry — A comparison, *Erzmetall,* 34, 588, 1981 (in German).

48. **Zelentsova, L. V., Yudelevich, I. G., and Chanysheva, T. A.,** Use of atomic absorption spectrometry for the analysis of highly pure substances. Analysis of highly pure indium and gallium, *Izv. Sib. Otd. Akad. Nauk SSSR Ser. Khim. Nauk,* 1, 132, 1982.

49. **Aller, A. J.,** Solvent extraction-flame spectrometric methods for the determination of indium in aluminum alloys, *Anal. Chim. Acta,* 134, 293, 1982.

50. **Jackwerth, E. and Salewski, S.,** Contribution to the multielement preconcentration from pure cadmium, *Z. Anal. Chem.,* 310, 108, 1982 (in German).

51. **Ondov, J. M. and Biermann, A. H.,** Effects of Particle-Control Devices on Atmospheric Emissions of Minor and Trace Elements from Coal Combustion, EPA-600/9-80-039d, Symp. Transfer Util. Part. Control. Technol., 2nd., PB 81-122228, U.S. Environmental Protection Agency, Office of Research and Development, Corvallis, Ore., 1980, 454-85; CA 96, 90897g, 1982.

AUTHOR INDEX

SUBJECT INDEX (INDIUM)

THALLIUM, Tl (ATOMIC WEIGHT 204.37)

Thallium occurs in igneous rocks, pyrites, lead and zinc ores, and some rare minerals. It is also found in small quantities in the earth crust. Thallium and its compounds are highly toxic, therefore, these are chiefly used in the manufacture of rodenticides and vermin poison. It is odorless and tasteless, giving no warning of its presence, therefore, adequate ventilation should be used while handling thallium and its compounds. Thallium is also used in photocells, IR detectors, preparation of artificial stones, optical glass of very high refracting index, and low-melting glasses.

All the materials containing thallium require its determination. Due to its toxic nature, its determination is performed in biological samples. Sample solution can be prepared by treatment with hydrochloric acid, nitric acid, or aqua regia and evaporation with sulfuric acid to fumes. Hard-to-dissolve samples should be fused with sodium carbonate.

In air-acetylene no serious interferences have been reported. A partial ionization effect can be controlled by the addition of 0.1% or more potassium as chloride, when a nitrous oxide-acetylene flame is used.

Standard Solution

To prepare 1000 mg/ℓ of thallium solution, dissolve 1.303 g of thallium nitrate ($TlNO_3$) in 20 mℓ of 1% (v/v) nitric acid and dilute to 1 ℓ with deionized water.

Instrumental Parameters

Wavelength	276.8 nm
Slit width	0.7 nm
Light source	Hollow cathode lamp
Flame	Air-acetylene, oxidizing (lean, blue)
Sensitivity	0.1—0.6 mg/ℓ
Detection limit	0.02 mg/ℓ
Optimum range	40 mg/ℓ
Secondary wavelengths and their sensitivities	
377.6 nm	0.3—1.6 mg/ℓ
238.0 nm	3.0—4.0 mg/ℓ
258.0 nm	10.0—13.0 mg/ℓ

REFERENCES

1. **Slavin, W., Sprague, S., Rieders, F., and Cordova, V.,** The determination of certain toxicological trace metals by atomic absorption spectrophotometry, *At. Absorpt. Newsl.,* 17, 1964.
2. **Mulford, C. E.,** Solvent extraction techniques for atomic absorption spectroscopy, *At. Absorpt. Newsl.,* 5, 88, 1966.
3. **Ivanov, N. P., Minervina, L. V., Baranov, S. V., Pofralidi, L. G., and Olikov, I. I.,** Tubes without electrodes with high frequency excitation of the spectrum of indium, gallium, bismuth, antimony, thallium, lead, magnesium, calcium and copper as a radiation source in atomic absorption analysis, *Zh. Anal. Khim.,* 21, 1129, 1966 (in Russian).
4. **Veenendaal, W. A. and Polak, H. L.,** The determination of thallium by atomic absorption spectrophotometry. I, *Anal. Chim. Acta,* 223, 17, 1966.
5. **Berman, E.,** Determination of cadmium, thallium and mercury in biological materials, *At. Absorpt. Newsl.,* 6, 57, 1967.
6. **Kartasheva, L. S., Kraulinya, E. K., and Liepa, S. A.,** Electrodeless high-frequency sources of thallium and indium resonance radiation, *Zh. Prikl. Spektrosk.,* 8, 206, 1968 (in Russian).
7. **Savory, J., Roszel, N. O., Mushak, P., and Sunderman, F. W., Jr.,** Measurements of thallium in biologic materials by atomic absorption spectrometry, *Am. J. Clin. Pathol.,* 50, 505, 1968.
8. **Browner, R. F., Dagnall, R. M., and West, T. S.,** Atomic fluorescence spectroscopy: atomic fluorescence and atomic absorption of thallium and mercury with electrodeless discharge tubes as sources, *Talanta,* 16, 75, 1969.
9. **Curry, A. S., Read, J. F., and Knott, A. R.,** Determination of thallium in biological material by flame spectrophotometry and atomic absorption, *Analyst (London),* 94, 744, 1969.
10. **Veenendaal, W. A. and Polak, M. L.,** Determination of thallium by atomic absorption spectrophotometry, *Spectrosc. Lett.,* 2, 173, 1969.
11. **Kambayashi, K.,** Measurement of thallium in urine by atomic absorption spectrophotometry and the effect of penicillamine upon thallotoxicosis, *Igaku To Seibutsugaku,* 80, 309, 1969.
12. **Kubasik, N. P. and Volosin, M. T.,** Simplified determination of urinary cadmium, lead and thallium with use of carbon rod atomization and atomic absorption spectrophotometry, *Clin. Chem. (Winston-Salem, N.C.),* 19, 954, 1973.
13. **Shkolnik, G. M. and Bevill, R. F.,** The determination of thallium in urine and plasma by Delves cup atomic absorption, *At. Absorpt. Newsl.,* 12, 112, 1973.
14. **Sighinolfi, G. P.,** Determination of thallium in geochemical reference samples by flameless atomic absorption spectrophotometry, *At. Absorpt. Newsl.,* 12, 136, 1973.
15. **Machata, G. and Bindes, R.,** Determination of traces of lead, thallium, zinc and cadmium in biological material by flameless atomic absorption spectrophotometry, *Z. Rechtsmed.,* 73, 29, 1973 (in German).
16. **Pille, P. and Boehmer, R. G.,** Some observations concerning thallium electrodeless discharge tubes, *Spectrosc. Lett.,* 6, 731, 1973.
17. **Talalaev, B. M., Mironova, O. N., and Brevnova, T. N.,** Atomic absorption determination of thallium using a filterless photometer, *Zh. Anal. Khim.,* 29, 249, 1974.
18. **Burke, K. E.,** Determination of microgram quantities of thallium in aluminum, iron and nickel-base alloys, *Appl. Spectrosc.,* 28, 234, 1974.
19. **Amore, F.,** Determination of cadmium, lead, thallium and nickel in blood by atomic absorption spectrometry, *Anal. Chem.,* 46, 1597, 1974.
20. **Fratta, M.,** Atomic absorption spectroscopy determination of ppb amounts of thallium in silicate rocks, *Can. J. Spectrosc.,* 19, 33, 1974.
21. **Langmyhr, F. J., Stubergh, J. R., Thomassen, Y., Hanssen, J. E., and Dolezal, J.,** Atomic absorption spectrometric determination of cadmium, lead, silver, thallium and zinc in silicate rocks by direct atomization from the solid state, *Anal. Chim. Acta,* 71, 35, 1974.
22. **Miksovsky, M. and Rubeska, I.,** Determination of antimony, indium and thallium by atomic absorption spectrometry, *Chem. Listy,* 68, 299, 1974 (in Czech).
23. **Welcher, G. G., Kriege, O. H., and Marks, J. Y.,** Direct determination of trace quantities of lead, bismuth, selenium, tellurium and thallium in high temperature alloys by non-flame atomic absorption spectrophotometry, *Anal. Chem.,* 46, 1227, 1974.
24. **Schrader, H. G.,** Thallium determination in lead, zinc, copper and calcium, *Metall,* 28, 705, 1974 (in German).
25. **Singh, N. P. and Joselow, M. M.,** Determination of thallium in whole blood by Delves cup atomic absorption spectrophotometry, *At. Absorpt. Newsl.,* 14, 42, 1975.
26. **Belyaev, Y. I., Oreshkin, V. N., and Vnukovskaya, G. L.,** Atomic absorption determination of element traces in rocks by using pulse thermal atomization of solid samples. III. Suppression of the nonresonance absorption in determination of cadmium, silver and thallium, *Zh. Anal. Khim.,* 30, 503, 1975.

27. **Fuller, C. W.,** The effect of acids on the determination of thallium by atomic absorption spectrometry with a graphite furnace, *Anal. Chim. Acta,* 81, 129, 1976.

28. **Muradov, V. G., Muradova, O. N., and Truzin, G. G.,** Direct determination thallium in metallic cadmium by flameless atomic absorption spectroscopic method, *Zh. Prikl. Spektrosk.,* 24, 5, 1976.

29. **Shaburova, V. P., Yudelevich, I. G., Seryakova, I. V., and Zolotov, Y. A.,** Extraction-atomic absorption determination of copper, silver and thallium in some metal halides, *Zh. Anal. Khim.,* 31, 255, 1976.

30. **Marks, J. Y., Welcher, G. G., and Spellman, R. J.,** Atomic absorption determination of lead, bismuth, selenium, tellurium, thallium and tin in complex alloys using direct atomization from metal chips in the graphite furnace, *Appl. Spectrosc.,* 31, 9, 1977.

31. **Wall, C. D.,** The determination of thallium in urine by atomic absorption spectroscopy and emission spectrography, *Clin. Chem. Acta,* 76, 259, 1977.

32. **Yudelevich, I. G. and Beizel, N. F.,** Determination of indium and thallium in semiconductor silicon by flameless atomic absorption spectrophotometry, *Izv. Sib. Otd. Akad. Nauk SSSR Ser. Khim. Nauk,* 4, 39, 1978.

33. **Heinrichs, H.,** Determination of bismuth, cadmium and thallium in 33 international standard reference rocks by fractional distillation combined with flameless atomic absorption spectrometry, *Z. Anal. Chem.,* 294, 345, 1979.

34. **Forrester, J. E., Lehecka, V., Johnston, J. R., and Ott, W. L.,** Direct determination of trace quantities of antimony, arsenic, bismuth, cadmium, lead, selenium, silver, tellurium and thallium in high-purity nickel by electrothermal atomic absorption spectrometry, *At. Absorpt. Newsl.,* 18, 173, 1979.

35. **Kujirai, O., Kobayashi, T., and Sudo, E.,** Rapid determination of trace quantities of thallium in heat-resisting cobalt and nickel alloys by graphite furnace atomic absorption spectrometry, *Z. Anal. Chem.,* 297, 398, 1979.

36. **Boyer, K. W., Capar, S. G., Jones, J. W., Suddendorf, R. F., and Forwalter, J.,** Multielement/ trace analysis identifies 48 elements in foods. Simultaneous element analysis and data reduction reaches parts per billion, *Food Process.,* 40, 72, 1979.

37. **Dittrich, K.,** Molecular absorption spectrometry with electrothermal volatilization in a graphite tube. III. A study of the determination of fluoride traces by aluminum fluoride, gallium fluoride, indium fluoride and thallium fluoride molecular absorption, *Anal. Chim. Acta,* 111, 123, 1979.

38. **Korkisch, J. and Steffan, I.,** Determination of thallium in natural waters, *Int. J. Environ. Anal. Chem.,* 6, 111, 1979.

39. **Gomez Coedo, A., Lopez, M. T. D., and Sistiaga, J. M.,** Application of the techniques of atomic absorption and plasma spectrometry (ICP) to the determination of indium and thallium in zinc and its alloys, *Rev. Metal.,* 15, 379, 1979 (in Spanish).

40. **Baker, A. A., Headridge, J. B., and Nicholson, R. A.,** Determination of silver and thallium in nickel-base alloys by atomic absorption spectrometry with introduction of solid samples into an induction furnace, *Anal. Chim. Acta,* 113, 47, 1980.

41. **Morgan, J. M., McHenry, J. R., and Masten, L. W.,** Simultaneous determination of inorganic and organic thallium by atomic absorption analysis, *Bull. Environ. Contam. Toxicol.,* 24, 333, 1980.

42. **Aihara, M. and Kiboku, M.,** Determination of bismuth and thallium by atomic absorption spectrophotometry after extraction with potassium xanthate-methyl isobutyl ketone, *Bunseki Kagaku,* 29, 243, 1980.

43. **Chapman, J. F. and Leadbeatter, B. F.,** Determination of thallium in human hair by graphite furnace atomic absorption spectrophotometry, *Anal. Lett.,* 13, 349, 1980.

44. **Jones, E. A. and Lee, A. F.,** Determination of thallium in ores, concentrates and metals, *Natl. Inst. Metall. Repub. S. Afr. Rep.,* No. 2036, 1980.

45. **Franz, J. and Grubert, G.,** Results of an intercomparison analysis of thallium in surface water, *Z. Wasser Abwasser Forsch.,* 13, 138, 1980.

46. **Katskov, D. A., Grinshtein, L. I., and Kruglikova, L. P.,** Study of the evaporation of the metals indium, gallium, thallium, germanium, tin, lead, antimony, bismuth, selenium and tellurium from a graphite surface by the atomic absorption method, *Zh. Prikl. Spektrosk.,* 33, 804, 1980.

47. **Carpenter, R. C.,** Determination of cadmium, copper, lead and thallium in human hair and kidney tissue by flame atomic absorption spectrometry after enzymic digestion, *Anal. Chim. Acta,* 125, 209, 1981.

48. **Clark, J. R. and Viets, J. G.,** Multielement extraction system for the determination of 18 trace elements in geochemical samples, *Anal. Chem.,* 53, 61, 1981.

49. **Clark, J. R. and Viets, J. G.,** Back-extraction of trace elements from organo-metallic halide extracts for determination by flameless atomic absorption spectrometry, *Anal. Chem.,* 53, 65, 1981.

50. **Alt, F., Berndt, H., Messerschmidt, J., and Sommer, D.,** Determination of cadmium, lead and thallium in mineral raw materials after chemical preconcentration using different spectrometric methods, *Spektrometertagung (Votr.),* 13, 331, 1981.

51. **Sauer, K. H. and Eckhard, S.,** Traces of thallium in requisite materials, dust and iron materials, *Mikrochim. Acta Suppl.,* 9, 87, 1981.

52. **Keil, R.,** Trace determination of thallium in rocks by flame or flameless atomic absorption spectrophotometry following preconcentration by extraction, *Fresenius' Z. Anal. Chem.,* 309, 181, 1981.

53. **Berndt, H., Messerschmidt, J., Alt, F., and Sommer, D.,** Determination of thallium in minerals and coal by AAS (injection method, platinum loop method), *Z. Anal. Chem.,* 306, 385, 1981.

54. **Berndt, H. and Messerschmidt, J.,** Microanalytical method of flame atomic absorption and emission spectrometry (platinum loop method). II. Determination of arsenic, bismuth, cadmium, mercury, lead, antimony, selenium, tellurium, thallium, zinc, lithium, sodium, potassium, cesium, strontium and barium, *Spectrochim. Acta,* 36b, 809, 1981.

55. **Voskresenskaya, N. T., Pchelintseva, N. F., and Tsekhonya, T. I.,** Extraction-atomic absorption determination of thallium in sedimentary rocks, *Zh. Anal. Khim.,* 36, 667, 1981.

56. **Wronsky, R. and Weidhuener, J.,** Thallium intoxication and antidote therapy with reference to chemical detection by atomic absorption spectroscopy, *Aerztl. Lab.,* 27, 316, 1981 (in German).

57. **Elson, C. M. and Albuquerque, C. A. R.,** Determination of thallium in geological materials by extraction and electrothermal atomic absorption spectrometry, *Anal. Chim. Acta,* 134, 393, 1982.

58. **Heinrichs, H. and Keltsch, H.,** Determination of arsenic, bismuth, cadmium, selenium and thallium by atomic absorption spectrometry with a volatilization technique, *Anal. Chem.,* 54, 1211, 1982.

59. **Voskresenskaya, N. T.,** Thallium occurrence in halide salts, *Geokhimiya,* 3, 450, 1982.

60. **Cammann, K. and Anderson, J. T.,** Increased sensitivity and reproducibility through signal averaging in ranges near an instrumental limit of detection. Thallium and cadmium trace determinations in rock samples, *Z. Anal. Chem.,* 310, 45, 1982.

61. **Wachter, C. and Weisweller, W.,** Extraction of thallium and simultaneous extraction of lead, cadmium and thallium for their quantitative analysis using flame atomic absorption spectroscopy, *Mikrochim. Acta,* 1, 307, 1982.

62. **Iskowitz, J. M., Lee, J. J. H., Zeitlin, H., and Fernando, Q.,** Determination of thallium in deep-sea ferromanganese nodules, *Mar. Min.,* 3, 285, 1982.

63. **Slavin, W., Carnrick, G. R., and Manning, D. C.,** Graphite-tube effects on perchloric acid interferences on aluminum and thallium in the stabilized-temperature platform, *Anal. Chim. Acta,* 138, 103, 1982.

AUTHOR INDEX

SUBJECT INDEX (THALLIUM)

Group IVA Elements — C, Si, Ge, Sn, and Pb

CARBON, C (ATOMIC WEIGHT 12.001)

Carbon is one of the most abundant elements and occurs on a wider cosmic scale in the sun, stars, comets, and nebulae, etc. It is an element of prehistoric discovery and is widely distributed in nature. There are a million or more known organic compounds vital for life. It occurs free in nature in the form of amorphous, graphite, and diamond. One of its isotopes, ^{14}C, is used for dating archaeological specimens.

Carbon is used in the commercial production of cement, iron, dyes, food, fabric, drugs, plastics, perfumes, fibers, fuel, etc. Its determination is required in almost every carbon-containing product. In metallurgy its analysis is required in the alloys.

Carbon is generally determined as carbon dioxide by combustion methods. There is no direct method using an atomic absorption technique. There is one indirect method reported in the literature for carbon determination in organic compounds.

REFERENCES

1. **Sakla, A. B. and Shalaby, A. M.,** Microdetermination of carbon in organic compounds by oxygen flask combustion with atomic absorption spectrophotometric, gravimetric and titrimetric finishes, *Microchem. J.,* 24, 168, 1979.
2. **L'vov, B. V., Novotny, I., and Pelieva, L. A.,** Determination of the heat of formation of carbon molecules from absorption and emission of the swan band in a graphite furnace, *Zh. Prikl. Spektrosk.,* 32, 965, 1980.

SILICON, Si (ATOMIC WEIGHT 28.0855)

Silicon occurs abundantly in combined form. It is present in the sun and stars and is a principal component of a class of meteorites. Silicon doped with boron, arsenic, gallium, and phosphorus is used in solid state industries for electronic and space-age devices. It is also used to make concrete and bricks, refractory material for high-temperature work, enamels, pottery, and glass. Silicon is also important for plant and animal life.

The estimation of silicon is required in alloys, ores, minerals, rocks, soil, and water, etc. Sample preparation for the chemical analysis is categorized into (1) silicates decomposed by acids and (2) silicates not decomposed by acids. The second category of samples can be fused with carbonate or bicarbonate for complete dissolution.

The presence of hydrofluoric acid, boric acid, and potassium at a level of approximately 1% or greater has been reported to cause severe depression in silicon absorbance. Use of a slightly less reducing flame having a red feather height of 5 mm minimizes this effect and also minimizes the carbon build up, when flame atomic absorption method is used. Sensitivity of arsenic determination can be improved with the use of a hydride generation technique.

Standard Solution

To prepare 1000 mg/ℓ solution of silicon, dissolve 10.112 g of sodium meta silicate ($Na_2SiO_3 \cdot 9H_2O$) in 500 mℓ of deionized water, add 10 mℓ of concentrated hydrochloric acid, and dilute to 1 ℓ with deionized water.

Instrumental Parameters

Wavelength	251.6 nm
Slit	0.2 nm
Light source	Hollow cathode lamp
Flame	Nitrous oxide-acetylene, reducing (rich, red)
Sensitivity	0.8—2.0 mg/ℓ
Detection limit	0.06 mg/ℓ
Optimum range	150 mg/ℓ

Secondary wavelengths and their sensitivities

250.7 nm	2.0—6.0 mg/ℓ
252.8 nm	2.0—6.11 mg/ℓ
251.9 nm	3.0 mg/ℓ
251.4 nm	3.0 mg/ℓ
252.4 nm	3.0—7.0 mg/ℓ
221.7 nm	3.0—7.5 mg/ℓ
221.1 nm	6.0—14.0 mg/ℓ
221.7 nm	7.5 mg/ℓ
288.2 nm	13.0—37.0 mg/ℓ
220.8 nm	24.0 mg/ℓ

REFERENCES

1. **Walsh, A.,** Atomic absorption spectroscopy, in *Proc. 10th Coll. Spectroscopicum Internationale,* Margoshes, M. and Lippincott, E. R., Eds., Spartan Books, Rochelle Park, N.J., 1963.
2. **Capacho-Delgado, L. and Manning, D. C.,** The determination by atomic absorption spectroscopy of several elements including silicon, aluminum and titanium in cement, *Analyst (London),* 92, 553, 1967.
3. **Kirkbright, G. F., Smith, A. M., and West, T. S.,** An indirect sequential determination of phosphorus and silicon by atomic absorption spectrophotometry, *Analyst (London),* 92, 411, 1967.
4. **McAuliffe, J. J.,** A method for determination of silicon in cast iron and steel by atomic absorption spectrometry, *At. Absorpt. Newsl.,* 6, 69, 1967.
5. **Katz, A.,** The direct and rapid determination of alumina and silica in silicate rocks and minerals by atomic absorption spectroscopy, *Am. Mineral.,* 53, 283, 1968.
6. **Hurford, T. R. and Boltz, D. F.,** Indirect ultraviolet spectrophotometric and atomic absorption spectrometric methods for determination of phosphorus and silicon by heteropoly chemistry of molybdate, *Anal. Chem.,* 40, 379, 1968.
7. **LaFlamme, Y.,** Determination of free silica in soils by atomic absorption spectrophotometry, *At. Absorpt. Newsl.,* 7, 101, 1968.
8. **Mario, E. and Gerner, R. E.,** Determination of silicon in commercial hand lotion by atomic absorption spectroscopy, *J. Pharm. Sci.,* 57, 1243, 1968.
9. **Paralusz, C. M.,** Trace organic silicon analysis using atomic absorption spectroscopy, *Appl. Spectrosc.,* 22, 520, 1968.
10. **Gomez Coedo, A. and Jiminez Seco, J. L.,** Determination of Fe, Mn, Ca, Mg, Al, and Si in iron ores by atomic absorption, *Rev. Metal.,* 4, 58, 1968 (in Spanish).
11. **Price, W. J. and Roos, J. T. H.,** The determination of silicon by atomic absorption spectrophotometry with particular reference to steel, cast iron aluminum alloys and cement, *Analyst (London),* 93, 709, 1968.
12. **Gomez Coedo, A., Teresa Dorado, A., and Jiminez Seco, J. L.,** Determination of silicon in Al and its alloys by atomic absorption, *Rev. Metal.,* 4, 629, 1968 (in Spanish).
13. **Tumanov, A. A. and Petukhova, V. G.,** Determination of traces of Al and Si in their oxides and of aluminum in metallic chromium, *Zavod. Lab.,* 35, 654, 1969 (in Russian).
14. **Yuan, T. L. and Breland, H. L.,** Evaluation of atomic absorption methods for determinations of Al, Fe and Si in clay and soil extracts, *Soil Sci. Soc. Am. Proc.,* 33, 868, 1969.
15. **Knight, D. M. and Pyzyna, M. K.,** Determination of Cu, Cr, Co, Mn, Ni, Si, W and V in tool steel by atomic absorption spectrometry, *At. Absorpt. Newsl.,* 8, 129, 1969.
16. **Miller, J. R., Helprin, J. J., and Finlayson, J. S.,** Silicon lubricant flushed from disposable syringes. Determination by atomic absorption spectrophotometry, *J. Pharm. Sci.,* 58, 455, 1969.
17. **Morrow, R. W., Dean, J. A., Schults, W. O., and Guerin, M. R.,** Silicon specific detector based on interfacing a gas chromatograph and a flame emission or atomic absorption spectrometer, *J. Chromatogr. Sci.,* 7, 572, 1969.
18. **Campbell, D. E.,** Determination of silicon in aluminum alloys by atomic absorption spectroscopy, *Anal. Chim. Acta,* 46, 31, 1969.
19. **Dagnall, R. M., Kirkbright, G. F., West, T. S., and Wood, R.,** The determination of silicon by flame photometry and atomic fluorescence spectroscopy with a separated nitrous oxide-acetylene flame, *Anal. Chim. Acta,* 47, 407, 1969.
20. **Ramakrishna, T. V., Robinson, J. W., and West, P. W.,** Determination of P, As or Si by atomic absorption spectrometry of molybdenum heteropoly acids, *Anal. Chim. Acta,* 45, 43, 1969.
21. **Ferris, A. P., Jepson, W. B., and Shapland, R. C.,** Evaluation and correction of interference between Al, Si, and Fe in atomic absorption spectrophotometry, *Analyst (London),* 95, 574, 1970.
22. **Prey, V., Teichmann, H., and Bichler, D.,** Determination of silicon in organic compounds with the aid of atomic absorption, *Mikrochim. Acta,* 1, 138, 1970.
23. **Salama, C. and Dunn, R.,** Determining trace amounts of silicon using disposable plastic syringes, *At. Absorpt. Newsl.,* 9, 52, 1970.
24. **Govindaraju, K.,** Use of complexing agents to put silicates in solution and separation by ion-exchange resins, *Ed. CNRS Paris,* 269, 1970 (in French).
25. **Trudell, L. A. and Boltz, D. F.,** Indirect atomic absorption spectrometric method for the determination of silicon, *Mikrochim. Acta,* 1970, 1220, 1970.
26. **Thormahlen, D. J. and Frank, E. H.,** The determination of silicon in niobium bearing and various other alloys by atomic absorption spectroscopy, *At. Absorpt. Newsl.,* 10, 63, 1971.
27. **Nakahara, T., Munemori, M., and Musha, S.,** Determination of silicon in some metallurgical materials by atomic absorption spectrophotometry, *Bull. Univ. Osaka Perfect. Ser. A,* 20, 169, 1971.
28. **Looyenga, R. W. and Huber, C. O.,** Determination of silicate in waste waters by atomic absorption inhibition titration, *Anal. Chem.,* 43, 498, 1971.

29. **Gamble, E. E. and Daniels, R. B.,** Iron and silica in water, acid ammonium oxalate and dithionite extracts of some North Carolina coastal plain soils, *Soil Sci. Soc. Am. Proc.,* 36, 939, 1972.

30. **Musil, J. and Halirova, D.,** Atomic absorption spectrophotometry in metallurgical analysis. I. Determination of silicon in ferrovanadium, *Hutn. Listy,* 27, 437, 1972 (in Czech).

31. **Guest, R. J.,** The Use of Flame Procedures for the Analysis of Minerals, Ores and Electric Furnace Slags. I. Sample Dissolution and Atomic Absorption Procedures for Use in the Determination of Silicon. A review, Mines Branch Tech. Bull. TB, Canadian Department of Energy, Mines and Resources, Ottawa, Canada, 1972, 149.

32. **Torok, I. and Varju, M.,** Atomic absorption spectrometry and its application in silicate analysis, *Epitoanyag,* 24, 241, 1972.

33. **Govindaraju, K. and L'homel, N.,** Direct and indirect atomic absorption determination of silica in silicate rock samples, *At. Absorpt. Newsl.,* 11, 115, 1972.

34. **Jeanroy, E.,** Total analysis of natural silicates by atomic absorption spectrophotometry. Application to soils and their constituents, *Chim. Anal. (Paris),* 54, 159, 1972.

35. **Rooney, R. C. and Pratt, C. G.,** Determination of tungsten and silicon in highly alloyed materials by atomic absorption spectroscopy, *Analyst (London),* 97, 400, 1972.

36. **Lin, C. I. and Huber, C. O.,** Determination of phosphate, silicate and sulfate in natural and waste water by atomic absorption inhibition titration, *Anal. Chem.,* 44, 2200, 1972.

37. **Legrand, G., Louvrier, J., Musikas, N., and Voinovich, I. A.,** Determination of silica, alumina, ferric oxide, lime and magnesia in cements by X-ray fluorescence and atomic absorption spectrometry, *Analusis,* 2, 9, 1973.

38. **Brivot, E., Cohort, I., Legrand, G., Louvrier, J., and Voinovich, I. A.,** Comparative determination by atomic absorption spectrometry and by spectrophotocolorimetry of nickel, copper, chromium, manganese, aluminum and silicon in certain steels, *Analusis,* 2, 570, 1973.

39. **Lerner, L. A., Tikhomirova, E. I., and Shostak, R. V.,** Determination of total silicon content in soils by the atomic absorption, *Pochvovedenie,* 104, 1973.

40. **Manoliu, C., Tomi, B., Daescu, A., and Petruc, F.,** Continuous determination of the total silicon dioxide in silicate solutions by atomic absorption spectrometry, *Rev. Chim.,* 24, 639, 1973.

41. **Stahlavska, A., Prokopova, H., and Tuzar, M.,** Use of spectral-analytical methods in drug analysis. Determination of Bi, Al, Mg, Ca, Ti and Si in antiacids by atomic absorption spectrophotometry, *Pharmazie,* 29, 140A, 1974.

42. **Peetre, I. B. and Smith, B. E. F.,** Sensitivity of determination of organosilicon and tin compounds by atomic absorption spectroscopy, *Mikrochim. Acta,* 2, 301, 1974.

43. **Manoliu, C., Tomi, B., and Petruc, F.,** Determination of silicon, aluminum and sodium in a molecular sieve by emission and atomic absorption flame photometry, *Rev. Chim.,* 24, 991, 1974.

44. **Guest, R. J. and Macpherson, D. R.,** The use of flame procedures in metallurgical analysis. I. Determination of silicon in sulfide and silicate materials, *Anal. Chim. Acta,* 71, 233, 1974.

45. **Hurtubise, R. J.,** Determination of silicon in streptomycin by atomic absorption, *J. Pharm. Sci.,* 63, 1128, 1974.

46. **Radecki, A., Lamparczyk, H., Grybowski, J., and Halkiemicz, J.,** Acid interferences in indirect determination of silicon by atomic spectroscopy, *Spectrosc. Lett.,* 7, 627, 1974.

47. **Urbain, H. and Carret, G.,** Spectral interference of vanadium in the atomic absorption spectroscopic determination of silicon, *Analusis,* 3, 110, 1975.

48. **Burdo, R. A. and Wise, W. M.,** Determination of silicon in glasses and minerals by atomic absorption spectrometry, *Anal. Chem.,* 47, 2360, 1975.

49. **Gill, R. C. O. and Kronberg, B. I.,** The precise determination of silicon in fluoborate solutions by atomic absorption spectrophotometry, *At. Absorpt. Newsl.,* 14, 157, 1975.

50. **Luzar, O. and Sliva, V.,** Determination of calcium oxide, magnesium oxide, alumina, silica, iron, copper, zinc, lead, cadmium, sodium oxide and potassium oxide in iron ores and agglomerates, *Hutn. Listy,* 30, 55 1975 (in Czech).

51. **Ortner, H. M. and Kantuscher, E.,** Metal salt impregnation of graphite tube for improvement of silicon determination by atomic absorption spectrometry, *Talanta,* 22, 581, 1975 (in German).

52. **Hocquellet, P. and Labevrie, N.,** Thermometric atomization of silicon, germanium, tin and lead on tantalum carbide. Application to atomic absorption spectrometry, *Analusis,* 3, 505, 1975.

53. **Bigois, M. and Levy, R.,** Microdetermination of silicon in organo silicon compounds by chlorocarbopyrolysis and atomic absorption measurement, *Talanta,* 23, 119, 1976.

54. **Kato, K.,** Atomic absorption spectrophotometric determination of total silicon carbide, *At. Absorpt. Newsl.,* 15, 4, 1976.

55. **Bosch-Arino, F., Bosch-Reig, F., and Hernandez-Martinez, V.,** Analysis of silicates. Determination of silica by distillation and atomic absorption spectrophotometry, *Quim. Anal.,* 30, 82, 1976.

56. **Gabrovski, I., Chekhlarova, I., and Veselinova, E.,** Indirect flame photometric determination of silicon in chromium ores, *Zavod. Lab.,* 42, 938, 1976.

57. **Musil, J. and Nehasilova, M.,** Interferences in the atomic absorption determination of silicon, *Talanta,* 23, 729, 1976.

58. **Sand, J. R. and Huber, C. O.,** Determination of silicate, phosphate and sulfate by calcium atomization inhibition titration, *Anal. Chim. Acta,* 87, 79, 1976.

59. **Chapman, J. F. and Dale, L. S.,** Improved sensitivity for boron and silicon in flame spectrometry by a fluoride evolution technique, *Anal. Chim. Acta,* 89, 363, 1977.

60. **DeVine, J. C. and Suhr, N. H.,** Determination of silicon in water samples, *At. Absorpt. Newsl.,* 16, 39, 1977.

61. **Tekula-Buxbaum, P.,** Determination of small quantities of aluminum and silicon in tungsten metal and tungsten oxides by atomic absorption spectrophotometry, *Mikrochim. Acta,* 1, 145, 1977.

62. **Walsh, J. N.,** Determination of silica in rocks and minerals by a combined gravimetric and atomic absorption spectrophotometric procedure, *Analyst (London),* 102, 51, 1977.

63. **Henry, C. D.,** Determination of silicon in soil extracts by atomic absorption, *At. Absorpt. Newsl.,* 16, 128, 1977.

64. **Kundu, M. K.,** Determination of silicones in vegetable oils by nonflame atomic absorption spectrophotometry, *Fette Seifen Anstrichm.,* 79, 170, 1977.

65. **Lo, D. B. and Christian, G. D.,** Studies of sensitivity and interferences in silicon determinations by flameless atomic absorption spectrometry, *Can. J. Spectrosc.,* 22, 45, 1977.

66. **Riddle, C. and Turek, A.,** An indirect method for the sequential determination of silicon and phosphorus in rock analysis by atomic absorption spectrometry, *Anal. Chim. Acta,* 92, 49, 1977.

67. **Tanaka, T., Kumamaru, T., and Yamamoto, Y.,** Determination of silicon by graphite atomic absorption spectrometry combined with molybdoheteropoly acid-methyl iso butyl ketone extraction, *Bunseki Kagaku,* 26, 519, 1977.

68. **Guest, R. J., Macpherson, D. R., and Pugliese, R. J.,** The determination of silicon in fluoride-bearing materials by atomic absorption spectrometry, *Anal. Chim. Acta,* 96, 185, 1978.

69. **Husain, D. and Norris, P. E.,** Collisional behavior of electronically excited silicon atoms $Si[^3p^2\,(^1D_2)]$, by atomic absorption spectroscopy, *Chem. Phys. Lett.,* 53, 474, 1978.

70. **Husain, D. and Norris, P. E.,** Kinetic study of ground state silicon atoms, $Si[^3p^2\,(^3P_j)]$, *J. Chem. Soc. Faraday Trans.,* 2, 74, 93, 1978.

71. **Husain, D. and Norris, P. E.,** Kinetic study of ground state silicon atoms, $Si[^3p^2(^3P_j)]$, by atomic absorption spectroscopy, *J. Chem. Soc. Faraday Trans.,* 2, 106, 1978.

72. **Abbey, S. and Maxwell, J. A.,** A critical comment on: an indirect method for the sequential determination of silicon and phosphorus in rock analysis by atomic absorption spectrometry, *Anal. Chim. Acta,* 99, 397, 1978.

73. **Husain, D. and Norris, P. E.,** Collisional quenching of electronically excited silicon atoms, $Si[^3p^2(^1D_2)]$ by atomic absorption spectroscopy, *J. Chem. Soc. Faraday Trans.,* 2, 1483, 1978.

74. **Parker, R. D.,** Determination of organo silicon compounds in water by atomic absorption spectroscopy, *Z. Anal. Chem.,* 292, 362, 1978.

75. **Riddle, C. and Turek, A.,** An indirect method for the sequential determination of silicon and phosphorus in rock analysis by atomic absorption spectrometry. Reply to the critical comment by S. Abbey and J. A. Maxwell, *Anal. Chim. Acta,* 99, 398, 1978.

76. **Tsuchitani, Y., Harada, K., Saito, K., Muramatsu, N., and Uematsu, K.,** Determination of organo silicone in sewage sludge by atomic absorption spectrometry, *Bunseki Kagaku,* 27, 343, 1978.

77. **Lo, D. B. and Christian, G. D.,** Microdetermination of silicon in blood, serum, urine and milk using furnace atomic absorption spectrometry, *Microchem. J.,* 23, 481, 1978.

78. **Purushottam, A. and Naidu, P. P.,** Rapid atomic absorption spectrophotometric method for estimation of "reactive" and total silica, iron, aluminum and titanium in bauxite, *Curr. Sci.,* 47, 88, 1978.

79. **Terashima, S.,** Atomic absorption determination of manganese, iron, copper, nickel, cobalt, lead, zinc, silicon, aluminum, calcium, magnesium, sodium, potassium, titanium and strontium in manganese nodules, *Chishitsu Chosasho Geppo,* 29, 401, 1978 (in Japanese).

80. **Gregorczyk, S. and Wycislik, A.,** Chemical analysis of high-alloy manganese-aluminum steels by atomic absorption spectrophotometry, *Wiad. Hutn.,* 35, 261, 1979 (in Polish).

81. **Startseva, E. A., Popova, N. M., Khrapai, V. P., and Yudelevich, I. G.,** Indirect atomic absorption of phosphorus and silicon in pure silver, *Izv. Sib. Otd. Akad. Nauk SSSR Ser. Khim. Nauk,* 6, 139, 1979.

82. **Baucells, M., Lacort, G., and Roura, M.,** Determination of major and trace elements in marine sediments of Costa Brava by emission and atomic absorption spectroscopy, *Invest. Pesq.,* 43, 551, 1979.

83. **Das, A. K. and Chakrabarty, A. K.,** Estimation of silicon in minerals by atomic absorption spectrophotometry, *Fert. Technol.,* 16, 61, 1979.

84. **Malhotra, P. D., Naidu, P. P., and Rao, T. J.,** Method for decomposition of chromite for analysis by atomic absorption spectrophotometry, *Curr. Sci.,* 48, 584, 1979.

85. **Bower, N. W. and Ingle, J. D.,** Precision of flame atomic absorption spectrometric measurements of aluminum, chromium, cobalt, silicon, europium, lead, manganese, nickel, potassium, selenium, titanium and vanadium, *Anal. Chem.,* 51, 72, 1979.

86. **Rewa, J. A. and Henn, E. L.,** Determination of trace silica in industrial process waters by flameless atomic absorption spectrometry, *Anal. Chem.,* 51, 452, 1979.

87. **LaBrecque, J. J.,** Decomposition and determination of aluminum and silicon in Venezuelan laterites by atomic absorption spectroscopy, *Chem. Geol.,* 26, 321, 1979.

88. **Fabbri, B.,** Determination of the chemical composition of alumina-silicate materials: measurement by atomic absorption spectrophotometry, *Ceramurgia,* 9, 57, 1979.

89. **Gregorczyk, S. and Wycislik, A.,** Determination of silicon in high-temperature nickel-base alloys by atomic absorption spectrometry, *Chem. Anal.,* 24, 1071, 1979.

90. **Frech, W. and Cedergren, A.,** Investigations of reactions involved in flameless atomic absorption procedures. I. A theoretical and experimental study of factors influencing the determination of silicon, *Anal. Chim. Acta,* 113, 227, 1980.

91. **Brenner, I. B., Watson, A. E., Russell, G. M., and Goncalves, M.,** New approach to determination of major and minor constituents in silicate and phosphate rocks, *Chem. Geol.,* 28, 321, 1980.

92. **Clegg, J. B., Grainger, F., and Gale, I. G.,** Quantitative measurement of impurities in gallium arsenide, *J. Mater. Sci.,* 15, 747, 1980.

93. **Lieu, V. T. and Woo, D. H.,** Application of flame and graphite furnace atomic absorption spectroscopy to the determination of metals in oil-field injection water and its suspended solids, *At. Spectrosc.,* 1, 149, 1980.

94. **Mueller-Vogt, G. and Wendl, W.,** Analysis of DOPANT concentrations of silicon and iron in lithium niobate (V) single crystals by atomic absorption spectroscopy, *Mater. Res. Bull.,* 15, 1461, 1980.

95. **Nadkarni, R. A.,** Multitechnique multielemental analysis of coal and fly ash, *Anal. Chem.,* 52, 929, 1980.

96. **Simmons, M. S.,** Routine determination of particulate silica in water, *Anal. Lett.,* 13a, 67, 1980.

97. **Balaes, G. and Robert, R. V. D.,** Analysis by atomic absorption spectrophotometry of activated charcoal, *Natl. Inst. Metall. Repub. S. Afr. Rep.,* No. 2060, 1980.

98. **Dyulgerova, R. B., Zhechev, D. Z., and Oavova, T. P.,** Construction of new hollow cathode lamps for atomic absorption determination of molybdenum, iron and silicon, *Elektropromst. Priborostr.,* 15, 110, 1980 (in Russian).

99. **Lu, Y. Z., Liang, F. X., and Bai, Y. L.,** Rapid analysis of silicate rocks by fusion with boron oxide-lithium carbonate, *Ti Chiu Hua Hsueh,* 3, 282, 1980 (in Chinese).

100. **Tardon, S.,** Rapid determination of small quantities of inorganic constituents of suspended dust and dust fallouts, *Chem. Prum.,* 30, 195, 1980 (in Czech).

101. **Pletneva, T. I.,** Atomic absorption determination of trace silicon contaminants in barium nitrate and sodium chloride, *Khim. Promst. Ser. Reakt. Osobo Christ. Veshchestva,* 1, 36, 1981.

102. **Pletneva, T. I.,** AA determination of trace contaminants in potassium and sodium nitrates, *Khim. Promst. Ser. Reakt. Osobo Christ. Veshchestva,* 1, 39, 1981.

103. **Mueller-Vogt, G. and Wendl, W.,** Reaction kinetics in the determination of silicon by graphite furnace atomic absorption spectrometry, *Anal. Chem.,* 53, 651, 1981.

104. **Lythgoe, D. J.,** Method for improving the determination of silicon by atomic absorption spectrometry using a tantalum-coated carbon furnace, *Analyst (London),* 106, 743, 1981.

105. **Young, R. S.,** Analysis of nickel refinery slimes and residues, *Talanta,* 28, 25, 1981.

106. **Devyatykh, G. G., Rudnevskii, N. K., Demarin, V. T., Krylov, V. A., and Sklemina, L. V.,** Chromatography-atomic absorption determination of trichlorosilane and trimethylchlorosilane impurities in silicon tetrachloride, *Zh. Anal. Khim.,* 36, 2335, 1981.

107. **Domingues, A.,** Technology for pressure decomposition in silicate analysis by atomic absorption, *Ceramica,* 27, 1, 1981 (in Portuguese).

108. **Murase, I. and Kamiya, M.,** Atomic absorption spectrometric method for determination of silicon in high-silicon aluminum alloys, *Gifu-ken Kinzoku Shikenjo Gyomu Hokoku,* 11, 1981.

109. **Balaes, G. E., Dixon, K., Russell, G. M., and Wall, G. J.,** Analysis of activated carbon, as used in the carbon-in-pulp process, for gold and eight other constituents, *S. Afr. J. Chem.,* 35, 4, 1982.

110. **Dobbie, J. W. and Smith, M. J. B.,** Silicon content of body fluids, *Scott. Med. J.,* 27, 17, 1982.

111. **Hicks, D. G., O'Reilly, J. E., and Koppenaal, D. W.,** On the correlation of the weight percent ash of coals and their silicon content, *Fuel,* 61, 150, 1982.

112. **Kauffman, R. E., Saba, C. S., Rhine, W. E., and Eisentraut, K. J.,** Quantitative multielement determination of metallic wear species in lubricating oils and hydraulic fluids, *Anal. Chem.,* 54, 975, 1982.

113. **Leong, A. S. Y., Disney, A. P. S., and Gove, D. W.,** Spallation and migration of silicone from blood-pump tubings in patients on hemodialysis, *N. Engl. J. Med.,* 306, 135, 1982.

114. **Verbeek, A. A., Mitchell, M. C., and Ure, A. M.,** Analysis of small samples of rock and soil by atomic absorption and emission spectrometry after lithium metaborate fusion/nitric acid dissolution, *Anal. Chim. Acta,* 135, 215, 1982.
115. **Goldschmidt, A. and Kittl, P.,** Study on morphology of liquid phase evaporation product of Portland cement paste, *Ceramica (Sao Paolo),* 26, 1, 1980 (in Portuguese); CA 96, 109170p, 1982.
116. **Jenke, D. R. and Diebold, F. E.,** Characterization of phosphorite ores, *Anal. Chem.,* 54, 1008, 1982.
117. **Dobbie, J. W. and Smith, M. J. B.,** Silicate nephrotoxicity in the experimental animal: the missing factor in analgesic nephropathy, *Scott. Med. J.,* 27, 10, 1982.

AUTHOR INDEX

SUBJECT INDEX (SILICON)

GERMANIUM, Ge (ATOMIC WEIGHT 72.59)

Germanium is found usually as sulfide in minerals. It is commercially obtained from the zinc industry or flue dusts of coal-burning power plants. Germanium in a highly pure form is a very good semiconductor material. It is used in the construction of rectifiers, photodiodes, IR spectroscopes, wide-angle camera lenses, and other microscope objectives. It is also used as an alloying agent, catalyst, and as phosphor in fluorescent lamps. Due to its activity towards certain bacteria, it is used as a chemotherapeutic agent in medicine.

Sample solutions are best prepared by dissolution in nitric-sulfuric acids mixture or aqua regia with gentle heating. Fusion methods are employed wherever necessary. Since platinum is attacked by germanium and some of its alloys and compounds, use of a platinum crucible for fusion is not advised. No chemical interferences have been reported in the nitrous oxide-acetylene flame when atomic absorption method is used for germanium determination.

Standard Solution

To prepare 1000 mg/ℓ solution, dissolve 1.0000 g of metallic germanium in 10 mℓ of aqua regia with gentle heating or in 5 mℓ of concentrated hydrofluoric acid. Add nitric acid dropwise until complete dissolution. Dilute with deionized water to 1 ℓ. Use of hydrochloric acid may cause loss of germanium due to volatilization. Store the solution in a polyethylene bottle.

Instrumental Parameters

Wavelength	265.1 nm
Slit	0.2 nm
Light source	Hollow cathode lamp
Flame	Nitrous oxide-acetylene, reducing (rich, red)
Sensitivity	0.8—2.2 mg/ℓ
Detection limit	0.15 mg/ℓ
Optimum range	100 mg/ℓ

Secondary wavelengths and their sensitivities

271.0 nm	1.5—5.0 mg/ℓ
259.2 nm	1.8—5.0 mg/ℓ
275.5 nm	2.1—6.1 mg/ℓ
269.1 nm	3.0—8.6 mg/ℓ
303.9 nm	15.0—170.0 mg/ℓ

REFERENCES

1. **Yanagisawa, M., Suzuki, M., and Takeuchi, T.,** Determination of germanium in synthetic fibers by solvent extraction and atomic absorption spectrophotometry, *Anal. Chim. Acta,* 46, 152, 1969.
2. **Pollock, E. N.,** Gallium and germanium in limonite by atomic absorption, *At. Absorpt. Newsl.,* 10, 77, 1971.
3. **Matsuo, T., Shida, J., and Kudo, S.,** Indirect determination of germanium by atomic absorption spectrometry, *Bunseki Kagaku,* 22, 1009, 1973.
4. **Johnson, D. J., West, T. S., and Dagnall, R. M.,** Determination of germanium by atomic absorption spectrometry with a graphite tube atomizer, *Anal. Chim. Acta,* 67, 79, 1973.
5. **Thompson, K. C. and Thomerson, D. R.,** Atomic absorption studies on the determination of antimony, arsenic, bismuth, germanium, lead, selenium, tellurium and tin by utilizing the generation of covalent hydrides, *Analyst (London),* 99, 595, 1974.
6. **Hocquellet, P. and Labevrie, M.,** Thermoelectric atomization of silicon, germanium, tin and lead on tantalum carbide. Application to atomic absorption spectrometry, *Analusis,* 3, 505, 1975.
7. **Rozenblum, V.,** Indirect determination of picogram amounts of germanium in pure water by flameless atomic absorption (Mo) spectroscopy, *Microchem. J.,* 21, 82, 1976.
8. **Chowdhury, N. A. and Husain, D.,** Collisional quenching of electronically excited germanium atoms $\{(Ge[4_p^2\,(^1S_0)])\}$ by atomic absorption spectroscopy, *J. Chem. Soc. Faraday Trans.,* 2, 73, 1805, 1977.
9. **Chowdhury, N. A. and Husain, D.,** Collision behavior of germanium $(4^3\,P_{0,\,1,\,2})$ by atomic absorption spectroscopy, *J. Photochem.,* 7, 41, 1977.
10. **Shimomura, S., Sakurai, H., Morita, H., and Mino, Y.,** Determination of germanium by atomic absorption spectrometry after solvent extraction-enhancement of sensitivity by a nebulizer effect, *Anal. Chim. Acta,* 96, 69, 1978.
11. **Ohta, K. and Suzuki, M.,** Atomic absorption spectrometry of germanium with a tungsten electrothermal atomizer, *Anal. Chim. Acta,* 104, 293, 1979.
12. **Palliere, M. and Gernez, G.,** Apparatus for the production of volatile hydrides for arsenic, antimony, bismuth, tin, germanium, selenium and tellurium determination by atomic absorption spectrometry, *Analusis,* 7, 46, 1979.
13. **Schnepfe, M. M.,** Germanium contents of USGS standard rocks by flameless atomic absorption, *Geostandards Newsl.,* 3, 93, 1979.
14. **Broeckx, J., Clauws, P., Van den Steen, K., and Vernik, J.,** Zeeman effect in the excitation spectra of shallow acceptors in germanium: experimental, *J. Phys. C,* 12, 4061, 1979.
15. **Epting, M. A., Sweigart, J. R., and Nixon, E. R.,** Absorption and emission spectra of matrix-isolated germanium telluride, *J. Mol. Spectrosc.,* 78, 277, 1979.
16. **Mino, Y., Shimomura, S., and Ota, N.,** Determination of germanium in different media by atomic absorption spectrometry with electrothermal atomization, *Anal. Chim. Acta,* 107, 253, 1979.
17. **Boyer, K. W., Capar, S. G., Jones, J. W., Suddendorf, R. F., and Forwalter, J.,** Multielement/trace analysis identifies 48 elements in foods. Simultaneous element analysis and data reduction reaches parts per billion, *Food Process.,* 40, 72, 1979.
18. **Hahn, M. H., Mulligan, K. J., Jackson, M. E., and Caruso, J. A.,** Sequential determination of arsenic, selenium, germanium and tin as their hydrides by gas-solid chromatography with an atomic absorption detector, *Anal. Chim. Acta,* 118, 115, 1980.
19. **Mino, Y., Ota, N., Sakao, S., and Shimomura, S.,** Determination of germanium in medicinal plants by atomic absorption spectrometry with electrothermal atomization, *Chem. Pharm. Bull.,* 28, 2687, 1980.
20. **Studnicki, M.,** Determination of germanium, vanadium and titanium by carbon furnace atomic absorption spectrometry, *Anal. Chem.,* 52, 1762, 1980.
21. **Katskov, D. A., Grinshtein, I. L., and Kruglikova, L. P.,** Study of the evaporation of the metals indium, gallium, thallium, germanium, tin, lead, antimony, bismuth, selenium and tellurium from a graphite surface by the atomic absorption method, *Zh. Prikl. Spektrosk.,* 33, 804, 1980.
22. **Zakhariya, A. N., Olenovich, N. L., and Dramitskaya, R. M.,** Determination of germanium by an atomic absorption spectrometric method, *Zh. Prikl. Spektrosk.,* 33, 612, 1980.
23. **Castillo, J. R., Lanaja, J., Belarra, M. A., and Aznarez, J.,** Flame atomic absorption determination of germanium in lignite ash with extraction of $GeCl_4$ into n-hexane, *At. Spectrosc.,* 2, 159, 1981.
24. **Andreae, M. O. and Froelich, P. N.,** Determination of germanium in natural waters by graphite furnace atomic absorption spectrometry with hydride generation, *Anal. Chem.,* 53, 287, 1981.
25. **Jin, K., Terada, H., and Taga, M.,** Determination of germanium by atomic absorption spectrometry following volatile hydride generation, *Bull. Chem. Soc. Jpn.,* 54, 2934, 1981.
26. **DeCarlo, E. H., Zeitlin, H., and Fernando, Q.,** Simultaneous separation of trace levels of germanium, antimony arsenic and selenium from an acid matrix by adsorbing colloid flotation, *Anal. Chem.,* 53, 1104, 1981.

27. **Ikeda, M., Nishibe, J., and Nakahara, T.,** Study of some fundamental conditions for the determination of germanium by continuous hydride generation-atomic absorption spectrophotometry, *Bunseki Kagaku,* 30, 548, 1981.
28. **Castillo, J. R., Lanaza, J., and Aznarez, J.,** Determination of germanium in coal ashes by hydride generation and flame atomic absorption spectrophotometry, *Analyst (London),* 107, 89, 1982.
29. **Paik, N. H., Park, M. K., Choi, S. H., Moon, D. C., and Kang, T. L.,** Studies of germanium in herbal drugs. II. Extracting effect of germanium of various solvents in zingiberis ehizome, *Soul Taehakkyo Yakhak Nonmunjip,* 5, 75, 1980 (in Korean); CA 96, 74708r, 1982.

AUTHOR INDEX

SUBJECT INDEX (GERMANIUM)

TIN, Sn (ATOMIC WEIGHT 118.69)

Tin occurs as sulfide and oxide in rocks and minerals. It is used as a coating on other metals to prevent corrosion. Its alloys such as soft solder, type metal, bell metal, fusible metal, pewter, bronze, Babbitt metal, die casting alloys, and phosphor bronze are very important for various industries. Tin is also used in calico printing, frost-free windshields, panel lighting, and superconductive magnets. Tin determination is required in the analyses of ores, dross, ashes, dust, plating materials, alloys, foodstuff, iron and steel, and general products.

In the decomposition of the sample, special precautions must be observed to prevent loss of tin due to volatilization. Covered acid digestion with hydrochloric acid in the presence of sulfuric acid is the recommended method for dissolution. Hard-to-dissolve samples can be fused with alkali or peroxide.

In air-acetylene or air-hydrogen or argon-hydrogen flames, many cationic and anionic interferences have been reported. Specifically the alkali and alkaline earth metals affect the absorbance in any concentration tending toward a constant elevation or depression of signal. Copper at a concentration of 200 to 1000 mg/ℓ has been reported to cause a 20% increase in tin absorbance, while a copper concentration of 2000 mg/ℓ or more produced only 10% increase in tin absorbance. Cobalt causes a continuously increasing elevation in the absorbance with increasing concentration, while zinc causes a continuously increasing depression in tin absorbance with the increasing zinc concentration. Elevation or depression of tin absorbance in the presence of aluminum and titanium depends markedly on the burner height. Aluminum interference becomes negligible at higher regions of the flame and titanium interference in the lower regions of the flame becomes minimum or negligible. Phosphoric and sulfuric acids have also been reported to cause depression in tin absorbance. No interferences have been observed in the nitrous oxide-acetylene flame.

Standard Solution

To prepare 1000 mg/ℓ solution, dissolve 1.0000 g of tin metal in 100 mℓ of concentrated hydrochloric acid and dilute to 1 ℓ with deionized water. All further dilutions of the stock solutions should be done with 10% (v/v) hydrochloric acid to prevent tin precipitation.

Instrumental Parameters

Wavelength	235.5 nm
Slit	0.4 nm
Light source	Hollow cathode lamp
Flame	Nitrous oxide-acetylene, reducing (rich and red)
Sensitivity	1.2 mg/ℓ
Detection limit	0.03 mg/ℓ
Optimum range	100 mg/ℓ

Secondary wavelengths and their sensitivities

235.5 nm	1.2—2.2 mg/ℓ
286.3 nm	1.4—3.2 mg/ℓ
270.6 nm	2.0—4.0 mg/ℓ
303.4 nm	3.0—5.0 mg/ℓ
219.9 nm	4.0—7.3 mg/ℓ
254.7 nm	4.0—9.4 mg/ℓ
233.5 nm	5.0—9.2 mg/ℓ
300.9 nm	5.0—9.2 mg/ℓ
266.1 nm	22.0—37.0 mg/ℓ

REFERENCES

1. **Slavin, W., Sprague, S., and Manning, D. C.,** Detection limits in analytical atomic absorption spectrophotometry, *At. Absorpt. Newsl.,* 18, 1964.
2. **Means, E. A. and Ratcliffe, D.,** Determination of wear metals in lubricating oils by atomic absorption spectroscopy, *At. Absorpt. Newsl.,* 4, 174, 1965.
3. **Sattur, T. W.,** Routine atomic absorption analysis on non-flame alloys and plant intermediates, *At. Absorpt. Newsl.,* 5, 37, 1966.
4. **Vollmer, J.,** Molten-tin hollow cathode lamps, *At. Absorpt. Newsl.,* 5, 35, 1966.
5. **Simpson, G. R. and Blay, R. A.,** Rapid method for determination of the metals Cu, Zn, Sn, Fe and Ca in foodstuffs by atomic absorption spectroscopy, *Food Trade Rev.,* August 1966.
6. **Slavin, S. and Slavin, W.,** Fully automatic analysis of used aircraft oils, *At. Absorpt. Newsl.,* 5, 106, 1966.
7. **Dagnall, R. M., Thompson, K. C., and West, T. S.,** The use of microwave excited electrodeless discharge tubes as spectral sources in atomic absorption spectroscopy, *At. Absorpt. Newsl.,* 6, 117, 1967.
8. **Bowman, J. A.,** The determination of tin ores and concentrates by atomic absorption spectrophotometry in the nitrous oxide-acetylene flame, *Anal. Chim. Acta,* 42, 285, 1968.
9. **Schallis, J. E. and Kahn, H. L.,** The determination of tin in lubricating oils with a nitrous oxide-acetylene flame, *At. Absorpt. Newsl.,* 7, 84, 1968.
10. **Harrison, W. W. and Juliano, P. A.,** Effects of organic solvents on tin absorbance in an air-hydrogen flame, *Anal. Chem.,* 41, 1016, 1969.
11. **McCracken, J. D., Vecchione, M. C., and Longo, S. L.,** Ion exchange separation of traces of tin, calcium and zinc from copper and their determination by atomic absorption spectrophotometry, *At. Absorpt. Newsl.,* 8, 102, 1969.
12. **Price, W. J. and Roos, J. T. H.,** Analysis of fruit by atomic absorption spectrophotometry. Determination of iron and tin in canned juice, *J. Sci. Food Agric.,* 20, 437, 1969.
13. **Burke, K. E.,** Study of the scavanger properties of manganese (IV) oxide with atomic absorption spectrometry. Determination of microgram quantities of Sb, Bi, Pb, and Sn in nickel, *Anal. Chem.,* 42, 1536, 1970.
14. **Juliano, P. O. and Harrison, W. W.,** Atomic absorption interferences of tin, *Anal. Chem.,* 42, 84, 1970.
15. **Levine, J. R., Moore, S. G., and Levine, S. L.,** Effect of potassium on determination of tin by atomic absorption spectrophotometry, *Anal. Chem.,* 42, 412, 1970.
16. **Moldan, B., Rubeska, I., Miksovsky, M., and Huka, M.,** Determination of tin in geological materials by atomic absorption spectrophotometry, *Anal. Chim. Acta,* 52, 91, 1970.
17. **Vickers, T. J., Cottrell, C. R., and Breakey, D. W.,** Atomization and excitation processes for tin in turbulent diffusion hydrogen-air flames, *Spectrochim. Acta,* 25b, 437, 1970.
18. **Zook, E., Greene, F., and Morris, E.,** Nutrient composition of selected wheats and wheat products. Distribution of manganese, copper, nickel, zinc, magnesium, lead, tin, cadmium, chromium and selenium as determined by atomic absorption spectroscopy and colorimetry, *Cereal Chem.,* 47, 720, 1970.
19. **Pearlman, R. S., Hefferren, J. J., and Lyon, H. W.,** Determination of tin in enamel and dentin by atomic absorption spectroscopy, *J. Dent. Res.,* 49, 1437, 1970.
20. **Clauss, C., Laugel, P., and Hasselmann, M.,** Quick determination of the tin content in tin plate by using simultaneously electrolytes and atomic absorption, *Chim. Anal. (Paris),* 53, 102, 1971.
21. **Mastelerz, M. and Raczynska, D.,** Quantitative determination of tin in slags, flue dusts and zinc concentrates by atomic absorption spectrophotometry, *Chem. Anal. (Warsaw),* 16, 85, 1971.
22. **Matsuo, T., Shida, J., and Nakamura, C.,** Determination of tin in lead alloys by atomic absorption spectrometry, *Bunseki Kagaku,* 20, 697, 1971.
23. **Kono, T.,** Determination of tin by atomic absorption spectroscopy, *Bunseki Kagaku,* 20, 552, 1971.
24. **Aue, W. A.,** A tin-sensitive hydrogen flame detector, *J. Chromatogr.,* 70, 158, 1972.
25. **Burke, K. E.,** Determination of microgram amounts of antimony, bismuth, lead and tin in aluminum, iron and nickel-base alloys by non-aqueous atomic absorption spectroscopy, *Analyst (London),* 97, 19, 1972.
26. **Nakahara, T., Munemori, M., and Musha, S.,** Atomic absorption spectrometric determination of tin in premixed inert gas (entrained air)- hydrogen flame, *Anal. Chim. Acta,* 62, 267, 1972.
27. **Gouin, J. U., Holt, J. L., and Miller, R. E.,** Determination of tin and antimony in type metal using atomic absorption spectrophotometry, *Anal. Chem.,* 44, 1042, 1972.
28. **Headridge, J. B. and Sowerbutts, A.,** Determination of tin in steels by solvent extraction followed by atomic absorption spectrophotometry, *Analyst (London),* 97, 422, 1972.
29. **Engberg, A.,** A comparison of a spectrophotometric (Quercetin) method and an atomic absorption method for the determination of tin in food, *Analyst (London),* 98, 137, 1973.
30. **Williams, A. I.,** The determination of bis(tri-n-butyl-tin)oxide and di-n-butyltin oxide in preserved softwood by atomic absorption spectrophotometry and polarography, *Analyst (London),* 98, 233, 1973.

31. **George, G. M., Albrecht, M. A., Frahm, L. J., and McDonnel, J. P.,** Atomic absorption spectrophotometric determination of dibutyl tin dilaurate in finished area, *J. Assoc. Off. Anal. Chem.,* 56, 1480, 1973.
32. **Woidich, H. and Pfannhauser, W.,** Distribution of tin between solid and liquid contents of some cans of vegetables and fruits, *Dtsch. Lebensm. Rundsch.,* 69, 224, 1973.
33. **Woidich, H. and Pfannhauser, W.,** Comparison of three methods for the determination of tin in canned food, *Z. Lebensm. Unters. Forsch.,* 151, 114, 1973 (in German).
34. **Faye, G. H., Bowman, W. S., and Sutarno, R.,** Result of an inter-laboratory program for the determination of tin in a standard reference ore: a cuvet, *Anal. Chim. Acta,* 67, 202, 1973.
35. **Quarrell, T. M., Powell, R. J. W., and Cluley, H. J.,** Determination of tin and antimony in lead alloy for cable sheathing by atomic absorption spectroscopy, *Analyst (London),* 98, 443, 1973.
36. **Sato, N., Tsuruta, K., Kamada, I., Narita, S., and Abukawa, H.,** Determination of dissolved tin in canned foods by atomic absorption spectrophotometry, *J. Food Hyg. Soc. Jpn.,* 14, 245, 1973 (in Japanese).
37. **Everett, G. L., West, T. S., and Williams, R. W.,** The determination of tin by carbon filament atomic absorption spectrometry, *Anal. Chim. Acta,* 70, 291, 1974.
38. **Mensik, J. D. and Seidemann, H. J., Jr.,** Determination of tin in mineralized rocks and ores by atomic absorption spectrophotometry, *At. Absorpt. Newsl.,* 13, 8, 1974.
39. **Peetre, I. B. and Smith, B. E. F.,** Sensitivity of determination of organosilicon and tin compounds by atomic absorption spectroscopy, *Mikrochim. Acta,* 2, 301, 1974.
40. **Rubeska, I.,** Tin atomization in hydrogen supported flames, *Spectrochim. Acta,* 29b, 263, 1974.
41. **Umland, F. and Schumacher, E.,** Highly sensitive indirect determination of tin by atomic absorption spectrometry by the mercury cold vapor method, *Z. Anal. Chem.,* 269, 367, 1974.
42. **Thompson, K. C. and Thomerson, D. R.,** Atomic absorption studies on the determination of antimony, arsenic, bismuth, germanium, lead, selenium, tellurium and tin by utilizing the generation of covalent hydrides, *Analyst (London),* 99, 595, 1974.
43. **Thornton, K. and Burke, K. E.,** Modification to the extraction-atomic absorption method for the determination of antimony bismuth, lead and tin, *Analyst (London),* 99, 469, 1974.
44. **Brown, R. and Husain, D.,** Collisional quenching of electronically excited tin atoms Sn (S^1D_2) by time resolved atomic absorption spectroscopy, *Int. J. Chem. Kinet.,* 7, 77, 1975.
45. **Condylis, A. and Mejean, B.,** Recent applications of atomic absorption to metallurgical analysis. Determination of low concentrations of the elements aluminum, barium, titanium, tin, antimony and tantalum, *Analusis,* 3, 94, 1975.
46. **Harrington, D. E. and Bramstedt, W. R.,** The determination of tin, antimony and tantalum in the presence of precious metals by atomic absorption spectroscopy, *At. Absorpt. Newsl.,* 14, 36, 1975.
47. **Ratcliffe, D. B., Byford, C. S., and Osman, P. B.,** The determination of arsenic, antimony and tin in steels by flameless atomic absorption spectrometry, *Anal. Chim. Acta,* 75, 457, 1975.
48. **Schmidt, F. J., Royer, J. L., and Muir, S. M.,** Automated determination of arsenic, selenium, antimony, bismuth and tin by atomic absorption utilizing sodium borohydride reduction, *Anal. Lett.,* 8, 123, 1975.
49. **L'vov, B. V., Kruglikova, L. P., Polzik, L. K., and Katskov, D. A.,** Theory of flame atomic absorption analysis. V. Behavior of tin in the hydrogen-air flame on the basis of a thermodynamic calculation of temperature and composition of the combustion products, *Zh. Anal. Khim.,* 30, 1045, 1975.
50. **Meranger, J. C.,** A rapid screening method for the determination of di-(n-octyl) tin stabilizers in alcoholic beverages, using a heated graphite atomizer, *J. Assoc. Off. Anal. Chem.,* 58, 1143, 1975.
51. **Ostroumenko, P. P. and Eremenko, A. M.,** Measurement of relative values of oscillator strengths in spectra of iron and tin atoms by the absorption method, *Opt. Spektrosk.,* 39, 413, 1975.
52. **Terashima, S.,** Determination of microamounts of tin in silicates by atomic absorption spectrometry with an argon-hydrogen flame, *Bunseki Kagaku,* 24, 319, 1975.
53. **Barnett, W. B. and McLaughlin, E. A.,** The atomic absorption determination of antimony, arsenic, bismuth, cadmium, lead and tin in iron, copper and zinc alloys with graphite furnace, *Anal. Chim. Acta,* 80, 285, 1975.
54. **Hocquellet, P. and Labevrie, N.,** Thermoelectric atomization of silicon, germanium, tin and lead on tantalum carbide. Application to atomic absorption spectrometry, *Analusis,* 3, 505, 1975.
55. **Musil, J.,** Determination of small amounts of lead or tin in ferrous metals and nickel by atomic absorption spectrometry, *Hutn. Listy,* 30, 292, 1975 (in Czech).
56. **Acatini, C., deBerman, S. N., Colombo, O., and Fondo, O.,** Determination of silver, copper, lead, tin, antimony iron, zinc, magnesium, potassium, and manganese in canned tomatoes by atomic absorption spectrophotometry, *Rev. Asoc. Bioquim. Argent.,* 40, 175, 1975 (in Spanish).
57. **Shelton, B. J.,** Determination of silver, selenium, tellurium, antimony, tin, lead and arsenic in anode sludges, *Natl. Inst. Metall. Repub. S. Afr. Rep.,* No. 1771, 1975.
58. **Bedard, M. and Kerbyson, J. D.,** Determination of arsenic, selenium, tellurium and tin in copper by hydride evolution atomic absorption spectrophotometry, *Can. J. Spectrosc.,* 21, 64, 1976.

59. **Christmann, D. R. and Ingle, J. D., Jr.,** Cold vapor mercury-coupled atomic absorption determination of tin, *Anal. Chim. Acta,* 86, 285, 1976.

60. **Vijan, P. and Chan, C. Y.,** Determination of tin by gas phase atomization and atomic absorption spectrometry, *Anal. Chem.,* 48, 1788, 1976.

61. **Guimont, J., Bouchard, A., and Pichette, M.,** Determination of tin in sediments by atomic absorption spectrophotometry, *Talanta,* 23, 62, 1976 (in French).

62. **Joshi, M. M. and Gopal, R.,** Spectroscopic investigation of flame test for tin and bismuth, *Indian J. Pure Appl. Phys.,* 14, 325, 1976.

63. **Thompson, K. C. and Godden, R. G.,** The application of a wide slot nitrous oxide-nitrogen-acetylene burner for the atomic absorption spectrophotometric determination of aluminum, arsenic and tin in steels by the single-pulse nebulization technique, *Analyst (London),* 101, 96, 1976.

64. **Welsch, E. P. and Chao, T. T.,** Determination of trace amounts of tin in geological materials by atomic absorption spectrometry, *Anal. Chim. Acta,* 82, 337, 1976.

65. **Garcia Olmedo, R., Garcia Puertas, P., Salesa Perez, M., Masoud, T. A., and Rubio, F. J.,** Determination of tin in fruit juices by atomic absorption spectrophotometry, *An. Bromatol.,* 28, 1, 1976 (in Spanish).

66. **Wehrer, C., Thiersault, J., and Laugel, P.,** Determination of tin in canned fruits and vegetables by atomic absorption spectrometry, *Ind. Aliment. Agric.,* 93, 1439, 1976 (in French).

67. **Marks, J. Y., Welcher, G. G., and Spellman, R. J.,** Atomic absorption determination of lead, bismuth, selenium, tellurium, thallium and tin in complex alloys using direct atomization from metal chips in the graphite furnace, *Appl. Spectrosc.,* 31, 9, 1977.

68. **Condylis, A. and Hocquaux, H.,** Atomic absorption spectrophotometric determination of arsenic, selenium and tin in metallurgical mixtures, *Analusis,* 5, 228, 1977.

69. **Del Monte, M. G. and Luperi, N.,** Determination of tin in plain carbon, stainless and high-speed steels by atomic absorption spectrophotometry with electrothermal atomization, *Analyst (London),* 102, 489, 1977.

70. **Hocquellet, P. and Labevrie, N.,** Determination of tin in foods by flameless atomic absorption, *At. Absorpt. Newsl.,* 16, 24, 1977.

71. **Trachman, H. L., Tyberg, A. J., and Branigan, P. D.,** Atomic absorption spectrometric determination of subpart per million quantities of tin in extracts and biological materials with a graphite furnace, *Anal. Chem.,* 49, 1090, 1977.

72. **Clay, A. F.,** Rapid determination of antimony in lead-tin base materials by atomic absorption spectrophotometry, *Lab. Pract.,* 26, 690, 1977.

73. **Dojmovic, M.,** Determination of bismuth, mercury, antimony, tin, tellurium and lead in aqueous solutions by flameless atomic absorption spectroscopy, *GIT Fachz. Lab.,* 21, 861, 1977.

74. **Boldyreva, N. N. and Malakov, V. V.,** Atomic absorption determination of tin in tin-platinum catalysts, *Metody Anal. Kontrolya Proizvod. Khim. Promst.,* 11, 30, 1977.

75. **Varnes, A. W. and Gaylor, V. F.,** Determination of dioctyltin stabilizers in food-stimulating solvents by atomic absorption spectrometry with electrothermal atomization, *Anal. Chim. Acta,* 101, 393, 1978.

76. **Matsuzaki, J., Miyoshi, H., and Takeshita, R.,** Determination of tin as stannane by atomic absorption spectrophotometry with use of an argon-hydrogen system, *Eisei Kagaku,* 24, 299, 1978 (in Japanese).

77. **Rees, D. I.,** A simple procedure for the determination of arsenic and tin in food by atomic absorption spectrophotometry, *J. Assoc. Publ. Anal.,* 16, 71, 1978.

78. **Amakawa, E., Ohnishi, K., Taguchi, N., and Seki, H.,** Determination of tin in soft drinks by flameless atomic absorption spectroscopy using a method of coprecipitation with zirconium hydroxide, *Bunseki Kagaku,* 27, 81, 1978.

79. **Boiteau, H. L. and Metayer, C.,** Microdetermination of lead, cadmium, zinc and tin in biological materials by atomic absorption spectrometry after mineralization and extraction, *Analusis,* 6, 350, 1978.

80. **Evans, W. H., Jackson, F. J., and Dellar, D.,** Evaluation of a method for determination of total antimony, arsenic, and tin in food stuffs using measurement by atomic absorption spectrophotometry with atomization in a silica tube using the hydride generation technique, *Analyst (London),* 104, 16, 1979.

81. **Frotzsche, H., Wegscheider, W., Knapp, G., and Ortner, H. M.,** A sensitive atomic absorption spectrometric method for the determination of tin with atomization from impregnated graphite surfaces, *Talanta,* 26, 219, 1979.

82. **Nyagah, C. G. and Wandiga, S. O.,** Use of complexing ligands in the determination of antimony and tin by atomic absorption spectrometry, *Talanta,* 26, 333, 1979.

83. **Palliere, M. and Gernez, G.,** Apparatus for the production of volatile hydrides for arsenic, antimony, bismuth, tin, germanium, selenium and tellurium determination by atomic absorption spectrometry, *Analusis,* 7, 46, 1979.

84. **Pyen, G. and Fishman, M.,** Automated determination of tin in water, *At. Absorpt. Newsl.,* 18, 34, 1979.

85. **Elkins, E. R. and Sulek, A.,** Atomic absorption determination of tin in foods: collaborative study, *J. Assoc. Off. Anal. Chem.,* 62, 1050, 1979.

86. **Hodge, V. F., Seidel, S. L., and Goldberg, E. D.,** Determination of tin (IV) and organo tin compounds in natural waters, coastal sediments and macro algae by atomic absorption spectrometry, *Anal. Chem.,* 51, 1256, 1979.

87. **Kojima, S.,** Separation of organo tin compounds by using the difference in partition behavior between hexane and methanolic buffer solution. I. Determination of butyl tin compounds in textiles by graphite furnace atomic absorption spectrophotometry, *Analyst (London),* 104, 660, 1979.

88. **Masunaga, K., Okada, M., and Miyagawa, H.,** Determination of trace tin by combined thorium hydroxide coprecipitation-flameless atomic absorption spectrophotometry, *Nippon Kagaku Kaishi,* 8, 1050, 1979.

89. **Nakashima, S.,** The separation of tin from water and sea water by flotation and the determination of tin by atomic absorption spectrophotometry following stannane generation, *Bull. Chem. Soc. Jpn.,* 52, 1844, 1979.

90. **Ohta, K. and Suzuki, M.,** Atomic absorption spectrometry of tin with electrothermal atomization in a molybdenum microtube, *Anal. Chim. Acta,* 107, 245, 1979.

91. **Sano, M., Furukawa, M., Kouri, M., and Tomota, I.,** Flameless atomic absorption spectrophotometric determination of vendex an inorganic tin miticide, in apples, oranges and tea leaves, *J. Assoc. Off. Anal. Chem.,* 62, 764, 1979.

92. **Tominaga, M. and Umezaki, Y.,** Determination of submicrogram amounts of tin by atomic absorption spectrometry with electrothermal atomization, *Anal. Chim. Acta,* 110, 55, 1979.

93. **Boyer, K. W., Capar, S. G., Jones, J. W., Suddendorf, R. F., and Forwalter, J.,** Multielement/trace analysis identifies 48 elements in foods. Simultaneous element analysis and data reduction reaches parts per billion, *Food Process.,* 40, 72, 1979.

94. **Graf-Harsanyi, E., Polos, L., and Pungor, E.,** Investigation of the atomization processes of tin in various atomizers and of the interference by copper, *Acta Chim. Acad. Sci. Hung.,* 101, 139, 1979.

95. **Welz, B. and Melcher, M.,** Hydride atomic absorption spectroscopic technique in trace analysis, *Labor Prixis,* 3, 41, 1979 (in German).

96. **Nakashima, S.,** Determination of tin by long absorption cell atomic absorption spectrophotometry following stannane generation, *Ber. Ohara Inst. Landwirtsch. Biol. Okayama Univ.,* 17, 187, 1979.

97. **Fricke, F. L., Robbins, W. B., and Caruso, J. A.,** Trace element analysis of food and beverages by atomic absorption spectrometry, *Prog. Anal. At. Spectrosc.,* 2, 185, 1979.

98. **Miyagawa, H.,** Determination of trace tin by a combined coprecipitation flameless atomic absorption spectrophotometry, *Nenpo-Fukui-Ken Kogyo Shikenjo,* 53, 168, 1979 (in Japanese).

99. **Hahn, M. H., Mulligan, K. J., Jackson, M. E., and Caruso, J. A.,** Sequential determination of arsenic, selenium, germanium and tin as their hydrides by gas-solid chromatography with an atomic absorption detector, *Anal. Chim. Acta,* 118, 115, 1980.

100. **Hall, A.,** Determination of tin content of some geological materials by atomic absorption spectrophotometry, *Chem. Geol.,* 30, 135, 1980.

101. **Knezevic, G. and Hueppe, K.,** Digestion methods for determination of lead, iron and tin in canned soups by atomic absorption spectrophotometry, *Dtsch. Lebensm. Rundsch.,* 76, 50, 1980.

102. **Sterritt, R. M. and Lester, J. N.,** Determination of silver, cobalt, manganese, molybdenum and tin in sewage sludge by a rapid electrothermal atomic absorption spectroscopic method, *Analyst (London),* 105, 616, 1980.

103. **Hon, P. K., Lau, O. W., Cheung, W. C., and Wong, W. C.,** The atomic absorption spectrometric determination of arsenic, bismuth, lead, antimony, selenium and tin with a flame heated silica T-tube after hydride generation, *Anal. Chim. Acta,* 115, 355, 1980.

104. **Itsuki, K. and Ikeda, T.,** Atomic absorption spectrometry of tin using graphite furnace with the aid of lanthanum, *Bunseki Kagaku,* 29, 309, 1980.

105. **Zakhariya, A. N., Olenovich, N. L., and Khutornoi, A. M.,** Atomic absorption determination of tin, lead, antimony and bismuth in copper products, *Ukr. Khim. Zh.,* 46, 421, 1980.

106. **Brown, J. R., Saba, C. S., Rhine, W. E., and Eisentraut, K. J.,** Particle size independent spectrometric determination of wear metals in aircraft lubricating oils, *Anal. Chem.,* 52, 2365, 1980.

107. **Clegg, J. B., Grainger, F., and Gale, I. G.,** Quantitative measurement of impurities in gallium arsenide, *J. Mater. Sci.,* 15, 747, 1980.

108. **Donaldson, E. M.,** Determination of tin in ores, iron, steel and nonferrous alloys by atomic absorption spectrophotometry after separation by extraction as the iodide, *Talanta,* 27, 499, 1080.

109. **Subramanian, K. S. and Sastri, V. S.,** A rapid hydride evolution electrothermal atomic absorption method for the determination of tin in geological materials, *Talanta,* 27, 469, 1980.

110. **Vickrey, T. M., Howell, H. E., Harrison, G. V., and Ramelow, G. J.,** Post column digestion methods for liquid chromatography-furnace atomic absorption speciation of organo lead and organo tin compounds, *Anal. Chem.,* 52, 1743, 1980.

111. **Chrichton, T. J., Farr, J. P. G., Russon, J., Saremi, M., Pountney, J., Bentley, A. J., and Earwaker, L. G.,** Analysis of annealed electrotinned steel, *Anal. Proc. (London),* 17, 479, 1980.

112. **Glenc, T., Jurczyk, J., and Robosz-Kabza, A.,** Determination of tin and lead in transformer carbon and alloy steels in the range from 0.0005 to 0.04% by atomic absorption method with electrothermal atomization, *Chem. Anal.,* 25, 515, 1980.

113. **Hughes, M. J.,** Analysis of Roman tin and pewter ingots, *Occas. Pap. British Museum,* 17, 41, 1980.

114. **Katskov, D. A., Grinshtein, I. L., and Kruglikova, L. P.,** Study of the evaporation of the metals indium, gallium, thallium, germanium, tin, lead, antimony, bismuth, selenium and tellurium from a graphite surface by the atomic absorption method, *Zh. Prikl. Spektrosk.,* 33, 804, 1980.

115. **Lindh, U., Brune, D., Nordberg, G., and Wester, P. O.,** Levels of antimony, arsenic, cadmium, copper, lead, mercury, selenium, silver, tin, and zinc in bone tissue of industrially exposed workers, *Sci. Total Environ.,* 16, 109, 1980.

116. **Norman, E. A., Orlova, E. S., and Shestakova, A. I.,** Determination of arsenic, tin and antimony impurities in steel by atomic absorption with electrothermal atomization, *Zavod. Lab.,* 46, 1108, 1980.

117. **Matsubara, K., Ota, N., Taniguchi, M., and Narita, K.,** Atomic absorption spectrophotometric determination of small amounts of arsenic, tin and aluminum in steels by the hydride generation method and the injection method, *Trans. Iron Steel Inst. Jpn.,* 20, B406, 1980.

118. **Vickrey, T. M., Harrison, G. V., and Ramelow, G. J.,** Treated graphite surfaces for determination by graphite furnace atomic absorption spectrometry, *Anal. Chem.,* 53, 1573, 1981.

119. **Carleer, R., Francois, J. P., and Van Poucke, L. C.,** Determination of the main, minor and trace elements in lead/tin based solder by atomic absorption spectrophotometry, *Bull. Soc. Chim. Belg.,* 90, 357, 1981.

120. **Clark, J. R. and Viets, J. G.,** Multielement extraction system for the determination of 18 trace elements in geochemical samples, *Anal. Chem.,* 53, 61, 1981.

121. **Clark, J. R. and Viets, J. G.,** Back-extraction of trace elements from organometallic-halide extracts for determination by flameless atomic absorption spectrometry, *Anal. Chem.,* 53, 65, 1981.

122. **Kahn, H. L., Cristiano, L. C., Dulude, G. R., and Sotera, J. J.,** Automated hydride generation, *Am. Lab.,* 13, 136, 138, 141, 1981.

123. **Welz, B., Grobenski, Z., and Melcher, M.,** Determination of antimony, arsenic, selenium, tellurium, bismuth and tin using the hydride-AAS technique, *Spektrometertagung (Votr.),* 13, 337, 1981.

124. **Ogihara, K., Chiba, M., and Kikuchi, M.,** Determination of plasma tin by flameless atomic absorption spectrometry, *Sangyo Igaku,* 23, 420, 1981.

125. **Takahashi, K., Minami, S., and Ohyagi, Y.,** Determination of copper and tin in antifouling coatings for ship bottom by atomic absorption spectrophotometry, *Shikizai Kyokaishi,* 54, 606, 1981.

126. **Guo, X. and Wang, S.,** A new type of double capillary nebulizer used in atomic absorption analysis, *Fen Hsi Hua Hsueh,* 9, 258, 1981.

127. **Dabeka, R. W. and McKenzie, A. D.,** Atomic absorption spectrometric determination of tin in canned foods using nitric acid-hydrochloric acid digestion and nitrous oxide-acetylene flame, *J. Assoc. Off. Anal. Chem.,* 64, 1297, 1981.

128. **Adelman, H., Jenniss, S. W., and Katz, S. A.,** Interlaboratory analysis of sewage sludge, *Am. Lab.,* 31, 1981.

129. **Gladwell, D., Thompson, M., and Wood, S. J.,** The determination of cassiterite tin in soils and sediments by an atomic absorption-volatile hydride method, *J. Geochem. Explor.,* 16, 41, 1981.

130. **Young, R. S.,** Analysis of nickel refinery slimes and residues, *Talanta,* 28, 25, 1981.

131. **Welz, B. and Melcher, M.,** Determination of antimony, arsenic, bismuth, selenium, tellurium and tin in metallurgical samples using the hydride AA technique. I. Analysis of low alloy steels, *Spectrochim. Acta,* 36b, 439, 1981.

132. **Vickrey, T. M., Harrison, G. V., and Ramelow, G. R.,** Treated graphite surfaces for determination of tin by graphite furnace atomic absorption spectrometry, *Anal. Chem.,* 53, 1573, 1981.

133. **Ikeda, M., Nishibe, J., and Nakahara, T.,** Study of some fundamental conditions for the determination of antimony, tin, selenium and tellurium with continuous hydride generation-atomic absorption spectrometry, *Bunseki Kagaku,* 30, 545, 1981.

134. **Burns, D. T., Glockling, F., and Harriott, M.,** Investigations of the determination of tin tetraalkyls and alkyl tin chlorides by atomic absorption spectrometry after separation by gas-liquid or high performance liquid-liquid chromatography, *Analyst (London),* 106, 921, 1981.

135. **Brodie, K. G. and Rowland, J. J.,** Trace analysis of arsenic, lead and tin, *Eur. Spectrosc. News,* 36, 41, 1981.

136. **Senesi, N. and Polemio, M.,** Trace element addition to soil by application of NPK fertilizers, *Fert. Res.,* 2, 289, 1981.

137. **Kaiser, M. L., Koirtyohann, S. R., Hinderberger, E. J., and Taylor, H. E.,** Reduction of matrix interferences in furnace atomic absorption with the L'vov platform, *Spectrochim. Acta,* 36b, 773, 1981.

138. **Koroscetz, F., Hoke, E., and Grasserbauer, M.,** Diffusion processes in electrodeposited lead-tin-copper bearing overlays, *Mikrochim. Acta Suppl.,* 9, 139, 1981 (in German).

139. **Jones, E. A. and Dixon, K.,** The separation of trace elements in manganese dioxide, *Natl. Inst. Metall. Repub. S. Afr. Rep.,* No. 2131, 1981.

140. **Korolev, N. V., Rodionova, M. S., Belova, I. V., and Kovalenko, N. K.,** Comparative evaluation of trace element content in fungicidal coatings, *Opt. Mikh. Promst.,* 11, 28, 1981 (in German).

141. **Takahashi, Y., Shiozawa, Y., Kawai, T., Shimizu, K., Kato, H., and Hasegawa, J.,** Elements dissolved from three types of amalgams in synthetic saliva, *Aichi Gakuin Daigaku Shigakkaishi,* 19, 107, 1981.

142. **Monteil, A. and Welte, B.,** Comparison of the different methods of attack during the determination of micropollutants in the sediments, Congr. Mediterr. Ing. Quim., (Actas), 2nd C20-1/C20-13, F.O.I.M., Barcelona, Spain, 1981 (in French).

143. **Iwai, H., Wada, O., and Arakawa, Y.,** Determination of tin- and di- and monobutyltin and inorganic tin in biological materials and some aspects of their metabolism in rats, *J. Anal. Toxicol.,* 5, 300, 1981.

144. **Mendiola Ambrioso, J. M. and Gonzalez Lopez, A.,** Analysis for tin in PVC composites by atomic absorption spectrophotometry, *Rev. Plast. Mod.,* 41, 550, 1981 (in Spanish).

145. **Marr, I. L. and Anwar, J.,** Microdetermination of tin in organotin compounds by flame-emission and atomic absorption spectrophotometry, *Analyst (London),* 107, 260, 1982.

146. **Thibaud, Y.,** Analysis of tin by atomic absorption spectrophotometry with electrothermal furnace, *Rev. Trav. Inst. Peches Marit.,* 44, 349, 1980 (in French); CA 97, 34245b, 1982.

147. **Voronkova, M. A., Antonova, E. A., and Gorshkov, V. V.,** Determining Tin and Molybdenum, USSR, SU 903,297 (Cl. C01G19/00), 07 Feb. 1982; appl. 80/100,525, July 24, 1980.

148. **Kauffman, R. E., Saba, C. S., Rhine, W. E., and Eisentraut, K. J.,** Quantitative multielement determination of metallic wear species in lubricating oils and hydraulic fluids, *Anal. Chem.,* 54, 975, 1982.

149. **Tsukahara, I. and Yamamoto, T.,** Determination of tin in copper, lead, nickel, selenium and aluminum metals and in copper and lead alloys by flame atomic absorption spectrometry after extraction of trioctyl-methylammonium hexachlorostannate (IV), *Anal. Chim. Acta,* 135, 235, 1982.

150. **Nadkarni, R. A.,** Applications of hydride generation-atomic absorption spectrometry to coal analysis, *Anal. Chim. Acta,* 135, 363, 1982.

151. **Harley, M. L.,** Determination for tin in geochemical samples by nonaqueous atomic absorption spectrophotometry, *At. Spectrosc.,* 3, 76, 1982.

152. **Matsumoto, K., Nishio, M., Misaki, Y., and Terada, K.,** Decomposition of tin (IV) oxide, antimony (III) oxide and bismuth (III) oxide by fusion with ammonium iodide and its application for analysis of environmental samples, *Bunseki Kagaku,* 31, 141, 1982.

153. **Chau, Y. K., Wong, P. T. S., and Bengert, G. A.,** Determination of methyltin (IV) and tin (IV) species in water by gas chromatography-atomic absorption spectrophotometry, *Anal. Chem.,* 54, 246, 1982.

154. **Graham, P. P., Bittel, R. J., Bovard, K. P., Lopez, A., and Williams, H. L.,** Mineral element composition of bovine spleen and separated spleen components, *J. Food Sci.,* 47, 720, 1982.

155. **Chan, C. Y. and Baig, M. W. A.,** Semiautomatic method for determination of total tin in rocks, *Anal. Chim. Acta,* 136, 413, 1982.

156. **Rayson, G. D. and Holcombe, J. A.,** Tin atom formation in a graphite furnace atomizer, *Anal. Chim. Acta,* 136, 249, 1982.

157. **Terashima, S.,** Determination of trace amounts of tin in seventy three geochemical reference samples by atomic absorption spectrometry, *Geostandards Newsl.,* 6, 77, 1982.

AUTHOR INDEX

SUBJECT INDEX (TIN)

LEAD, Pb (ATOMIC WEIGHT 207.2)

Lead occurs in the combined state in a large number of minerals and is believed to be the oldest metal associated with the planet Saturn. It has been used since ancient times. Lead is very resistant to corrosion and therefore is used in the manufacture of containers for corrosive liquids and lead pipes. It is also used in storage batteries, cable coverings, plumbing, ammunitions, and in the manufacture of lead tetraethyl, which is used as an antiknocking compound in gasoline. More recent uses are in radiation shields for X-ray and nuclear reactors and vibration absorbing pads for high-speed machinery. Lead compounds are used in paint pigments, drying agents for oils, for producing crystal glass, plumber's cement, and as a protective coating. Lead is also used in medicine.

The determination of lead is required in a large number of commercially important products such as lead mattes, slags, insecticides, paint pigments, alloys and ores, etc. Lead is a cumulative poison and is analyzed in food products, environmental samples, soil, water, and biological products for its toxicity.

Sample preparation is very important for lead analysis. Dissolution is done by using dilute nitric acid and expelling the acid by heating with sulfuric acid. This method may cause lower results. Fusion with sodium carbonate and precipitation of lead as lead sulfide are the routine methods. Lead sulfide precipitate is filtered, washed, and dissolved in dilute acid.

In the atomic absorption method of analysis, large excess of other metals such as 10,000 mg/ℓ iron may cause interference. No other cation interferences are reported in air-acetylene flame. Interferences due to acetate, carbonate, fluoride, iodide, and phosphate concentrations when present in greater amounts than that of lead, suppress absorbance significantly. These interferences can be reduced by preparing all the solutions including blank solutions in 0.1 M EDTA.

Standard Solution

To prepare 1000 mg/ℓ lead solution, dissolve 1.0000 g of metallic lead or 1.598 g of lead nitrate ($Pb(NO_3)_2$) in 50 mℓ of 1:1 nitric acid and dilute to 1 ℓ with deionized water.

Instrumental Parameters

Wavelength	217.0 nm
Slit	0.7 nm
Light source	Hollow cathode lamp
Flame	Air-acetylene, oxidizing (lean, blue)
Sensitivity	0.1—0.4 mg/ℓ
Detection limit	0.03 mg/ℓ
Optimum range	20 mg/ℓ

Secondary wavelengths and their sensitivities

283.3 nm	0.2—0.5 mg/ℓ
261.4 nm	3.5—11.0 mg/ℓ
202.2 nm	5.0—7.1 mg/ℓ
368.4 nm	5.0—27.0 mg/ℓ
205.3 nm	5.4—34.0 mg/ℓ
364.0 nm	67 mg/ℓ

REFERENCES

1. **Zeeman, P. B. and Butler, L. R. P.,** Determination of lead in wine by the method of atomic absorption, *Tegnikon,* 13, 96, 1960.
2. **Elwell, W. T. and Gidley, J. A. F.,** The determination of lead in copper-base alloys and steel by atomic absorption spectrophotometry, *Anal. Chim. Acta,* 24, 71, 1961.
3. **Robinson, J. W.,** Determination of lead in gasoline atomic absorption spectroscopy, *Anal. Chem.,* 32, 17A, 1961.
4. **Willis, J. B.,** Determination of lead in urine by atomic absorption spectroscopy, *Nature (London),* 191, 381, 1961.
5. **Elwell, W. T. and Gidley, J. A. F.,** *Atomic Absorption Spectrophotometry,* Rev. 2nd ed., Pergamon Press, New York, 1961, 1962.
6. **Schuler, V. C. O., Jansen, A. V., and James, G. S.,** The development of atomic absorption methods for the determination of silver, copper, iron, lead and zinc in high purity gold and the role of organic additives, *J. S. Afr. Inst. Min. Metal.,* 62, 807, 1962.
7. **Willis, J. B.,** Determination of lead and other heavy metals in urine by atomic absorption spectroscopy, *Anal. Chem.,* 34, 614, 1962.
8. **Zeeman, P. B. and Butler, L. R. P.,** The determination of lead, copper and zinc in wines by atomic absorption spectroscopy, *Appl. Spectrosc.,* 16, 120, 1962.
9. **Sprague, S. and Slavin, W.,** Determination of the metal content of lubricating oils by atomic absorption spectrophotometry, *At. Absorpt. Newsl.,* 12, 1963.
10. **Slavin, W., Sprague, S., Rieders, F., and Cordova, V.,** The determination of certain toxicological trace metals by atomic absorption spectrophotometry, *At. Absorpt. Newsl.,* 17, 1964.
11. **Berman, E.,** The determination of lead in blood and urine by atomic absorption spectrophotometry, *At. Absorpt. Newsl.,* 3, 111, 1964.
12. **Dagnall, R. M. and West, T. S.,** Observations on the atomic absorption spectroscopy of lead in aqueous solution, in organic extracts and in gasoline, *Talanta,* 11, 1553, 1964.
13. **Sprague, S. and Slavin, W.,** Determination of very small amounts of copper and lead in KCl by organic extraction and atomic absorption spectrophotometry, *At. Absorpt. Newsl.,* 20, 1964.
14. **Musha, S., Munemori, M., and Nakanishi, Y.,** Determination of zinc, lead and calcium in poly (vinyl chloride) by atomic absorption spectrophotometry, *Bunseki Kagaku,* 13, 330, 1964.
15. **Novoselov, V. A. and Aidarov, T. K.,** Spectrographic determination of the trace elements Ag, Cu, Pb, Bi, Cd and Al in solutions by using a hollow cathode source, *Tr. Khim. Khim. Tekhnol.,* 108, 1964.
16. **Strasheim, A., Norval, E., and Butler, L. R. P.,** The atomic absorption determination of traces of lead in fish flour, *J. S. Afr. Chem. Inst.,* 17, 55, 1964.
17. **Trent, D. J.,** The determination of lead in gasoline by atomic absorption spectroscopy, *At. Absorpt. Newsl.,* 4, 348, 1965.
18. **Slavin, W. and Manning, D. C.,** Performance of lead hollow cathode lamps for atomic absorption spectroscopy, *Appl. Spectrosc.,* 19, 65, 1965.
19. **Biechler, D. G.,** Determination of trace copper, lead, zinc, cadmium, nickel and iron in industrial waste waters by atomic absorption spectrometry after ion exchange concentration on Dowex A-1, *Anal. Chem.,* 37, 1054, 1965.
20. **Bradfield, E. G. and Osborne, M.,** Application of atomic absorption spectroscopy to the determination of Zn, Fe, Cu, Mn and Pb in bottled ciders, *Long Ashton Agric. Hort. Res. Stn. Univ. Bristol Annu. Rep.,* 157, 1965.
21. **Razumov, V. A., Utkina, T. P., and Aidarov, T. K.,** Atomic absorption determination of lead in biological liquids, *Zh. Anal. Khim.,* 20, 1371, 1965 (in Russian).
22. **Bordonali, C., Biancifiori, M. A., and Besazza, G.,** Determination of traces of metals in sodium by atomic absorption (spectrometry). Determination of copper, manganese, iron, lead, nickel and zinc, *Chim. Ind. (Milan),* 47, 397, 1965.
23. **Dagnall, R. M., West, T. S., and Young, P.,** Determination of trace amounts of lead in steels, brass and bronze alloys by atomic absorption spectrometry, *Anal. Chem.,* 38, 358, 1966.
24. **Sprague, S. and Slavin, W.,** A simple method for the determination of lead in blood, *At. Absorpt. Newsl.,* 5, 9, 1966.
25. **Chakrabarti, C. L., Robinson, J. W., and West, P. W.,** The atomic absorption spectroscopy of lead, *Anal. Chim. Acta,* 34, 269, 1966.
26. **Jaworowski, R. and Weberling, R. P.,** Spectral interference, *At. Absorpt. Newsl.,* 5, 125, 1966.
27. **Sattur, T. W.,** Routine atomic absorption analyses on non-ferrous alloys and plant intermediates, *At. Absorpt. Newsl.,* 5, 37, 1966.
28. **Slavin, S. and Slavin, W.,** Fully automatic analysis of used aircraft oils, *At. Absorpt. Newsl.,* 5, 106, 1966.

29. **Takeuchi, T., Suzuki, M., and Yanagisawa, M.,** Some observations on the determination of metals by atomic absorption spectroscopy combined with extractions, *Anal. Chim. Acta,* 36, 258, 1966.

30. **Wilson, H. W.,** Note on the determination of lead in gasoline by atomic absorption spectrometry, *Anal. Chim. Acta,* 38, 921, 1966.

31. **Ivanov, N. P., Minervina, L. V., Baranov, S. V., Pofralidi, L. G., and Olikov, I. I.,** Tubes without electrodes with high frequency excitation of the spectrum of In, Ga, Bi, Sb, Tl, Pb, Mg, Ca and Cu as a radiation source in atomic absorption analysis, *Zh. Anal. Khim.,* 21, 1129, 1966 (in Russian).

32. **Kolb, B., Kemmer, G., Schleser, F. H., and Wiedeking, E.,** Element-specific detection of gas chromatographically separated metal compounds by atomic absorption spectroscopy. Determination of alkylleads in gasoline, *Z. Anal. Chem.,* 221, 166, 1966 (in German).

33. **Spitzer, H.,** Atomic absorption spectrophotometry for Cu, Pb, Fe, Mn, Ni, and Ca in zinc oxide, *Z. Erzbergbau Metallhuettenwes.,* 19, 567, 1966 (in German).

34. **Krumpack, J.,** Semiquantitative analysis by atomic absorption, *At. Absorpt. Newsl.,* 6, 20, 1967.

35. **Scott, T. C., Roberts, E. D., and Cain, D. A.,** Determination of minor constituents in ferrous materials by atomic absorption spectrophotometry, *At. Absorpt. Newsl.,* 6, 1, 1967.

36. **Chakrabarti, C. L.,** Determination of lead in aqueous and organic media by atomic absorption spectroscopy, *Appl. Spectrosc.,* 21, 160, 1967.

37. **Kopito, L., Byers, R. K., and Shwachman, H.,** Lead in hair of children with chronic lead poisoning, *N. Engl. J. Med.,* 276, 949, 1967.

38. **Kopito, L. and Shwachman, H.,** Determination of lead in urine by atomic absorption spectroscopy using coprecipitation with bismuth, *J. Lab. Clin. Med.,* 70, 326, 1967.

39. **Thilliez, G.,** Determination of traces of lead in air by atomic absorption spectrometry, *Anal. Chem.,* 39, 427, 1967.

40. **Bazhov, A. S.,** Atomic absorption determination of lead in mineral material, *Zh. Anal. Khim.,* 23, 1640, 1968.

41. **Berman, E., Valavanis, V., and Dubin, A.,** A micromethod for determination of lead in blood, *Clin. Chem. (Winston-Salem, N.C.),* 14, 239, 1968.

42. **Bober, A. and Mills, A. L.,** Determination of lead crystals by atomic absorption spectrometry, *Appl. Spectrosc.,* 22, 62, 1968.

43. **Fishman, M. J. and Midgett, M. R.,** Extraction techniques for the determination of cobalt, nickel and lead in fresh water by atomic absorption, *Adv. Chem. Ser.,* 73, 1968.

44. **Hessel, D. W.,** A simple and quantitative determination of lead in blood, *At. Absorpt. Newsl.,* 7, 55, 1968.

45. **Jordan, J.,** The determination of lead in food grade phosphates and phosphoric acid by atomic absorption spectroscopy, *At. Absorpt. Newsl.,* 7, 48, 1968.

46. **Roosels, D. and Vanderkeel, J. V.,** An atomic absorption determination of lead in urine after extraction with dithizone, *At. Absorpt. Newsl.,* 7, 9, 1968.

47. **Selander, S. and Cramer, K.,** Determination of lead in urine by atomic absorption spectrophotometry, *Br. J. Ind. Med.,* 25, 139, 1968.

48. **Devoto, G.,** Determination of lead in urine and blood by atomic absorption spectrophotometry, *Boll. Sco. Ital. Biol. Sper.,* 44, 421, 1968 (in Italian).

49. **Nagura, M. and Iida, C.,** Atomic absorption spectrophotometric determination of cobalt, nickel, lead and copper in silicates with the absorption tube technique, *Bunseki Kagaku,* 17, 1513, 1968.

50. **Torres, F.,** The Quantitative Determination of Lead, Cadmium and Indium in Microgram Amounts by Atomic Absorption Spectrophotometry, SL-TM-68-4, Sandia Laboratories, Albuquerque, N. M., 1968.

51. **Lehnert, G., Schaller, K. H., and Szadkowski, D.,** Rapid and reliable method for the determination of lead in small quantities of blood, *Z. Klin. Chem. Klin. Biochem.,* 7, 310, 1969 (in German).

52. **Segal, R. J.,** Non-specificity of urinary lead measurements by atomic absorption spectroscopy, *Clin. Chem. (Winston-Salem, N.C.),* 15, 1124, 1969.

53. **Brimhall, W. H.,** Measurement of lead isotopes by differential atomic absorption, *Anal. Chem.,* 41, 1349, 1969.

54. **Burnham, C. D., Moore, C. E., Kanabrocki, E., and Hattori, D. M.,** Determination of lead in airborne particulates in Chicago and Cook County, Ill. by atomic absorption spectroscopy, *Environ. Sci. Technol.,* 3, 472, 1969.

55. **Dalton, E. F. and Malanoski, A. J.,** Atomic absorption analysis of copper and lead in meat and meat products, *J. Assoc. Off. Anal. Chem.,* 52, 1035, 1969.

56. **Donovan, P. P. and Feeley, D. T.,** Method for the determination of lead in blood by atomic absorption spectrophotometry, *Analyst (London),* 94, 871, 1969.

57. **Einarsson, O. and Lindstedt, G.,** A non-extraction atomic absorption method for the determination of lead in blood, *Scand. J. Clin. Lab. Invest.,* 23, 367, 1969.

58. **Endo, V., Hata, T., and Nakahara, V.,** Determination of titanium, vanadium, nickel, chromium, lead and bismuth in iron ore by atomic absorption method, *Bunseki Kagaku,* 18, 883, 1969.

59. **Farrelly, R. D. and Pybus, J.,** Measurement of lead in blood and urine by atomic absorption spectrophotometry, *Clin. Chem. (Winston-Salem, N.C.),* 15, 566, 1969.

60. **Hall, J. M. and Woodward, C.,** Interference in the atomic absorption determinations of lead in copper-based materials, *Spectrosc. Lett.,* 2, 113, 1969.

61. **Hoover, W. L., Reagor, J. C., and Garner, J. C.,** Extraction and atomic absorption analysis of lead in plant and animal products, *J. Assoc. Off. Anal. Chem.,* 52, 708, 1969.

62. **Kettner, H.,** Comparison of the determination of lead in precipitated dust by dithizone and atomic absorption spectrophotometric methods, *J. Water Earth Air Hyg.,* 29, 55, 1969.

63. **Kirdhhof, H.,** Determination of isotope ratio for lead samples by atomic absorption, *Spectrochim. Acta,* 24b, 235, 1969.

64. **Searle, B., Chan, W., Jensen, C., and Davidow, B.,** Determination of lead in paint scrapings by atomic absorption, *At. Absorpt. Newsl.,* 8, 126, 1969.

65. **Tanaka, T. and Iida, C.,** Atomic absorption spectrophotometric determination of lead by absorption tube, *Bunseki Kagaku,* 18, 492, 1969.

66. **Zurlo, N., Griffini, A. M., and Colombo, G.,** Determination of lead in urine by atomic absorption spectrophotometry after coprecipitation with thorium, *Anal. Chim. Acta,* 47, 203, 1969.

67. **Baranova, S. V., Ivanov, N. P., Pofralidi, L. G., Knyazev, V. V., Talalaev, B. M., and Vasilev, E. N.,** Electrodeless lamps with high frequency excitation of the spectrum as a radiation source in atomic absorption analysis. Iodide lamps for the atomic absorption determination of silver, lead and iron, *Zh. Anal. Khim.,* 24, 1649, 1969 (in Russian).

68. **Burke, K. E.,** Study of the scavanger properties of manganese (IV) oxide with atomic absorption spectrometry determination of microgram quantities of antimony, bismuth, lead and tin in nickel, *Anal. Chem.,* 42, 1536, 1970.

69. **Burnham, C. D., Moore, C. E., Kowalski, T., and Krasniewski, J.,** A detailed study of lead determination in airborne particulates over Morton Grove, Ill., by atomic absorption spectroscopy, *Appl. Spectrosc.,* 24, 411, 1970.

70. **Delves, H. T.,** A micro-sampling method for the rapid determination of lead in blood by atomic absorption spectrophotometry, *Analyst (London),* 95, 431, 1970.

71. **Emmermann, R. and Luecke, W.,** Determination of trace amounts of lead, zinc and silver in soil samples by atomic absorption spectrometry using a tantalum sampling-boat, *Z. Anal. Chem.,* 248, 325, 1970 (in German).

72. **Gandrud, R. and Marshall, J. C.,** The determination of arsenic, lead, nickel and zinc in copper by atomic absorption spectrophotometric method, *Appl. Spectrosc.,* 24, 367, 1970.

73. **Kahn, H. L. and Sebesteyen, J. S.,** The determination of lead in blood and urine by atomic absorption spectrophotometry with the sampling boat system, *At. Absorpt. Newsl.,* 9, 33, 1970.

74. **Kono, T.,** Determination of lead by atomic absorption spectroscopy, *Bunseki Kagaku,* 19, 1032, 1970.

75. **Loftin, H. P., Christian, C. M., and Robinson, J. W.,** The continuous determination of lead in air, *Spectrosc. Lett.,* 3, 161, 1970.

76. **Matsumoto, I., Takabayashi, T., and Nakamura, I.,** Atomic absorption spectrometric determination of microamounts of lead in iron oxides, *Bunseki Kagaku,* 19, 771, 1970.

77. **Moldan, B., Rubeska, I., and Miksovsky, M.,** The determination of lead in silicate rocks by atomic absorption spectrophotometry with long path absorption tubes, *Anal. Chim. Acta,* 50, 342, 1970.

78. **Motto, H. L., Daines, R. H., Chilko, D. M., and Motto, C. K.,** Lead in soils and plants: its relationship to traffic volume and proximity to highways, *Environ. Sci. Technol.,* 4, 231, 1970.

79. **Nix, J. and Goodwin, T.,** The simultaneous extraction of Fe, Mn, Cu, Co, Ni, Cr, Pb and Zn from natural water for determination by atomic absorption spectroscopy, *At. Absorpt. Newsl.,* 9, 119, 1970.

80. **Page, A. L. and Ganje, T. J.,** Accumulations of lead in soils for regions of high and low motor vehicle traffic density, *Environ. Sci. Technol.,* 4, 140, 1970.

81. **Westerlund-Helmerson, U.,** Determination of lead and cadmium in blood by a modification of the Hessel method, *At. Absorpt. Newsl.,* 9, 133, 1970.

82. **Hermon, S. E. and Rennie, R. J.,** Rapid determination of lead and bismuth in aluminum alloys either separately or in the presence of each other using atomic absorption spectrophotometry, *Metallurgia,* 82, 201, 1970.

83. **Bazhov, A. S. and Zherebenko, A. V.,** Spectral characteristics of high frequency electrodeless lamps containing vapors of lead salts, *Zh. Prikl. Spektrosk.,* 12, 760, 1970.

84. **Hofton, M. E. and Hubbard, D. P.,** Determination of trace amounts of lead in high alloy steels by solvent extraction and atomic absorption spectroscopy, *Anal. Chim. Acta,* 52, 425, 1970.

85. **Keppler, J. F., Maxfield, M. E., Moss, W. D., Tieten, G., and Lynch, A. L.,** Inter-laboratory evaluation of the reliability of blood-lead analyses, *Am. Ind. Hyg. Assoc. J.,* 31, 412, 1970.

86. **Kinnen, J. and Wennerstrand, B.,** Determination of lead in copper and copper-base alloys, *Metallurgia,* 82, 81, 1970.

87. **Lorimier, D. F. and Fernandez, J. G.,** Determination of lead in urine by atomic absorption spectrophotometry, *Helv. Chim. Acta,* 53, 1990, 1970 (in French).
88. **Zook, E., Greene, F., and Morris, E.,** Nutrient composition of selected wheats and wheat products. Distribution of manganese, copper, nickel, zinc, magnesium, lead, tin, cadmium, chromium and selenium as determined by atomic absorption spectroscopy and colorimetry, *Cereal Chem.,* 47, 720, 1970.
89. **Arroyo, M. and Soldevilla, L.,** Blood lead concentration in general population, *Med. Segur. Trab.,* 69, 15, 1970 (in Spanish).
90. **Dancheva, R.,** Atomic absorption determination of lead, *Khim. Ind. (Sofia),* 43, 127, 1971.
91. **Den-Tonkela, W. A. M. and Bikker, A. M.,** Microdetermination of lead on tapes of an AISI automatic air sampler by atomic absorption spectroscopy, *Atmos. Environ.,* 5, 353, 1971.
92. **Fernandez, F. J. and Kahn, H. L.,** The determination of lead in blood by atomic absorption spectrophotometry with the "Delves Sampling Cup" technique, *At. Absorpt. Newsl.,* 10, 1, 1971.
93. **Fletcher, K.,** Direct determination of lead in plant materials by atomic absorption spectrophotometry, *J. Sci. Food Agric.,* 22, 260, 1971.
94. **Gambrell, J. W.,** The determination of lead and zinc as principal constituents by atomic absorption spectroscopy, *At. Absorpt. Newsl.,* 10, 81, 1971.
95. **Hwang, J. Y.,** Lead analysis in air particulate samples by atomic absorption spectrometry, *Can. Spectrosc.,* 16, 43, 1971.
96. **Hwang, J. Y., Ullucci, P. A., Smith, S. B., Jr., and Malenfant, A. L.,** Microdetermination of lead in blood by flameless atomic absorption spectrometry, *Anal. Chem.,* 43, 1319, 1971.
97. **John, M. K.,** Lead contamination of some agricultural soils in Western Canada, *Environ. Sci. Technol.,* 5, 1199, 1971.
98. **Kashiki, M., Yamazoe, S., and Oshima, S.,** Determination of lead in gasoline by atomic absorption spectroscopy, *Anal. Chim. Acta,* 53, 95, 1971.
99. **Kisfaludi, G., Henry, C., and Jourain, J. L.,** Determination of lead in water by atomic absorption flame spectrophotometry, *Chim. Anal. (Paris),* 53, 388, 1971 (in French).
100. **Kisfakudi, G. and Linhof, M.,** Determination of traces of lead in cast irons and steels by atomic absorption flame spectrometry, *Anal. Chim. Acta,* 53, 83, 1971 (in French).
101. **Luecke, W. and Emmermann, R.,** The application of the boat technique for lead, zinc, silver and cadmium in soil samples, *At. Absorpt. Newsl.,* 10, 45, 1971.
102. **Lyons, H. and Quinn, F. E.,** Measurement of lead in biological materials by combined anion exchange chromatography and atomic absorption spectrophotometry, *Clin. Chem. (Winston-Salem, N.C.),* 17, 152, 1971.
103. **Matousek, J. P. and Stevens, B. J.,** Biological applications of the carbon rod atomizer in atomic absorption spectroscopy. Preliminary studies on magnesium, iron, copper, lead and zinc in blood and plasma, *Clin. Chem. (Winston-Salem, N.C.),* 17, 363, 1971.
104. **Matsumoto, I., Okamoto, M., and Kanda, M.,** Rapid determination of trace amounts of lead in bismuth-oxychloride by atomic absorption spectrophotometry, *Bunseki Kagaku,* 20, 287, 1971.
105. **Okamoto, M., Kanda, M., Matsumoto, I., and Miya, Y.,** Fast analysis of trace amounts of lead in cosmetics by atomic absorption spectrophotometry, *J. Soc. Cosmet. Chem.,* 22, 589, 1971.
106. **Omang, S. H.,** Determination of lead in air by flameless atomic absorption spectrophotometry, *Anal. Chim. Acta,* 55, 439, 1971.
107. **Prakash, N. J. and Harrison, W. W.,** Simple demountable hollow cathode tube for the analysis of solutions. Application to lead in biological materials, *Anal. Chim. Acta,* 53, 421, 1971.
108. **Thomas, B. G.,** Determination of silver, lead and zinc in high grade ores, *At. Absorpt. Newsl.,* 10, 73, 1971.
109. **Weissberg, J. B., Lipschutz, F., and Oski, F. A.,** δ-Aminolevulinic acid dehydratase activity in circulating blood cells. A sensitive laboratory test for the detection of childhood lead poisoning, *N. Engl. J. Med.,* 284, 565, 1971.
110. **Yamamoto, Y., Kumamaru, T., Hayashi, Y., and Kanke, M.,** Determination of a trace amounts of cadmium, zinc, lead and copper in water by atomic absorption spectrophotometry, *Bunseki Kagaku,* 30, 347, 1971.
111. **Yeager, D. W., Cholak, J., and Henderson, E. W.,** Determination of lead in biological and related material by atomic absorption spectrophotometry, *Environ. Sci. Technol.,* 5, 1020, 1971.
112. **Zinterhofer, L. J. M., Jatlow, P. I., and Fappiano, A.,** Atomic absorption determination of lead in blood and urine in the presence of EDTA, *J. Lab. Clin. Med.,* 78, 664, 1971.
113. **Dollefeld, E.,** Micromethod for determination of the lead content of blood and urine by atomic absorption spectrophotometry, *Aerztl. Lab.,* 17, 369, 1971 (in German).
114. **Hislop, J. S. and Parker, A.,** Lead in Human Tissue: Literature Survey with Particular Reference to the Determination of Lead in Human Rib, AERE-R6987, U.K. Atomic Energy Authority, London, 1971.

115. **Uny, G., Mathien, C., Tardif, J. P., and Van Danh, T.,** Determination of zinc, iron and lead in high purity cobalt, *Anal. Chim. Acta,* 53, 109, 1971 (in French).

116. **Toma, O. and Crisan, T.,** Determination of copper, tin, lead, zinc, nickel and iron in bronze by atomic absorption spectrophotometry, *Metallurgia,* 23, 715, 1971 (in Romanian).

117. **Arroyo, M. and Ximinez, L.,** Analytical determination of lead in blood by atomic absorption spectro-photometry (direct micro method with the Delves system), *Med. Trop.,* 47, 221, 1971 (in Spanish).

118. **Cernik, A. A. and Sayers, M. H. P.,** Determination of lead in capillary blood using a paper punched disc atomic absorption technique. Application to the supervision of lead workers, *Br. J. Ind. Med.,* 28, 392, 1971.

119. **John, P.,** Determination of lead in petrol by atomic absorption spectrophotometry, *Spectrovision,* 26, 1971.

120. **Kanke, M., Hayashi, Y., Kumamaru, T., and Yamamoto, Y.,** Determination of ppb levels of cadmium, zinc, lead and copper extracted as their dithizonates into nitrobenzene by atomic absorption spectropho-tometry, *Nippon Kagaku Zasshi,* 92, 983, 1971.

121. **Bohnstedt, U.,** Determination of traces of lead, bismuth, zinc and cadmium in alloys: an example of the use of combined analytical procedures, *DEW Tech. Ber.,* 11, 101, 1971 (in German).

122. **Bratzel, M. P., Jr. and Chakrabarti, C. L.,** Determination of lead in petroleum and petroleum products by atomic absorption spectrometry with a carbon rod atomizer, *Anal. Chim. Acta,* 61, 25, 1972.

123. **Briggs, D.,** Population differentiation in *Marchantia polymorphia* L. in various lead pollution levels, *Nature (London),* 238, 166, 1972.

124. **Gajan, R. J. and Larry, D.,** Determination of lead in fish by atomic absorption spectrophotometry and by polarography. I. Development of the methods, *J. Assoc. Off. Anal. Chem.,* 55, 727, 1972.

125. **Gajan, R. J. and Larry, D.,** Determination of lead in fish by atomic absorption spectrophotometry and by polarography. II. Collaborative study, *J. Assoc. Off. Anal. Chem.,* 55, 733, 1972.

126. **Hauser, T. R., Hinners, T. A., and Kent, J. L.,** Atomic absorption determination of cadmium and lead in whole blood by a reagent-free method, *Anal. Chem.,* 44, 1819, 1972.

127. **Hoover, W. L.,** Collaborative study of a method for determining lead in plant and animal products, *J. Assoc. Off. Anal. Chem.,* 55, 737, 1972.

128. **Husain, D. and Littler, J. G. F.,** Kinetic study of electronically excited lead atoms, lead 6 1 D 2 by atomic absorption spectroscopy using attentuation of resonance radiation at lambda 373.995 nm, lead 7S 3P 2 O from 6P 2 1 D 2, *Chem. Phys. Lett.,* 16, 145, 1972.

129. **Needleman, H. L., Tuncay, O. C., and Shapiro, I. M.,** Lead levels in deciduous teeth of urban and suburban American children, *Nature (London),* 235, 111, 1972.

130. **Norval, E. and Butler, L. R. P.,** Determination of lead in blood by atomic absorption with the high temperature graphite tube, *Anal. Chim. Acta,* 58, 47, 1972.

131. **Sellers, N. G.,** Ion exchange separation and determination of lead in steel by atomic absorption, *Anal. Chem.,* 44, 410, 1972.

132. **Sorensen, L. L. C. and Nobe, K.,** Effects of lead on oxidation activity of copper oxide-alumina catalyst, *Environ. Sci. Technol.,* 6, 239, 1972.

133. **Campbell, K. and Palmer, J.,** Determination of trace amounts of lead in petroleum products by iodine mono chloride extraction and atomic absorption analysis, *Inst. Pet. J.,* 58, 193, 1972.

134. **Fine, P. R., Thomas, C. W., Suhs, R. H., Cohnberg, R. E., and Flashner, B. A.,** Pediatric blood levels, a study of 14 Illinois cities of intermediate population, *JAMA,* 221, 1475, 1972.

135. **Burke, K. E.,** Determination of microgram amounts of antimony, bismuth, lead and tin in aluminum, iron and nickel-base alloys by nonaqueous atomic absorption spectroscopy, *Analyst (London),* 97, 19, 1972.

136. **Ediger, R. D. and Coleman, R. L.,** A modified Delves cup atomic absorption procedure for the deter-mination of lead in blood, *At. Absorpt. Newsl.,* 11, 33, 1972.

137. **Gregor, C. D.,** Solubilization of lead in lake and reservoir sediments by NTA, *Environ. Sci. Technol.,* 6, 278, 1972.

138. **Kahn, H. L., Fernandez, F. J., and Slavin, S.,** The determination of lead and cadmium in soils and leaves by atomic absorption spectroscopy, *At. Absorpt. Newsl.,* 11, 42, 1972.

139. **Kubasik, N. P., Volosin, M. T., and Murray, M. H.,** Carbon rod atomizer applied to measurement of lead in whole blood by atomic absorption spectrophotometry, *Clin. Chem. (Winston-Salem, N.C.),* 18, 410, 1972.

140. **Moyers, J. L., Zoller, W. H., Duce, B. A., and Hoffman, G. L.,** Gaseous bromine and particulate lead, vanadium and bromine in a polluted atmosphere, *Environ. Sci. Technol.,* 6, 68, 1972.

141. **Husain, D. And Littler, J. G.,** Collisional quenching of electronically excited lead atoms 6 (1) D (2) by time resolved atomic absorption spectroscopy using attenuation of resonance radiation, *J. Chem. Soc. Faraday Trans.,* 2, 2110, 1972.

142. **Ishizuka, T., Sunahara, H., and Tanaka, K.,** Determination of copper, iron, lead and zinc in yttrium oxide and yttrium oxide sulfide by atomic absorption spectrometry, *Bunseki Kagaku,* 21, 847, 1972.

143. **Joselow, M. M. and Bogden, J. D.,** A simplified micro method for collection and determination of lead in blood using a paper disk-in Delves cup technique, *At. Absorpt. Newsl.,* 11, 99, 1972.

144. **Kubasik, N. P. and Volosin, M. T.,** Concentrations of lead in capillary blood of newborns, *Clin. Chem. (Winston-Salem, N.C.),* 18, 1415, 1972.

145. **Lee, J. A.,** Lead pollution from a factory manufacturing anti-knock compounds, *Nature (London),* 238, 165, 1972.

146. **McDuffie, B.,** Rapid screening of pencil paint for lead by a combustion atomic absorption technique, *Anal. Chem.,* 44, 1551, 1972.

147. **Mitchell, D. G., Ryan, F. J., and Aldous, K. M.,** The precise determination of lead in whole blood by solvent extraction-atomic absorption spectrometry, *At. Absorpt. Newsl.,* 11, 120, 1972.

148. **Olsen, E. D. and Jatlow, P. I.,** An improved Delves cup atomic absorption procedure for determination of lead in blood and urine, *Clin. Chem. (Winston-Salem, N.C.),* 18, 1312, 1972.

149. **Pagenkopf, G. K., Neuman, D. R., and Woodriff, R.,** Determination of lead in fish by furnace atomic absorption, *Anal. Chem.,* 44, 2248, 1972.

150. **Pueschel, S. M., Kopito, L., and Schwachman, H.,** Children with an increased lead burden. A screening and follow up study, *JAMA,* 222, 462, 1972.

151. **Rabinowitz, M. B. and Wetherill, G. W.,** Identifying sources of lead contamination by stable isotope technique, *Environ. Sci. Technol.,* 6, 705, 1972.

152. **Rajegowda, B. K., Glass, L., and Evans, H. E.,** Lead concentrations in the newborn infant, *J. Pediatr.,* 80, 116, 1972.

153. **Renshaw, G. D., Pounds, C. A., and Pearson, E. F.,** Variation in lead concentration along single hairs as measured by nonflame atomic absorption spectrophotometry, *Nature (London),* 238, 162, 1972.

154. **Rosen, J. F. and Trinidad, E. E.,** The micro determination of blood lead in children by flameless atomic absorption: the carbon rod atomizer, *J. Lab. Clin. Med.,* 80, 567, 1972.

155. **Seeley, J. L., Dick, D., Arvik, J. H., Zimdahl, R. L., and Skogerboe, R. K.,** Determination of lead in soil, *Appl. Spectrosc.,* 26, 456, 1972.

156. **Shiriashi, N., Masegawa, T., Hisayuki, T., and Takahashi, H.,** Dissolution of methyl isobutyl ketone in various salt solutions. Determination of cadmium and lead by atomic absorption, *Bunseki Kagaku,* 21, 705, 1972.

157. **Tsukahara, I. and Yamamoto, Y.,** Determination of microamounts of lead in copper, nickel and aluminum metals and in copper base alloys by solvent extraction and atomic absorption spectrometry, *Anal. Chem. Acta,* 61, 33, 1972.

158. **Woodriff, R. and Lech, J. F.,** Determination of trace lead in the atmosphere by furnace atomic absorption, *Anal. Chem.,* 44, 1323, 1972.

159. **Blumenthal, S., Davidow, B., Harris, D., and Oliver-Smith, F.,** A comparison between two diagnostic tests for lead poisoning, *Am. J. Public Health,* 62, 1060, 1972.

160. **Cheremisinoff, P. N. and Habib, Y. H.,** Cadmium, chromium, lead and mercury: plenary account for water pollution. I. Occurrence, toxicity and detection, *Water Sewage Works,* 119, 73, 1972.

161. **Debras-Guedon, J., Draignaud, M., and Boix, A.,** Contribution of atomic absorption spectrometry to the problem of soluble lead in glazed tableware, *Bull. Soc. Fr. Ceram.,* 53, 1972.

162. **Friswell, N. J. and Jenkins, D. R.,** Identification of lead compounds in flames and determination of the lead oxide bond energy, *Combust. Flame,* 19, 197, 1972.

163. **Mack, D. and Berg, H.,** Determination of lead in vegetables and fruits, including wine grapes, by atomic absorption spectrophotometry, *Dtsch. Lebensm. Rundsch.,* 68, 262, 1972.

164. **Vens, M. D. and Lauwerys, R.,** Simultaneous determination of lead and cadmium in blood and urine by coupling ion exchange resin chromatography and atomic absorption spectrophotometry, *Arch. Mal. Prof.,* 33, 97, 1972 (in French).

165. **Olzhataev, B. A., Koifman, M. D., Shugurova, Z. V., and Nedugova, G. A.,** Atomic absorption determination of lead in products of the flotation beneficiation of lead-zinc ores, *Dokl. Akad. Nauk Uzb. SSR,* 29, 20, 1972.

166. **Zuber, R.,** Determination of lead in plant material with use of atomic absorption spectrophotometry, *Mitt. Geb. Lebensmittelunters. Hyg.,* 63, 229, 1972 (in German).

167. **Arroyo, M. and de Salamanca, R. E.,** Lead in normal rats, *Med. Segur. Trab.,* 80, 15, 1972 (in Spanish).

168. **Damiani, P.,** Determination of lead in cosmetible vegetables by atomic absorption spectrophotometry, *Ind. Aliment. Pinerolo,* 11, 106, 1972 (in Italian).

169. **de Salamanca, R. E. and Arroyo, M.,** Experimental intoxication of lead in rats, *Med. Segur. Trab.,* 80, 25, 1972 (in Spanish).

170. **de Salamanca, R. E., Arroyo, M., and Alonso, J.,** Properties of free erythrocytes in individual laboratory workers in contact with lead, *Med. Segur. Trab.,* 80, 19, 1972 (in Spanish).

171. **Ishizuka, T., Sunahara, H., and Tanaka, K.,** Determination of copper, iron, lead and zinc in yttrium oxide and yttrium oxide sulfide by atomic absorption spectrometry, *Nagoyo Kogyo Gijutsu Shikensho Hokoku,* 21, 385, 1972.

172. **Schulz-Baldes, M.,** Toxicity and accumulation of lead in the common mussel *mytilus edulis* in laboratory experiment, *Mar. Biol.,* 16, 226, 1972 (in German).

173. **Toma, O. and Crisan, T.,** Atomic absorption determination of zinc and traces of iron, lead, sodium, nickel, calcium and magnesium in saturated zinc chloride solutions, *Rev. Chim.,* 23, 189, 1972 (in Romanian).

174. **Fey, R. and Becker, G.,** Release of lead from ceramic glazes and decorations of utensils: analysis and evaluation for food regulations, *Z. Lebensm. Unters. Forsch.,* 150, 87, 1972 (in German).

175. **Holroyd, P. M. and Snodin, D. J.,** Determination of lead and cadmium in tapwater, rain water, milk and urine, *J. Assoc. Public Anal.,* 10, 110, 1972.

176. **Steele, T. W., Mallet, R. C., Pearton, D. C. G., and Ring, E. J.,** Determination of atomic absorption spectrophotometry of platinum, palladium, rhodium, ruthenium, gold, silver and lead in prills, *Lab. Meth.* No. 78/4, 1972.

177. **Toma, O. and Crisan, T.,** Determination of the concentration of zinc and traces of iron, lead, sodium, nickel, calcium and magnesium in saturated zinc chloride solution with atomic absorption spectrophotometry, *Chim. Anal.,* 2, 189, 1972.

178. **Piscator, M. and Lind, B.,** Cd, Zn, Cu and Pb in human renal cortex, *Arch. Environ. Health,* 24, 426, 1972.

179. **Arroyo, M., de Salamanca, R. E., and Mas, V.,** Correlation between the values of blood and urinary lead in the analysis of control of individuals with work exposure, *Med. Segur. Trab.,* 21, 32, 1973 (in Spanish).

180. **Boppel, B.,** Lead contents of foodstuffs. II. Fruit juices, soft drinks and mineral waters, *Z. Lebensm. Unters. Forsch.,* 153, 345, 1973.

181. **Brockhaus, A.,** Quantitative detection of lead, cadmium and zinc in blood and organs, *Staub Reinhalt. Luft,* 33, 437, 1973 (in German).

182. **Chao, T. T. and Sanzolone, R. F.,** The atomic absorption spectrophotometric determination of microgram levels of cobalt, nickel, copper, lead and zinc in soil and sediment extracts containing large amounts of manganese and iron, *J. Res. U.S. Geol. Surv.,* 1, 681, 1973.

183. **Henn, E. L.,** Determining lead contents of paints, *Paint Varn. Prod.,* 63, 29, 1973.

184. **Renshaw, G. D., Pounds, C. A., and Pearson, E. F.,** Determination of lead and copper in hair by nonflame atomic absorption spectrophotometry, *J. Forensic Sci.,* 18, 143, 1973.

185. **Hwang, J. Y., Ullucci, P. A., and Mokeler, C. J.,** Direct flameless atomic absorption determination of lead in blood, *Anal. Chem.,* 45, 795, 1973.

186. **Jackwerth, E., Hoehn, R., and Koos, K.,** Trace enrichment by partial dissolution of the matrix in presence of mercury. Determination of bismuth, copper, lead, nickel, silver, gold and palladium in high purity cadmium by atomic absorption spectrometry, *Z. Anal. Chem.,* 264, 1, 1973 (in German).

187. **Mansell, R. E. and Hiller, T. A.,** Application of the PARR acid digestion bomb to decomposition of tetraethyl lead, *Anal. Chem.,* 45, 975, 1973.

188. **Murthy, L., Menden, E. E., Eller, P. M., and Petering, H. G.,** Atomic absorption determination of zinc, copper, cadmium and lead in tissues solubilized by aqueous tetra methyl ammonium hydroxide, *Anal. Biochem.,* 53, 365, 1973.

189. **Nakagawa, R.,** Atomic absorption analysis of zinc, lead and cadmium in hot spring waters, *Nippon Kagaku Kaishi,* 514, 1973.

190. **Okusu, H., Ueda, Y., Ota, K., and Kawano, K.,** Determination of cadmium, zinc, copper and lead in waste water by atomic absorption spectrometry, *Bunseki Kagaku,* 22, 84, 1973.

191. **Purdue, L. J., Enrione, R. E., Thompson, R. J., and Bonfield, B. A.,** Determination of organic and total lead in the atmosphere by atomic absorption spectrometry, *Anal. Chem.,* 45, 527, 1973.

192. **Renshaw, G. D., Pounds, C. A., and Pearson, E. F.,** The quantitative estimation of lead, antimony and barium in gunshot residues by non-flame atomic absorption spectrophotometry, *At. Absorpt. Newsl.,* 12, 55, 1973.

193. **Rose, G. A. and Willden, E. G.,** An improved method for the determination of whole blood lead by using an atomic absorption technique, *Analyst (London),* 98, 243, 1973.

194. **Sorensen, J. R. J., Levin, L. S., and Petering, H. G.,** Cadmium, copper, lead, mercury and zinc concentrations in the hair of individuals living in the United States, *Interface,* 2, 17, 1973.

195. **Supp, G. R., Gibbs, I., and Juszli, M.,** Determination of lead in petroleum additives using atomic absorption spectrophotometry, *At. Absorpt. Newsl.,* 12, 66, 1973.

196. **Bertinuson, J. R. and Clark, C. S.,** The contribution to lead content of soils from urban housing, *Interface,* 2, 6, 1973.

197. **Buchauer, M. J.,** Contamination of soil and vegetation near a zinc smelter by zinc, cadmium, copper and lead, *Environ. Sci. Technol.,* 7, 131, 1973.

198. **Cernik, A. A.,** Some observations on the filter paper punched disk method for the determination of lead in capillary blood, *At. Absorpt. Newsl.,* 12, 42, 1973.

199. **Fernandez, F. J.,** Some observations on the determination of lead in blood with the Delves cup method, *At. Absorpt. Newsl.,* 12, 70, 1973.

200. **Gross, S. B., Pfitzer, E. A., and Kehoe, R. A.,** Human body burden of lead: preliminary observations on the relationship with age, *Interface,* 2, 18, 1973.
201. **Hicks, J. M., Gutierrez, A. N., and Worthy, B. E.,** Evaluation of the Delves micro system for blood lead analysis, *Clin. Chem. (Winston-Salem, N.C.),* 19, 322, 1973.
202. **Holak, W.,** Determination of traces of lead and cadmium in foods by atomic absorption spectrophotometry using the "sampling boat," *At. Absorpt. Newsl.,* 12, 63, 1973.
203. **Arroyo, M.,** Analytical determination of lead in urine by atomic absorption spectrophotometry. Direct micromethod using the Delves system, *Rev. Clin. Esp.,* 128, 119, 1973 (in Spanish).
204. **Arroyo, M., de Salamanca, R. E., and Piga, A.,** Normal levels of blood lead, *Med. Segur. Trab.,* 81, 27, 1973 (in Spanish).
205. **Blakeley, J. H., Manson, A., and Zatka, V. J.,** Improvements in the manganese dioxide collection of trace lead and bismuth in nickel, *Anal. Chem.,* 45, 1941, 1973.
206. **Bolter, E., Hemphill, D., Wixson, B., Chen, R., and Butherus, D.,** Geochemical and vegetation studies of trace substances from lead smelting, in *Trace Substances in Environmental Health,* Vol. 4, Hemphill, D. D., Ed., University of Missouri, Columbia, 1973, 79.
207. **Chauvin, J. V., Newton, M. P., and Davis, D. G.,** The determination of lead and nickel by atomic absorption spectrometry with a flameless wire loop atomizer, *Anal. Chim. Acta,* 65, 291, 1973.
208. **Chisolm, J. J., Jr.,** Management of increased lead absorption and lead poisoning in children, *N. Engl. J. Med.,* 289, 1016, 1973.
209. **Crecelius, E. A. and Piper, D. Z.,** Particulate lead contamination recorded in sedimentary cores from Lake Washington, Seattle, *Environ. Sci. Technol.,* 7, 1053, 1973.
210. **Crisler, J. P., Lao, N. T., Tang, L. C., Serrano, B. A., and Shields, A.,** A micro-sampling method for the determination of blood lead, *Microchem. J.,* 18, 77, 1973.
211. **Favretto, L., Pertoldi Marletta, G., and Favretto Gabrielli, L.,** Determination of lead in grapes exposed to automobile exhausts gases, *At. Absorpt. Newsl.,* 12, 101, 1973.
212. **Fiorino, J. A., Moffitt, R. A., Woodson, A. L., Gajan, R. J., Huskey, G. E., and Scholz, R. G.,** Determination of lead in evaporated milk by atomic absorption spectrophotometry and anodic stripping voltammetry. Collaborative study, *J. Assoc. Off. Anal. Chem.,* 56, 1246, 1973.
213. **Gish, C. D. and Christensen, R. E.,** Cadmium, nickel, lead and zinc in earthworms from roadside soil, *Environ. Sci. Tech.,* 7, 1060, 1973.
214. **Heinemann, G.,** Atomic absorption spectrometric determination of lead in blood and urine using the Delves sampling system, *Z. Klin. Chem. Klin. Biochem.,* 11, 197, 1973 (in German).
215. **Henn, E. L.,** Rapid determination of lead in paint by atomic absorption utilizing the Delves cup technique, *At. Absorpt. Newsl.,* 12, 109, 1973.
216. **Hislop, J. S., Parker, A., Spicer, G. S., and Webb, M. S. W.,** Determination of Lead in Human Rib Bone (Comparison of Methods), AERE-R7321, U.K. Atomic Energy Authority, London, 1973.
217. **Hohnadel, D. C., Sunderman, F. W., Jr., Nechay, M. W., and McNeeley, M. D.,** Atomic absorption spectrometry of nickel, copper, zinc and lead in sweat collected from healthy subjects during sauna bathing, *Clin. Chem. (Winston-Salem, N.C.),* 19, 1288, 1973.
218. **Janssens, M. and Dams, R.,** Determination of lead in atmospheric particulates by flameless atomic absorption spectrometry with a graphite tube, *Anal. Chim. Acta,* 65, 41, 1973.
219. **Klevay, L. M.,** Hair as a biopsy material. III. Assessment of environmental lead exposure, *Arch. Environ. Health,* 26, 169, 1973.
220. **Krinitz, B. and Franco, V.,** Collaborative study of an atomic absorption method for the determination of lead and cadmium extracted from glazed ceramic surfaces, *J. Assoc. Off. Anal. Chem.,* 56, 869, 1973.
221. **Kubasik, N. P. and Volosin, M. T.,** Simplified determination of urinary cadmium, lead and thallium with use of carbon rod atomizer and atomic absorption spectrophotometry, *Clin. Chem. (Winston-Salem, N.C.),* 19, 954, 1973.
222. **Kubo, Y., Nakazawa, N., and Sato, M.,** Determination of trace amounts of lead, chromium and copper in sea water by solvent extraction-atomic absorption spectrophotometry, *Sekiyu Gakkai Shi,* 16, 588, 1973.
223. **Lamm, S., Cole, B., Glynn, K., and Ullman, W.,** Lead content of milks fed to infants 1971-1972, *N. Engl. J. Med.,* 289, 574, 1973.
224. **Lin-Fu, J. S.,** Vulnerability of children to lead exposure and toxicity (first of two parts), *N. Engl. J. Med.,* 289, 1229, 1973.
225. **Lin-Fu, J. S.,** Vulnerability of children to lead exposure and toxicity (second of two parts), *N. Engl. J. Med.,* 289, 1289, 1973.
226. **Manning, D. C.,** Atomic absorption analysis by flameless sampling: lead in milk, *Am. Lab.,* 5, 39, 1973.
227. **Marumo, Y., Oikawa, T., and Niwaguchi, T.,** Determination of lead in biological materials by atomic absorption spectrophotometry, *Bunseki Kagaku,* 22, 1024, 1973.
228. **Matousek, J. P. and Brodie, K. G.,** Direct determination of lead airborne particulates by nonflame atomic absorption, *Anal. Chem.,* 45, 1606, 1973.

229. **Olsen, R. D. and Sommerfeld, M. R.**, A technique for extraction and storage of water samples for Mn, Cd, and Pb determination by atomic absorption spectroscopy, *At. Absorpt. Newsl.*, 12, 165, 1973.

230. **Pinkerton, C., Hammer, D. I., Bridbord, K., Creason, J. P., Kent, J. L., and Murthy, G. K.**, Human milk as a dietary source of cadmium and lead, in *Trace Substances in Environmental Health*, Vol. 6, Hemphill, D. D., Ed., University of Missouri, Columbia, 1973, 39.

231. **Robbins, W. K.**, Determination of lead in gasoline by heated vaporization atomic absorption spectrometry, *Anal. Chim. Acta*, 65, 285, 1973.

232. **Roques, Y. and Mathieu, J.**, Rapid determination of lead ultratraces in atmospheric particulates matter by flameless atomic absorption spectrophotometry, *Analusis*, 2, 481, 1973.

233. **Roschnik, R. K.**, Determination of lead in foods by atomic absorption spectrophotometry, *Analyst (London)*, 98, 596, 1973.

234. **Sorensen, J. R. J., Melby, E. G., Nord, P. J., and Petering, H. G.**, Interferences in the determination of metallic elements in human hair. Evaluation of Zn, Cu, Pb, and Cd using atomic absorption spectrophotometry, *Arch. Environ. Health*, 27, 36, 1973.

235. **Snodin, D. J.**, A rapid method for the determination of lead in fruit juice and other beverages, *J. Assoc. Public. Anal.*, 11, 47, 1973.

236. **Sauerhoff, M. W. and Michaelson, I. A.**, Hyperactivity and brain catecholamines in lead-exposed developing rats, *Science*, 182, 1022, 1973.

237. **Snodin, D. J.**, Determination of lead and cadmium in baby foods, *J. Assoc. Public Anal.*, 11, 112, 1973.

238. **Cohen, C. J., Bowers, G. N., and Lepow, M. L.**, Epidemiology of lead poisoning: a comparison between urban and rural children, *J. Am. Med. Soc.*, 26, 1430, 1973.

239. **Kiboku, M. and Aihara, M.**, Atomic absorption spectrophotometric determination of microamounts of lead by using solvent extractions with potassium ethyl xanthate-methyl isobutyl ketone, *Bunseki Kagaku*, 22, 1581, 1973.

240. **Kozlov, M. G., Mileshina, S. A., and Startsev, G. P.**, Absorption spectrum of lead vapor in ultraviolet and Schumann regions, *Opt. Spectrosc.*, 35, 89, 1973.

241. **Langmyhr, F. J., Thomassen, Y., and Massoumi, A.**, Atomic absorption spectrometric determination of lead and cadmium in silicon, ferrosilicon and ferromanganese by direct atomization from the solid state, *Anal. Chim. Acta*, 67, 460, 1973.

242. **Machata, G. and Bindes, R.**, Determination of traces of lead, thallium, zinc and cadmium in biological material by flameless atomic absorption spectrophotometry, *Z. Rechtsmed.*, 73, 29, 1973 (in German).

243. **Okuneva, G. A. and Tarasevich, N. I.**, Atomic absorption determination of iron and lead in brasses, *Vestn. Mosk. Univ. Khim.*, 14, 600, 1973 (in Russian).

244. **Ronco, A. E. and Merodio, J. C.**, Determination of lead in silicate rocks by atomic absorption spectrometry, *An. Asoc. Quim. Argent.*, 61, 219, 1973.

245. **Shigematsu, T., Matsui, M., Fujino, O., and Kinoshita, K.**, Determination of lead by atomic absorption spectrometry with a carbon tube atomizer, *Nippon Kagaku Kaishi*, 2123, 1973.

246. **Smit, A. and Backe-Hansen, K.**, Determination of contamination of zinc compounds by lead and cadmium, *Acta Pharm. Suec.*, 10, 254, 1973.

247. **Hantzsch, S., Kaffanke, K., and Nietruch, F.**, Lead determination in street dust by atomic absorption spectrophotometry, *Staub Reinhalt. Luft*, 33, 34, 1973.

248. **Oelschlaeger, W. and Frankel, E.**, Determination of lead in plant and animal tissue and in mineral feed additives by atomic absorption spectrophotometry, *Landwirtsch. Forsch.*, 26, 281, 1973.

249. **Reusmann, G. and Westphalen, J.**, Automated determination of lead, zinc and cadmium in vegetable material, *Staub Reinhalt. Luft*, 33, 435, 1973 (in German).

250. **Hahn, C.**, Determination of lead, cadmium and zinc in fired glazes and engobe decorations using atomic absorption spectroscopy, *Silik. J.*, 12, 38, 1973.

251. **Machata, G.**, Blood lead level of the Viennese population, *Wien Klin. Wochenschr.*, 85, 216, 1973 (in German).

252. **Frech, W. and Cedergren, A.**, Investigations of reactions involved in flameless atomic absorption procedures. II. An experimental study of the role of hydrogen in eliminating the interference from chlorine in the determination of lead in steel, *Anal. Chim. Acta*, 82, 93, 1973.

253. **Kubota, J., Mills, E. L., and Oglesby, R. T.**, Lead, Cd, Zn, Cu and Co in streams and lake waters of Cayuga Lake Basin, New York, *Environ. Sci. Technol.*, 8, 243, 1974.

254. **Langmyhr, F. J., Thomassen, Y., and Massoumi, A.**, Atomic absorption spectrometric determination of copper, lead, cadmium and manganese in pulp and paper by the direct atomization technique, *Anal. Chim. Acta*, 68, 305, 1974.

255. **Porter, W. K., Jr.**, Collaborative study of the determination of lead in alkyd and latex paints, *J. Assoc. Off. Anal. Chem.*, 57, 614, 1974.

256. **Rattonetti, A.**, Determination of soluble cadmium, lead, silver and indium in rain water and steam water with the use of flameless atomic absorption, *Anal. Chem.*, 46, 739, 1974.

257. **Shaw, F. and Ottaway, J. M.,** Determination of trace amounts of lead in steel and cast iron by atomic absorption spectrometry with the use of carbon furnace atomization, *Analyst (London),* 99, 184, 1974.

258. **Taylor, W., Molyneux, M. K. B., and Blackaddes, E. S.,** Lead over-absorption in a population of oxygas burners, *Nature (London),* 247, 53, 1974.

259. **Evenson, M. A. and Pendergast, D. D.,** Rapid ultramicro direct determination of erythrocyte lead concentration by atomic absorption spectrophotometry with the use of a graphite tube furnace, *Clin. Chem. (Winston-Salem, N.C.),* 20, 163, 1974.

260. **Haelen, P., Cooper, G., and Pampel, C.,** The determination of lead in evaporated milk by delves cup atomic absorption spectrophotemetry, *At. Absorpt. Newsl.,* 13, 1, 1974.

261. **Heistand, R. N. and Shaner, W. C.,** Automated atomic absorption determination of lead in gasoline, *At. Absorpt. Newsl.,* 13, 65, 1974.

262. **Kashiki, M., Yamazoe, S., Ikeda, N., and Oshima, S.,** Determination of lead in gasoline by atomic absorption spectrometry with a carbon rod atomizer, *Anal. Lett.,* 7, 53, 1974.

263. **Kopito, L. E., Davis, M. A., and Schwachman, H.,** Sources of error in determining lead in blood by atomic absorption spectrophotometry, *Clin. Chem. (Winston-Salem, N.C.),* 20, 205, 1974.

264. **Kubasik, N. P. and Volosin, M. T.,** Use of the carbon rod atomizer for direct analysis of lead in blood, *Clin. Chem. (Winston-Salem, N.C.),* 20, 300, 1974.

265. **Anderson, W. N., Broughton, P. M. G., Dawson, J. B., and Fisher, G. W.,** Evaluation of some atomic absorption systems for the determination of lead in blood, *Clin. Chim. Acta,* 50, 129, 1974.

266. **Boppel, B.,** Lead contents of food stuffs. I. On the analysis of lead in foodstuffs, *Z. Anal. Chem.,* 268, 114, 1974 (in German).

267. **Bratzel, M. P. and Reed, A. J.,** Microsampling of blood lead analysis, *Clin. Chem. (Winston-Salem, N.C.),* 20, 217, 1974.

268. **Childs, E. A. and Gaffke, J. N.,** Organic solvent extraction of lead and cadmium from aqueous solutions for atomic absorption spectrophotometric measurement, *J. Assoc. Off. Anal. Chem.,* 57, 360, 1974.

269. **Childs, E. A. and Gaffke, J. N.,** Possible interference in the measurement of lead and cadmium from elements found in fish muscle, *J. Assoc. Off. Anal. Chem.,* 57, 365, 1974.

270. **Cooke, R. E., Glynn, K. L., Ullman, W. W., Lurie, N., and Lepow, M.,** Comparative study of a micro scale test for lead in blood for use in mass screening programs, *Clin. Chem. (Winston-Salem, N.C.),* 20, 582, 1974.

271. **Thompson, K. C. and Thomerson, D. R.,** Atomic absorption studies on the determination of antimony, arsenic, bismuth, germanium, lead, selenium, tellurium and tin by utilizing the generation of covalent hydrides, *Analyst (London),* 99, 595, 1974.

272. **Thornton, K. and Burke, K. E.,** Modification to the extraction-atomic absorption method for the determination of antimony, bismuth, lead and tin, *Analyst (London),* 99, 469, 1974.

273. **Riner, J. C., Wright, F. C., and McBeth, C. A.,** A technique for determining lead in feces of cattle by flameless atomic absorption spectrophotometry, *At. Absorpt. Newsl.,* 13, 129, 1974.

274. **Rosen, J. F., Zarate-Salvador, C., and Trinidad, E. E.,** Plasma lead levels in normal and lead-intoxicated children, *J. Pediatr.,* 84, 45, 1974.

275. **Sapek, A.,** Atomic absorption spectrometric determination of lead, nickel and cobalt in soil extracts, *Chem. Anal.,* 19, 687, 1974.

276. **Shaw, F. and Ottaway, J. M.,** Determination of trace amounts of lead in high purity copper and copper alloys by atomic absorption spectrometry with graphite furnace atomization, *At. Absorpt. Newsl.,* 13, 77, 1974.

277. **Smulevicz, J. J.,** Lead lines at the Iliac Crest and "Early" diagnosis of lead poisoning, *Am. J. Med. Sci.,* 267, 49, 1974.

278. **Addis, G. and Moore, M.,** Lead levels in the water of suburban Glasgow, *Nature (London),* 252, 120, 1974.

279. **Amore, F.,** Determination of cadmium, lead, thallium and nickel in blood by atomic absorption spectrometry, *Anal. Chem.,* 46, 1597, 1974.

280. **Asokan, S. K.,** Experimental lead cardiomyopathy: myocardial structural changes in rats given small amounts of lead, *J. Lab. Clin. Med.,* 84, 20, 1974.

281. **Brachaczek, W. W., Butler, J. W., and Pierson, W. R.,** Acid enhancement of zinc and lead atomic absorption signals, *Appl. Spectrosc.,* 28, 585, 1974.

282. **Brady, D. V., Montalvo, J. G., Jr., Jung, J., and Curran, R. A.,** Direct determination of lead in plant leaves via graphite furnace atomic absorption, *At. Absorpt. Newsl.,* 13, 118, 1974.

283. **Caprio, R. J., Margulis, H. L., and Joselow, M. M.,** Lead absorption in children and its relationship to urban traffic densities, *Arch. Environ. Health,* 28, 195, 1974.

284. **Cernik, A. A.,** Lead in blood — the analysis problem, *Chem. Br.,* 10, 58, 1974.

285. **Chisholm, J. J., Jr.,** Lead in red blood cells and plasma, *J. Pediatr.,* 84, 163, 1974.

286. **Damiani, M., Tamba, M. G., and Catano, M.,** Determination of lead in stainless steel by atomic absorption spectroscopy, *Talanta,* 21, 601, 1974.

287. **Ealy, J., Bolton, N. E., McElheny, R. J., and Morrow, R. W.,** Determination of lead in whole blood by graphite furnace atomic absorption spectrophotometry, *Am. Ind. Hyg. Assoc. J.,* 35, 566, 1974.

288. **Everson, R. J. and Parker, H. E.,** Effect of hydrogen ion concentration on the determination of lead by solvent extraction and atomic absorption spectrophotometry, *Anal. Chem.,* 46, 1966, 1974.

289. **Gershanik, J. J., Brooks, G. G., and Little, J. A.,** Fetal blood lead values in a rural area, *J. Pediatr.,* 84, 112, 1974.

290. **Goto, T.,** Atomic absorption spectrochemical determination of lead extracted with high molecular weight amines, *Bunseki Kagaku,* 23, 1165, 1974.

291. **Govindaraju, K., Mevelle, G., and Chouard, C.,** Solid sampling atomic absorption determination of lead in rock samples using the iron screw rod technique, *Anal. Chem.,* 46, 1672, 1974.

292. **Huffman, P. L. and Caruso, J. A.,** Analysis of lead in evaporated milk by flameless atomic absorption spectroscopy, *J. Agric. Food Chem.,* 22, 824, 1974.

293. **Jensen, F. O., Dolezal, J., and Langmyhr, F. J.,** Atomic absorption spectrometric determination of Cd, Pb and Zn in salts or salt solutions by hanging mercury drop electrodeposition and atomization in a graphite furnace, *Anal. Chim. Acta,* 72, 245, 1974.

294. **Kapur, J. K. and West, T. S.,** Determination of lead in instant coffee and tea powders by carbon filament atomic absorption spectrometry, *Anal. Chim. Acta,* 73, 180, 1974.

295. **Korkisch, J. and Gross, H.,** Atomic absorption determination of lead in geological materials, *Talanta,* 21, 1025, 1974.

296. **Krinitz, B.,** Effect of light on the extraction of lead and cadmium from glazed ceramic surfaces, *J. Assoc. Off. Anal. Chem.,* 57, 966, 1974.

297. **Lech, J. F., Siemer, D., and Woodriff, R.,** Determination of lead in atmospheric particulates by furnace atomic absorption, *Environ. Sci. Technol.,* 8, 840, 1974.

298. **Langmyhr, F. J., Sundli, A., and Jonsen, J.,** Atomic absorption spectrometric determination of cadmium and lead in dental material by atomization directly from the solid state, *Anal. Chim. Acta,* 73, 81, 1974.

299. **Langmyhr, F. J., Stubergh, J. R., Thomassen, Y., Hanssen, J. E., and Dolezal, J.,** Atomic absorption spectrometric determination of Cd, Pb, Ag, Th and Zn in silicate rocks by direct atomization from the solid state, *Anal. Chim. Acta,* 71, 35, 1974.

300. **Markus, J. R.,** Atomic absorption determination of lead in apples, *J. Assoc. Off. Anal. Chem.,* 57, 970, 1974.

301. **Mitchell, D. G., Aldous, K. M., and Ward, A. F.,** Determination of lead in pencil paint using a microsampling cup atomic absorption technique, *At. Absorpt. Newsl.,* 13, 121, 1974.

302. **Balraadjsing, B. D.,** The determination of total lead in soil by atomic absorption spectrophotometry, *Commun. Soil Sci. Plant Anal.,* 5, 25, 1974.

303. **Browne, R. C., Ellis, R. W., and Weightman, D.,** Interlaboratory variation in measurement of blood levels, *Lancet,* 2, 1112, 1974.

304. **Ebert, J. and Jungmann, H.,** Rapid and sensitive determination of lead in urine by flameless atomic absorption spectrometry, *Z. Anal. Chem.,* 272, 287, 1974.

305. **Welcher, G. G., Kriege, O. H., and Marks, J. Y.,** Direct determination of trace quantities of Pb, Bi, Se, Te and Tl in high temperature alloys by non-flame atomic absorption spectrophotometry, *Anal. Chem.,* 46, 1227, 1974.

306. **Yasuda, S. and Kakiyama, H.,** Determination of trace amounts of copper and lead in water by flameless atomic absorption spectroscopy, *Bunseki Kagaku,* 23, 670, 1974.

307. **Fletcher, G. E. and Collins, A. G.,** Atomic absorption methods of analysis of oilfield brines: Ba, Ca, Cu, Fe, Pb, Li, Mg, Mn, K, Na, Sr and Zn, *Rep. Invest. U.S. Bur. Mines,* RI 7861, 1974.

308. **Ford, A., Young, B., and Meloan, C.,** Determination of lead in organic coloring dyes by atomic absorption spectroscopy, *J. Agric. Food Chem.,* 22, 1034, 1974.

309. **Joselow, M. M. and Singh, N. P.,** Microanalysis for lead in blood with built-in contamination control, *Am. Ind. Hyg. Assoc. J.,* 35, 793, 1974.

310. **Stringer, C. A., Jr., Zingaro, R. A., Creech, B., and Kolar, F. L.,** Lead concentration in human lung samples, *Arch. Environ. Health,* 29, 268, 1974.

311. **Stulikova, M. and Adam, J.,** Determination of small amounts of lead in uranium, *Talanta,* 21, 1203, 1974.

312. **Szivos, K., Polos, L., Feher, I., and Pungor, E.,** Atomic absorption method for the determination of lead in air, *Period. Polytech. Chem. Eng.,* 18, 281, 1974.

313. **Thiersault, J., Gandhour, M., and Laugel, P.,** Rapid determination of lead in tin plating, *Ann. Falsif. Expert. Chim.,* 67, 27, 1974 (in French).

314. **Cernik, A. A.,** Determination of blood lead by atomic absorption spectrophotometry using a four millimeter paper punched disc and carbon sampling cup technique, *Br. J. Ind. Med.,* 31, 239, 1974.

315. **Harrison, R. M., Perry, R., and Slater, D. H.,** An adsorption technique for the determination of organic lead in street air, *Atmos. Environ.,* 8, 1187, 1974.

316. **Kondrachoff, W. and Pujade-Renaud, J. M.,** Determination of the content of soluble lead in lead chromate pigments, *Double Liaison,* 21, 593, 1974 (in French).

317. **Reith, J. F., Engelsma, J., and Van Ditmarsch, M.,** Lead and zinc contents of food and diets in the Netherlands, *Z. Lebensm. Unters. Forsch.,* 156, 271, 1974.

318. **Stoeppler, M., Backhaus, F., Dahl, R., Dumont, M., Hagedorn-Goetz, H., Hilpert, K., Klahre, P., Rutzel, H., Valenta, P., and Nurnberg, H. W.,** Determination of lead and cadmium in biological matrices, in Proc. Int. Symp. on Recent Advances in the Assessment of the Health Effects of Environmental Pollution, Paris, France, June 24 to 28, 1974.

319. **Beyer, M. E. and Bond, A. M.,** Simultaneous determination of cadmium, copper, lead and zinc in lead and zinc concentrates by ac polarographic methods. Comparison with atomic absorption spectrometry, *Anal. Chim. Acta,* 75, 409, 1975.

320. **Coker, D. T.,** Determination of individual and total diet alkyls in gasoline by a simple rapid gas chromatography-atomic absorption spectrometry technique, *Anal. Chem.,* 47, 386, 1975.

321. **Day, J. P., Hart, M., and Robinson, M. S.,** Lead in urban street dust, *Nature (London),* 253, 343, 1975.

322. **Fernandez, F. J.,** Micromethod for lead determination in whole blood by atomic absorption with use of the graphite furnace, *Clin. Chem. (Winston-Salem, N.C.),* 21, 558, 1975.

323. **Gegiou, D. and Botsivali, M.,** Atomic absorption spectrophotometric determination of lead in beverages and fruit juices and of lead extracted by their action on glazed ceramic surfaces, *Analyst (London),* 100, 234, 1975.

324. **Hermann, P.,** Determination of lead and zinc in powders by atomic absorption, *Spectrochim. Acta,* 30b, 15, 1975 (in German).

325. **Holak, W.,** Analysis of paints for lead by atomic absorption spectrometry, *Anal. Chim. Acta,* 74, 216, 1975.

326. **Kilroe-Smith, T. A.,** Linear working graphs in blood lead determination with the Beckman flameless atomic absorption cuvet, *Clin. Chem. (Winston-Salem, N.C.),* 21, 630, 1975.

327. **Korkisch, J. and Sorio, A.,** Use of ion exchange for the determination of trace elements in natural waters. V. Lead, *Talanta,* 22, 273, 1975 (in German).

328. **Landrigan, P. J., Gehlbach, S. H., Rosenblum, B. F., Shoults, J. M., Candelaria, R. M., Barthel, W. F., Liddle, J. A., Smrek, A. L., Staehling, N. W., and Sanders, J. F.,** Epidemic lead absorption near an ore smelter, *N. Engl. J. Med.,* 292, 123, 1975.

329. **Lauwerys, R., Buchet, J. P., Roels, H., Berlin, A., and Smeets, J.,** Intercompany ion program of lead, mercury, and cadmium analysis in blood, urine and aqueous solutions, *Clin. Chem. (Winston-Salem, N.S.),* 21, 551, 1975.

330. **Lakasiewicz, R. J., Bereus, P. H., and Buell, B. E.,** Rapid determination of lead in gasoline by atomic absorption spectrometry in the nitrous oxide-hydrogen flame, *Anal. Chem.,* 47, 1045, 1975.

331. **Marcus, M., Hollander, M., Lucas, R. E., and Pfeiffer, N. C.,** Microscale blood lead determination in screening: evaluation of factors affecting results, *Clin. Chem. (Winston-Salem, N.C.),* 21, 553, 1975.

332. **McCorriston, L. L. and Ritchie, R. K.,** Determination of lead in gasoline by atomic absorption spectrometry using a total consumption burner, *Anal. Chem.,* 47, 1137, 1975.

333. **Maxfield, M. E., Stopps, G. J., Barnes, J. R., Snee, R. D., Finan, M., and Azar, A.,** Recovery of blood lead concentration and of red cell δ-aminolevulinic acid dehydrase activity in dogs following return to normal diets after 75 weeks of lead feeding, *Am. Ind. Hyg. Assoc. J.,* 36, 193, 1975.

334. **Murphy, J. and Stockton, H.,** Magnesium interference in the atomic absorption determination of lead, *At. Absorpt. Newsl.,* 14, 40, 1975.

335. **Posma, F. D., Balke, J., Herber, R. F. M., and Stuik, E. J.,** Microdetermination of cadmium and lead in whole blood by flameless atomic absorption spectrometry using carbon-tube and carbon-cup as simple cell and comparison with flame studies, *Anal. Chem.,* 47, 834, 1975.

336. **Rendall, R. E. G., Baily, P., and Soskolne, C. L.,** The effect of particle size on absorption of inhaled lead, *Am. Ind. Hyg. Assoc. J.,* 36, 207, 1975.

337. **Siemer, D. D. and Stone, R. W.,** Analytical potential of nonresonance line flameless atomic absorption spectrometry for lead determination, *Appl. Spectrosc.,* 29, 240, 1975.

338. **Waldron, H. A.,** Lead levels in blood of residents near the M6-A38(M) interchange, Birmingham, *Nature (London),* 253, 345, 1975.

339. **Wilkinson, D. R. and Palmer, W.,** Lead in teeth as a function of age, *Am. Lab.,* 7, 67, 1975.

340. **Baily, P. and Kilroe-Smith, T. A.,** Effect of sample preparation on blood lead values, *Anal. Chim. Acta,* 77, 29, 1975.

341. **Bowen, C. B. and Foote, H.,** Total Lead in Gasoline Determined by Atomic Absorption Spectrophotometry, IP 74-010, Institute of Petroleum, London, 1975.

342. **Barlow, P. J. and Khera, A. K.,** Sample preparation using tissue solubilization by solvene-350 TM for lead determinations by graphite furnace atomic absorption spectrophotometry, *At. Absorpt. Newsl.,* 14, 149, 1975.

343. **Briese, L. A. and Giesy, J. P.,** Determination of lead and cadmium associated with naturally occurring organics extracted from surface waters, using flameless atomic absorption, *At. Absorpt. Newsl.,* 14, 133, 1975.

344. **Byr'ko, V. M., Prishchepov, L. F., and Shikheeva, I. A.,** Extraction-atomic absorption determination of lead in electrolytes of nickel production, *Zavod. Lab.,* 41, 525, 1975.

345. **Campbell, W. C. and Ottaway, J. M.,** Determination of lead in carbonate rocks by carbon furnace atomic absorption spectrometry after dissolution in nitric acid, *Talanta,* 22, 729, 1975.

346. **Frech, W.,** Rapid determination of lead in steel by flameless atomic absorption spectrometry, *Anal. Chim. Acta,* 77, 43, 1975.

347. **Fung, H., Yaffe, S. J., Mattar, M. E., and Lanigham, M. C.,** Blood and salivary lead levels in children, *Clin. Chim. Acta,* 61, 423, 1975.

348. **Garnys, V. P. and Matousek, J. P.,** Correction for spectral interference with determination of lead in blood by nonflame atomic absorption spectrometry, *Clin. Chem. (Winston-Salem, N.C.),* 21, 891, 1975.

349. **Garnys, V. P. and Smyths, L. E.,** Fundamental studies on improvement of precision and accuracy in flameless atomic absorption spectroscopy using the graphite tube atomizer. Lead in whole blood, *Talanta,* 22, 881, 1975.

350. **Gomez Coedo, A. and Dorado Lopez, M. T.,** Determination of low contents of lead in aluminum and its alloys, *Rev. Metal.,* 11, 61, 1975 (in Spanish).

351. **Hancock, S. and Slater, A.,** Specfic method for the determination of trace concentrations of tetramethyl and tetraethyl lead vapors in air, *Analyst (London),* 100, 422, 1975.

352. **Hankin, L., Heichel, G. H., and Botsford, R. A.,** Lead in pet foods and processed organ meats. A human problem?, *JAMA,* 231, 484, 1975.

353. **Henderson, R. W. and Andrews, D.,** Lead extraction from aluminum, *Bull. Environ. Contam. Toxicol.,* 13, 330, 1975.

354. **Huffman, H. L., Jr. and Caruso, J. A.,** Preliminary study on the effect of time on apparent lead content of evaporated milk as determined by non-flame atomic absorption spectrometry, *Talanta,* 22, 871, 1975.

355. **Jackson, K. W., Fuller, T. D., Mitchell, D. G., and Aldous, K. M.,** A rapid microsampling-cup atomic absorption procedure for the determination of lead in urine, *At. Absorpt. Newsl.,* 14, 121, 1975.

356. **Kanda, M., Hori, Y., and Matsumoto, I.,** Rapid determination of microamounts of lead in titanium dioxide by atomic absorption spectrophotometry using a graphite atomizer, *Bunseki Kagaku,* 24, 299, 1975.

357. **Kirk, M., Perry, E. G., and Arritt, J. M.,** Separation and atomic absorption measurement of trace amounts of lead, silver, zinc, bismuth and cadmium in high nickel alloys, *Anal. Chim. Acta,* 80, 163, 1975.

358. **Kono, T. and Nemori, A.,** Extraction of copper, cadmium, lead, silver and bismuth with iodide methyl isobutyl ketone in atomic absorption spectrophotometric analysis, *Bunseki Kagaku,* 24, 419, 1975.

359. **Korkisch, J. and Gross, H.,** Analysis of nuclear raw materials. VIII. Atomic absorption spectrophotometric determination of lead in tri-uranium octoxide and yellow cake samples, *Mikrochim. Acta,* 2, 413, 1975.

360. **Korkisch, J. and Sorio, A.,** Determination of cadmium, copper and lead in natural waters after anion exchange separation, *Anal. Chim. Acta,* 76, 393, 1975.

361. **Lau, O. W. and Li, K. L.,** Determination of lead and cadmium in paint by atomic absorption spectrophotometry by the delves microsampling technique, *Analyst (London),* 100, 430, 1975.

362. **Luzar, O. and Sliva, V.,** Determination of calcium oxide, magnesium oxide, alumina, silica, iron, copper, zinc, lead, cadmium, sodium oxide and potassium oxide in iron ores and agglomerates, *Hutn. Listy,* 30, 55, 1975 (in Czech).

363. **Mack, D.,** Determination of lead in wine and juices by flameless atomic absorption, *Dtsch. Lebensm. Rundsch.,* 71, 1975 (in German).

364. **Nakahara, T. and Musha, S.,** Chemical interference effects in the atomic absorption spectrometric determination of lead with premixed inert gas (entrained air)-hydrogen flames, *Appl. Spectrosc.,* 29, 352, 1975.

365. **Noller, B. N. and Bloom, H.,** Determination of atmospheric particulate lead using low volume sampling and nonflame atomic absorption, *Atmos. Environ.,* 9, 505, 1975.

366. **Robinson, J. W., Rhodes, L., and Wolcott, D. K.,** Determination and identification of molecular lead pollutants in the atmosphere, *Anal. Chim. Acta,* 78, 474, 1975.

367. **Robinson, J. W., Vadaurreta, L. E., Wolcott, D. K., Goodbread, J. P., and Kiesel, E.,** Metal specific atomic absorption detector for gas chromatography. Its use in the determination of lead alkyls in gasoline, *Spectrosc. Lett.,* 8, 491, 1975.

368. **Tarasevich, N. I., Kozyreva, G. V., and Portugal'skaya, Z. P.,** Extraction-atomic absorption determination of indium, bismuth and lead trace impurities in rocks and soils, *Vestn. Mosk. Univ. Khim.,* 16, 241, 1975.

369. **Vandeberg, J. T., Swafford, H. D., and Scott, R. W.,** Determination of low concentrations of lead in paint by atomic absorption spectroscopy, *J. Paint Technol.,* 57, 84, 1975.

370. **Yamamoto, Y., Kumamaru, T., Kamada, T., Tanaka, T., and Kawabe, M.,** Determination of the ppb level of cadmium, lead and copper in water by a carbon-tube flameless atomic absorption spectrophotometry, *Nippon Kagaku Kaishi,* 5, 836, 1975.

371. **Barnett, W. B. and McLaughlin, E. A.,** The atomic absorption determination of antimony, arsenic, bismuth, cadmium, lead and tin in iron, copper and zinc alloys with the graphite furnace, *Anal. Chim. Acta,* 80, 285, 1975.

372. **Behne, D., Braetter, P., and Wolters, W.,** Determination of lead in biological materials using flameless atomic absorption spectrometry, *Z. Anal. Chem.,* 277, 355, 1975.

373. **Block, C.,** Determination of lead in coal and coal ashes by flameless atomic absorption spectrometry, *Anal. Chim. Acta,* 80, 369, 1975.

374. **Coles, L. E., Bishop, J. R., Cassidy, W., Greenfield, S., Hill, W. H., Hoodless, R. A., King, E. E. J., Lambie, D. A., Liebmann, H., Milton, R. F., Sheppard, W. L., Watson, C. A., Shallis, P. W., and Wilson, J. J.,** Determination of small amounts of lead in organic matter by atomic absorption spectrometry, *Analyst (London),* 100, 899, 1975.

375. **Eaton, D. F., Fowles, G. W. A., Thomas, M. W., and Turnbull, G. B.,** Chromium and lead in colored printing inks used for children's magazines, *Environ. Sci. Technol.,* 9, 768, 1975.

376. **Hocquellet, P. and Labevrie, N.,** Thermoelectric atomization of silicon, germanium, tin and lead on tantalum carbide. Application to atomic absorption spectrometry, *Analusis,* 3, 505, 1975.

377. **Huntzicker, J. J., Friedlander, S. K., and Davidson, C. I.,** Material balance for automobile-emitted lead in Los Angeles Basin, *Environ. Sci. Technol.,* 9, 448, 1975.

378. **Hunt, D. C.,** Screening technique for the presence of lead and cadmium in solid samples by atomic absorption spectrophotometry, *Lab. Pract.,* 24, 411, 1975.

379. **Iosof, V., Mihalka, S., and Colios, E.,** Determination of copper, lead and zinc in mineral products by atomic absorption spectrophotometry, *Rev. Chim.,* 26, 680, 1975.

380. **Juselius, R. E., Lupovich, P., and Moriarty, R.,** Sampling problems in the microdetermination of blood lead, *Clin. Toxicol.,* 8, 1975.

381. **Musil, J.,** Determination of small amounts of lead or tin in ferrous metals and nickel by atomic absorption spectrometry, *Hutn. Listy,* 30, 292, 1975 (in Czech).

382. **Nishishita, T., Yamazoe, S., Mallett, W. R., Kashiki, M., and Oshima, S.,** Determination of possible olefin interference in the analysis of lead in gasoline by atomic absorption spectroscopy, *Anal. Lett.,* 8, 849, 1975.

383. **Stoeppler, M., Kuppers, G., Matthes, W., Rutzel, H., and Valenta, P.,** On the accuracy of lead and cadmium determination in human blood, Int. Conf. on Heavy Metals in the Environment, Toronto, October 27 to 31, 1975.

384. **Volosin, M. T., Kubasik, N. P., and Sine, H. E.,** Use of the carbon rod atomizer for analysis of lead in blood. Three methods compared, *Clin. Chem. (Winston-Salem, N.C.),* 21, 1986, 1975.

385. **Yamamoto, Y., Kumamaru, T., Kamada, T., and Tanaka, T.,** Determination of parts per 10^9 levels of cadmium, lead and copper in water by carbon tube flameless atomic absorption spectrophotometry combined with ammonium pyrrolidine-1-carbodithionate-isobutyl methyl ketone extraction, *Eisei Kagaku,* 21, 71, 1975.

386. **Acatini, C., deBerman, S. N., Colombo, C., and Fondo, O.,** Determination of silver, copper, lead, tin, antimony, iron, calcium, zinc, magnesium, potassium and manganese in canned tomatoes by atomic absorption spectrophotometry, *Revta Asoc. Bioquim. Argent.,* 40, 175, 1975 (in Spanish).

387. **Haller, H. E. and Mack, D.,** Determination of lead in wine by flameless atomic absorption spectrometry, *Dtsch. Lebensm. Rundsch.,* 71, 430, 1975 (in German).

388. **Lee, Y. S., Lao, N. T., and Crisler, J. P.,** Micromethod for lead in canned baby foods, such as juices, *Microchem. J.,* 20, 319, 1975.

389. **Vondenhoff, T.,** Determination of Pb, Cd, Cu and Zn in plant and animal material by atomic absorption in a flame and in a graphite tube after sample decomposition by the Schoniger technique, *Mitt. GDCh Fachgruppe Lebensmittelchem. Gerichtl. Chem.,* 29, 341, 1975 (in German).

390. **Brovko, I. A., Sultanov, M., and Nazarov, S. N.,** Microdetermination of atmospheric lead by extraction-atomic absorption, *Tr. Sredneaz. Reg. Nauchno Issled. Gidrometeorol. Inst.,* 35, 227, 1975 (in Russian).

391. **Favretto, L., Pertoldo Marletta, G., and Favretto Gabrielli, L.,** Lead in the Environment. A Basis for Choice, Indus. Graf. Del Bianco S.R.L., 1975 (in Italian).

392. **Ranfft, K.,** Determination of lead and cadmium in feeding stuffs by absorption spectroscopy, *Landwirtsch. Forsch.,* 31, 135, 1975 (in German).

393. **Shelton, B. J.,** Determination of silver, selenium, tellurium, antimony, tin, lead and arsenic in anode sludges, *Natl. Inst. Metall. Repub. S. Afr. Rep.,* No. 1771, 1975.

394. **Andersson, A.,** Interferences from sulfur in the determination of lead with the graphite furnace, *At. Absorpt. Newsl.*, 15, 71, 1976.

395. **Edington, D. N. and Robbins, J. A.,** Records of lead deposition in Lake Michigan sediments since 1800, *Environ. Sci. Technol.*, 10, 266, 1976.

396. **Frech, W., Lundren, G., and Lunner, S. E.,** Routine use of flameless AAS for the determination of lead in high and low alloy steels, *At. Absorpt. Newsl.*, 15, 57, 1976.

397. **Frech, W. and Cedergren, A.,** Investigations of reactions involved in flameless atomic absorption procedures. I. Application of high-temperature equilibrium calculations to a multicomponent system with special reference to the interference from chlorine in the flameless atomic absorption method for lead in steel, *Anal. Chim. Acta,* 82, 83, 1976.

398. **Graef, V.,** Atomic absorption spectrometric determination of lead in beard hair, *J. Clin. Chem. Clin. Biochem.*, 14, 181, 1976.

399. **Kilroe-Smith, T. A.,** Linearization of calibration curves with the HGA-72 flameless cuvette for the determination of lead in blood, *Anal. Chim. Acta,* 82, 421, 1976.

400. **Kovar, K. A., Lautenschlaeger, W., and Seidel, R.,** Limit test determination of heavy metals (Pb, As) in drugs by atomic absorption, *Dtsch. Apoth. Ztg.*, 115, 1855, 1976 (in German).

401. **Krinitz, B. and Holak, W.,** Collaborative study of effect of light on cadmium and lead leaching from ceramic glazes, *J. Assoc. Off. Anal. Chem.*, 59, 158, 1976.

402. **Kubota, M., Golightly, D. W., and Mavrodineanu, R.,** A rapid wire loop method for determination of lead in paint by atomic absorption spectrometry, *Appl. Spectrosc.*, 1976.

403. **Morris, N. M., Clarke, M. A., Tripp, V. W., and Carpenter, F. G.,** Determination of lead, cadmium and zinc in sugar, *J. Agric. Food Chem.*, 24, 45, 1976.

404. **Ohta, K. and Suzuki, M.,** Some observations on the atomization of lead with a metal micro-tube atomizer, *Anal. Chim. Acta,* 83, 381, 1976.

405. **Ooghe, W. and Kastelijn, H.,** Determination of traces of lead and cadmium in apertitifs by flameless atomic absorption spectrophotometry, *Ann. Falsif. Expert. Chim.*, 69, 351, 1976 (in French).

406. **Singh, N. P., Thind, I. S., Vitale, L. F., and Pawlow, M.,** Lead contents of tissues of baby rats born of and nourished by lead-poisoned mothers, *J. Lab. Clin. Med.*, 87, 273, 1976.

407. **Thompson, K. C. and Godden, R. G.,** A simple method for monitoring excessive levels of lead in whole blood using atomic absorption spectrophotometry and a rapid, direct nebulization technique, *Analyst (London)*, 101, 174, 1976.

408. **Tsutsumi, C., Koizumi, H., and Yoshikawa, S.,** Atomic absorption spectrophotometric determination of lead, cadmium and copper in foods by simultaneous extraction of the iodides with methyl isobutyl ketone, *Bunseki Kagaku*, 25, 150, 1976.

409. **Zook, E. G., Powell, J. J., Hackley, B. M., Emerson, J. A., Brooks, J. R., and Knobl, G. M., Jr.,** National marine fisheries service preliminary survey of selected seafoods for mercury, lead, cadmium, chromium and arsenic content, *J. Agric. Food Chem.*, 24, 47, 1976.

410. **Akama, Y., Nakai, T., and Kawamura, F.,** Extraction and atomic absorption spectrophotometric determination of lead with 4-benzoyl-3-methyl-1-phenyl-5-pyrazolone, *Bunseki Kagaku*, 25, 496, 1976.

411. **Chau, Y. K., Wong, P. T., and Goulden, P. D.,** Gas chromatography-atomic absorption spectrometry for the determination of tetra alkyl lead compounds, *Anal. Chim. Acta,* 85, 421, 1976.

412. **Coughtrey, P. J. and Martin, M. H.,** A comment on the analysis of biological materials for lead using atomic absorption spectroscopy, *Chemosphere,* 5, 183, 1976.

413. **Favretto, L.,** Determination of lead in seaweed at the nanogram level using a microsampling system, *At. Absorpt. Newsl.*, 15, 98, 1976.

414. **Hadeishi, T. and McLaughlin, R. D.,** Zeeman atomic absorption determination of lead with a dual chamber furnace, *Anal. Chem.*, 58, 1009, 1976.

415. **Jacobsen, J. and Snowdon, C. T.,** Increased lead ingestion in calcium-deficient monkeys, *Nature (London)*, 262, 51, 1976.

416. **Jedrzejewska, H. and Malusecka, M.,** Interference effects of some carriers on the atomic absorption of Cr, Mn, Sb, Fe, Bi and Pb in an air-acetylene flame, *Chem. Anal.*, 21, 585, 1976.

417. **Joensson, H.,** Determination of lead and cadmium in milk with modern analytical methods, *Lebensm. Unters. Forsch.*, 160, 1, 1976.

418. **Kinnison, R. R.,** Lead and its biological impacts on man, *Environ. Sci. Technol.*, 10, 644, 1976.

419. **Koizumi, H. and Yasuda, K.,** Determination of lead, cadmium and zinc using the Zeeman effect in atomic absorption spectrometry, *Anal. Chem.*, 48, 1178, 1976.

420. **Magyar, B. and Vonmont, H.,** Combined use of atomic absorption and X-ray fluorescence for the determination of lead in atmospheric dust, *Z. Anal. Chem.*, 280, 115, 1976.

421. **Minkkinen, P.,** Lead, chromium and barium in Finnish colored magazines, *Kemia-Kemi,* 6, 282, 1976.

422. **Natelson, E. A. and Fred, H. L.,** Lead poisoning from cocktail glasses. Observations on two patients, *JAMA*, 236, 2527, 1976.

423. **Pakalns, P. and Farrar, Y. J.,** Effect of fats, mineral oils and creosote on the extraction-atomic absorption determination of copper, iron, lead and manganese in water, *Water Res.*, 10, 1027, 1976.

424. **Prishchep, N. N., Guchkova, A. N., Pilipenko, A. T., and Ol'khovich, P. F.,** Atomic absorption determination of lead in alkalis, *Ukr. Khim. Zh.*, 42, 524, 1976.

425. **Sanzolone, R. F. and Chao, T. T.,** Atomic absorption spectrometric determination of copper, zinc and lead in geological materials, *Anal. Chim. Acta*, 86, 163, 1976.

426. **Satake, M., Asano, T., Takagi, Y., and Yonekubo, T.,** Determination of copper, zinc, lead, cadmium and manganese in brackish and coastal waters by the combination of chelating ion exchange separation and atomic absorption spectrophotometry, *Nippon Kagaku Kaishi*, 5, 762, 1976.

427. **Sato, A. and Saitoh, N.,** Flameless atomic absorption spectroscopic determination of arsenic, chromic and lead in sea water by use of coprecipitation with zirconium hydroxide, *Bunseki Kagaku*, 25, 663, 1976.

428. **Scott, D. R., Hemphill, D. C., Holboke, L. E., Long, S. J., Loseke, W. A., Pranger, L. J., and Thompson, R. J.,** Atomic absorption and optical emission analysis of NASN atmospheric particulates samples for lead, *Environ. Sci. Technol.*, 10, 877, 1976.

429. **Solomon, R. L. and Hartford, J. W.,** Lead and cadmium in dusts and soils in a small urban community, *Environ. Sci. Technol.*, 10, 773, 1976.

430. **Stegvic, K., Mikalsen, G., Ophus, E. M., and Mylius, E. A.,** Determination of lead in human lungs by direct flameless atomic absorption analysis of small tissue samples, *Bull. Environ. Contam. Toxicol.*, 15, 734, 1976.

431. **Toda, S., Fuwa, K., Badlaender, P., and Vallee, B. L.,** Determination of lead in MIBK extracts using a modified long path atomic absorption cell, *Spectrosc. Lett.*, 9, 225, 1976.

432. **Torsi, G. and Tessari, G.,** Time-resolved distribution of atoms in flameless spectrometry: lead release in hydrogen atmosphere, *Anal. Chem.*, 48, 1318, 1976.

433. **Watkins, D., Corbyons, T., Bradshaw, J., and Winefordner, J.,** Determination of lead in confection wrappers by atomic absorption spectrometry, *Anal. Chim. Acta*, 85, 403, 1976.

434. **Wood, E. J., Gonzalez, R., Blanco, J. A., and Rucci, A. O.,** Determination of lead in propellant samples by atomic absorption spectrophotometry, *Talanta*, 23, 473, 1976.

435. **Broeckaert, J. A.,** Application of hollow cathode excitation coupled to vidicon detection to the simultaneous multielement determination of toxic elements in airborne dust. A unique sampling-analysis procedure for lead and cadmium, *Bull. Soc. Chim. Belg.*, 85, 755, 1976.

436. **Ott, R. E.,** Determination of submicro concentrations of lead in water by atomic absorption, *Tr. Tallin. Politekh. Inst.*, 397, 79, 1976 (in Russian).

437. **Petrino, P., Cas, M., and Estienne, J.,** Application of physico-chemical methods to the determination of lead in wine, *Ann. Falsif. Expert. Chim.*, 69, 87, 1976.

438. **Vijan, P. N. and Wood, G. R.,** Semiautomated determination of lead by hydride generation and atomic absorption spectrophotometry, *Analyst (London)*, 101, 966, 1976.

439. **Wall, G.,** Determination by atomic absorption spectrophotometry of copper, lead and zinc in sulfide concentrates, *Natl. Inst. Metall. Repub. S. Afr. Rep.*, No. 1798, 1976.

440. **Albert, R., Beigl, E., Kinzel, H., and Steiner, G. M.,** Micro-determination of lead in samples of biological material by flameless atomic absorption spectrophotometry, *Z. Pflanzenphysiol.*, 80, 43, 1976 (in German).

441. **Hoffman, H. E., Nathansen, H., and Altman, F.,** Determination of calcium, zinc and lead in unused lubricating oils with the AAS-1 atomic absorption spectrophotometer, *Jenaer Rundsch.*, 21, 302, 1976 (in German).

442. **Horak, O.,** Determination by flameless atomic absorption spectroscopy of lead and cadmium in grain and grass samples taken at different distances from motorways, *Landwirtsch. Forsch.*, 29, 289, 1976 (in German).

443. **Polos, L., Fodor, P., Szivos, K., Kantor, T., and Pungor, E.,** Determination of lead in air-borne dust by the atomic absorption techniques, *Hung. Sci. Instrum.*, 38, 45, 1976.

444. **Gherardi, S., Bigliardi, D., and Bellucci, G.,** Determination of lead in foodstuff using flameless atomic absorption. I. Influence of various ions, *Ind. Conserve*, 51, 273, 1976.

445. **Oudart, N., Guichard, C., and Delage, C.,** Determination of lead by atomic absorption spectrophotometry in relation to toxicological research, *J. Eur. Toxicol.*, 9, 69, 1976.

446. **Schuller, P. L. and Egan, H.,** Cadmium, Lead, Mercury and Methyl Mercury Compounds: Review of Methods of Trace Analysis and Sampling with Special Reference to Food, Rep. No. 92-5-1-00094-M-84, Food and Agriculture Organization, Rome, 1976.

447. **Brashnarova, A., Faitondzhiev, L., and Havezov, I.,** Effect of matrix components in the determination of lead in soils by atomic absorption spectrophotometry, *Z. Anal. Chem.*, 283, 303, 1977.

448. **Castellani, F., Riccioni, R., Gusteri, M., Bartocci, V., and Ceskon, P.,** Direct microdetermination of lead in wax crayons by flameless atomic absorption technique, *At. Absorpt. Newsl.*, 16, 57, 1977.

449. **Eller, P. M. and Haartz, J. C.,** A study of methods for the determination of lead and cadmium, *J. Am. Ind. Hyg. Assoc.,* 38, 116, 1977.

450. **Favretto, Gabreilli, L., Pertoldo Marletta, G., and Favretto, L.,** Rapid determination of particulate lead in sea water by a reagent-free microsampling system, *At. Absorpt. Newsl.,* 16, 4, 1977.

451. **Florence, T. M. and Batley, G. E.,** Determination of the chemical forms of trace metals in natural waters with special reference to copper, lead, cadmium and zinc, *Talanta,* 24, 151, 1977.

452. **Frech, W. and Cedergren, A.,** Investigations of reactions involved in flameless atomic absorption procedures. III. A study of factors influencing the determination of lead in strong sodium chloride solutions, *Anal. Chim. Acta,* 88, 57, 1977.

453. **Heffron, C. L., Reid, J. T., Furr, A. K., Parkinson, T. F., King, J. M., Bache, C. A., St. John, L. E., Jr., Gutenmann, W. H., and Lisk, D. J.,** Lead and other elements in sheet fed colored magazines and newsprint, *J. Agric. Food Chem.,* 25, 657, 1977.

454. **Hodges, D. J.,** Observations on the direct determination of lead in complex matrixes by carbon furnace atomic absorption spectrophotometry, *Analyst (London),* 102, 66, 1977.

455. **Marks, J. Y., Welcher, G. G., and Spellman, R. J.,** Atomic absorption determination of lead, bismuth, selenium, tellurium, thallium and tin in complex alloys using direct atomization from metal chips in the graphite furnace, *Appl. Spectrosc.,* 31, 9, 1977.

456. **Mitchell, D. G., Mills, W. N., Ward, A. F., and Aldous, K. M.,** Determination of cadmium, copper and lead in sludges by microsampling cup atomic absorption spectrometry in a nitrous oxide-acetylene flame, *Anal. Chim. Acta,* 90, 275, 1977.

457. **Pakalns, P. and Farrar, Y. J.,** The effect of surfactants on the extraction-atomic absorption spectrophotometric determination of copper, iron, manganese, lead, nickel, zinc, cadmium and cobalt, *Water Res.,* 11, 145, 1977.

458. **Prager, M. J. and Graves, D.,** Comparison of X-ray fluorescence, atomic absorption and gravimetric methods for the determination of lead in glassware, *J. Assoc. Off. Anal. Chem.,* 60, 609, 1977.

459. **Sirota, G. R. and Uthe, J. F.,** Determination of tetraalkyllead compounds in biological materials, *Anal. Chem.,* 49, 823, 1977.

460. **Thompson, K. C., Wagstaff, K., and Wheatstone, K. C.,** Method for the minimization of matrix interferences in the determination of lead and cadmium in nonsaline waters by atomic absorption spectrophotometry using electrothermal atomization, *Analyst (London),* 102, 310, 1977.

461. **Wegscheider, W., Knapp, G., and Spitzy, H.,** Statistical investigations of interferences in graphite furnace atomic absorption spectrometry. III. Lead, *Z. Anal. Chem.,* 283, 183, 1977.

462. **Woodis, T. C., Jr., Hunter, G. B., and Johnson, F. J.,** Statistical studies of matrix effects on the determination of cadmium and lead in fertilizer and plant tissue by flameless atomic absorption spectrometry, *Anal. Chim. Acta,* 90, 127, 1977.

463. **Yasuda, S. and Kakiyama, H.,** Vaporization of atoms and molecules during heating of cadmium, lead and zinc salts in a carbon tube atomizer, *Anal. Chim. Acta,* 89, 369, 1977.

464. **Capar, S. G.,** Atomic absorption spectrophotometric determination of lead, cadmium, zinc and copper in clams and oysters: collaborative study, *J. Assoc. Off. Anal. Chem.,* 60, 1400, 1977.

465. **Fjerdingstad, E., Danscher, G., and Fjerdingstad, E. J.,** Changes in zinc and lead content of rat hippocampus and whole brain following intravital dithizone treatment as determined by flameless atomic absorption spectrophotometry, *Brain Res.,* 130, 369, 1977.

466. **Fuller, C. W.,** The effect of graphite tube condition on the determination of lead in the presence of magnesium chloride by electrothermal atomic absorption spectrometry, *At. Absorpt. Newsl.,* 16, 106, 1977.

467. **Inoue, S., Yotsuyanagi, T., Sasaki, M., and Aomura, K.,** Atomic absorption method for the determination of trace amounts of cadmium, copper, lead and zinc by ion-pair extraction of dimercaptomaleonitrile complexes, *Bunseki Kagaku,* 26, 550, 1977.

468. **Kanuti, R., Balsiger, C., and Schlatter, C.,** Trace metal analysis in the parts per billion range in biological material; e.g., the determination of lead in blood by atomic absorption spectrometry with a graphite tube, *Mitt. Geb. Lebensmittelunters. Hyg.,* 68, 78, 1977.

469. **Lester, J. N., Harrison, R. M., and Perry, R.,** Rapid flameless atomic absorption analysis of the metallic content of sewage sludges. I. Lead, cadmium and copper, *Sci. Total Environ.,* 8, 153, 1977.

470. **McLaren, J. W. and Wheeler, R. C.,** Double peaks in the atomic absorption determination of lead using electrothermal atomization, *Analyst (London),* 102, 542, 1977.

471. **Menden, E. E., Brockman, D., Choudhury, H., and Petering, H. G.,** Dry ashing of animal tissues for atomic absorption spectrometric determination of zinc, copper, cadmium, lead, iron, manganese, magnesium and calcium, *Anal. Chem.,* 49, 1644, 1977.

472. **Mueller, U., Hauser, E., Kappeler, A., Merk, E., Steiner, K., and Windemann, K.,** Collaborative test on determination of lead in foods by atomic absorption spectrophotometry, *Mitt. Geb. Lebensmittelunters. Hyg.,* 68, 126, 1977 (in German).

473. **Brodie, K. G. and Stevens, B. J.**, Measurement of whole blood lead and cadmium at low levels using an automatic sampler dispenser and furnace atomic absorption, *J. Anal. Toxicol.*, 1, 282, 1977.

474. **Weiner, J.**, Determination of lead in beer by atomic absorption spectroscopy, *J. Inst. Brew.*, 83, 82, 1977.

475. **Zimdahl, R. L. and Skogerboe, R. K.**, Behavior of lead in soil, *Environ. Sci. Technol.*, 11, 1202, 1977.

476. **Zawadzka, H., Baralkiewicz, D., and Elbanowska, H.**, Flame atomic absorption spectrometric determination of cobalt, cadmium, lead, nickel, copper and zinc in natural waters, *Chem. Anal.*, 22, 913, 1977.

477. **Cirlin, E. H. and Housely, R. M.**, A flameless atomic absorption study of the volatile trace metal lead in lunar samples, *Proc. Lunar Sci. Conf.*, Geochim. Cosmochim. Acta, 2 (Suppl. 8), 3931, 1977.

478. **Cohen, E. S. and Kurchatova, G.**, Determination of lead and cadmium traces in atomospheric air by flame and flameless atomic absorption, *Dokl. Bolg. Akad. Nauk*, 30, 1439, 1977.

479. **Delves, H. T.**, Analytical techniques for blood-lead measurements, *J. Anal. Toxicol.*, 1, 261, 1977.

480. **Del-Castiho, P. and Herber, R. F.**, The rapid determination of Cd, Pb, Cu and Zn in whole blood by atomic absorption spectrometry with electrothermal atomization improvements in precision with a peak-shape monitoring devices, *Anal. Chim. Acta*, 94, 269, 1977.

481. **Johansson, K., Frech, W., and Cedergren, A.**, Investigations of reactions involved in flameless atomic absorption procedures. VI. A study of some factors influencing the determination of lead in sulfate matrices, *Anal. Chim. Acta*, 94, 245, 1977.

482. **Kalman, S. M.**, The pathophysiology of lead poisoning: a review and a case report, *J. Anal. Toxicol.*, 1, 277, 1977.

483. **Kozusnikova, J. and Kolarova, A.**, Rapid determination of low contents of lead in steels by the method of flameless atomic absorption spectrometry, *Chem. Listy*, 71, 1287, 1977.

484. **Pellerin, F. and Goulle, J. P.**, Detection and rapid determination of cadmium, copper, lead and zinc in dyes and antioxidants authorized for use in drugs and foodstuffs, *Ann. Pharm. Fr.*, 35, 189, 1977.

485. **Shamsipoor, M. and Wahdat, F.**, Determination of lead in plants, *Z. Anal. Chem.*, 288, 191, 1977.

486. **Zmudzki, J.**, Determination of lead in biological samples by atomic absorption spectrophotometry, *Med. Water*, 33, 179, 1977 (in Polish).

487. **Barlow, P. J.**, Microdetermination of lead and cadmium in pasturized market milks by flameless atomic absorption spectroscopy using a base digest, *J. Dairy Res.*, 44, 377, 1977.

488. **Dujmovic, M.**, Determination of bismuth, mercury, antimony, tin, tellurium and lead in aqueous solutions by flameless atomic absorption spectroscopy, *GIT Fachz. Lab.*, 21, 861, 1977.

489. **Kronstein, M.**, Atomic absorption analysis of organolead toxicants released from antifouling paints, *Mod. Paint Coat.*, 67, 57, 1977.

490. **Sabet, S., Ottaway, J. M., and Fell, G. S.**, Comparison of the Delves cup and carbon furnace atomization used in atomic absorption spectrometry for the determination of lead in blood, *Proc. Anal. Div. Chem. Soc.*, 14, 300, 1977.

491. **Favretto, L., Marletta, G. P., and Gabrielli, L. F.**, Rapid determination of lead in atmospheric particulated matter by atomic absorption spectroscopy with a microsampling system, *Ann. Chim.*, 67, 377, 1977.

492. **Harrison, R. M. and Laxen, D. P. H.**, Comparative study of methods for analysis of total lead in soils, *Water Air Soil Pollut.*, 8, 387, 1977.

493. **Kujirai, O., Kobayashi, T., and Sudo, E.**, Determination of trace quantities of lead and bismuth in heat-resisting alloys by atomic absorption spectrometry with heated graphite atomizer, *Trans. Jpn. Inst. Met.*, 18, 775, 1977.

494. **Alt, F.**, A simple, fast and reliable determination of lead in blood by atomic absorption spectrometry, *Z. Anal. Chem.*, 290, 108, 1978.

495. **Cross, P. J. and Husain, D.**, Reactions of ground state lead atoms Pb ($6^3 P_0$) with alkyl bromides studied by atomic absorption spectroscopy, *J. Photochem.*, 8, 183, 1978.

496. **Geladi, P. and Adams, F.**, The determination of cadmium copper, iron, lead and zinc in aerosols by atomic absorption spectrometry, *Anal. Chim. Acta*, 96, 229, 1978.

497. **Girgis-Takla, F. and Chroneos, I.**, Determination of lead in plastic containers for pharmaceutical products by atomic absorption spectrophotometry using a carbon rod atomizer, *Analyst (London)*, 103, 122, 1978.

498. **Hudnik, V., Gomiscek, S., and Gorenc, B.**, The determination of trace metals in mineral waters. I. Atomic absorption spectrometric determination of Cd, Co, Cr, Cu, Ni and Pb by electrothermal atomization after concentration by coprecipitation, *Anal. Chim. Acta*, 98, 39, 1978.

499. **Ikeda, M., Kaneko, I., Watanabe, T., Ishihara, N., and Miura, T.**, An automated system for the determination of lead in blood, manganese in urine and nickel in waste water, *J. Am. Ind. Hyg. Assoc.*, 39, 226, 1978.

500. **Jackson, K. W., Marczak, E., and Mitchell, D. G.**, Rapid determination of lead in biological tissues by microsampling cup atomic absorption spectrometry, *Anal. Chim. Acta*, 97, 37, 1978.

501. **Lerner, L. A., Sedykh, E. M., and Igishina, E. V.**, Determination of lead content in soils by the method of nonflame atomic absorption spectroscopy, *Pochvovedenie*, 2, 115, 1978.

502. **Manning, D. C. and Slavin, W.,** Determination of lead in a chloride matrix, *At. Absorpt. Newsl.,* 17, 43, 1978.

503. **Masunaga, K., Okada, M., and Miyagawa, H.,** Determination of trace lead by a combined coprecipitation-flameless atomic absorption spectrophotometry, *Nippon Kagaku Kaishi,* 3, 395, 1978.

504. **Nise, G. and Vesterberg, O.,** Blood lead determination by flameless atomic absorption spectroscopy, *Clin. Chim. Acta,* 84, 129, 1978.

505. **Pickford, C. J. and Rossi, G.,** Determination of lead in atmospheric particulates using an automated atomic absorption spectrophotometric system with electrothermal atomization, *Analyst (London),* 103, 341, 1978.

506. **Siemer, D. D. and Wei, H.,** Determination of lead in rocks and glasses by temperature controlled graphite cup atomic absorption spectrometry, *Anal. Chem.,* 50, 147, 1978.

507. **Ure, A. M., Hernandez-Artiga, M. P., and Mitchell, M. C.,** A carbon rod atomizer for the determination of cadmium and lead in plant materials and soil extracts. III. Simultaneous determination of cadmium by atomic fluorescence and lead by atomic absorption spectrometry, *Anal. Chim. Acta,* 96, 37, 1978.

508. **Watling, R. J.,** The use of a slotted tube for the determination of lead, zinc, cadmium, bismuth, cobalt, manganese and silver by atomic absorption spectrometry, *Anal. Chim. Acta,* 97, 395, 1978.

509. **Andrews, D. G., Aziz-Alrahman, A. M., and Headridge, J. G.,** Determination of lead in iron and steel by atomic absorption spectrophotometry with the introduction of solid samples into an induction furnace, *Analyst (London),* 103, 909, 1978.

510. **Boiteau, H. L. and Metayer, C.,** Microdetermination of lead, cadmium, zinc and tin in biological materials by atomic absorption spectrometry after mineralization and extraction, *Analusis,* 6, 350, 1978.

511. **Elson, C. M., Dostal, J., Hynes, D. L., and de Albuquerque, C. A. R.,** Silver, cadmium, and lead contents of some rock reference samples, *Geostandards Newsl.,* 11, 121, 1978.

512. **Evans, W. H., Read, J. I., and Lucas, B. E.,** Evaluation of a method for the determination of total cadmium, lead and nickel in foodstuffs using measurement by flame atomic absorption spectrophotometry, *Analyst (London),* 103, 580, 1978.

513. **Fernandez, F. J.,** Automated microdetermination of lead in blood, *At. Absorpt. Newsl.,* 17, 115, 1978.

514. **Farris, F. F., Poklis, A., and Griesmann, G. E.,** Atomic absorption spectroscopic determination of lead extracted from acid-solubilized tissues, *J. Assoc. Off. Anal. Chem.,* 61, 660, 1978.

515. **Hoenig, M. and Vanderstappen, R.,** Determination of cadmium, copper, lead, zinc and manganese in plants by flame atomic absorption spectroscopy. Mineralization effects, *Analusis,* 6, 312, 1978.

516. **Krinitz, B.,** Rapid screening field test for detecting cadmium and lead extracted from glazed ceramic dinnerware: collaborative study, *J. Assoc. Off. Anal. Chem.,* 61, 1124, 1978.

517. **Langmyhr, F. J. and Kjuus, I.,** Direct atomic absorption spectrometric determination of cadmium, lead and manganese in bone and of lead in ivory, *Anal. Chim. Acta,* 100, 139, 1978.

518. **Manning, D. C. and Slavin, W.,** Determination of lead in a chloride mixture with a graphite furnace, *Anal. Chem.,* 50, 1234, 1978.

519. **Maruta, T., Minegishi, K., and Sudoh, G.,** Atomic absorption spectrometric determination of amounts of copper, manganese, lead and chromium in cements by direct atomization in a carbon furnace, *Yogyo Kyokai Shi,* 86, 532, 1978.

520. **Moore, M. R., Campbell, B. C., Meredith, P. A., Beattie, A. D., Goldberg, A., and Campbell, D.,** The association between lead concentrations in teeth and domestic water lead concentrations, *Clin. Chim. Acta,* 87, 11, 1978.

521. **Robbins, W. B. and Caruso, J. A.,** Determination of lead and cadmium in normal human lung tissue by flameless atomic absorption spectroscopy, *Spectrosc. Lett.,* 11, 333, 1978.

522. **Schmidt, W. and Dietl, F.,** Enrichment and determination of lead in digested soils and sediments and in soil and sediment extracts using flame atomic absorption, *Z. Anal. Chem.,* 291, 213, 1978.

523. **Stoeppler, M., Brandt, K., and Rains, T. C.,** Contributions to automated trace analysis. II. Rapid method for the automated determination of lead in whole blood by electrothermal atomic absorption spectrophotometry, *Analyst (London),* 103, 714, 1978.

524. **Sulek, A. M., Elkins, E. R., and Zink, E. W.,** Lead in evaporated milk by anodic stripping voltammetry and atomic absorption spectrophotometry: cooperative interlaboratory study, *J. Assoc. Off. Anal. Chem.,* 61, 931, 1978.

525. **Therrell, B. L., Drosche, J. M., and Dziuk, T. W.,** Analysis for lead in undiluted whole blood by tantalum ribbon atomic absorption spectrophotometry, *Clin. Chem. (Winston-Salem, N.C.),* 24, 1182, 1978.

526. **Viets, J. G.,** Determination of silver, bismuth, cadmium, copper, lead and zinc in geologic materials by atomic absorption spectrometry with tricaprylyl methyl ammonium chloride, *Anal. Chem.,* 50, 1097, 1978.

527. **Wettern, M.,** Determination of lead in small samples of the marine diatom *skeletonama costatum* (grev.) cleve by flameless atomic absorption spectrometry after hydrofluoric acid decomposition, *Z. Anal. Chem.,* 292, 278, 1978.

528. **Zaguzin, V. P., Karmanova, N. G., and Pogrebnyak, Y. F.,** Use of the graphite capsule-flame atomizer for determining copper and lead in powdered rock samples, *Zh. Prikl. Spektrosk.,* 28, 963, 1978.

529. **Akama, Y., Naka, H., Nakai, T., and Kawamura, F.,** Extraction and atomic absorption spectrophotometric determination of lead with 4-capryl-3-methyl-1-phenyl-5-pyrazolone, *Bunseki Kagaku,* 27, 680, 1978.

530. **Alt, F. and Massmann, H.,** Determination of lead in blood by atomic absorption spectrometry, *Spectrochim. Acta,* 33b, 337, 1978.

531. **Cammann, K.,** A critical comparison between flameless atomic absorption spectroscopy and an improved electrochemical anodic stripping technique in the case of a rapid trace determination of lead in geological samples, *Z. Anal. Chem.,* 293, 97, 1978.

532. **Feinberg, M. and Ducauze, C.,** Mineralization of lead and cadmium in animal tissue with regard to their determination by atomic absorption spectrometry, *Bull. Soc. Chim. Fr.,* 11-12, 419, 1978.

533. **Jin, K., Taga, M., Yoshida, H., and Hikime, S.,** Determination of trace amount of lead by atomic absorption spectrometry after hydride evolution process. I. Sensitivity enhancement effect by the combined use of dichromate hydrogen peroxide, peroxodisulfate or permanganate with sodium borohydride in atomic absorption spectrometry of lead, *Bunseki Kagaku,* 27, 759, 1978.

534. **Kantor, T., Fodor, P., and Pungor, E.,** Determination of traces of lead, cadmium and zinc in copper by an arc-nebulization and flame atomic absorption technique, *Anal. Chim. Acta,* 102, 15, 1978.

535. **Loos-Neskovic, C., Fedoroff, M., and Revel, G.,** Determination of lead in high purity silver. Comparison between photon activation and atomic absorption spectrometry, *Radiochem. Radioanal. Lett.,* 36, 13, 1978.

536. **McDonald, C., Mahayani, M. M., and Kanjo, M.,** Solvent extraction studies of lead using alamine 336 and aliquot 336-S, *Sep. Sci. Technol.,* 13, 429, 1978.

537. **Minochkhina, L. N., Gazarov, R. A., Fadeeva, V. I., and Zorov, N. B.,** Extraction atomic absorption determination of palladium and lead in catalysts, *Mosk. Univ. Ser. 2 Khim.,* 19, 619, 1978.

538. **Minochkhina, L. N., Gazarov, R. A., Fadeeva, V. I., Zorov, N. B., Bezlepkin, A. I., and Panchishnyi, V. I.,** Determination of lead and palladium in catalysts of the complete combustion of exhaust gases by an atomic absorption method, *Zavod. Lab.,* 44, 1323, 1978.

539. **Schetze, I. and Mueller, W.,** Determination of trace metals in dietary fats and emulsifiers by flameless atomic absorption spectrometry. I. Determination of bound heavy metals; copper, iron, nickel, zinc, lead and cadmium in dietary fats, *Nahrung,* 22, 777, 1978.

540. **Holm, J.,** Simplified digestion method and measuring technique for determining lead, cadmium and arsenic in animal tissues by atomic absorption spectrometry, *Fleischwirtschaft,* 58, 745, 1978 (in German).

541. **Koops, J. and Westerbeck, D.,** Determination of lead and cadmium in pasteurized liquid milk by flameless atomic absorption spectrophotometry, *Neth. Milk Dairy J.,* 32, 149, 1978.

542. **Vackova, M. and Zemberyova, M.,** Determination of Cu, Pb, Co, Ni and Zn in ferromanganese by atomic absorption spectrometry, *Hutn. Listy,* 33, 890, 1978.

543. **Viala, A., Cano, J. P., Gouezo, F., Sauve, J. M., Bourbon, P., and Mallet, B.,** A methodical approach for determining lead, cadmium and chromium in atmospheric dusts by flameless atomic absorption spectrometry, *Pollut. Atmos.,* 20, 115, 1978 (in French).

544. **Terashima, S.,** Atomic absorption determination of Mn, Fe, Cu, Ni, Co, Pb, Zn, Si, Al, Ca, Mg, Na, K, Ti and Sr in manganese nodules, *Chishitsu Chosasho Geppo,* 29, 401, 1978 (in German).

545. **McKown, M. M., Tschrin, C. R., and Lee, P. P. F.,** Investigation of Matrix Interferences for Atomic Absorption Spectroscopy of Sediments, Rep. No. EPA-600/7-78/085, U.S. Environmental Protection Agency, Office of Research and Development, Corvallis, Ore., 1978.

546. **Allain, P. and Mauras, Y.,** Microdetermination of lead and cadmium in blood and urine by graphite furnace atomic absorption, *Clin. Chim. Acta,* 91, 41, 1979.

547. **Berenguer, V., Guinon, J. L., and De la Guardia, M.,** Rapid determination of lead in petrol by atomic absorption spectrometry of emulsified samples, *Z. Anal. Chem.,* 294, 416, 1979.

548. **Boone, J., Hearn, T., and Lewis, S.,** Comparison of interlaboratory results for blood lead with results from a definitive method, *Clin. Chem. (Winston-Salem, N.C.),* 25, 389, 1979.

549. **Bower, N. W. and Ingle, J. D.,** Precision of flame atomic absorption spectrometric measurements of Al, Cr, Co, Eu, Pb, Mn, Ni, K, Se, Si, Ti and V, *Anal. Chem.,* 51, 72, 1979.

550. **Chow, C.,** Determination of lead in columbite concentrates by atomic absorption spectrometry after sulfide separation, *Analyst (London),* 104, 154, 1979.

551. **Fetterolf, D. D. and Syty, A.,** Sample preparation and the determination of lead in chewing gum by nonflame atomic absorption spectrometry, *J. Agric. Food Chem.,* 27, 377, 1979.

552. **Fischbein, A., Rice, C., Sarzoki, L., Kon, S. H., Petrocci, M., and Selikoff, I. J.,** Exposure to lead in firing ranges, *JAMA,* 241, 1141, 1979.

553. **Hirao, Y., Fukumota, K., Sugisaki, H., and Kimura, K.,** Determination of lead in seawater by furnace atomic absorption spectrometry after concentration with yield tracer, *Anal. Chem.,* 51, 651, 1979.

554. **Iu, K. L., Pulford, I. D., and Duncan, H. J.,** Determination of cadmium, cobalt, copper, nickel and lead in soil extracts by dithizone extraction and atomic absorption spectrometry with electrothermal atomization, *Anal. Chim. Acta,* 106, 319, 1979.

555. **Lundberg, E. and Frech, W.,** Direct determination of trace metals in solid samples by atomic absorption spectrometry with electrothermal atomizers. I. Investigations of homogeneity for lead and antimony in metallurgical materials, *Anal. Chim. Acta,* 104, 67, 1979.

556. **Lundberg, E. and Frech, W.,** Direct determination of trace metals in solid samples by atomic absorption spectrometry with electrothermal atomizers. II. Determination of lead in steels and nickel-base alloys, *Anal. Chim. Acta,* 104, 75, 1979.

557. **Meranger, J. C., Subramanian, K. S., and Chalifoux, C.,** A national survey for Cd, Cr, Cu, Pb, Zn, Ca and Mg in Canadian drinking water supplies, *Environ. Sci. Technol.,* 13, 707, 1979.

558. **Ng, S. S. M. and Bhattacharya, P. K.,** Determination of microquantities of lead in ferrous sulfate, zinc oxide, manganese sulfate and iron powder by using solvent extraction and flameless atomic absorption spectroscopy, *J. Assoc. Off. Anal. Chem.,* 62, 473, 1979.

559. **Nuhfer, E. B. and Romanosky, R. R.,** Determination of lead in earth materials from lithium metaborate flux-fusion dissolutions, *At. Absorpt. Newsl.,* 18, 8, 1979.

560. **Okuneva, G. A., Gibalo, I. M., Kuzyakov, Y. Y., Zheleznova, A. A., and Shashova, M. V.,** Effect of cations during atomic absorption determination of Fe, Mn, Ni, Pb and Zn in copper-based alloys, *Zh. Anal. Khim.,* 34, 661, 1979.

561. **Radziuk, B., Thomassen, Y., Van Loon, J. C., and Chau, Y. K.,** Determination of alkyl lead compounds in air by gas chromatography and atomic absorption spectrometry, *Anal. Chim. Acta,* 105 255, 1979.

562. **Ihida, M., Ishii, T., and Ohnishi, R.,** Precision on the determination of trace elements in coal, *Am. Chem. Soc. Div. Fuel Chem.,* 24, 262, 1979.

563. **Anonymous,** Determination of Trace Metals in Liquid Coke-Oven Effluents by Atomic Absorption Spectrophotometry, Carbonization Res. Rep. 76, British Carbonization Research Association, Chesterfield, Derbyshire, 1979, 17.

564. **Fonds, A. W., Kempf, T., Minderhound, A., and Sonneborn, M.,** Heavy metal content of various kinds of water, *Reinhalt Wassers,* 78, 1979 (in German).

565. **Slavin, W. and Manning, D. C.,** Reduction of matrix interferences for lead determination with the L'vov platform and the graphite furnace, *Anal. Chem.,* 51, 261, 1979.

566. **Vratkovskaya, S. V. and Pogrebnyak, Y. F.,** Determination of copper, lead and zinc in slightly mineralized water by flame atomic absorption spectrophotometry, *Zh. Anal. Khim.,* 34, 759, 1979.

567. **Weisel, C. P., Fasching, J. L., Piotrowicz, S. R., and Duce, R. A.,** A modified standard addition method for determining Cd, Pb, Cu and Fe in seawater derived samples by atomic absorption spectroscopy, *Adv. Chem. Ser.,* 172, 134, 1979.

568. **Arpadjan, S. and Alexandrova, I.,** Determination of trace elements (Pb, Bi, Cu, Zn and Ag) in high-purity tin by flame atomic absorption spectrophotometry, *Z. Anal. Chem.,* 298, 159, 1979.

569. **Armannsson, H.,** Dithizone extraction and flame atomic absorption spectrometry for the determination of Cd, Zn, Pb, Cu, Ni, Co and Ag, in seawater and biological tissues, *Anal. Chim. Acta,* 110, 21, 1979.

570. **Aruscavage, P. J. and Campbell, E. Y.,** The determination of lead in 13 USGS rocks, *Talanta,* 26, 1052, 1979.

571. **Bengtsson, M., Danielsson, L. G., and Magnusson, B.,** Determination of Cd and Pb in seawater after extraction using electrothermal atomization. Minimization of interferences from excavated sea salts, *Anal. Lett.,* 12, 1367, 1979.

572. **Berndt, H. and Messerschmidt, J.,** A microanalytical method using flame atomic absorption and emission spectrometry (Platinum Loop Method). I. Basic method and application to the determination of lead and cadmium in drinking water, *Spectrochim. Acta,* 34b, 241, 1979.

573. **Bozsai, G. and Csanady, M.,** Systematic investigations on the heavy metal pollution (cadmium, lead, copper, zinc, chromium and barium) of drinking water using atomic absorption spectrometric analysis, *Z. Anal. Chem.,* 297, 370, 1979.

574. **Debeka, R. W.,** Graphite furnace atomic absorption spectrometric determination of lead and cadmium in foods after solvent extraction and stripping, *Anal. Chem.,* 51, 902, 1979.

575. **Faitondzhiev, L. P. and Braschnarova, A. G.,** Determination of lead in earth by atomic absorption spectrophotometry, *Landwirtsch. Forsch.,* 32, 1, 1979.

576. **Forrester, J. E., Lehecka, V., Johnston, J. R., and Ott, W. L.,** Direct determination of trace quantities of Sb, As, Bi, Cd, Pb, Se, Ag, Te and Tl in high-purity nickel by electrothermal atomic absorption spectrometry, *At. Absorpt. Newsl.,* 18, 73, 1979.

577. **Goto, T., Hirayama, K., and Unohara, N.,** Atomic absorption spectrophotometry of lead after thiosulfato complex solvent extraction with trioctyl methyl ammonium chloride, *Bunseki Kagaku,* 28, 432, 1979.

578. **Hageman, L. R., Nichols, J. A., Viswanadham, P., and Woodriff, R.,** Comparative interference study for atomic absorption lead determination using a constant temperature vs. a pulsed-type atomizer, *Anal. Chem.,* 51, 1406, 1979.

579. **Heinrichs, H.,** Determination of lead in geological and biological materials by graphite furnace atomic absorption spectrometry, *Z. Anal. Chem.,* 295, 355, 1979.

580. **Hirata, S.,** Determination of lead in Bay sediments by atomic absorption spectrometry with graphite furnace, *Bunseki Kagaku,* 28, 503, 1979.

581. **Kharlamov, I. P., Eremina, G. V., Niemark, V. Y., and Belkova, G. V.,** Determination of lead impurity in alloys by the flameless atomic absorption spectrophotometric method, *Zavod. Lab.,* 45, 391, 1979.

582. **Ohta, K. and Suzuki, M.,** Determination of lead in water by atomic absorption spectrometry with electrothermal atomization, *Z. Anal. Chem.,* 298, 140, 1979.

583. **Pleban, P. A. and Pearson, K. H.,** Determination of lead in whole blood and urine using Zeeman effect flameless atomic absorption spectroscopy, *Anal. Lett.,* 12, 935, 1979.

584. **Baily, P., Norval, E., Kilroe-Smith, T. A., Skikne, M. I., and Rollin, H. B.,** The application of metal-coated graphite tubes to the determination of trace metals in biological materials. I. The determination of lead in blood using a tungsten coated graphite tube, *Microchem. J.,* 24, 107, 1979.

585. **Boyer, K. W., Capar, S. G., Jones, J. W., Suddendorf, R. F., and Forwalter, J.,** Multielement/trace analysis identifies 48 elements in foods. Simultaneous element analysis and data reduction reaches parts per billion, *Food Process.,* 40, 72, 1979.

586. **Flanjak, J. and Lee, H. Y.,** Trace metal content of livers and kidneys of cattle, *J. Sci. Food Agric.,* 30, 503, 1979.

587. **Giuffre, G. P. and Litman, R.,** Comparison of elemental uptake by pine trees in varied environments, *J. Environ. Sci. Health,* A14, 365, 1979.

588. **Sefzic, E.,** Determination of As, Pb, Cd, Cr and Se with respect to the requirements of drinking water regulations by means of flameless atomic absorption spectroscopy, *Vom Wasser,* 50, 285, 1979 (in German).

589. **Tominaga, M. and Umezaki, Y.,** Suppression of interferences in the determination of lead and cadmium by graphite furnace atomic absorption spectrometry, *Bunseki Kagaku,* 28, 347, 1979.

590. **Tsalev, D. and Petrova, V.,** Hexamethylene ammonium hexamethylene dithiocarbamate-n-butyl acetate as an extractant of Bi, Cd, Co, Hg, In, Pb and Pd from acidic media, *Dokl. Bolg. Akad. Nauk,* 32, 911, 1979.

591. **Tasushida, T. and Takeo, T.,** Direct determination of copper, lead and cadmium in tea infusions by flameless atomic absorption spectrometry, *Agric. Biol. Chem.,* 43, 1347, 1979.

592. **Baeckman, S. and Karlsson, R. W.,** Determination of lead, bismuth, zinc, silver and antimony in steel and nickel-base alloys by atomic absorption spectrophotometry using direct atomization of solid samples in a graphite furnace, *Analyst (London),* 104, 1017, 1979.

593. **Lagesson, V. and Andrasko, L.,** Direct determination of lead and cadmium in blood and urine by flameless atomic absorption spectrophotometry, *Clin. Chem. (Winston-Salem, N.C.),* 25, 1948, 1979.

594. **Long, S. J., Suggs, J. C., and Walling, J. F.,** Lead analysis of ambient air particulates: interlaboratory evaluation of EPA lead reference method, *J. Air Pollut. Control. Assoc.,* 29, 28, 1979.

595. **Mamontova, S. A., and Pchelintseva, N. F.,** Extraction-atomic absorption determination of lead and cadmium in natural waters, suspensions and sediments, *Zh. Anal. Khim.,* 34, 2231, 1979.

596. **Markunas, L. D., Barry, E. F., Giuffre, G. P., and Litman, R.,** Improved procedure for determination of lead in environmental samples by atomic absorption spectroscopy, *J. Environ. Sci. Health,* A14, 501, 1979.

597. **Rezchikov, V. G. and Usvatov, V. A.,** Atomic absorption determination of copper, lead and bismuth in silver nitrate, *Zavod. Lab.,* 45, 1112, 1979.

598. **Ruttner, O. and Jarc, H.,** Residue studies in Austria. I. Investigations of the lead, cadmium and chromium content of meat from Upper Austrian cattle, *Wein Tieraerztl. Monatsschr.,* 66, 259, 1979.

599. **Sedykh, E. M., Belyaev, Y. I., and Ozhegov, P. I.,** Study of the mechanism of atomization of lead compounds during atomic absorption determination of lead in a graphite furnace, *Zh. Anal. Khim.,* 34, 1984, 1979.

600. **Tsujii, K., Kuga, K., Murayama, S., and Yasuda, M.,** Evaluation of high-frequency discharge lamps for atomic absorption and atomic fluorescence spectrometry of cadmium, lead and zinc, *Anal. Chim. Acta,* 111, 103, 1979.

601. **Baucells, M., Lacort, G., and Roura, M.,** Determination of major and trace elements in marine sediments of Costa Brava by emission and atomic absorption spectroscopy, *Invest. Pesqu.,* 43, 551, 1979.

602. **Berta, E. and Olah, J.,** Survey of heavy metal content of waste water sludges in Hungary, *Hidrol. Kozl.,* 59, 469, 1979.

603. **Branca, P. and Maina, E.,** Contents of lead, copper, zinc, manganese and iron in products preserved in tin-plated containers, *Boll. Chim. Unione Ital. Lab. Prov. Parte Sci.,* 5, 593, 1979.

604. **Carelli, G., Rimatori, V., and Sperduto, B.,** Determination of lead in blood by wet ashing, solvent extraction and flameless atomic absorption spectrophotometry, *Med. Lav.,* 70, 313, 1979.

605. **Colovos, G., Eaton, W. S., Ricci, G. E., Shepard, L. S. and Wang, H.,** Collaborative Testing of NIOSH Atomic Absorption Method, DHEW (NIOSH) Publ. 79-144, U.S. Department of Health, Education and Welfare, Washington, D.C., 1979.

606. **Dolske, D. A. and Sievering, H.,** Trace element loading of Southern Lake Michigan by dry deposition of atmospheric aerosols, *Water Air Soil Pollut.,* 12, 485, 1979.

607. **El-Enany, F. F., Mahmoud, K. F., and Varma, M. M.,** Single organic extraction for determination of ten heavy metals in sea water, *J. Water Pollut. Control. Fed.,* 51, 2545, 1979.

608. **Fujino, O., Matsui, M., and Shigematsu, T.,** Determination of trace metals in natural water by atomic absorption spectrometry. Preconcentration of metals by diethyldithiocarbamate-di-isobutyl ketone solvent extraction, *Mizu Shori Gijutsu,* 20, 201, 1979 (in Japanese).

609. **Jackson, C. J.,** Atomatic monitor for lead emissions from stacks: design philosophy and preliminary evaluation, *J. Autom. Chem.,* 1, 267, 1979.

610. **Kobayashi, Y., Tanabe, T., and Nakamoto, Y.,** Determination of lead in aromatic oils by flameless atomic absorption spectrometry, *Aromatikkusu,* 31, 228, 1979.

611. **Palacios Remondo, J. and Ramirez Diaz, S.,** Lead and cadmium levels in liver of monogastric and polygastric animals determined by atomic absorption spectrophotometry, *Dev. Agroquim. Tecnol. Aliment.,* 19, 279, 1979.

612. **Chowdhury, A. N. and Das, A. K.,** Determination of traces of lead and cadmium in iron ores by nonaqueous atomic absorption spectrometry, *Indian J. Chem. Soc.,* A17, 536, 1979.

613. **Fricke, F. L., Robbins, W. B., and Caruso, J. A.,** Trace element analysis of food and beverages by atomic absorption spectrometry, *Prog. Anal. At. Spectrosc.,* 2, 185, 1979.

614. **Ward, N. I. and Brooks, R. R.,** Lead levels in wool as an indication of lead in blood of sheep exposed to automotive emissions, *Bull. Environ. Contam. Toxicol.,* 21, 403, 1979.

615. **Shan, X. and Ni, Z.,** Matrix modification for the determination of lead in seawater using graphite furnace atomic absorption spectrometry, *Environ. Sci.,* 1, 24, 1980 (in Chinese).

616. **Vijan, P. N. and Sadana, R. S.,** Determination of lead in drinking waters by hydroxide generation and atomic absorption spectroscopy and three other methods, *Talanta,* 27, 321, 1980.

617. **Zakhariya, A. N., Olenovich, N. L., and Khutornoi, A. M.,** Atomic absorption determination of tin, lead, antimony and bismuth in copper products, *Ukr. Khim. Zh.,* 46, 421, 1980.

618. **Amiard, J. C., Amiard-Triquet, C., Meranger, C., Marchand, J., and Ferre, R.,** Study of transfer of cadmium, lead, copper and zinc in estuarine food chains. I. Study in the Loire estuary during summer 1978, *Water Res.,* 14, 665, 1980.

619. **Ash, C. P. J. and Lee, D. L.,** Lead, cadmium, copper and iron in earthworms from roadside sites, *Environ. Pollut. Ser. A Ecol. Biol.,* 22, 59, 1980.

620. **Mitcham, R. P.,** Determination of lead in drinking water by atomic absorption spectrophotometry using an electrically heated graphite furnace and an ammonium tetramethlene dithiocarbamate extraction technique, *Analyst (London),* 105, 43, 1980.

621. **Pachuta, D. G. and Love, L. J.,** Determination of lead in urban particulates by microsampling cup atomic absorption spectrometry, *Anal. Chem.,* 52, 444, 1980.

622. **Pederson, B., Willems, M., and Storgaard-Joergensen, S.,** Determination of copper, lead, cadmium, nickel and cobalt in EDTA extracts of soil by solvent extraction and graphite furnace atomic absorption spectrophotometry, *Analyst (London),* 105, 119, 1980.

623. **Petrov, I. I., Tsalev, D. L., and Barsev, A. I.,** Atomic absorption spectrometric determination of cadmium, cobalt, copper, manganese, nickel, lead and zinc in acetate soil extracts, *At. Spectrosc.,* 1, 47, 1980.

624. **Polo Diez, L., Hernandez Mendez, J., and Pedraz Penalava, F.,** Analytical applications of emulsions: determination of lead in gasoline by atomic absorption spectrophotometry, *Analyst (London),* 105, 37, 1980.

625. **Harbach, D., Diehl, H., Timm, J., and Huntemann, D.,** Comparison of three sample preparation methods for the determination of lead in fruit juices by flameless atomic absorption, *Z. Anal. Chem.,* 301, 215, 1980.

626. **Hon, P. K., Lau, O. W., Cheung, W. C., and Wong, M. C.,** The atomic absorption spectrometric determination of As, Bi, Pb, Sb, Se and Sn with a flame-heated silica T-tube after hydride generation, *Anal. Chim. Acta,* 115, 355, 1980.

627. **Kowalska, A., Kedziora, M., and Kedziora, A.,** Determination of lead in graphite by flameless atomic absorption and solid sampling techniques, *At. Spectrosc.,* 1, 33, 1980.

628. **Legotte, P. A., Rosa, W. C., and Sutton, D. C.,** Determination of cadmium and lead in urine and other biological samples by graphite furnace atomic absorption spectrometry, *Talanta,* 27, 39, 1980.

629. **Callio, S.,** Lead by flameless atomic absorption with phosphate matrix modification, *At. Spectrosc.,* 1, 80, 1980.

630. **Chakrabarti, C. L., Wan, C. C., and Li, W. C.,** Direct determination of traces of copper, zinc, lead, cobalt, iron and cadmium in bovine liver by graphite furnace atomic absorption spectrometry using the solid sampling and the platform technique, *Spectrochim. Acta*, 35b, 93, 1980.

631. **DeJonghe, W., Chakraborti, D., and Adams, F.,** Graphite furnace atomic absorption spectrometry as a metal-specific detection system for tetraalkyl lead compounds separated by gas-liquid chromatography, *Anal. Chim. Acta*, 115, 89, 1980.

632. **Favretto-Gabreilli, L., Pertoldi Marletta, G., and Favretto, L.,** Determination of lead in mussels by atomic absorption spectrophotometry and solid microsampling, *At. Spectrosc.*, 1, 35, 1980.

633. **Feinberg, M. and Ducauze, C.,** High temperature ashing of foods for atomic absorption spectrometric determination of lead, cadmium and copper, *Anal. Chem.*, 52, 207, 1980.

634. **Knight, M. J.,** Comparison of Four Digestion Procedures not Requiring Perchloric Acid for the Trace Element Analysis of Plant Material, Rep. ANL/LRP-TM-18, Argonne National Laboratory, Argonne, Ill., 1980, 31; available from the NTIS, Springfield, Va.

635. **Orpwood, B.,** Use of Chelating Ion-Exchange Resins for the Determination of Trace Metals in Drinking Waters, TR 153 Water Research Center, Medmenham Laboratory, Medmenhem, Marlow, Bucks., England, 1980.

636. **Farag, R. S., El-Aassar, S. T., Mostafa, M. A., and Abdel Rahim, E. A.,** Use of atomic absorption analysis of mineral content of certain animal hairs in relation to sex (I), *J. Drug Res.*, 12, 217, 1980.

637. **Hoshino, Y., Utsunomiya, T., and Fukui, K.,** Graphite furnace atomic absorption spectrometry utilizing selective concentration onto tungsten wire, *Eng. Mater. Tokyo Inst. Technol.*, 5, 109, 1980.

638. **Caristi, C., Cimino, G., and Ziino, M.,** Heavy metal pollution. II. Determination of trace heavy elements in lemon, orange and mandarin juices by flameless atomic absorption spectroscopy, *Essenz. Deriv. Agrum.*, 50, 165, 1980.

639. **Inhat, M., Gordon, A. D., Gaynor, J. D., Berman, S. S., Desauliners, A., Stoeppler, M., and Valenta, P.,** Interlaboratory analysis of natural fresh waters for Cu, Zn, Cd and Pb, *Int. J. Environ. Anal. Chem.*, 8, 259, 1980.

640. **Chakrabarti, C. L., Wan, C. C., and Li, W. C.,** Atomic absorption spectrometric determination of cadmium, lead, zinc, copper, cobalt and iron in oyster tissue by direct atomization from the solid state using the graphite furnace platform technique, *Spectrochim. Acta*, 35b, 547, 1980.

641. **Saito, S. and Kamoda, M.,** Sample preparation by dry ashing prior to the determination of metals in sugars, *Seito Gijutsu Kenkyu Kaishi*, 29, 36, 1980 (in Japanese).

642. **Fink, L. K., Jr., Harris, A. B., and Schick, L. L.,** Trace Metals in Suspended Particulates, Biota and Sediments of the St. Croix, Narraguagus and Union Estuaries and the Goose Cove Region of Penobscot Bay, Rep. W80-04809, OWRT-A-041-ME (I), 1980; available from the NTIS, Springfield, Va; *Govt. Rep. Announce. Index (U.S.)*, 80, 3372, 1980.

643. **Tursunov, A. T., Brovko, I. A., and Nazarov, S. N.,** Use of a "furnace-flame" system for the atomic absorption determination of some trace elements in soil extracts, *Uzb. Khim. Zh.*, 5, 5, 1980.

644. **Hirokawa, K. and Takada, K.,** Determination of trace lead in a very small amount of steel, copper and copper alloys by atomic absorption spectrometry using direct atomization of solid sample in a graphite furnace, *Bunseki Kagaku*, 29, 675, 1980.

645. **Akama, Y., Nakai, T., and Kawamura, F.,** Determination of traces of Co, Cu, Mn, Ni and Pb in solar salt by AA spectrometry combined with extraction, *Nippon Kaishi Gakkaishi*, 34, 196, 1980 (in Japanese).

646. **Takiyama, K., Ishii, Y., and Yoshimura, I.,** Determination of metals in vegetables by ashing in a teflon crucible, *Mukogawa Joshi Daigaku Kiyo Shokumotsu-hen*, 28, F15, 1980 (in Japanese).

647. **Aungst, B. J., Dolce, J., and Fung, H.,** Solubilization of rat whole blood and erythrocytes for automated determination of lead using atomic absorption spectrophotometry, *Anal. Lett.*, 13, 347, 1980.

648. **Barudi, W. and Bielig, H. J.,** Heavy metal content of vegetables which grow above ground and fruits, *Z. Lebensm. Unters. Forsch.*, 170, 254, 1980.

649. **Beyer, N. W. and Moore, J.,** Lead residues in Eastern tent caterpillars (*Malacosoma americanum*) and their host plant (*Prunus serotinas*) close to a major highway, *Environ. Entomol.*, 9, 10, 1980.

650. **Birch, J., Harrison, R. M., and Laxen, D. P. H.,** Specific method for 24-48 hour analysis of tetraalkyl lead in air, *Sci. Total Environ.*, 14, 31, 1980.

651. **Birnie, S. E. and Noden, F. G.,** Determination of tetramethyl and tetraethyl lead vapors in air following collection on a glass fiber-iodized carbon filter disk, *Analyst (London)*, 105, 110, 1980.

652. **Borriello, R. and Sciaudone, G.,** Zinc, copper, iron and lead in bottled and canned beer by atomic absorption spectroscopy, *At. Spectrosc.*, 1, 131, 1980.

653. **Colella, M. B., Siggia, S., and Barnes, R. M.,** Poly (acrylamidoxime) resin for determination of trace metals in natural waters, *Anal. Chem.*, 52, 2347, 1980.

654. **Danielson, L. G.,** Cadmium, cobalt, copper, iron, lead, nickel and zinc in Indian Ocean water, *Mar. Chem.*, 8, 199, 1980.

655. **DeJonghe, W. R., Chakraborti, D., and Adams, F. C.,** Sampling of tetraalkyl lead compounds in air for determination by gas chromatography-atomic absorption spectrometry, *Anal. Chem.,* 52, 1974, 1980.

656. **Gorshkov, V. V., Orlova, L. P., and Voronkova, M. A.,** Preconcentration and atomic absorption determination of cadmium and lead in natural materials, *Zh. Anal. Khim.,* 35, 1277, 1980.

657. **Halliday, M. C., Houghton, C., and Ottaway, J. M.,** Direct determination of lead in polluted sea water by carbon furnace atomic absorption spectrometry, *Anal. Chim. Acta,* 119, 67, 1980.

658. **Holak, W.,** Analysis of foods for lead, cadmium, copper, zinc, arsenic and selenium using closed system sample digestion: collaborative study, *J. Assoc. Off. Anal. Chem.,* 63, 485, 1980.

659. **Horler, D. N. H., Barber, J., and Barringer, A. R.,** Multielemental study of plant surface particles in relation to geochemistry and biogeochemistry, *J. Geochem. Explor.,* 13, 41, 1980.

660. **Hubert, J., Candelaria, R. M., and Applegate, H. G.,** Determination of lead, zinc, cadmium and arsenic in environmental samples, *At. Spectrosc.,* 1, 90, 1980.

661. **Jin, K. and Taga, M.,** Determination of trace amounts of lead by atomic absorption spectrometry after hydride evolution process. II. Fundamental conditions for the determination of lead and application to rocks, steel and water, *Bunseki Kagaku,* 29, 522, 1980.

662. **Knezevic, G. and Hueppe, K.,** Digestion methods for determination of lead, iron and tin in canned soups by atomic absorption spectrophotometry, *Dtsch. Lebensm. Rundsch.,* 76, 50, 1980.

663. **Magnusson, B. and Westerlund, S.,** Determination of cadmium, copper, iron, nickel, lead and zinc in Baltic Sea water, *Mar. Chem.,* 8, 231, 1980.

664. **Mendiola, Ambrosio, J. M., Gonzalez, A., and Arribas Jimeno, S.,** Analysis of lead in PVC compounds by atomic absorption spectrophotometry, *Afinidad,* 37, 39, 1980.

665. **Morse, R. A. and Lisk, D. J.,** Elemental analysis of honeys from several nations, *Am. Bee J.,* 120, 592, 1980.

666. **Nadkarni, R. A.,** Multitechnique multielemental analysis of coal and fly ash, *Anal. Chem.,* 52, 929, 1980.

667. **Pande, J. and Das, S. M.,** Metallic contents in water and sediments of Lake Nainital, India, *Water Air Soil Pollut.,* 13, 3, 1980.

668. **Pande, S. P. and Pendharkar, A. V.,** Some analytical aspects of lead and cadmium determination in drinking water by atomic absorption spectrophotometry, *J. Inst. Chem.,* 52, 141, 1980.

669. **Poldoski, J. E.,** Determination of lead and cadmium in fish and clam tissue by an atomic absorption spectrometry with a molybdenum and lanthanum treated pyrolytic atomizer, *Anal. Chem.,* 52, 1147, 1980.

670. **Rohbock, E., Georgii, H. W., and Mueller, J.,** Measurements of gaseous lead alkyls in polluted atmospheres, *Atmos. Environ.,* 14, 89, 1980.

671. **Schramel, P., Wolf, A., Seif, R., and Klose, B. J.,** New device for ashing of biological materials under pressure, *Z. Anal. Chem.,* 302, 62, 1980.

672. **Sefflova, A. and Komarek, J.,** Determination of lead by atomic absorption spectrometry with electrothermal atomization, *Chem. Listy,* 74, 971, 1980.

673. **Sinex, S. A., Cantillo, A. Y., and Helz, G. R.,** Accuracy of acid extraction methods for trace metals in sediments, *Anal. Chem.,* 52, 2342, 1980.

674. **Abo-Rady, M. D. K.,** Aquatic macrophytes as indicator for heavy metal pollution in the river Seine, *Arch. Hydrobiol.,* 89, 387, 1980 (in German).

675. **Cool, M., Marcoux, F., Paulin, A., and Mehra, M. C.,** Metallic contaminants in street soils of Moncton, New Brunswick, Canada, *Bull. Environ. Contam. Toxicol.,* 25, 409, 1980.

676. **Cruz, R. B., Lorouso, C., George, S., Thomassen, Y., Kinrade, J. D., Butler, L. R., Lye, J., and Van Loon, J. C.,** Determination of total organic solvent extractable, volatile and tetraalkyl lead in fish, vegetation, sediment and water samples, *Spectrochim. Acta,* 35b, 775, 1980.

677. **Davison, W.,** Ultratrace analysis of soluble zinc, cadmium, copper and lead in Windermere Lake water using anodic stripping voltammetry and atomic absorption spectroscopy, *Freshwater Biol.,* 10, 223, 1980.

678. **Denton, J. E., Potter, G. D., and Santolucito, J. A.,** Comparison of skull and femur lead levels in adult rats, *Environ. Res.,* 23, 264, 1980.

679. **Diehl, H., Harbach, D., and Timm, J.,** Design and Evaluation of Atomic Absorption Spectrometry Studies with the Addition Method, *Math.-Arbeitspap. No. 22,* University of Bremen, 1980 (in German).

680. **Eiden, C. A., Jewell, C. A., and Wightman, J. P.,** Interaction of lead and cadmium with chitin and chitosan, *J. Appl. Polym. Sci.,* 25, 1587, 1980.

681. **Filkova, L.,** Reproducibility, accuracy and detection limit in determination of lead in blood and hair by flameless atomic absorption spectrophotometry, *Chem. Listy,* 74, 533, 1980 (in Czech).

682. **Foldes, W., Kosa, F., Viragos-Kis, E., Rengei, B., and Ferke, A.,** Atomic absorption spectrophotometric study of contents of inorganic substances from skeletons for determination of time of burial in the earth, *Arch. Kriminol.,* 166, 105, 1980 (in German).

683. **Glenc, T., Jurczyk, J., and Robosyj-Kabza, A.,** Determination of tin and lead in transformer, carbon and alloy steels in the range of 0.0005 to 0.04% by atomic absorption method with electrothermal atomization, *Chem. Anal.,* 25, 515, 1980.

684. **Harrison, R. M. and Laxen, D. P. H.,** Physicochemical speciation of lead in drinking water, *Nature (London),* 286, 791, 1980.

685. **Hartman, V. E. M. and Cavillini, Z. T. L.,** Determination of lead and zinc in Brazilian oils of sicilian lemon and orange limonene, *An. Acad. Bras. Cienc.,* 52, 311, 1980 (in Portuguese).

686. **Itokawa, H., Watanabe, K., Tazaki, T., Hayashi, T., and Hayashi, Y.,** Quantitative analysis of metals in crude drugs, *Shoyakugaku Zasshi,* 34, 155, 1980 (in Japanese).

687. **Jenniss, S. W., Katz, S. A., and Mount, T.,** Comparison of sample preparation methods for the determination of cadmium and lead in sewage sludges by AAS, *Am. Lab.,* 12, 18, 1980.

688. **Katskov, D. A., Grinshtein, I. L., and Kruglikova, L. P.,** Study of the evaporation of the metals In, Ga, Tl, Ge, Sn, Pb, Sb, Bi, Se and Te from a graphite surface by the atomic absorption method, *Zh. Prikl. Spektrosk.,* 33, 804, 1980.

689. **Kosa, F., Foldes, V., Viragos-Kis, E., Rengei, B., and Ferke, A.,** Atomic absorption spectrophotometric study on content of inorganic substances in fetal bone for the determination of age, *Arch. Kriminol.,* 166, 44, 1980 (in German).

690. **Kruse, R.,** Multiple determination of lead and cadmium in fish by electrothermal AAS after wet ashing in commercially available teflon beakers, *Z. Lebensm. Unters. Forsch.,* 171, 261, 1980 (in German).

691. **Lindh, U., Brune, D., Nordberg, G., and Wester, P. O.,** Levels of antimony, arsenic, cadmium, copper, lead, mercury, selenium, silver, tin and zinc in bone tissue of industrially exposed workers, *Sci. Total Environ.,* 16, 109, 1980.

692. **Maeda, T., Nakagawa, M., Kawakatsu, M., and Tanimoto, Y.,** Determination of metallic elements in serum and urine by flame and graphite furnace atomic absorption spectrophotometry, *Shimadzu Hyoron,* 37, 1980 (in Japanese).

693. **Ogihara, K., Seki, H., and Nagase, K.,** Comparison of the method of bottom sediment analysis and the method of the agricultural soil pollution prevention law, *Nagano-ken Eisei Kogai Kenkyusho Kenkyu Hokoku,* 2, 88, 1980 (in Japanese).

694. **Oreshkin, V. N., Belyaev, Y. I., Tatsii, Y. G., and Vnukovakaya, G. L.,** Direct simultaneous determination of cadmium, lead and silver in sea water and eolian suspended matter by flameless atomic absorption, *Okeanologiya,* 20, 736, 1980 (in Russian).

695. **Posta, J.,** Determination of the optimum parameters of a combined atomic absorption spectrophotometric plus arc flame method for the analysis of floating dust samples, *Hung. Sci. Instrum.,* 47, 33, 1980.

696. **Riandey, C., Gavinelli, R., and Pinta, M.,** Effect of heating rate on electrothermal atomization in atomic absorption spectrometry: application to the volatile elements cadmium and lead, *Spectrochim. Acta,* 35b, 765, 1980.

697. **Schmidt, W. and Dietl, F.,** Determination of lead in digested soils and sediments and in soil and sediment extracts by flameless atomic absorption with zirconium-coated graphite tubes, *Z. Anal. Chem.,* 303, 385, 1980.

698. **Sadykh, E. M., Belyazev, Y. I., and Sorokina, E. V.,** Matrix effect during electrothermal atomic absorption determination of silver, tellurium, lead, cobalt, nickel in materials of complex composition, *Zh. Anal. Khim.,* 35, 2162, 1980 (in Russian).

699. **Sadykh, E. M., Belyaev, Y. I., and Sorokina, E. V.,** Elimination of matrix effects in electrothermal atomic absorption determination of silver, lead, cobalt, nickel and tellurium in samples of complicated composition, *Zh. Anal. Khim.,* 35, 2348, 1980 (in Russian).

700. **Sourova, J. and Capkova, A.,** Determination of trace elements in water with a high iron content by AAS, *Vodni Hospod.,* 30b, 133, 1980 (in Czech).

701. **Stein, V. B., Canelli, E., and Richards, A. H.,** Simplified determination of cadmium, lead and chromium in estuarine waters by flameless atomic absorption, *Int. J. Environ. Anal. Chem.,* 8, 99, 1980.

702. **Stendal, H.,** Leaching studies for the determination of copper, zinc, lead, nickel and cobalt in geological materials by atomic absorption spectrophotometry, *Chem. Erde,* 39, 276, 1980.

703. **Tsushida, T. and Takeo, T.,** Determination of copper, lead and cadmium in tea by graphite furnace atomic absorption spectrophotometry, *Nippon Shokuhin Kogyo Gakkaishi,* 27, 585, 1980 (in Japanese).

704. **Tursunov, A. T., Brovko, I. A., and Nazarov, S. N.,** Use of a furnace-flame system for the atomic absorption determination of some trace elements in soil extracts, *Uzb. Khim. Zh.,* 5, 5, 1980 (in Russian).

705. **Van Willis, W., El-Ahraf, A., Vinjamoori, D. V., and Aref, K.,** Analysis of animal feed ingredients and soil amendment products produced from beef cattle manure for selected trace metals using atomic absorption spectrophotometry, *J. Food Prot.,* 43, 834, 1980.

706. **Jaros, J. and Radil, J.,** Use of Dithiocarbamates for Determining Selected Zirconium Impurities by Atomic Absorption Spectrometry, Rep. UJP-496, 1980, 38 (in Czech).

707. **Thamzil, L., Suwirma, S., and Surtipanti, S.,** Determination of heavy metals in the stream of Sunter River, *Majalah BATAN,* 13, 41, 1980 (in Indonesian).

708. **Suwirma, S., Surtipanti, S., and Thamzil, L.,** Determination of heavy metals mercury, lead, cadmium, chromium, copper and zinc in fish, *Majalah BATAN,* 13, 9, 1980 (in Indonesian).

709. **Chakrabarti, C. L., Wan, C. C., Hamed, W. A., and Bartels, P. C.,** Trace element determination by capacitive discharge atomic absorption spectrometry, *Nature (London),* 288, 246, 1980.

710. **Noller, B. N. and Bloom, H.,** Application of graphite furnace atomic absorption spectrometry to the determination of metals in air particulates, *Clean Air (Melbourne),* 14, 9, 1980.

711. **Alt, F., Berndt, H., Messerschmidt, J., and Sommer, D.,** Determination of cadmium, lead and thallium in mineral raw materials after chemical preconcentration using different spectrometric methods, *Spektrometertagung (Votr.),* 13, 331, 1981.

712. **Behari, J. R.,** Determination of lead in blood, *Int. J. Environ. Anal. Chem.,* 10, 149, 1981.

713. **Mueller, H. and Siepe, V.,** Quantitative determination of arsenic, lead, cadmium, mercury and selenium in foods by flameless atomic absorption spectrophotometry, *Dtsch. Lebensm. Rundsch.,* 77, 392, 1981.

714. **Magnusson, B. and Westerlund, S.,** Solvent extraction procedures combined with back-extraction for trace metal determinations by atomic absorption spectrometry, *Anal. Chim. Acta,* 131, 63, 1981.

715. **Halls, D. J.,** Application of graphite furnace atomic absorption spectrometry in clinical analysis, *Anal. Proc. (London),* 18, 344, 1981.

716. **Mitchell, G. E.,** Trace metal levels in Queensland dairy products, *Aust. J. Dairy Technol.,* 36, 70, 1981.

717. **Chakraborti, D., Jiang, S. G., Surkijn, P., DeJonghe, W., and Adams, F.,** Determination of tetra-alkyllead compounds in environmental samples by gas chromatography-graphite furnace atomic absorption spectrometry, *Anal. Proc. (London),* 18, 347, 1981.

718. **Brunner, W., Heckner, H., and Sansoni, B.,** Fully automatic simultaneous operation of several atomic absorption spectrometers in the analytical service laboratory of a research center by use of a process computer, *Spektrometertagung (Votr.),* 13, 99, 1981 (in German).

719. **Ranchet, J., Menissier, F., Lamathe, J., and Voinovitch, I.,** Interlaboratory comparison of the determination of cadmium, chromium, copper and lead by flameless atomic absorption spectrometry, *Bull. Liaison Lab. Ponts Chaussees,* 114, 81, 1981 (in French).

720. **Meranger, J. C., Hollebone, B. R., and Blanchette, G. A.,** Effects of storage times, temperatures and container types on the accuracy of atomic absorption determination of cadmium, copper, mercury, lead and zinc in whole hepatinized blood, *J. Anal. Toxicol.,* 5, 33, 1981.

721. **Meranger, J. C., Subramanian, K. S., and Chalifoux, C.,** Survey for Cd, Co, Cr, Cu, Ni, Pb, Zn, Ca and Mg, in Canadian drinking water supplies, *J. Assoc. Off. Anal. Chem.,* 64, 44, 1981.

722. **Pakalns, P.,** Effect of surfactants on mixed chelate extraction-atomic absorption spectrophotometric determination of copper, nickel, iron, cobalt, cadmium, zinc and lead, *Water Res.,* 15, 7, 1981.

723. **Pleban, P. A., Kerkay, J., and Pearson, K. H.,** Polarized Zeeman effect flameless atomic absorption spectrometry of cadmium, copper, lead and manganese in human kidney cortex, *Clin. Chem. (Winston-Salem, N.C.),* 27, 68, 1981.

724. **Salmon, S. G., Davis, R. H., and Holcombe, J. A.,** Time shifts and double peaks for lead caused by chemisorbed oxygen in electrothermally heated graphite atomizers, *Anal. Chem.,* 53, 324, 1981.

725. **Fernandez, F. J., Beaty, M. M., and Barnett, W. B.,** Use of the L'vov platform for furnace atomic absorption applications, *At. Spectrosc.,* 2, 16, 1981.

726. **Hinderberger, E. J., Kaiser, M. L., and Koirtyohann, S. R.,** Furnace atomic absorption analysis of biological samples using the L'vov platform and matrix modification, *At. Spectrosc.,* 2, 1, 1981.

727. **Hodges, D. J. and Skelding, D.,** Determination of lead in urine by atomic absorption spectroscopy with electrothermal atomization, *Analyst (London),* 106, 299, 1981.

728. **Laxen, D. P. and Harrison, R. M.,** Cleaning methods for polythene containers prior to the determination of trace metals in fresh water samples, *Anal. Chem.,* 53, 345, 1981.

729. **Martynyuk, T. G., Sevast'yanova, N. I., Nikolaeva, V. A., and Mukhtarov, E. I.,** Flameless atomic absorption method for determining the lead and cadmium content of poultry products, *Vopr. Pitan.,* 2, 16, 1981.

730. **Baker, A. A. and Headridge, J. B.,** Determination of bismuth, lead and tellurium in copper by atomic absorption spectrometry with introduction of solid samples into an induction furnace, *Anal. Chim. Acta,* 125, 93, 1981.

731. **Bertenshaw, M. P., Gelsthorpe, D., and Wheatstone, K. C.,** Determination of lead in drinking water by atomic absorption spectrophotometry with electrothermal atomization, *Analyst (London),* 106, 23, 1981.

732. **Carpenter, R. C.,** Determination of cadmium, copper, lead and thallium in human liver and kidney tissue by flame atomic absorption spectrometry after enzymic digestion, *Anal. Chim. Acta,* 125, 209, 1981.

733. **Clark, J. R. and Viets, J. G.,** Multielement extraction system for the determination of 18 trace elements in geochemical samples, *Anal. Chem.,* 53, 61, 1981.

734. **Clark, J. R. and Viets, J. G.,** Back-extraction of trace elements from organometallic-halide extracts for determination by flameless atomic absorption spectrometry, *Anal. Chem.,* 53, 65, 1981.

735. **deHaas, E. J. M. and deWolff, F. A.,** Microassay of lead in blood with an improved procedure for silanization of reagent tubes, *Clin. Chem. (Winston-Salem, N.C.),* 27, 205, 1981.

736. **Paschal, D. C. and Bell, C. J.,** Improved accuracy in the determination of blood lead by electrothermal atomic absorption, *At. Spectrosc.,* 2, 146, 1981.

737. **Smith, R.,** Use of a hydride-generation technique for the atomic absorption determination of lead in drinking water, *At. Spectrosc.,* 2, 155, 1981.

738. **Wandiga, S. O.,** The concentrations of zinc, copper, lead, manganese, nickel and fluoride in rivers and lakes of Kenya, *Sinet,* 3, 67, 1981.

739. **Kimura, M. and Kawanami, K.,** Separation and preconcentration of trace amounts of several metals in sodium perchlorate using activated carbon as a collector, *Nippon Kagaku Kaishi,* 1, 1, 1981 (in Japanese).

740. **Capel, I. D., Pinnock, M. H., Dorrell, H. M., Williams, D. C., and Grant, E. C. G.,** Comparison of concentrations of some trace, bulk and toxic metals in the hair of normal and dyslexic children, *Clin. Chem. (Winston-Salem, N.C.),* 27, 879, 1981.

741. **Cowgill, U. M.,** The chemical composition of bananas. Market basket values, *Biol. Trace Elem. Res.,* 3, 33, 1981.

742. **Carleer, R., Francois, J. P., and Van Poucke, L. C.,** Determination of the main, minor and trace elements in lead-tin based solders by atomic absorption spectrometry, *Bull. Soc. Chim. Belg.,* 90, 357, 1981.

743. **Shevchuk, I. A., Dovzhenko, N. P., and Kravtsova, Z. N.,** Atomic absorption determination of lead and bismuth in steels using ion-exchange chromatography, *Ukr. Khim. Zh.,* 47, 773, 1981 (in Russian).

744. **Pihlaja, H.,** Determination of traces of metals in Finnish margarines by the flameless atomic absorption spectrophotometry method, *Fette Seifen Anstrichm.,* 83, 294, 1981.

745. **Rasmussen, L.,** Determination of trace metals in sea water by Chelax-100 or solvent extraction techniques and atomic absorption spectrometry, *Anal. Chim. Acta,* 125, 117, 1981.

746. **Slovak, Z., Docekal, B., and Bohumil, N.,** Determination of trace metals in aluminum oxide by electrothermal atomic absorption spectrometry with direct injection of aqueous suspensions, *Anal. Chim. Acta,* 129, 263, 1981.

747. **Bewers, J. M., Dalziel, J., Yeats, P. A., and Barron, J. L.,** An intercalibration for trace metals in sea water, *Mar. Chem.,* 10, 173, 1981.

748. **Aksyuk, A. F., Merzlyakova, N. M., and Tarkhova, L. P.,** Determination of trace elements in desalinated sea water by an atomic absorption method, *Gig. Sanit.,* 5, 72, 1981.

749. **Clanet, F., Deloncle, R., and Popoff, G.,** Chelating resin catcher for capture, preconcentration and determination of toxic metal traces (Zn, Cd, Hg and Pb) in waters, *Water Res.,* 15, 591, 1981.

750. **Henrion, G., Gelbrecht, J., Hoffmann, T., and Marquardt, D.,** AAS trace determination in tungsten after preconcentration on chelate resin WOFATIT MC50, *Z. Anal. Chem.,* 21, 192, 1981 (in German).

751. **Petrov, I., Tsalev, D., and Vasileva, E.,** Pulse nebulization atomic absorption spectrometry after preconcentration from acidic media, *Dokl. Bolg. Akad. Nauk,* 34, 679, 1981; CA 95, 146378n, 1981.

752. **Amini, M. K., Defreese, J. D., and Hathaway, L. R.,** Comparison of three instrumental spectroscopic techniques for elemental analysis of Kansas shales, *Appl. Spectrosc.,* 35, 497, 1981.

753. **Farmer, J. G.,** The analytical chemist in studies of metal pollution in sediment cores, *Anal. Proc. (London),* 18, 249, 1981.

754. **Krinitz, B. and Tepedino, N.,** Lead in preserved duck eggs: field screening test and confirmation and quantitation by atomic absorption spectrophotometry and anodic stripping voltammetry, *J. Assoc. Off. Anal. Chem.,* 64, 1014, 1981.

755. **Berndt, H. and Messerschmidt, J.,** *O,O*-Diethyl dithiophosphate for trace enrichment on activated carbon. I. Analysis of high purity gallium-aluminum determination of element traces by flame AAS (injection method and loop AAS), *Fresenius' Z. Anal. Chem.,* 308, 104, 1981.

756. **Julshamn, K.,** Studies on major and minor elements in molluscs, in Western Norway. VII. The contents of 12 elements including copper, zinc, cadmium and lead in common mussel (*mytilus adulis*) and brown seaweed (*ascophyllum nodusum*) relative to the distance from the industrial sites in Sorfjordan, Inner Hardangefjord, *Fiskeridir. Skr. Ser. Ernaer.,* 1, 267, 1981.

757. **Sugisaki, H., Nakamura, H., Hirao, Y., and Kimura, K.,** Determination of lead in geostandard rocks by electrothermal atomic absorption spectrometry after isolation of lead with yield monitoring, *Anal. Chim. Acta,* 125, 203, 1981.

758. **Torsi, G., Desimoni, E., Palmisano, F., and Sabbatini, L.,** Determination of lead in seawater by electrothermal atomic absorption spectrometry after electrolytic accumulation on a glassy carbon atomizer, *Anal. Chim. Acta,* 124, 143, 1981.

759. **Wittmers, L. W., Jr., Alich, A., and Aufderheide, A. C.,** Lead in bone. I. Direct analysis for lead in milligram quantities of bone ash by graphite furnace atomic absorption spectroscopy, *Am. J. Clin. Pathol.,* 75, 80, 1981.

760. **Farmer, J. G. and Gibson, M. J.,** Direct determination of cadmium, chromium, copper and lead in silicious standard reference materials from a fluoboric acid matrix by graphite furnace atomic absorption spectrometry, *At. Spectrosc.,* 2, 176, 1981.

761. **Sturgeon, R. B., Berman, S. S., Willie, S. N., and Desauliners, J. A. H.,** Preconcentration of trace elements from sea water with silica-immobilized 8-hydroxy quinoline, *Anal. Chem.,* 53, 2337, 1981.

762. **Adelman, H., Jenniss, S. W., and Katz, S. A.,** Interlaboratory analysis of sewage sludge, *Am. Lab.,* 31, December 1981.

763. **Koster, P. B., Raats, P., Hibbert, D., Phillipson, R. T., Schiweck, H., and Steinle, G.,** Collaborative study on the determination of trace elements in dried sugar beet pulp and molasses. III. Lead, *Int. Sugar J.,* 83, 291, 1981.

764. **Koster, P. B., Raats, P., Hibbert, D., Phillipson, T. R., Schiweck, H., and Steinle, G.,** Collaborative study on the determination of trace elements in dried sugar pulp and molasses. III. Lead, *Zuckerindistrie (Berlin),* 106, 895, 1981 (in German).

765. **Mattsson, P., Albanus, L., and Frank, A.,** Cadmium and some other elements in liver and kidney from moose *(alces alces), Var Foeda,* 33, 335, 1981.

766. **Wesenberg, G. B. R., Fosse, G., and Rasmusson, P.,** The effect of graded doses of cadmium on lead, zinc, copper content of target and indicator organs in rats, *Int. J. Environ. Stud.,* 17, 191, 1981.

767. **Blakimore, W. M. and Billedeau, S. M.,** Analysis of laboratory animal feed for toxic and essential elements by atomic absorption and inductively coupled argon plasma emission spectrometry, *J. Assoc. Off. Anal. Chem.,* 64, 1284, 1981.

768. **Haluska, M. and Smrhova, A.,** Study of the mechanical state of a device by means of lubricating oils, *Ropa Uhlie,* 23, 549, 1981 (in Czech).

769. **Shaburova, V. P., Ruchkin, E. D., Yudelevich, I. G., and Kruglikov, L. M.,** Atomic absorption method of analysis for piezoceramic materials of the lead zirconate titanate system, *Izv. Sib. Otd. Akad. Nauk SSSR Ser. Khim. Nauk,* 5, 95, 1981.

770. **Fukushima, I.,** A Study on Monitoring Methodology for Trace Elements in the Environment Through Chemical Analysis of Hair. Part of a Coordinated Program on Nuclear Methods in Health-Related Monitoring of Trace Element Pollutants, Rep. IAEA-R-2480/F, International Atomic Energy Agency, 1981; INIS Atomindex 12, Abstr. No. 626438, 1981.

771. **Denton, G. R. W. and Burdon-Jones, C.,** Influence of temperature and salinity on the uptake, distribution and depuration of mercury, cadmium and lead by the black-lip oyster *Saccostrea echinata, Mar. Biol. (Berlin),* 64, 317, 1981.

772. **Berndt, H. and Messerschmidt, J.,** Microanalytical method of flame atomic absorption and emission spectrometry (Platinum Loop Method). II. Determination of arsenic, bismuth, cadmium, mercury, lead, antimony, selenium, tellurium, thallium, zinc, lithium, sodium, potassium, cesium, strontium and barium, *Spectrochim. Acta,* 36b, 809, 1981.

773. **Bye, R.,** Improvement of the sensitivity of the "sampling boat" technique in atomic absorption spectrometry. Determination of silver, lead and cadmium, *Z. Anal. Chem.,* 306, 30, 1981.

774. **Faitondzhiev, L.,** Comparison of the determination of lead in soils by atomic absorption and by dithizone spectrophotometry, *Dokl. Bolg. Akad. Nauk,* 34, 75, 1981.

775. **Favretto, L., Pertoldi-Marletta, G., and Gabrielli-Favretto, L.,** Atomic absorption spectroscopic determination of lead in biological materials by direct solid microsampling, *Mikrochim. Acta,* 1, 387, 1981.

776. **Genc, O., Akman, S., Ozdural, A. R., Ates, S., and Balkis, T.,** Theoretical analysis of atom formation-time curves for the HGA-74 furnace. II. Evaluation of the atomization mechanisms for manganese, chromium and lead, *Spectrochim. Acta,* 36b, 163, 1981.

777. **Ikeda, M., Jiro, N., and Nakahara, T.,** Investigation of some fundamental conditions for the determination of arsenic, lead and tellurium and continuous hydride generation-atomic absorption spectrometry, *Bunseki Kagaku,* 30, 368, 1981.

778. **Iwata, Y., Matsumoto, K., Haraguchi, H., Fuwa, K., and Okamoto, K.,** Proposed certified reference material for pond sediment, *Anal. Chem.,* 53, 1136, 1981.

779. **Jackson, K. W., Ebdon, L., Webb, D. C., and Cox, A. G.,** Determination of lead in vegetation by a rapid microsampling cup atomic absorption procedure with solid sample introduction, *Anal. Chim. Acta,* 128, 67, 1981.

780. **Jastrow, J. D., Zimmerman, C. A., Dvorak, A. J., and Hinchman, R. R.,** Plant growth and trace element uptake on acidic coal refuse amended with lime or fly ash, *J. Environ. Qual.,* 10, 154, 1981.

781. **Jiang, S. G., Chakraborti, D., DeJonghe, W., and Adams, F.,** Atomic absorption spectrophotometric determination of volatile organolead compounds in the atmosphere, *Anal. Chem.,* 305, 177, 1981.

782. **Katz, S. A., Jenniss, S. W., Mount, T., Tout, R. E., and Chatt, A.,** Comparison of sample preparation methods for the determination of metals in sewage sludges by flame atomic absorption spectrometry, *Int. J. Environ. Anal. Chem.,* 9, 209, 1981.

783. **Kurfuerst, U. and Rues, B.,** Determination of heavy metals (lead, cadmium, mercury) in settling sludge using Zeeman AAS without chemical decomposition, *Z. Anal. Chem.,* 308, 1, 1981.

784. **Matsumoto, K.,** Determination of lead oxide in and on the surface of high-purity lead metal, *Z. Anal. Chem.,* 305, 370, 1981.

785. **Messman, J. D. and Rains, T. C.,** Determination of tetraalkyllead compounds in gasoline by liquid chromatography-atomic absorption spectrometry, *Anal. Chem.,* 53, 1632, 1981.

786. **Sheng-jun, M., Bin Shi-yan, Y., Wen-sheng, Z., and Yong-mao, L.,** Atomic absorption determination of high concentrations of calcium, magnesium, lead and zinc in ores, *Fen Hsi Hua Hsueh,* 9, 12, 1981 (in Chinese).

787. **Newton, J. T.,** Rapid determination of antimony, barium and lead in gunshot residue via automated atomic absorption spectrophotometry, *J. Forensic Sci.,* 26, 302, 1981.

788. **Wolff, E. W., Landy, M. P., and Peel, D. A.,** Preconcentration of cadmium, copper, lead and zinc in water at the 10^{-12}g/g level by absorption onto tungsten wire followed by flameless atomic absorption spectrometry, *Anal. Chem.,* 53, 1566, 1981.

789. **Wunderlich, E.,** Determination of lead and cadmium in biological material, *Erzmetall,* 34, 577, 1981.

790. **Jathar, V. S., Pendharkar, P. B., Pandey, V. K., Raut, S. J., and Parameswaran, M.,** Intake of lead through food in India, *Bull. Radiat. Prot.,* 4, 3, 1981.

791. **Jackson, K. F., Benedik, J. E., and Birholz, F. A.,** Metals distribution in shale oil fractions, *Oil Shale Symp. Proc.,* 14, 75, 1981.

792. **Burba, P. and Schaefer, W.,** Atomic absorption spectrometric determination (flame-AAS) of heavy metals in bacterial leach liquors (jarosite) after their analytical separation on cellulose, *Erzmetall,* 34, 582, 1981.

793. **Moreiskaya, L. V., Nemodruk, A. A., Simonova, E. T., Emel'yanova, T. S., and Raevich, T. L.,** Substoichiometry in the atomic absorption spectrometric method. Determination of high component content, *Zavod. Lab.,* 47, 38, 1981 (in Russian).

794. **Frech, W., Lundberg, E., and Barbooti, M. M.,** Direct determination of trace metals in solid samples by atomic absorption spectrometry with electrothermal atomizers. IV. Interference effects in the determination of lead and bismuth in steels, *Anal. Chim. Acta,* 131, 45, 1981.

795. **Satsmadjis, J. and Voutsinou-Taliadouri, F.,** Determination of trace metals at concentrations above the linear calibration range by electrothermal atomic absorption spectrometry, *Anal. Chim. Acta,* 131, 83, 1981.

796. **Khalighie, J., Ure, A. M., and West, T. S.,** Atom-trapping atomic absorption spectrometry of arsenic, cadmium, lead, selenium and zinc in air-acetylene and air-propane flames, *Anal. Chim. Acta,* 131, 27, 1981.

797. **Dorner, W. G.,** Trace analysis of lead and its compounds, *Labor Praxis,* 5, 1042, 1981 (in Russian).

798. **Suzuki, M., Ohta, K., Yamakita, T., and Katsuno, T.,** Electrothermal atomization with a metal microtube in atomic absorption spectrometry, *Spectrochim. Acta,* 36b, 679, 1981.

799. **Rains, T. C.,** Determination of aluminum, barium, calcium, lead, magnesium and silver in ferrous alloys by atomic emission and atomic absorption spectrometry, *ASTM STP,* 747, 43, 1981.

800. **Koster, P. B., Raats, P., Hibbert, D., Phillipson, R. T., Schiweck, H., and Steinle, G.,** Collaborative study on the determination of trace elements in dried sugar beet pulp and molasses. III. Lead, *Sucr. Belge,* 100, 333, 1981 (in French).

801. **Elson, C. M., Bem, E. M., and Ackman, R. G.,** Determination of heavy metals in a menhaden oil after refining and hydrogenation using several analytical methods, *J. Am. Oil Chem. Soc.,* 58, 1024, 1981.

802. **Pilipenko, A. T. and Samchuk, A. I.,** Extraction-atomic absorption determination of trace elements in natural waters, *Khim. Tekhnol. Vody,* 3, 343, 1981.

803. **Kovarskii, N. Y., Kovekovdova, L. T., Pryazhevskaya, I. S., Belenkii, V. S., Shapovalova, E. N., and Popkova, S. M.,** Preconcentration of trace elements from water by electrodeposited magnesium hydroxide, *Zh. Anal. Khim.,* 36, 2264, 1981.

804. **Subramanian, K. S. and Meranger, J. C.,** A rapid electrothermal atomic absorption spectrophotometric method for cadmium and lead in human whole blood, *Clin. Chem. (Winston-Salem, N.C.),* 27, 1866, 1981.

805. **Brodie, K. G. and Rowland, J. J.,** Trace analysis of arsenic, lead and tin, *Eur. Spectrosc. News.,* 36, 41, 1981.

806. **Xia, L.,** Flame atomic absorption spectrophotometric determination of copper, cadmium, lead and zinc in tea leaves, *Fenxi Huaxue,* 9, 498, 1981 (in Chinese).

807. **Foerster, M. and Lieser, K. H.,** Determination of traces of heavy metals in inorganic salts and organic solvents by energy-dispersive X-ray fluorescence analysis of flameless atomic absorption spectrometry after enrichment on a cellulose exchanger, *Fresenius' Z. Anal. Chem.,* 309, 355, 1981.

808. **Monteil, A. and Welte, B.,** Comparison of the different methods of attack during the determination of micropollutants in the sediments, *Congr. Mediterr. Ing. Quim. (Actas), F.O.I.M.,* Barcelona, Spain, 2nd C20-13, 1981 (in French).

809. **Karai, I., Fukumoto, K., and Horiguchi, S.,** Improvement in the atomic absorption determination of lead in blood, *J. Appl. Toxicol.,* 1, 295, 1981.

810. **Kruse, R.,** Experiences in the trace analysis of mercury, lead and cadmium in fish using modern methods of determination, *Fisch. Umwelt,* 10, 19, 1981.

811. **Okamoto, A., Ohmori, M., and Ishibashi, T.,** Analysis of the trace elements in natural food dyes, *Annu. Rep. Osaka City Inst. Public Health Environ. Sci.,* 43, 98, 1981.

812. **Mishima, M., Hoshiai, T., Watanuki, T., and Sugawara, K.,** Ion-exchange and atomic absorption spectrometric determination of trace metals in human teeth, *Koshu Eiseiin Kenkyu Hokoku*, 30, 40, 1981.

813. **Suddendorf, R. F., Gajan, R. J., and Boyer, K. W.,** Utilization of atomic spectroscopy for the determination of lead and cadmium in zinc salts used as food additives, in *Dev. At. Plasma Spectrochem. Anal.,* Proc. Int. Winter Cong. 1980, Barnes, R. M., Ed., Heyden, London, 1981, 706.

814. **Amakawa, E., Ohonishi, K., Seki, H., and Matsumoto, M.,** Determination of lead in soft drinks by flameless atomic absorption spectroscopy using a method of coprecipitation with zirconium hydroxide, *Tokyo Toritsu Eisei Kenkyusho Kenkyu Nempo*, 32, 199, 1981.

815. **Koroschetz, F., Hoke, E., and Grasserbauer, M.,** Diffusion processes in electrodeposited lead-tin-copper bearing overlays, *Mikrochim. Acta Suppl.*, 9, 139, 1981.

816. **Fugas, M.,** Metals in airborne atmospheric particles. Review, *Zast. Atmos.*, 9, 13, 1981.

817. **Marlier-Geets, O., Heck, J. P., Barideau, L., and Rocher, M.,** Method for Determination of Heavy Metals in Sewage Sludge, Their Distribution According to Their Origin and Their Concentration Variation with Time, Rep. EUR 7076, Comm. Eur. Communities, Charct., Treat Use Sewage Sludge, Serv. Sci. Sol., Fac. Sci. Agrono. Etat, Gembloux, Belgium, 1981, 284 (in French).

818. **Hirano, K., Iida, K., Shimada, T., Iguchi, K., and Nagasaki, Y.,** Study on practical application of monitoring method of total amounts of heavy metals in waste water on adsorption of chelating resins and ion-exchange resins, *Zenkoku Kogaiken Kaishi*, 6, 9, 1981.

819. **Inhat, M.,** Analytical approach to the determination of copper, zinc, cadmium and lead in natural fresh waters, *Int. J. Environ. Anal. Chem.*, 10, 217, 1981.

820. **Berndt, H. and Messerschmidt, J.,** A microanalytical method of flame atomic absorption and emission spectrometry (platinum loop method). II. Determination of As, Bi, Cd, Hg, Pb, Sb, Se, Te, Tl, Zn, Li, Na, K, Ce, Sr and Ba, *Spectrochim. Acta*, 36b, 809, 1981.

821. **Grossi, G., Piazzi, S., Perugini, D., and Paolini, M.,** A simplified method for determining urinary and hematic lead by atomic absorption spectrophotometry, *Laboratorio (Milan)*, 8, 179, 1981 (in Italian).

822. **Borg, H., Edin, A., Holm, K., and Skoeld, E.,** Determination of metals in fish livers by flameless atomic absorption spectroscopy, *Water Res.*, 15, 1291, 1981.

823. **Costantini, S.,** Application of anodic voltammetry to direct determination of lead and cadmium in powdered milk for zootechnical use, *Riv. Soc. Ital. Sci. Aliment.*, 10, 239, 1981.

824. **Wen, J. and Geng, T.,** Determination of available cadmium, lead, copper and zinc in calcareous soil, *Fenxi Huaxue*, 9, 565, 1981.

825. **Senesi, N. and Polemio, M.,** Trace element addition to soil by application of NPK fertilizers, *Fert. Res.*, 2, 289, 1981.

826. **Simon, J. and Liese, T.,** Trace determination of lead and cadmium in bones. Comparison of atomic absorption spectrometry and inverse polarography, *Fresenius' Z. Anal. Chem.*, 309, 383, 1981.

827. **Ferenc, H.,** Determination of lead in blood by the atomic absorption spectrophotometric method, *Munkavedelem*, 27, 51, 1981 (in Hungarian).

828. **Fosse, G., Justesen, N. P. B., and Wesenberg, G. B. R.,** Microstructure and chemical composition of fossil mammalian teeth, *Calcif. Tissue Int.*, 33, 521, 1981.

829. **Kaiser, M. L., Koirtyohann, S. R., Hinderberger, E. J., and Taylor, H. E.,** *Spectrochim. Acta*, 36b, 773, 1981.

830. **Rohbock, E.,** The effect of airborne heavy metals on automobile passengers in Germany, *Environ. Int.*, 5, 133, 1981 (in English).

831. **Kanikawa, M., Kojima, S., and Nakamura, A.,** Analysis of lead in paints, paint scrapings and stationeries used by children, *Eisei Kagaku*, 27, 391, 1981.

832. **Borg, H.,** Trace Metals in Natural Waters. An Analytical Interpretation, Rapp.-Naturvardsverket (Swed.), SNV PM 1463, Forskningssekretariatet, Statens Naturvaardsverk, Solna, Sweden, 32pp., 1981 (in Swedish).

833. **Ueta, T., Mori, K., Nishida, S., and Taketishi, Y.,** Hygienic chemical studies on dental treatment materials. I. Content of cadmium, lead and arsenic in dental casting alloys and acrylic dental resins, *Tokyo Toritsu Eisei Kenkyusho Kenkyu Nempo*, 32, 110, 1981.

834. **Ikeda, S., Nagashima, C., and Ishikawa, T.,** A rapid determination of lead in blood by flameless atomic absorption spectrophotometer, *Tokyo Toritsu Eisei Kenkyusho Kenkyu Nempo*, 32, 227, 1981 (in Japanese).

835. **Suwirma, S., Surtipanti, S., and Yatim, S.,** Determination of mercury, lead, cadmium and chromium in several sea fishes, *Majalah BATAN*, 14, 2, 1981 (in Indonesian).

836. **Greenwood, R.,** Distribution of heavy metals (copper, lead and zinc) in rural areas of Rio de Janero (Brazil), *Rev. Bras. Geoscience*, 11, 98, 1981 (in Portuguese).

837. **Balaes, G. E. E. and Robert, R. V. D.,** Determination by atomic absorption spectrophotometry of impurities in manganese dioxide, *Natl. Inst. Metall. Repub. S. Afr. Rep.*, No. 2094, 1981.

838. **Batley, G. E.,** Insitu electrodeposition for the determination of lead and cadmium in sea water, *Anal. Chim. Acta*, 124, 121, 1981.

839. **Borg, H., Edin, A., Holm, K., and Skoeld, E.,** Determination of metals in fish livers by flameless atomic absorption spectroscopy, *Water Res.,* 15, 1291, 1981.

840. **Boyer, K. W., Jones, J. W., Linscott, D., Wright, S. K., Stroube, W., and Cunningham, W.,** Trace element levels in tissues from cattle fed a sewage sludge amended diet, *J. Toxicol. Environ. Health,* 8, 281, 1981.

841. **Brodie, K. G. and Rowland, J. J.,** Trace analysis of arsenic, lead and tin, *Eur. Spectrosc. News,* 36, 41, 1981.

842. **Brzozowska, B. and Zawadzka, T.,** Determination of lead, cadmium, zinc and copper in vegetable products by atomic absorption spectrophotometry, *Rocz. Panst. Zakl. Hig.,* 32, 9, 1981.

843. **Christensen, E. R. and Nan-kwang, C.,** Fluxes of arsenic, lead, zinc and cadmium to Green Bay and Lake Michigan sediments, *Environ. Sci. Technol.,* 15, 553, 1981.

844. **Debus, H., Hanle, W., Scharmann, A., and Wirz, P.,** Trace element detection by forward scattering using a continuum light source, *Spectrochim. Acta,* 36b, 1015, 1981.

845. **Dorner, W. G.,** Trace analysis of lead and its compounds, *Labor Praxis,* 5, 1042, 1981.

846. **Elson, C. M., Bem, E. M., and Ackman, R. G.,** Determination of heavy metals in a menhaden oil after refining and hydrogenation using several analytical methods, *J. Am. Oil Chem. Soc.,* 58, 1024, 1981.

847. **Henrion, G., Gelbrecht, J., Hoffmann, T., and Marquardt, D.,** Atomic absorption spectrophotometric determination of trace metal impurities in tungsten after separation on WOFATIT MC 50 chelating resin, *Z. Chem.,* 21, 192, 1981 (in German).

848. **Knutti, R.,** Matrix effects and matrix modifications in graphite furnace atomic absorption spectrophotometry as exemplifed by determination of lead in urine, *Mitt. Geb. Lebensmittelunters. Hyg.,* 72, 183, 1981.

849. **Nagy, S. and Rouseff, R. L.,** Lead contents of commercially canned single strength orange juice stored at various temperatures, *J. Agric. Food Chem.,* 29, 889, 1981.

850. **Rohbock, E. and Schmidt, G.,** Comparison of the aromatic and lead content of European gasolines, *Environ. Technol. Lett.,* 2, 263, 1981 (in German).

851. **Shaburova, V. P., Ruchkin, E. D., Yudelevich, I. G., and Kruglov, L. M.,** Atomic absorption method of analysis for pieziceramic materials of the lead zirconate titanate system, *Izv. Sib. Otd. Akad. Nauk SSSR Ser. Khim. Nauk,* 5, 95, 1981.

852. **Tunstall, M., Berndt, H., Sommer, D., and Ohls, K.,** Direct determination of trace elements in aluminum with flame atomic absorption spectrometry and ICP emission spectrometry — a comparison, *Erzmetall,* 34, 588, 1981 (in German).

853. **Viala, A., Goueze, F., Mallet, B., Fondarai, J., Cano, J. P., Sauve, J. M., Grimaldi, F., and Deturmeny, E.,** Determination of four trace metals (lead, cadmium, chromium and zinc) in atmospheric dust in Marseille between 1977-1979, *Pollut. Atmos.,* 91, 107, 1981 (in French).

854. **Wesenberg, G. B. R., Fosse, G., and Rasmussen, P.,** Effect of graded doses of cadmium on lead, zinc and copper content of target and indicator organs in rats, *Int. J. Environ. Stud.,* 17, 191, 1981.

855. **Brzozowska, B. and Zawadzka, T.,** Atomic absorption spectrophotometry in determination of lead, cadmium, zinc and copper in meat products, *Rocz. Panstv. Zakl. Hig.,* 32, 323, 1981 (in Polish).

856. **Uchino, E., Tsuzuki, T., Inoue, K., and Kishi, R.,** Concentration of iron, copper, zinc and lead in three tissues of young (22 days old) and adult (290 days old) rats, *Hokkaidoritsu Eisei Kenkyushoho,* 31, 117, 1981.

857. **Taliadouri-Voutsinou, F.,** Trace metals in marine organisms from the Saronikos Gulf (Greece), *J. Etud. Pollut. Mar. Mediterr.,* 5, 275, 1981.

858. **Hobart, E. W., Van Dien, J. E., and Green, M. H.,** Comparison of atomic absorption and X-ray fluorescence methods of analysis for lead, *Electron Microsc. X-ray Appl. Environ. Occup. Health Anal.,* 2, 51, 1981.

859. **Uysal, H.,** Levels of trace elements in some food-chain organisms from the Aegean coasts, *J. Etud. Mar. Mediterr.,* 5, 503, 1981.

860. **Haq, I. and Khan, C.,** Hazards of a traditional eye cosmetic-Surma, *J. Pak. Med. Assoc.,* 32, 7, 1982.

861. **Erzinger, J. and Puchelt, H.,** Methods for the determination of trace elements in geological materials, *Erzmetall,* 35, 173, 1982.

862. **Merchandise, P., Olie, J. L., Robbe, D., and Legret, M.,** Trace metal determination in river sediments and sewage sludges. Inter-laboratory comparison of extraction techniques, *Environ. Technol. Lett.,* 3, 157, 1982.

863. **Jackwerth, E. and Salewski, S.,** Contribution to the multielement preconcentration from pure cadmium, *Z. Anal. Chem.,* 310, 108, 1982 (in German).

864. **Lendermann, B. and Hundeshagen, D.,** Use of multielement standards for calibration in water analysis by AAS, *Z. Anal. Chem.,* 310, 415, 1982 (in German).

865. **Sakata, M. and Shimoda, O.,** Simple and rapid method for determination of lead and cadmium in sediment by graphite atomic absorption spectrometry, *Water Res.,* 16, 231, 1982.

866. **Kuga, K.,** Determination of exhausting gas from flameless atomizer in atomic absorption spectrometry. An estimation of chloride interference phenomenon in lead determination, *Bunseki Kagaku,* 31, 27, 1982.

867. **Brodie, K. G. and Routh, M. W.,** Trace metal analysis of biological samples with the GTA-95 graphite tube atomizer, *Varian Instrum. Appl.,* 16, 18, 1982.

868. **Ristow, R. and Bernau, M.,** Reproducibility and comparability of various atomic absorption spectrophotometric methods to determine lead in wine, *Dtsch. Lebensm. Rundsch.,* 78, 125, 1982.

869. **Abbey, S.,** Matrix interferences in graphite furnace atomic absorption spectrometry by capacitive discharge heating. Comments, *Anal. Chem.,* 54, 136, 1982.

870. **Chakrabarti, C. L., Wan, C. C., Hamed, H. A., and Bartels, P. C.,** Matrix interferences in graphite furnace atomic absorption spectrometry by capacitive discharge heating. Reply to comments, *Anal. Chem.,* 54, 137, 1982.

871. **Durando, M. L. and Aragon, S. R.,** Atmospheric lead in downtown Guatemala City, *Environ. Sci. Technol.,* 16, 20, 1982.

872. **Danielsson, L. G., Magnusson B., and Zhang, K.,** Matrix interference in the determination of trace metals by graphite furnace AAS after Chelax-100 preconcentration *At. Spectrosc.,* 3, 39, 1982.

873. **Bae, E. S. and Kim, Y. W.,** Mean concentration of lead in coal and fuel oil, *Kolyo Uikitae Chapchi,* 11, 21, 1980 (in Korean); CA 96, 71580g, 1982.

874. **Ough, C. S., Crowell, E. A., and Benz, J.,** Metal content of California wines, *J. Food Sci.,* 47, 825, 1982.

875. **Cirlin, E. H. and Housley, R. M.,** Distribution and evolution of zinc, cadmium and lead in Apollo 16 regoloth samples and the average uranium-lead ages of the parent rock, *Geochim. Cosmochim. Acta Suppl.,* 16, 529, 1982.

876. **Bertenshaw, M. P., Gelsthorpe, D., and Wheatstone, K. C.,** Reduction of matrix interferences in the determination of lead in aqueous samples by atomic absorption spectrophotometry with electrothermal atomization with lanthanum pretreatment, *Analyst (London),* 197, 163, 1982.

877. **Yamagata, N.,** Interlaboratory comparison study on the reliability of environmental analyses. Soil and sediment 1978-80, *Bunseki Kagaku,* 31, T1, 1982.

878. **Shamrai, Z. Y., Sergeeva, K. I., Naftal, I. P., and Dymova, M. S.,** Atomic absorption analysis of copper-manganese alloys, *Zavod. Lab.,* 48, 73, 1982 (in German).

879. **Akama, Y., Nakai, T., and Kawamura, F.,** Determination of lead in high purity cadmium by solvent extraction and atomic absorption spectrometry, *Fresenius' Z. Anal. Chem.,* 310, 429, 1982.

880. **Fischer, W. R. and Fechter, H.,** Analytical determination and fractionation of copper, zinc, lead, cadmium, nickel and cobalt in soils and underwater soils, *Z. Pflanzenernaer. Bodenkd.,* 145, 151, 1982 (in German).

881. **Bush, B., Doran, D. R., and Jackson, K. W.,** Evaluation of erythrocyte protoporphyrin and zinc protoporphyrin as micro screening procedures for lead poisoning detection, *Ann. Clin. Biochem.,* 19, 71, 1982.

882. **May, T. W. and Brumbaugh, W. G.,** Matrix modifier and L'vov platform for elimination of matrix interferences in the analysis of fish tissues for lead by graphite furnace atomic absorption spectrometry, *Anal. Chem.,* 54, 1032, 1982.

883. **Torsi, G. and Desimoni, E.,** Electrostatic accumulation furnace for electrothermal atomic spectrometry: a new apparatus for the determination of metals in the atmosphere, EUR 7624, 1982.

884. **Schilcher, H.,** Residues and impurities in medicinal plants and drug preparations. XIX. Report: on the determination of value and quality control of drugs, *Plant Med.,* 44, 65, 1982 (in German).

885. **Herber, R. F. M. and Van Deyck, W.,** On the optimization of blood lead standards in electrothermal atomization atomic absorption spectrometry, *Clin. Chim. Acta,* 120, 313, 1982.

886. **Bruhn, F., Carlos, R., Aldo, E., Navarrete, A., and Bravo, C.,** Method for the direct determination of lead in blood by graphite furnace atomic absorption spectroscopy, *Bol. Soc. Chil. Quim.,* 27, 346, 1982 (in Spanish).

887. **Hausknecht, K. A., Ryan, E. A., and Leonard, L. P.,** Determination of lead in paint chips using a modified ashing procedure and atomic absorption spectrophotometry, *At. Spectrosc.,* 3, 53, 1982.

888. **Kojima, I., Iida, C., and Yamasaki, K.,** Determination of five elements in a milligram amount of lead glasses by flame atomic absorption and emission spectrometry using a one-drop method, *Bunseki Kagaku,* 31, E167, 1982.

889. U.S. Environmental Protection Agency, Standards of performance for new stationary sources; lead-acid battery manufacture, *Fed. Regist.,* 47, 16564, April 16, 1982.

890. **Xu, H., Xiao, R., and Qiu, S.,** Determination of total lead in sewage using the ring oven technique, *Huanjing Kexue,* 3, 57, 1982.

891. **Lazareva, E. A. and Ryabinin, A. I.,** Atomic absorption determination of nanogram concentrations of lead in Black Sea waters, *Metod. Anal. Mor. Vod. Tr. Sov. Bolg. Sotrudnichestva,* L., 66, 1981 (in Russian); from *Ref. Zh. Khim.,* Abstr. No. 7G173, 1982.

892. **Brodie, K. and Routh, M. W.,** Trace metal analysis of biological samples with the GTA-95 graphite tube atomizer, *Varian Instrum. Appl.,* 16, 18, 1982.

893. **Baldini, M., Grossi, M., Micco, C., and Stacchini, A.,** Presence of metal in cereal. II. Contamination of 1978 Italian rice, *Riv. Soc. Ital. Sci. Aliment.,* 11, 23, 1982.

894. **Sturgeon, R. E., Desauliners, J. A. H., Berman, S. S., and Russell, D. S.,** Determination of trace metals in estuarine sediments by graphite furnace-atomic absorption spectrometry, *Anal. Chim. Acta,* 134, 283, 1982.

895. **Olney, C. E., Schauer, P. S., McLean, S., Lu, Y., and Simpson, K. L.,** International study on Artemia. VIII. Comparison of the chlorinated hydrocarbons and heavy metals in five different strains of newly hatched Artemia and a laboratory-reared marine fish, Brine Shrimp *Artemia,* in *Proc. Int. Symp.,* Universa Press, Wetteren, Belgium, 1979, 3, 343; CA 96, 81005n, 1982.

896. **Collins, B. J. and Sells, N. J.,** Lead contamination associated with snowmobile trails, *Environ. Res.,* 27, 159, 1982.

897. **Jacyszyn, K., Walas, J., Malinowski, A., Latkowski, T., and Cwynar, L.,** Concentration of heavy metals in pregnant women, *Zentralbl. Gynaekol.,* 104, 117, 1982.

898. Technical Assistance Report No. TA-80-79-753 at U.S. Postal Inspection Service, Atlanta, Georgia, Salisbury Rep., NIOSH-HE/TA-80-79-753: Order No. PB82-103185, 1980; available from the NTIS, Springfield, Va.; *Govt. Rep. Announce. Index (U.S.),* 82, 44, 1982.

899. **Jonas, D. M. and Parker, L. R., Jr.,** The effect of pH on the chelation of lead with ammonium pyrrolidine-dithiocarbamate for atomic absorption spectrometry, *Anal. Chim. Acta,* 134, 389, 1982.

900. **Hejtmancik, M. R., Jr., Dawson, E. B., and Williams, B. J.,** Tissue distribution of lead in rat pups nourished by lead-poisoned mothers, *J. Toxicol. Environ. Health,* 9, 77, 1982.

901. **Brueggemeyer, T. W. and Caruso, J. A.,** Determination of lead in aqueous samples as the tetramethyl derivative by atomic absorption spectrometry, *Anal. Chem.,* 54, 872, 1982.

902. **Kauffman, R. E., Saba, C. S., Rhine, W. E., and Eisentraut, K. J.,** Quantitative multielement determination of metallic wear species in lubricating oils and hydraulic fluids, *Anal. Chem.,* 54, 975, 1982.

903. **Sakla, A. B., Badran, A. H., and Shalaby, A. M.,** Determination of elements by atomic absorption spectrometry after destruction of blood in the oxygen flask, *Mikrochim. Acta,* 1, 483, 1982.

904. **Berndt, H., Guecer, S., and Messerschmidt, J.,** Direct determination of lead in urine with a new micromethod of flame atomic absorption spectroscopy (Loop AAS), *J. Clin. Chem. Clin. Biochem.,* 20, 85, 1982.

905. **Ranchet, J., Menissier, F., Lamathe, J., and Voinovitch, I.,** Interlaboratory comparison: the determination of cadmium, chromium, copper and lead in standard solutions by flameless atomic absorption spectrometry, *Analusis,* 10, 71, 1982.

906. **Nadkarni, R. A.,** Application of hydride generation-atomic absorption spectrometry to coal analysis, *Anal. Chim. Acta,* 135, 363, 1982.

907. **May, T. W. and Brumbaugh, W. G.,** Matrix modifier and L'vov platform for elimination of matrix interferences in the analysis of fish tissue for lead by graphite furnace atomic absorption spectrometry, *Anal. Chem.,* 54, 1032, 1982.

908. **Camara Rica, C. and Kirkbright, G. F.,** Determination of trace concentrations of lead and nickel in human milk by electrothermal atomization atomic absorption spectrophotometry and inductively coupled plasma emission spectroscopy, *Sci. Total Environ.,* 22, 193, 1982.

909. **Hoenig, M., Lima, C., and Dupire, S.,** Validity of atomic absorption spectrometric determinations of cadmium, cobalt, chromium, nickel and lead in animal tissues, particularly in fishes and their organs, *Analusis,* 10, 32, 1982.

910. **Wuensch, I.,** Rapid determination of lead in urine by AAS in the presence of CaEDTA, *Fresenius' Z. Anal. Chem.,* 310, 430, 1982.

911. **Landsberger, S., Jervis, R. E., Aufreiter, S., and Van Loon, J. C.,** The determination of heavy metals (aluminum, manganese, iron, nickel, copper, zinc, cadmium and lead) in urban snow using an atomic absorption graphite furnace, *Chemosphere,* 11, 237, 1982.

912. **Salmon, S. G. and Holcombe, J. A.,** Alteration of metal release mechanisms in graphite furnace atomizers by chemosorbed oxygen, *Anal. Chem.,* 54, 630, 1982.

913. **Smith, B. M. and Griffiths, M. B.,** Determination of lead and antimony in urine by atomic absorption spectroscopy with electrothermal atomization, *Analyst (London),* 107, 253, 1982.

914. **Wachter, C. and Weisweller, W.,** Extraction of thallium and simultaneous extraction of lead, cadmium and thallium for their quantitative analysis using flame atomic absorption spectroscopy, *Mikrochim. Acta,* 1, 307, 1982.

915. **Fernandez, F. J. and Hilligoss, D.,** An improved graphite furnace method for the determination of lead in blood using matrix modification and the L'vov platform, *At. Spectrosc.,* 3, 130, 1982.

916. **Djuvic, I., Djordjevic, V., and Radovic, N.,** Lead, cadmium, copper, arsenic and mercury in some additives (for meat), *Tehnol. Mesa,* 22, 355, 1981 (in English/Serbo-Croatian); CA 97, 54150v, 1982.

917. **Wittmann, Z.,** Determination of wear metals in used lubricating oils by atomic absorption spectrometry, *Acta Chim. Anal. Sci. Hung.,* 109, 295, 1982.

918. **Balkas, T., Tugrul, S., and Salihoglu, I.,** Trace metal levels in fish and crustacea from Northeastern Mediterranean Coastal waters, *Mar. Environ. Res.,* 6, 281, 1982.

919. **Maslowska, J. and Legedz, E.,** Comparison of methods for determination of trace lead amounts in dyes and finished products of the cosmetic industry, *Rocz. Panstw. Zakl. Hig.,* 32, 327, 1981; CA 97, 60800y, 1982.

920. **Jones, E. A.,** The separation and determination of trace elements in chromic oxide, *Rep. MINTEK,* Council for Mineral Technology, Randburg, South Africa, M15 1982.

921. **Brovko, I. A.,** Diphenylcarbazone as an extraction reagent in atomic absorption, *Deposited Doc. VINITI,* 2842-81, VINITI, Moscow, U.S.S.R., 1981 (in Russian); CA 97, 65661p, 1982.

AUTHOR INDEX

SUBJECT INDEX (PHYSICS)

SUBJECT INDEX (BIOLOGY)

Group VA Elements — N, P, As, Sb, and Bi

NITROGEN, N (ATOMIC WEIGHT 14.0067)

Nitrogen occurs free in air and in a combined state in nature. It can be found in plants, animals, protein-containing materials, fossil fuel, ammonia, and ammonium salts. Nitrogen is used in the manufacture of feed, fertilizers, ammonia, dyes, explosives, drugs, and other chemicals. Liquid nitrogen is used in space research and in the oil industry.

Usually the combined form of nitrogen is the main concern of an analyst. Determination of nitrogen is required in the analyses of nitrogen-containing organic compounds and evaluation of fertilizers, feed, hay, fodders, grain, sewage, water, and soil, etc.

Most of the nitrogen compounds are water soluble, but organic compounds require digestion. Atomic absorption method is the indirect procedure for the determination of nitrogen. Experimental details are not available.

REFERENCES

1. **Kumamaru, T., Tao, E., Okamoto, N., and Yamamoto, Y.,** A new method for the determination of nitrate ions by atomic absorption spectroscopy, *Bull. Chem. Soc. Jpn.,* 38, 2204, 1966 (in English).
2. **Houser, M. E. and Fauth, M. I.,** Indirect determination of nitrate, nitrite and nitro groups by atomic absorption spectrophotometry, *Microchem. J.,* 15, 1399, 1970.
3. **Sevcik, J.,** Selective detection of sulfur, chlorine and nitrogen with the help of the combination of flame ionization flame photometric detectors, *Chromatographia,* 4, 195, 1971.
4. **Hoshikawa, G. and Fudaho, Y.,** Atomic absorption determination of nitrate in plant material, *Kagawa Daigaku Nogakuba Gakujutsu Hokoku,* 27, 111, 1976 (in Japanese).
5. **Eklund, R. H. and Holcombe, J. A.,** Signal depression in electrothermal atomic absorption spectrometry by nitrate and sulfate ions, *Anal. Chim. Acta,* 109, 97, 1979.
6. **Watson, M. E.,** Interlaboratory comparison in the determination of nutrient concentrations of plant tissue, *Commun. Soil Sci. Plant Anal.,* 12, 661, 1981.
7. **Hassan, S. S. M.,** New atomic absorption spectrometric, potentiometric and polarographic method for the determination of nitrates, *Talanta,* 28, 89, 1981.
8. **Wieckowska, J.,** Determination of trace elements in oils, *Mikrochim. Acta,* 1, 195, 1982.

AUTHOR INDEX

SUBJECT INDEX

PHOSPHORUS, P (ATOMIC WEIGHT 30.9738)

Phosphorus occurs in nearly all igneous rocks and in combination in minerals. It is an essential component of living matter and is mostly found in teeth and bones. It is used in the manufacture of safety matches, steels, pyrotechnics, phosphor bronze, pesticides, smoke bombs, and tracer bullets. Its compounds are used as acids, cleaning agents, water softeners, fertilizers, chinaware, and special glasses.

Quantitative determination of total phosphorus is required in industrial waste, soil, fertilizers, water, insecticides, food stuffs, pharmaceuticals, biomaterials, commercial phosphorus salts, alloy steel, ferroalloys, slags, refractory materials, ores, rocks, minerals, additives, lubricants, and various other organic compounds.

Most of the phosphorus-containing samples can be dissolved in acids or fusion is used, wherever necessary. No interferences have been reported in the nitrous oxide-acetylene flame for the atomic absorption analysis of phosphorus.

Standard Solution

To prepare 50,000 mg/ℓ solution of phosphorus, dissolve 21.32 g of dibasis ammonium phosphate $(NH_4)_2HPO_4$ in water and dilute to 1ℓ with deionized water. Store it in polyethylene bottles only.

Instrumental Parameters

Wavelength	213.6 nm
Slit	0.2 nm
Light source	Hollow cathode lamp
Flame	Nitrous oxide-acetylene, reducing (rich and red)
Sensitivity	250—290 mg/ℓ
Detection limit	20 mg/ℓ
Optimum range	10,000 mg/ℓ
Secondary wavelength and its sensitivity	
214.9 nm	460—500 mg/mℓ

REFERENCES

1. **Walsh, A.,** Atomic absorption spectroscopy, in *Proc. 10th Coll. Spectroscopicum Int.,* Lippinocott, E.R. and Margoshes, M., Eds., Spartan Books, Rochelle Park, N.J., 1963.
2. **Kirkbright, G. F., Smith, A. M., and West, T. S.,** An indirect sequential determination of phosphorus and silicon by atomic absorption spectrophotometry, *Analyst (London),* 92, 411, 1967.
3. **Zaugg, W. S.,** Determination of phosphate in fresh and seawater by atomic absorption spectrophotometry, *At. Absorpt. Newsl.,* 6, 63, 1967.
4. **Kumamaru, T., Otani, Y., and Yamamoto, Y.,** A new method for the determination of phosphorus by atomic absorption spectrophotometry using molybdenum as a light source, *Chem. Soc. Jpn.,* 40, 429, 1967.
5. **Zaugg, W. S. and Knox, R. J.,** Determination of phosphate in biological materials and reaction mixtures by atomic absorption spectrophotometry, *Anal. Biochem.,* 20, 282, 1967.
6. **Singhal, K. C., Banerji, A.C., and Banerjee, B. K.,** Indirect method of estimation of phosphorus by atomic absorption spectrophotometry, *Technology,* 5, 117, 1968.
7. **Singhal, K. C. and Banerjee, B. K.,** Indirect estimation of phosphorus in rock phosphate by atomic absorption spectrophotometer using ion-exchange decomposition technique, *Technology,* 5, 239, 1968.
8. **Hurford, T. R. and Boltz, D. F.,** Indirect ultraviolet spectrophotometric and atomic absorption spectrometric methods for determination of phosphorus and silicon by heteropoly chemistry of molybdate, *Anal. Chem.,* 40, 379, 1968.
9. **Syty, A. and Dean, J. A.,** Determination of phosphorus and sulfur in fuel rich air-hydrogen flames, *Appl. Spectrosc.,* 7, 1331, 1968.
10. **Devoto, G.,** Indirect determination of phosphates in urine by atomic absorption spectrophotometry, *Boll. Soc. Ital. Biol. Sper.,* 44, 424, 1968 (in Italian).
11. **L'vov, B. V. and Khartsyzov, A. D.,** Atomic absorption determination of phosphorus in a graphite cuvet, *Zh. Prikl. Spektrosk.,* 11, 9, 1969.
12. **L'vov, B.V. and Khartsyzov, A. D.,** Atomic absorption determination of S, P, I, and Hg, from resonance lines in the vacuum ultraviolet spectral region, *Zh. Prikl. Spektrosk.,* 11, 413, 1969.
13. **Manning, D. C. and Slavin, S.,** The direct determination of phosphorus by atomic absorption spectroscopy, *At. Absorpt. Newsl.,* 8, 132, 1969.
14. **Ramakrishna, T. V., Robinson, J. W., and West, P. W.,** Determination of P, As, or Si by atomic absorption spectrometry of molybdenum heteropoly acids, *Anal. Chim. Acta,* 45, 43, 1969.
15. **Aldous, K. M., Dagnall, R. M., and West, T. S.,** Flame spectroscopic determination of S and P in organic and aqueous matrixes by using a simple flame photometer, *Analyst (London),* 95, 417, 1970.
16. **Berck, B., Westlake, W. F., and Gunther, F. A.,** Micro determination of phosphate by gas-liquid chromatography with micro coulometric, thermionic and flame photometric detection, *J. Agric. Food Chem.,* 18, 143, 1970.
17. **Kerber, J. D., Barnett, W. B., and Kahn, H. L.,** The determination of phosphorus by atomic absorption and flame emission spectroscopy, *At. Absorpt. Newsl.,* 9, 39, 1970.
18. **Morard, P. and Gullo, J. L.,** Mineralization of vegetable tissues for the determination of phosphorus, potassium, calcium, magnesium and sodium, *Ann. Agron.,* 21, 229, 1970 (in French).
19. **Linden, G. and Turk, S.,** Determination of phosphorus in biological media by atomic absorption spectrophotometry, *Chim. Anal.,* 53, 244, 1971.
20. **Melton, J. R., Hoover, W. L., Howard, P. A., and Green, V. S.,** Atomic absorption adaptation of the quinolinium molybdophosphate method for phosphorus, *J. Assoc. Off. Anal. Chem.,* 54, 373, 1971.
21. **Veillon, C. and Park, J. Y.,** Use of the salet phenomenon in the determination of sulfur and phosphorus in aqueous and organic samples, *Anal. Chim. Acta,* 60, 293, 1972.
22. **Wright, F. C. and Riner, J. C.,** Dry-ashing eggshells for determining calcium, magnesium and phosphorus, *J. Assoc. Off. Anal. Chem.,* 55, 662, 1972.
23. **Lin, C. I. and Huber, C. O.,** Determination of phosphate, silicate and sulfate in natural and waste water by atomic absorption inhibition titration, *Anal. Chem.,* 44, 2200, 1972.
24. **Toralballa, G. C., Spielholtz, G. I., and Steinberg, R. J.,** Determination of total phosphorus in detergents by atomic absorption spectroscopy, *Mikrochim. Acta,* 4, 484, 1972.
25. **Acuna, A. U., Husain, D., and Wiesenfeld, R. J.,** Kinetic study of electronically excited phosphorus atoms (3|2|D[J], 3|2|P[J]) by atomic absorption spectroscopy. II, *J. Chem. Phys.,* 58, 5272, 1973.
26. **Baker, D. E.,** A new approach to soil testing. II. Ionic equilibria involving H, K, Ca, Mg, Mn, Fe, Cu, Zn, Na, P and S, *Soil Sci. Soc. Am. Proc.,* 37, 537, 1973.
27. **Kirkbright, G. F. and Marshall, M.,** Direct determination of phosphorus by atomic absorption flame spectrometry, *Anal. Chem.,* 45, 1610, 1973.
28. **Baialardo, A. M. and Gomez Coedo, A.,** Determination of phosphorus in steel and cast iron by atomic absorption spectrophotometry, *Rev. Metal.,* 9, 35, 1973.

29. **Huber, C. O.,** Atomic Absorption Analysis of Phosphorus in Water, EPA 670/2-73-079, Environmental Protection Agency, Washington, D.C., 1973.

30. **Pleskach, L. I., Averin, U. D., and Fedorova, Z. V.,** Solvent extraction of alkali metal molybdophosphates as a method for increasing the sensitivity of the flame photometric determination of phosphorus, *Z. Anal. Khim.*, 28, 2378, 1973.

31. **Everett, G. L., West, T. S., and Williams, R. W.,** Trace analysis for sulfur and phosphorus in aqueous solution by a carbon filament atom reservoir technique, *Anal. Chim. Acta*, 68, 387, 1974.

32. **Kidani, Y., Takemura, H., and Koike, H.,** Indirect determination of phosphorus in organic compounds by atomic absorption spectrophotometry, *Bunseki Kagaku*, 23, 212, 1974.

33. **Evans, C. D., List, G. R., Beal, R. E., and Black, L. T.,** Iron and phosphorus contents of soybean oil from normal and damaged beans, *J. Am. Chem. Soc.*, 51, 444, 1974.

34. **Throneberry, G. O.,** Phosphorus and zinc measurements in kjeldahl digests, *Anal. Biochem.*, 60, 358, 1974.

35. **Ranjikar, K. P. and Townshend, A.,** Indirect pyrophosphate determination by atomic absorption using copper (II) and a liquid exchanger, *Anal. Lett.*, 7, 743, 1974.

36. **Stanton, R. E.,** Determination of calcium, cobalt, copper, iron, molybdenum, phosphorus, selenium and sulfur in plant material, *Lab. Pract.*, 23, 233, 1974.

37. **Suzuki, T., Morinaga, H., and Sasaki, A.,** Indirect determination of phosphorus in iron and steel by atomic absorption spectrometry, *Tetsu To Hagane*, 61, 1063, 1975.

38. **Rozenblum, V.,** Successive determination of picogram amounts of phosphorus and arsenic in pure water by indirect flameless atomic absorption (molybdenum) spectroscopy, *Anal. Lett.*, 8, 549, 1975.

39. **Subryan, V. L., Solomons, C. C., Gordon, S. G., Neldner, K. H., and Reeve, E. B.,** A micro method for the quantitative analysis of calcium, magnesium and phosphorus in human skin, *J. Lab. Clin. Med.*, 86, 1056, 1975.

40. **Campbell, D. R. and Seitz, W. R.,** Graphite oven flame analysis: application to phosphorus analysis, *Anal. Lett.*, 9, 543, 1976.

41. **Ediger, R. D.,** Standard conditions for the determination of phosphorus with the HGA graphite furnace, *At. Absorpt. Newsl.*, 15, 145, 1976.

42. **Gold'shtein, M. M. and Yudelevich, I. G.,** Indirect atomic absorption methods for determining fluorine, chlorine, iodine, phosphorus and sulfur, *Zh. Anal. Khim.*, 31, 801, 1976.

43. **Sand, J. R., Liu, J. H., and Huber, C. O.,** Determination of silicate, phosphate and sulfate by calcium atomization inhibition titration, *Anal. Chim. Acta*, 87, 79, 1976.

44. **Yudelevich, I. G., Buyanova, L. M., Bakhturova, N. F., and Korda, T. M.,** Indirect atomic absorption determination of phosphorus in silicon structures, *Izv. Sib. Otd. Akad. Nauk SSSR Ser. Khim. Nauk*, 5, 52, 1976.

45. **Husain, D. and Norris, P. E.,** Kinetic study of ground state phosphorus atoms, $P(3^4S_{3/2})$ by atomic absorption spectroscopy in the vacuum ultraviolet, *J. Chem. Soc. Faraday Trans.*, 73, 415, 1977.

46. **Inokuma, Y. and Endo, J.,** Determination of trace amounts of phosphorus in iron and steel by indirect atomic absorption spectrometric method combined with extraction of phosphomolybdic acid, *Tetsu To Hagana*, 63, 1026, 1977.

47. **Riddle, C. and Turek, A.,** An indirect method for the sequential determination of silicon and phosphorus in rock analysis by atomic absorption spectrometry, *Anal. Chim. Acta*, 92, 49, 1977.

48. **Whiteside, P. J. and Price, W. J.,** Determination of phosphorus in steel by atomic absorption spectrophotometry with electrothermal atomization, *Analyst (London)*, 102, 618, 1977.

49. **Mckown, M. M., Tschrin, C. R., and Lee, P. P. F.,** Investigation of Matrix Interferences for Atomic Absorption Spectroscopy of Sediments, Rep. No. EPA-600/7-78/085, U.S. Environmental Protection Agency, Office of Research and Development, Corvallis, Ore., 1978.

50. **Driscoll, D. J., Clay, D. A., Rogers, C. H., Jungers, R. H., and Butler, F. E.,** Direct determination of phosphorus in gasoline by flameless atomic absorption spectrometry, *Anal. Chem.*, 50, 767, 1978.

51. **Ediger, R. D., Knott, A. R., Peterson, G. E., and Beaty, R. D.,** The determination of phosphorus by atomic absorption using the graphite furnace, *At. Absorpt. Newsl.*, 17, 28, 1978.

52. **Prevot, A. and Gente-Jauniaux, M.,** Rapid determination of phosphorus in oils by flameless atomic absorption, *At. Absorpt. Newsl.*, 17, 1, 1978.

53. **Vigler, M. S., Strecker, A., and Varnes, A.,** Investigations of the determination of phosphorus in organic media by atomic absorption heated graphite atomizer, *Appl. Spectrosc.*, 32, 60, 1978.

54. **Abbey, S. and Maxwell, J. A.,** A critical comment on: "An indirect method for the sequential determination of silicon and phosphorus in rock analysis by atomic absorption spectrometry", *Anal. Chim. Acta*, 99, 397, 1978.

55. **L'vov, B. V. and Pelieva, L. A.,** Atomic absorption determination of phosphorus with an HGA atomizer using sample evaporation from a probe introduced into heated furnace, *Zh. Anal. Khim.*, 33, 1572, 1978.

56. **Rasmusson, B.,** Magnesium and phosphate in the serum of patients with medullary carcinoma of the thyroid, *Clin. Chim. Acta,* 89, 279, 1978.
57. **Riddle, C. and Turek, A.,** An indirect method for the sequential determination of silicon and phosphorus in rock analysis by atomic absorption spectrometry. Reply to the critical comment by Abbey, S. and Maxwell, J. A., *Anal. Chim. Acta,* 99, 398, 1978.
58. **Azuma, Y. and Aramaki, S.,** Direct determination of phosphorus by atomic absorption analysis, *Hokoku-Nara-Ken Kogyo Shikenjo,* 4, 51, 1978 (in Japanese).
59. **Hoft, D., Oxman, J., and Gurira, R. C.,** Direct determination of phosphorus in fertilizers by atomic absorption spectroscopy, *J. Agric. Food Chem.,* 27, 145, 1979.
60. **Janousek, L.,** Determination of phosphorus in iron and steel by indirect atomic absorption spectrometry and extraction of phosphovanadomolybdic acid, *Chem. Anal.,* 24, 59, 1979.
61. **Driedger, A. and Seitz, W. R.,** Flame photometry of phosphorus vaporized from a silica-treated graphite furnace, *Anal. Chem.,* 51, 1197, 1979.
62. **Gupta, P. K. and Sarkar, A. K.,** Quantitative estimation of phosphorus in water and effluents by atomic absorption method, *Indian J. Chem.,* 17a, 317, 1979.
63. **Boyer, K. W., Capar, S. G., Jones, J. W., Suddendorf, R. F., and Forwalter, J.,** Multielement/trace analysis identifies 48 elements in foods. Simultaneous element analysis and data reduction reaches parts per billion, *Food Process.,* 40, 72, 1979.
64. **Carmichael, N. G., Squibb, K. S., and Fowler, B. A.,** Metals in the molluscan kidney: a comparison of two closely related bivalve species (argopecten) using X-ray microanalysis and atomic absorption spectroscopy, *J. Fish. Res. Board Can.,* 36, 1149, 1979.
65. **Daidoji, H., Akai, Y., and Honma, A.,** Absorption and emission spectra of phosphorus and phosphorus monoxide, *Bunko Kenkyu,* 28, 77, 1979.
66. **Startseva, E. A., Popova, N. M., Khrapai, V. P., and Yudelevich, I. G.,** Indirect atomic absorption determination of phosphorus and silicon in pure silver, *Izv. Sib. Otd. Akad. Nauk SSSR Ser. Khim. Nauk,* 6, 139, 1979.
67. **Khavezov, I., Ruseva, E., and Iordanov, N.,** Flameless atomic absorption determination of phosphorus using zirconium carbide coated graphite atomizer tubes, *Z. Anal. Chem.,* 296, 125, 1979.
68. **Slikkerveer, F. J., Braad, A. A., and Hendrikse, P. W.,** Determination of phosphorus in edible oils by graphite furnace atomic spectrometry, *At. Spectrosc.,* 1, 30, 1980.
69. **Flannery, R. L. and Markus, D. K.,** Automated analysis of soil extracts for phosphorus, potassium, calcium and magnesium, *J. Assoc. Off. Anal. Chem.,* 63, 779, 1980.
70. **Haynes, R. J.,** Comparison of two modified kjeldahl digestion techniques for multielement plant analysis with conventional wet and dry ashing methods, *Soil Sci. Plant Anal.,* 11, 459, 1980.
71. **Horler, D. N. H., Barber, J., and Barringer, A. R.,** Multielemental study of plant surface particles in relation to geochemistry and biogeochemistry, *J. Geochem. Explor.,* 13, 41, 1980.
72. **Hilliard, E. P. and Smith, J. D.,** Minimum sample preparation for the determination of ten elements in pig feces and feeds by atomic absorption spectrophotometry and a spectrophotometric procedure for total phosphorus, *Analyst (London),* 104, 313, 1979.
73. **Morse, R. A. and List, D. J.,** Elemental analysis of honeys from several nations, *Am. Bee J.,* 120, 592, 1980.
74. **Nadkarni, R. A.,** Multitechnique multielemental analysis of coal and fly ash, *Anal. Chem.,* 52, 929, 1980.
75. **Persson, J. A. and Frech, W.,** Investigations of reactions involved in electrothermal atomic absorption procedures. VIIIA. A theoretical and experimental study of factors influencing the determination of phosphorus, *Anal. Chim. Acta,* 119, 75, 1980.
76. **El-Shaarawy, M. I.,** Phosphorus determination by atomic absorption spectrophotometry, *Ind. Aliment.,* 19, 477, 1980 (in Italian).
77. **Kosa, F., Foldes, V., Viragos-Kis, E., Rengei, B., and Ferke, A.,** Atomic absorption spectrophotometric study on content of inorganic substances in fetal bone for the determination of age, *Arch. Kriminol.,* 166, 44, 1980 (in German).
78. **Tardon, S.,** Rapid determination of small quantities of inorganic constituents of suspended dust and dust-fallout, *Chem. Prum.,* 30, 195, 1980 (in Czech).
79. **Watson, M. E.,** Interlaboratory comparison in the determination of nutrient concentrations of plant tissue, *Soil Sci. Plant Anal.,* 12, 601, 1981.
80. **Kuga, K. and Kitazume, E.,** Indirect determination of phosphorus by graphite furnace atomic absorption spectrometry of antimony and molybdenum, *Bunseki Kagaku,* 30, 164, 1981.
81. **Yudelevich, I. G., Buyanova, L. M., Beisel, N. F., and Kozhanova, L. A.,** Atomic absorption determination of concentration profiles of major compounds of $A_{III}B_V$ type semiconductor films, *At. Spectrosc.,* 2, 165, 1981.
82. **Bernal, J. L., del Nozal, M. J., Deban, L., and Aller, A. J.,** Indirect method for determining phosphorus in aluminum alloys by atomic absorption spectrometry, *Talanta,* 28, 469, 1981.

83. **Tekula-Buxbaum, P.,** Indirect sequential determination of phosphorus and arsenic in high-purity tungsten and its compounds by atomic absorption spectrophotometry, *Mikrochim. Acta,* 2, 183, 1981.
84. **Tittarelli, P. and Mascherpa, A.,** Liquid chromatography with graphite furnace atomic absorption spectrometric detector for speciation of organophosphorus compounds, *Anal. Chem.,* 53, 1466, 1981.
85. **Young, R. S.,** Analysis of nickel refinery slimes and residues, *Talanta,* 28, 25, 1981.
86. **Mimura, T. and Wakisaka, S.,** Indirect determination of small amounts of phosphorus and ATPase by flameless atomic absorption spectrometry, *Teikyo Igaku Zasshi,* 4, 143, 1981.
87. **Langmyhr, F. J. and Dahl, I. M.,** Atomic absorption spectrometric determination of phosphorus in biological materials, *Anal. Chim. Acta,* 131, 303, 1981.
88. **Kim, G. O. and Li, S. S.,** Study of analysis of pure metals by atomic absorption. II, *Punsok Hwahak,* 3, 8, 1981.
89. **Baliza, S. V. and Soledade, L. E. B.,** Application of atomic absorption in molecular analysis (spectrophotometry), *Congr. Annu. ABM,* 36th, Congress of the Brazilian Metals Association, Sao Paulo, 1981, 333.
90. **Boyer, K. W., Jones, J. W., Linscott, D., Wright, S. K., Stroube, W., and Cunningham, W.,** Trace element levels in tissues from cattle fed a sewage sludge amended diet, *J. Toxicol. Environ. Health,* 8, 281, 1981.
91. **Lee, C. Y., Parsons, G. F., and Downing, D. L.,** Effects of processing as amino acid and mineral contents of peas, *J. Food Sci.,* 47, 1034, 1982.
92. **Welz, B., Voellkopf, U., and Grobenski, Z.,** Determination of phosphorus in steel with a stabilized-temperature graphite furnace and Zeeman-corrected atomic absorption spectrometry, *Anal. Chim. Acta,* 136, 201, 1982.
93. **Graham, P. P., Bittel, R. J., Bovard, K. P., Lopez, A., and Williams, H. L.,** Mineral element composition of bovine spleen and separated spleen components, *J. Food Sci.,* 47, 720, 1982.

AUTHOR INDEX

SUBJECT INDEX

ARSENIC, As (ATOMIC WEIGHT 74.9216)

Arsenic has been known since the early Greeks and Romans. The metal and its compounds are used in alloys for hardening, insecticides, herbicides, silvicides, germicides, defoliants, dessicants, rodenticides, disinfectants, pigments, pharmaceuticals, and as a food additive in food for human consumption.

Due to the increased use of arsenic compounds and their known toxicity and carcinogenic properties, the trace level determination of arsenic in food products, agricultural products, and drinking water is a necessity. Several methods have been reported from time to time in literature, but in recent years atomic absorption method combined with hydride generation technique has taken a leading role to analyze the samples.

Preparation of sample solution is as important as the analysis. In any case arsenous solutions should not be boiled unless provisions are made to prevent its loss through volatilization. Arsenic trioxide is soluble in alkali carbonates and hydroxides, but dilute acids have no effect. Sulfides are insoluble in dilute acids, but are decomposed in concentrated strength. The recommended procedure for most of the samples is to treat the sample with aqua regia or fuming nitric acid or fusion with a mixture of sodium carbonate- sodium nitrate and addition of hydrochloric acid in the presence of an oxidizing agent to convert arsenic to its chlorides. All arsenites except the alkali arsenite require acids to have effective solution.

The major source of interferences in the atomic absorption method is molecular absorption of flame gases and species at the extreme UV region of the spectrum (193.7 and 197.2 nm). A sample with high total salt content, i.e., greater than 1%, will produce absorption at the 193.7 nm arsenic line, even when the element is not present in the sample solution. Background correction via a continuum source is strongly recommended. An argon-hydrogen flame is suitable for high solid and high acid concentration samples, but cannot be used with organic solvents. Improved sensitivity is obtained with this flame when used with the hydride generation technique. High concentrations of cobalt, copper, iron, lead, molybdenum, and nickel form compounds with arsenic which cannot be dissociated in the cool oxygen-hydrogen flame. Use of the hotter nitrous oxide-acetylene flame will eliminate the chemical interference with reduced sensitivity.

Standard Solution

To prepare 1000 mg/ℓ solution of arsenic, dissolve 1.3203 g of arsenious oxide (As_2O_3) in 25 mℓ of 20% (w/v) potassium hydroxide solution. Neutralize with 20% (v/v) sulfuric acid to a phenolphthalein end point and dilute it to 1 ℓ with 1% (v/v) sulfuric acid.

Instrumental Parameters

Wavelength	193.7 nm
Slit width	0.7 nm
Light source	Hollow cathode lamp/electrodeless discharge lamp
Flame	Air-acetylene, oxidizing (lean, blue)
Sensitivity	1 mg/ℓ
Detection limit	0.1 mg/ℓ
Optimum range	100 mg/ℓ

Secondary wavelengths and their sensitivities

189.0 nm	0.8 mg/ℓ
197.2 nm	1.0—2.0 mg/ℓ

REFERENCES

1. **Allan, J. E.,** Absorption flame photometry below 2000 Å, in 4th Australian Spectroscopy Conf., August 1963.
2. **Slavin, W., Sprague, S., and Manning, D. C.,** Detection limits in analytical atomic absorption spectrophotometry, *At. Absorpt. Newsl.,* 18, 1964.
3. **Mulford, C. E.,** Solvent extraction techniques for atomic absorption spectroscopy, *At. Absorpt. Newsl.,* 5, 88, 1966.
4. **Mulford, C. E.,** Low-temperature ashing for determination of volatile metals by atomic absorption spectroscopy, *At. Absorpt. Newsl.,* 5, 135, 1966.
5. **Dagnall, R. M., Thompson, K. C., and West, T. S.,** The use of microwave excited electrodeless discharge tubes as spectral sources in atomic absorption spectroscopy, *At. Absorpt. Newsl.,* 6, 117, 1967.
6. **Kahn, H. L. and Schallis, J. E.,** Improvement of detection limits of arsenic, selenium and other elements with an argon-hydrogen flame, *At. Absorpt. Newsl.,* 7, 5, 1968.
7. **Slavin, W.,** *Atomic Absorption Spectroscopy,* John Wiley & Sons, New York, 1968.
8. **Devoto, G.,** Indirect determination of arsenic in urine by atomic absorption spectrophotometry, *Boll. Soc. Ital. Biol. Sper.,* 44, 421, 1968 (in Italian).
9. **Smith, K. E. and Frank, C. W.,** Characterization of arsenic by atomic absorption spectroscopy in oxyacetylene flames, *Appl. Spectrosc.,* 22, 765, 1968.
10. **Holak, W.,** Gas sampling technique for arsenic determination by atomic absorption spectrophotometry, *Anal. Chem.,* 41, 1712, 1969.
11. **Kirkbright, G. F., Sargent, M., and West, T. S.,** The determination of arsenic and selenium by atomicabsorption spectroscopy in a nitrogen separated air-acetylene flame, *At. Absorpt. Newsl.,* 8, 34, 1969.
12. **Menis, O. S. and Rains, T. C.,** Determination of arsenic by atomic absorption spectrometry with an electrodeless discharge lamp as a source of radiation, *Anal. Chem.,* 41, 952, 1969.
13. **Ramakrishna, T. V., Robinson, J. W., and West, P. W.,** Determination of phosphorus, arsenic and silicon by atomic absorption spectrometry of molybdenum heteropoly acids, *Anal. Chim. Acta,* 45, 43, 1969.
14. **Lambert, M. J.,** The determination of copper, chromium and arsenic in preservation treated timber by the method of atomic absorption spectrophotometry, *J. Inst. Wood Sci.,* 27, 27, 1969.
15. **Ando, A., Suzuki, M., Fuwa, K., and Vallee, B. I.,** Atomic absorption of arsenic in nitrogen (entrained air)-hydrogen flames, *Anal. Chem.,* 41, 1974, 1969.
16. **Dean, J. A. and Fues, R. E.,** Determination of arsenic in organic compounds and other matrices by flame emission spectrophotometry, *Anal. Lett.,* 2, 105, 1969.
17. **Hill, U. T.,** Determination of arsenic in steel, iron ores and smelters by atomic absorption, *Am. Soc. Test. Mater.,* 443, 83, 1969.
18. **Gandrud, B. and Marshall, J. C.,** The determination of arsenic, lead, nickel and zinc in copper by atomic absorption spectrophotometry, *Appl. Spectrosc.,* 24, 367, 1970.
19. **Johns, P.,** The atomic absorption of arsenic and selenium, *Spectrovision,* 24, 6, 1970.
20. **Dalton, E. F. and Malonoski, A. J.,** Note on the determination of arsenic by atomic absorption by arsenic generation into an argon-hydrogen entrained air flame, *At. Absorpt. Newsl.,* 10, 92, 1972.
21. **Fernandez, F. J. and Manning, D. C.,** The determination of arsenic at sub-microgram levels by atomic absorption spectrophotometry, *At. Absorpt. Newsl.,* 10, 86, 1971.
22. **Kirkbright, G. F. and Ranson, L.,** Use of the nitrous oxide-acetylene flame for determination of arsenic and selenium by atomic absorption spectrometry, *Anal. Chem.,* 43, 1238, 1971.
23. **Madsen, R. E., Jr.,** Atomic absorption determination of arsenic subsequent to arsine reaction with 0.01 *M* silver nitrate, *At. Absorpt. Newsl.,* 10, 57, 1971.
24. **Manning, D. C.,** A high sensitivity arsenic-selenium sampling system for atomic absorption spectroscopy, *At. Absorpt. Newsl.,* 10, 123, 1971.
25. **Spielholtz, G. L., Toralballa, G. C., and Steinberg, R. J.,** Determination of arsenic in coal and in insecticides by atomic absorption spectroscopy, *Mikrochim. Acta,* 918, 1971.
26. **Yamamoto, Y., Kumamaru, T., Hayashi, Y., Kanke, M., and Matsui, A.,** Indirect atomic absorption determination of ppm levels of arsenic by combination of an MIBK extract of arsenomolybdic acid, *Talanta,* 19, 1633, 1972.
27. **Kirkbright, G. F., Ranson, L., and West, T. S.,** Use of a triple pass optical arrangement for the determination of arsenic and selenium by atomic absorption spectroscopy in an inert gas shielded nitrous oxide-acetylene flame, *Spectrosc. Lett.,* 5, 25, 1972.
28. **Williams, A. I.,** Use of atomic absorption spectrophotometry for the determination of copper, chromium and arsenic in preserved wood, *Analyst (London),* 97, 104, 1972.

29. **Yamamoto, Y., Kumamaru, T., Hayashi, Y., Kanke, M., and Matsui, A.,** Determination of arsenic in water colorimetry using arsenic-silver diethyldithiocarbamateburcine-chloroform system and atomic absorption spectrometry of molybdenum after the extraction of molybdoarsenic acid into MIBK, *Bunseki Kagaku,* 21, 379, 1972.

30. **Chu, R. C., Barron, G. P., and Baumgarner, P. A.,** Arsenic determination at sub-microgram levels by arsine evolution and flameless atomic absorption spectrophotometric technique, *Anal. Chem.,* 44, 1476, 1972.

31. **Leblanc, P. J. and Jackson, A. L.,** Dry ashing technique for the determination of arsenic in marine fish, *J. Assoc. Off. Anal. Chem.,* 56, 383, 1972.

32. **Hassan, S. S. M. and Eldesouky, M. H.,** Indirect microdetermination of arsenic compounds by atomic absorption spectroscopy, *Z. Anal. Chem.,* 259, 346, 1972.

33. **Schmidt, F. J. and Royer, J. L.,** Submicrogram determination of arsenic, selenium, antimony and bismuth by atomic absorption utilizing sodium borohydride reduction, *Anal. Lett.,* 6, 17, 1973.

34. **Singhal, S. P.,** Indirect atomic absorption spectrophotometric determination of arsenic, *Microchem. J.,* 18, 178, 1973.

35. **George, G. M., Frahm, L. J., and McDonnell, J. P.,** Dry ashing method for the determination of total arsenic in animal tissues: collaborative study, *J. Assoc. Off. Anal. Chem.,* 56, 793, 1973.

36. **Kan, K. T.,** An automated method for the determination of arsenic and antimony, *Anal. Lett.,* 6, 603, 1973.

37. **Melton, J. R., Hoover, W. L., Ayers, J. L., and Howard, P. A.,** Direct vaporization and qualification of arsenic from soils and water, *Soil Sci. Soc. Am. Proc.,* 37, 558, 1973.

38. **Nakamura, Y., Nagai, H., Kubota, D., and Himeno, S.,** Determination of arsenic in waste water and soils by atomic absorption spectrometry using the arsenic generation method, *Bunseki Kagaku,* 22, 1543, 1973.

39. **Nakahara, T., Nishino, H., Munemori, M., and Musha, S.,** Atomic absorption spectrophotometric determination of arsenic and selenium in premixed inert gas (entrained air)-hydrogen flames with a multiflame burner, *Bull. Chem. Soc. Jpn.,* 46, 1706, 1973.

40. **Yamamoto, Y., Kumamaru, T., Hayashi, Y., and Kamada, T.,** Rapid and sensitive atomic absorption determination of arsenic by the arsine-argon-hydrogen flame system with the use of a zinc powder tablet as the reductant, *Bull. Soc. Chem. Jpn.,* 46, 2604, 1973.

41. **Yamamoto, Y., Kumamaru, T., Hayashi, Y., and Kamada, T.,** Atomic absorption determination of ppb levels of arsenic in water by an arsine-argon-hydrogen flame system combined with use of zinc powder tablets, potassium iodide and stannous chloride as reductant, *Bunseki Kagaku,* 22, 876, 1973.

42. **Al-Sibaai, A. A. and Fogg, A. G.,** Stability of dilute standard solutions of antimony, arsenic, iron and rhenium used in colorimetry, *Analyst (London),* 98, 732, 1973.

43. **Coldwell, J. C., Lishka, R. J., and McFarren, E. F.,** Evaluation of a low-cost arsenic and selenium determination at microgram-per-liter levels, *J. Am. Water Works Assoc.,* 85, 731, 1973.

44. **Gomez Coedo, A. and Dorado, M. T.,** Determination of arsenic in different metal matrices by atomic absorption spectrometry, *Rev. Metal.,* 10, 355, 1974 (in Spanish).

45. **Baird, R. B. and Gabrielin, S. M.,** A tantalum foil lined graphite tube for the analysis of arsenic and selenium by atomic absorption spectroscopy, *Appl. Spectrosc.,* 28, 273, 1974.

46. **Barnett, W. B. and Kerber, J. D.,** The atomic absorption determination of arsenic and other "difficult" trace elements in metallurgical samples, *At. Absorpt. Newsl.,* 13, 56, 1974.

47. **Freeman, H. C. and Uthe, J. F.,** An improved hydride generation apparatus for determining arsenic and selenium by atomic absorption spectroscopy, *At. Absorpt. Newsl.,* 13, 75, 1974.

48. **Hoover, W. L., Melton, J. R., Howard, P. A., and Bassett, J. W., Jr.,** Atomic absorption spectrometric determination of arsenic, *J. Assoc. Off. Anal. Chem.,* 57, 18, 1974.

49. **Knudson, E. J. and Christian, G. D.,** A note on the determination of arsenic using sodium borohydride, *At. Absorpt. Newsl.,* 13, 74, 1974.

50. **Orheim, R. M. and Bovee, H. H.,** Atomic absorption determination of nanogram quantities of arsenic in biological media, *Anal. Chem.,* 46, 921, 1974.

51. **Robinson, J. W., Garcia, R. C., Hindman, G., and Slevin, P.,** Difficulties in the determination of arsenic by atomic absorption spectrometry, *Anal. Chim. Acta,* 69, 203, 1974.

52. **Tam, K. C.,** Arsenic in water by flameless atomic absorption spectrophotometry, *Sci. Technol.,* 8, 734, 1974.

53. **Terashima, S.,** Atomic absorption in analysis of microamounts of arsenic and antimony in silicates by the arsenic and stibine generation method, *Biokhimiya,* 39, 1331, 1974.

54. **Thompson, K. C. and Thomerson, D. R.,** Atomic absorption studies on the determination of antimony, arsenic, bismuth, germanium, lead, selenium, tellurium and tin by utilizing the generation of covalent hydrides, *Analyst (London),* 99, 595, 1974.

55. **Van Loon, J. C. and Brooker, E. J.**, A simplified method for determining arsenic and antimony by the hydride evolution-atomic absorption method, *Anal. Lett.*, 7, 505, 1974.

56. **Yasuda, S. and Kakiyama, H.**, Determination of arsenic and antimony by flameless atomic absorption spectroscopy using a carbon tube atomizer, *Bunseki Kagaku*, 23, 620, 1974.

57. **King, H. G. and Morrow, R. W.**, Determination of Arsenic and Selenium in Surface Water by Atomic Absorption to Support Environmental Monitoring Programs, Rep. Y-156, U.S. Atomic Energy Comm., Oak Ridge, Tenn., 1974.

58. **Vijan, P. N. and Wood, G. R.**, An automated submicrogram determination of arsenic in atmospheric particulate matter by flameless atomic absorption spectrophotometry, *At. Absorpt. Newsl.*, 13, 33, 1974.

59. **Elliott, S. C. and Loper, B. R.**, Improved absorption tube for arsenic determinations, *Anal. Chem.*, 46, 2256, 1974.

60. **Goulden, P. D. and Brooksbank, P.**, Automated atomic absorption determination of arsenic, antimony and selenium in natural waters, *Anal. Chem.*, 46, 1431, 1974.

61. **Kirkbright, G. F. and Wilson, P. J.**, Application of a demountable hollow cathode lamp as a source for the direct determination of sulfur, iodine, arsenic, selenium and mercury by atomic absorption flame spectrometry, *Anal. Chem.*, 46, 1414, 1974.

62. **Lunde, G. and Paus, P. E.**, Ethanol as a solvent in the determination of arsenic in marine extracts by atomic absorption spectrophotometry, *Anal. Lett.*, 7, 363, 1974.

63. **Reay, P. F.**, A routine method for the determination of arsenic in plants, sediments and natural waters, *Anal. Chim Acta*, 72, 145, 1974.

64. **Griffin, H. R., Hocking, M. B., and Lowery, D. G.**, Arsenic determination in tobacco by atomic absorption spectrometry, *Anal. Chem.*, 47, 229, 1975.

65. **Kamada, T., Kumamaru, T., and Yamamoto, Y.**, Rapid determination of trace amounts of arsenic (III, V), antimony (III, V) and selenium (IV) in water by atomic absorption spectrophotometry with a carbon tube atomizer, *Bunseki Kagaku*, 24, 89, 1975 (in Japanese).

66. **Oddo, N. and Vitalis, A.**, New instrumental systems in atomic absorption spectroscopy for the analysis of toxic elements. Determination of arsenic with more advanced sampling systems, *Chim. Ind.*, 57, 93, 1975.

67. **Ratcliffe, D. B., Byford, C. S., and Osman, P. B.**, The determination of arsenic, antimony and tin in steels by flameless atomic absorption spectrometry, *Anal. Chim. Acta*, 75, 457, 1975.

68. **Schmidt, F. J., Boyer, J. L., and Muir, S. M.**, Automated determination of arsenic, selenium, antimony, bismuth and tin by atomic absorption utilizing sodium borohydride reduction, *Anal. Lett.*, 8, 123, 1975.

69. **Fitchett, A. W., Daughtrey, E. H., and Mushak, P.**, Quantitative measurements of inorganic and organic arsenic by flameless atomic absorption spectrometry, *Anal. Chim. Acta,*79, 93, 1975.

70. **Frahm, L. J., Albrecht, M. E., and McDonnell, J. P.**, Atomic absorption spectrophotometric determination of 4-hydroxy-3-nitro benzene arsenic acid (roxarsone) in premixes, *J. Assoc. Off. Anal. Chem.*, 58, 945, 1975.

71. **Maruts, T. and Sudoh, G.**, Arsenic by atomic absorption spectrometry, *Anal. Chim. Acta*, 77, 37, 1975.

72. **Mesman, B. B. and Thomas, T. C.**, Two atomic absorption methods for the determination of submicrogram amounts of arsenic and selenium, *Anal. Lett.*, 8, 449, 1975.

73. **Rozenblum, V.**, Successive termination of picogram amounts of phosphorus and arsenic in pure water by indirect flameless atomic absorption (molybdenum).spectroscopy, *Anal. Lett.*, 8, 549, 1975.

74. **Woidich, H. and Pfannhauser, W.**, Determination of arsenic in biological material using flame atomic absorption spectroscopy, *Z. Anal. Chem.*, 276, 61, 1975.

75. **Crecetius, E. A., Bothner, M. H., and Carpenter, R.**, Geochemistries of arsenic, antimony, mercury and related elements in sediments of Puget Sound, *Environ. Sci. Technol.*, 9, 325, 1975.

76. **Barnett, W. B. and McLaughlin, E. A.**, The atomic absorption determination of antimony, arsenic, bismuth, cadmium, lead and tin in iron, copper and zinc alloys with the graphite furnace, *Anal. Chim. Acta*, 80, 285, 1975.

77. **Kurokawa, M., Kaneko, M., Nobuichi, N., Fukui, S., and Kanno, S.**, Hygiene chemistry studies on environmental pollution by arsenic and antimony. I. Determination of arsenic by zinc column arsine generator coupled to atomic absorption spectrophotometer, *Bunseki Kagaku*, 21, 77, 1975 (in Japanese).

78. **Gherardi, S., Dall'Aglio, G., and Versitano, A.**, Determination of arsenic in food products, *Ind. Conserve*, 50, 284, 1975 (in Italian).

79. **Shelton, B. J.**, Determination of silver, selenium, tellurium, antimony, tin, lead and arsenic in anode sludges, *Natl. Inst. Metall. Repub. S. Afr. Rep.*, No. 1771, 1975.

80. **Dixon, K., Mallett, R. C., and Kocaba, R.**, Determination by atomic absorption spectrophotometry of arsenic, selenium, tellurium, antimony and bismuth on ion-exchange resins, *Natl. Inst. Metall. Repub. S. Afr. Rep.*, No. 1689, 1975.

81. **Fiorino, J.A., Jones, J.W., and Capar, S. G.,** Sequential determination of arsenic, selenium, antimony and tellurium in foods via rapid hydride evolution and atomic absorption spectrometry, *Anal. Chem.,* 48, 120, 1976.

82. **Freeman, H. C., Uthe, J. F., and Flemming, B.,** A rapid and precise method for the determination of inorganic and organic arsenic with and without wet ashing using a graphite furnace, *At. Absorpt. Newsl.,* 15, 49, 1976.

83. **Johnson, C. A. and Lewin, J. F.,** The determination of some "toxic" metals in human liver as a guide to normal levels in New Zealand. II. Arsenic, mercury and selenium, *Anal. Chim. Acta,* 82, 79, 1976.

84. **Korn K.,** Determination of arsenic in leachates by flameless atomic absorption spectroscopy, *Z. Anal. Chem.,* 179, 288, 1976.

85. **Kovar, K. A., Lautenschlaeger, W., and Seidel, R.,** Limit test determination of heavy metals (Pb, As) in drugs by atomic absorption, *Dtsch. Apoth. Ztg.,* 115, 1855, 1976 (in German).

86. **Kunselman, G. C. and Huff, E. A.,** The determination of arsenic, antimony, selenium and tellurium in environmental water samples by flameless atomic absorption, *At. Absorpt. Newsl.,* 15, 29, 1976.

87. **Owens, J. W. and Gladney, E. S.,** The determination of arsenic in natural waters by flameless atomic absorption, *At. Absorpt. Newsl.,* 15, 47, 1976.

88. **Pierce, F. D. and Brown, H. R.,** Inorganic interference study of automated arsenic and selenium determination with atomic absorption spectrometry, *Anal. Chem.,* 48, 693, 1976.

89. **Pierce, F. D., Lamoreaux, T. C., Brown, H.R., and Fraser, R. S.,** An automated technique for the sub-microgram determination of selenium and arsenic in surface waters by atomic absorption spectroscopy, *Appl. Spectrosc.,* 30, 38, 1976.

90. **Siemer, D. D., Koteel, P., and Jariwala, V.,** Optimization of arsenic generation in atomic absorption arsenic determinations, *Anal. Chem.,* 48, 836,1976.

91. **Aggett, J. and Spell, A. C.,** Determination of arsenic (III) and total arsenic by atomic absorption spectroscopy, *Analyst (London),* 101, 341, 1976.

92. **Thompson, K. C. and Godden, R. G.,** The application of a wide slot nitrous oxide-nitrogen-acetylene burner for the atomic absorption spectrophotometric determination of aluminum, arsenic and tin in steels by the single-pulse nebulization technique, *Analyst (London),* 101, 96, 1976.

93. **Vijan, P. N., Rayner, A. C., Sturgis, D., and Wood, G. R.,** A semi-automated method for the determination of arsenic in soil and vegetation by gas phase sampling and atomic absorption spectrometry, *Anal. Chim. Acta,* 82, 329, 1976.

94. **Wanchope, R. D.,** Atomic absorption determination of trace quantities of arsenic. Application of a rapid arsine generation technique to soil, water and plant samples, *At. Absorpt. Newsl.,* 15, 64, 1976.

95. **Zook, E. G., Powell, J. J., Hackley, B. M., Emerson, J. A., Brooker, J. R., and Knobl, G. M., Jr.,** National marine fisheries service preliminary survey of selected seafoods for mercury, lead, cadmium, chromium and arsenic content, *J. Agric. Food Chem.,* 24, 47, 1976.

96. **Bedard, M. and Kerbyson, J. D.,** Determination of arsenic, selenium, tellurium and tin in copper by hydride evolution atomic absorption spectrophotometry, *Can. J. Spectrosc.,* 21, 64, 1976.

97. **Chambers, J. C. and McClellan, B. E.,** Enhancement of atomic absorption sensitivity for copper, cadmium, antimony, arsenic and selenium by means of solvent extraction, *Anal. Chem.,* 48, 2061, 1976.

98. **Corbin, D. R. and Barnard, W. M.,** Atomic absorption spectrophotometric determination of arsenic and selenium in water by hydride generation, *At. Absorpt. Newsl.,* 15, 116, 1976.

99. **Edmonds, J. S. and Francesconi, K. A.,** Estimation of methylated arsenicals by vapor generation atomic absorption spectrometry, *Anal. Chim.,* 48, 2019, 1976.

100. **Forehand, T. J., Depuy, A. E., and Tai, H.,** Determination of arsenic in sandy soils, *Anal. Chem.,* 48, 999, 1976.

101. **Hamner, R. M., Lechak, D. L., and Greenberg, P.,** Determination of silver, arsenic, bismuth, antimony, selenium and tellurium in chromium metal with flameless atomic absorption spectroscopy, *At. Absorpt. Newsl.,* 15, 122, 1976.

102. **Kaszerman, R. and Theurer, K.,** Effect of valence state on the determination of arsenic by flame atomic absorption, *At. Absorpt. Newsl.,* 15, 129, 1976.

103. **Kokot, M. L.,** A modification of a hydride generation apparatus for the rapid determination of arsenic in a large number of geochemical and other samples, *At. Absorpt. Newsl.,* 15, 105, 1976.

104. **Sato, A. and Saitoh, N.,** Flameless atomic absorption spectroscopic determination of arsenic, chromium and lead in sea water by use of coprecipitation with zirconium hydroxide, *Bunseki Kagaku,* 25, 663, 1976.

105. **Terashima, S.,** The determination of arsenic in rocks, sediments and minerals by arsine generation and atomic absorption spectrometry, *Anal. Chim. Acta,* 86, 43, 1976.

106. **Walsh, P. R., Fasching, J. L., and Duce, R. A.,** Losses of arsenic during the low temperature ashing of atmospheric particulate samples, *Anal. Chem.,* 48, 1012, 1976.

107. **Walsh, A., Fasching, J. L., and Duce, R. A.,** Matrix effects and their control during the flameless atomic absorption determination of arsenic, *Anal. Chem.,* 48, 1025, 1976.

108. **Yamamoto, Y. and Kamada, T.,** Fractional determination of ppb levels of arsenic (III) and arsenic (V) in water using graphite furnace atomic absorption spectrophotometry combined with ammonium pyrrolidine dithiocarbamate nitrobenzene extraction, *Bunseki Kagaku,* 25, 567, 1976.

109. **Yamamoto, Y. and Kumamaru, T.,** Comparative study of zinc tablet and sodium borohydride tablet reduction systems in the detemination of arsenic, antimony and selenium by atomic absorption spectrophotometry via their hydrides, *Z. Anal. Chem.,* 281, 353, 1976.

110. **Yamamoto, Y. and Kumamaru, T.,** Masking effect of potassium iodide on interferences in the determination of arsenic via the hydride by atomic absorption spectrophotometry using the sodium borohydride tablet as the reductant, *Z. Anal. Chem.,* 28, 139, 1976.

111. **Kamada, T.,** Selective determination of arsenic (III) and arsenic (V) with ammonium pyrrolidine dithiocarbamate, sodium diethyl dithiocarbamate and dithizone by means of flameless atomic absorption spectrophotometry with carbon tube atomizer, *Talanta,* 23, 835, 1976.

112. **Tiemann, F. and Maassen, J.,** Arsenic content of semitidal waters and reproducibility of determinations at microgram levels, *Z. Wasser Abwasser Forsch.,* 9, 42, 1976 (in German).

113. **Yamamoto, Y., Kumamaru, T., Edo, T., and Takemoto, J.,** Atomic absorption determination of ppb levels of arsenic in water by an arsine-nitrogen-hydrogen flame system combined with the use of sodium borohydride tablets and potassium iodide, *Bunseki Kagaku,* 25, 770, 1976.

114. **Aslin, G. E. M.,** Determination of arsenic and antimony in geological materials by flameless atomic absorption spectrometry, *J. Geochem. Explor.,* 6, 321, 1976.

115. **Andreae, M. C.,** Determination of arsenic species in natural waters, *Anal. Chem.,* 49, 820, 1977.

116. **Bower, N. W. and Ingle, J. D.,** Precision of flame atomic absorption measurements of arsenic, cadmium, calcium, copper, iron, magnesium, molybdenum, sodium and zinc, *Anal. Chem.,* 49, 574, 1977.

117. **Feldman, C.,** Determination of traces of arsenic in siliceous materials, *Anal. Chem.,* 49, 825, 1977.

118. **Gian, H. F. and Tong, S. L.,** Determination of traces of arsenic in water by arsine generation and radiometric analysis, *Anal. Chim. Acta,* 89, 151, 1977.

119. **Guimont, J., Pichette, M., and Rheaume, N.,** Determination of arsenic in water, rocks and sediments by atomic absorption spectrophotometry, *At. Absorpt. Newsl.,* 16, 53, 1977.

120. **Poldoski, J. E.,** Molecular spectral interference in the determination of arsenic by furnace atomic absorption, *At. Absorpt. Newsl.,* 16, 70, 1977.

121. **Aruscavage, P.,** Determination of arsenic, antimony and selenium in coal by atomic absorption spectrometry with graphite tube atomizer, *J. Res. U.S. Geol. Surv.,* 5, 405, 1977.

122. **Thompson, A. J. and Thorseby, P. A.,** Determination of arsenic in soil and plant materials by atomic absorption spectrophotometry with electrothermal atomization, *Analyst (London),* 102, 9, 1977.

123. **Condylis, A. and Hocquaux, H.,** Atomic absorption spectrophotometric determination of arsenic, selenium and tin in metallurgical matrixes, *Analusis,* 5, 228, 1977.

124. **Fishman, M. and Spencer, R.,** Automated atomic absorption spectrometric determination of total arsenic in water and stream bed materials, *Anal. Chem.,* 49, 1599, 1977.

125. **Inhat, M. and Miller, H. J.,** Analysis of foods for arsenic and selenium by acid digestion, hydride evolution atomic absorption spectrophotometry, *J. Assoc. Off. Anal. Chem.,* 60, 813, 1977.

126. **Inhat, M. and Miller, H. J.,** Acid digestion hydride evolution atomic absorption spectrophotometric method for determining arsenic and selenium in foods: collaborative study, I, *J. Assoc. Off. Anal. Chem.,* 60, 1414, 1977.

127. **Inokuma, Y. and Endo, J.,** Determination of trace amounts of arsenic in iron and steel by indirect atomic absorption spectrometric method combined with extraction of arsenomolybdic acid, *Tetsu To Hogane,* 63, 1581, 1977.

128. **Ishizaki, M.,** Determination of total arsenic in biological samples by flameless atomic absorption spectrometry using a carbon tube atomizer, *Bunseki Kagaku,* 26, 667, 1977.

129. **Kang, H. K. and Valentine, J. L.,** Acid interference in the determination of arsenic by atomic absorption spectrometry, *Anal. Chem.,* 49, 1829, 1977.

130. **Kneip, T. J.,** Arsenic, selenium, and antimony in urine and air. Analytical method by hydride generation and atomic absorption spectrometry, *J. Health Lab. Sci.,* 14, 53, 1977.

131. **May, I. and Greenland, L. P.,** Borohydride-flame determination of arsenic in phosphoric acid, *Anal. Chem.,* 49, 2376, 1977.

132. **Mullen, J. D.,** Determination of arsenic in high-purity copper by flameless atomic absorption spectrophotometry, *Talanta,* 24, 657, 1977.

133. **Pierce, F. D. and Brown, H. R.,** Comparison of inorganic interferences in atomic absorption spectrometric determination of arsenic and selenium, *Anal. Chem.,* 49, 1417, 1977.

134. **Rigin, V. I. and Verkhoturov, G. N.,** Atomic absorption determination of arsenic using prior electrochemical reduction, *Zh. Anal. Khim.,* 32, 1965, 1977.

135. **La Villa, F. and Oueraud, F.,** Determination of arsenic in catalytic reforming feedstocks by flameless atomic absorption, *Rev. Inst. Fr. Pet.,* 32, 413, 1977 (in French).

136. **Shaikh, A. U. and Tallman, D. E.,** Determination of submicrogram per liter quantities of arsenic in water by arsine generation followed by graphite furnace atomic absorption spectrometry, *Anal. Chem.,* 49, 1093, 1977.

137. **Siemer, D. D. and Koteel, P.,** Comparison of methods of hydride generation atomic absorption spectrometric arsenic and selenium determination, *Anal. Chem.,* 49, 1096, 1977.

138. **Siemer, D. D., Vitek, R. K., Koteel, P., and Houser, W. C.,** Determination of arsenic in beverages and foods by hydride generation atomic absorption spectroscopy, *Anal. Lett.,* 10, 357, 1977.

139. **Smith, R. G., Van Loon, J. C., Knechtel, J. R., Fraser, J. L., Pitts, A. E., and Hodges, A. E.,** A simple and rapid hydride generation-atomic absorption method for the determination of arsenic in biological, environmental and geological samples, *Anal. Chim. Acta,* 93, 61, 1977.

140. **Brodie, K. G.,** A comparative study: determining arsenic and selenium by atomic absorption spectrophotometry, *Int. Lab.,* 65, 68, 74, 1977.

141. **Howe, L. H.,** Trace Analysis of Arsenic by Colorimetry Atomic Absorption and Polarography, EPA-600/ 7-77-036, U.S. Environmental Protection Agency, Office of Research and Development, Corvallis, Ore., 1977.

142. **de Groot, G., Van Dijk, A., and Maes, R. A. A.,** Determination of arsenic in urine by atomic absorption spectrophotometry with the hydride generation technique, *Pharm. Week,* 112, 949, 1977 (in Dutch).

143. **Tam, K. H. and Conacher, H. B. S.,** Suitability of the dry ashing procedure for determination of mercury in fish, *J. Environ. Sci. Health,* 12b, 213, 1977.

144. **Welz, B.,** Arsenic determination by atomic absorption spectroscopy, *Chem. Ind.,* 59, 771, 1977.

145. **Yasui, A. and Tsutsumi, C.,** Adaptability of wet decomposition method to food samples for the determination of arsenic by arsine generation-atomic absorption spectrophotometry, *Bunseki Kagaku,* 26, 809, 1977.

146. **Holm, J.,** Simplified digestion method and measuring technique for determining lead, cadmium and arsenic in animal tissues by atomic absorption spectrometry, *Fleischwirtschaft,* 58, 745, 1978 (in German).

147. **Fukamachi, K. and Tokunaga, T.,** Determination of ppb of arsenic in water by atomic absorption spectrophotometry using a zinc column, *Eisei Kagaku,* 24, 265, 1978 (in Japanese).

148. **Reichert, J. K. and Gruber, H.,** Application of a technically simplified hydride system for arsenic and selenium determination by atomic absorption spectrophotometry, *Vom Wasser,* 51, 191, 1978.

149. **Crecelius, E. A.,** Modification of the arsenic speciation technique using hydride generation, *Anal. Chem.,* 50, 826, 1978.

150. **Fleming, D. E. and Taylor, G. A.,** Improvement in the determination of total arsenic by arsine generation and atomic absorption spectrophotometry using a flame heated silica furnace, *Analyst (London),* 103, 101, 1978.

151. **Haynes, B. W.,** Electrothermal atomic absorption determination of arsenic and antimony in combustible municipal solid waste, *At. Absorpt. Newsl.,* 17, 49, 1978.

152. **Wauchope, R. D.,** Selenium and arsenic levels in soybean from different production regions of the United States, *J. Agric. Food Chem.,* 26, 226, 1978.

153. **Agemian, H. and Cheam, V.,** Simultaneous extraction of mercury and arsenic from fish tissues and an automated determination of arsenic by atomic absorption spectrometry, *Anal. Chim. Acta,* 101, 193, 1978.

154. **Chrenekova, E. and Rusinova, N.,** Determination of arsenic in biological material by atomic absorption spectrophotometry, *Chem. Listy,* 72, 990, 1978.

155. **Denyszyn, R. B., Grohse, P. M., and Wagoner, D. E.,** Sampling and atomic absorption spectrometric determination of arsine at the 2 $\mu g/m^3$ level, *Anal. Chem.,* 50, 1094, 1978.

156. **Flanjak, J.,** Atomic absorption spectrometric determination of arsenic and selenium in offal and fish by hydride generation, *J. Assoc. Off. Anal. Chem.,* 61, 1299, 1978.

157. **Kirkbright, G. F. and Taddia, M.,** Application of masking agents in minimizing interferences from some metal ions in the determination of arsenic by atomic absorption spectrometry with the hydride generation technique, *Anal. Chim. Acta,* 100, 145, 1978.

158. **Nakashima, S.,** Flotation of submicrogram amounts of arsenic coprecipitated with iron (III) hydroxide from natural waters and determination of arsenic by atomic absorption spectrophotometry following hydride generation, *Analyst (London),* 103, 1031, 1978.

159. **Orlova, V. A., Malyutina, T. M., and Kirillova, T. I.,** Solvent extraction of arsenic from chloride-iodide solutions. Indirect extraction. Atomic absorption determination of arsenic in high-purity titanium chloride, *Zh. Anal. Khim.,* 33, 1961, 1978.

160. **Shaikh, A. U. and Tallman, D. E.,** Species-specific analysis for nanogram quantities of arsenic in natural waters by arsine generation followed by graphite furnace atomic absorption spectrometry, *Anal. Chim. Acta,* 98, 251, 1978.

161. **Hocquellet, P.,** Application of electrothermal atomization to the determination of arsenic, antimony, selenium and mercury by atomic absorption spectrometry, *Analusis,* 6, 426, 1978.

162. **Kaneko, E.,** Determination of ppb level arsenic in well water, *Bunseki Kagaku,* 27, 1250, 1978.

163. **Yasui, A., Tsutsumi, C., and Toda, S.,** Selective determination of inorganic arsenic (III), (V) and organic arsenic in biological materials by solvent extraction atomic absorption spectrophotometry, *Agric. Biol. Chem.,* 42, 2139, 1978.

164. **Yudelevich, I. G., Beizel, N. F., and Terent'eva, L. A.,** Atomic absorption determination of arsenic in layers on silicon structures and films, *Izv. Sib. Otd. Akad. Nauk SSR Ser. Khim. Nauk,* 4, 39, 1978 (in Russian).

165. **Rees, D. I.,** A simple procedure for the determination of arsenic and tin in food by atomic absorption spectrophotometry, *J. Assoc. Public Anal.,* 16, 71, 1978.

166. **Brodie, K. G.,** Analysis of arsenic and other trace elements by vapor generation, *Am. Lab.,* 11, 58, 1979.

167. **Evans, W. H., Jackson, F. J., and Dellar, D.,** Evaluation of a method for the determination of total antimony, arsenic and tin in food stuffs using measurement by atomic absorption spectrophotometry with atomization in a silica tube using the hydride generation technique, *Analyst (London),* 104, 6, 1979.

168. **Haynes, B. W.,** Arsenic, antimony, selenium and tellurium determination in high-purity copper by electrothermal atomization, *At. Absorpt. Newsl.,* 18, 46, 1979.

169. **Jackwerth, E., Willmer, P. G., Hohn, R., and Berndt, H.,** A simple accessory for the determination of mercury and the hydride-forming elements (As, Bi, Sb, Se and Te) using flameless atomic absorption spectroscopy, *At. Absorpt. Newsl.,* 18, 66, 1979.

170. **Likaits, E. R., Farrell, R. F., and Mackie, A. J.,** Determination of arsenic in blister copper and flue dust by atomic absorption spectroscopy, *At. Absorpt. Newsl.,* 18, 53, 1979.

171. **Lo, D. B. and Coleman, R. L.,** Determination of arsenic in animal tissues using graphite furnace atomic absorption, *At. Absorpt. Newsl.,* 18, 10, 1979.

172. **Milner, B. A., Whiteside, P. J., and Price, W. J.,** Determination of trace amounts of arsenic and antimony in zinc powder, *Analyst (London),* 104, 474, 1979.

173. **Nakashima, S.,** Selective determination of arsenic (III) and arsenic (V) by atomic absorption spectrophotometry following arsine generation, *Analyst (London),* 104, 172, 1979.

174. **Palliere, M. and Gernez, G.,** Apparatus for the production of volatile hydrides for As, Sb, Bi, Sn, Ge, Se and Te determination by atomic absorption spectrometry, *Analusis,* 7, 46, 1979.

175. **Peter, F., Growcock, G., and Strune, G.,** Determination of arsenic in urine by atomic absorption spectrometry with electrothermal atomization, *Anal. Chim. Acta,* 104, 177, 1979.

176. **Rubeska, I. and Hlavinkova, V.,** Determination of arsenic in rocks and soils by atomic absorption spectrophotometry using the MHS-1 automated hydride system, *At. Absorpt. Newsl.,* 18, 5, 1979.

177. **Stringer, C. E. and Attrep, M., Jr.,** Comparison of digestion methods for determination of organoarsenicals in waste water, *Anal. Chem.,* 51, 731, 1979.

178. **Berndt, H., Willmer, P. G., and Jackwerth, E.,** Determination of arsenic in lead and lead alloys, *Z. Anal. Chem.,* 296, 377, 1979 (in German).

179. **Forrester, J. E., Lehecka, V., Johnston, J. R., and Ott, W. L.,** Direct determination of trace quantities of Sb, As, Bi, Cd, Pb, Se, Ag, Te and thallium in high-purity nickel by electrothermal atomic absorption spectrometry, *At. Absorpt. Newsl.,* 18, 73, 1979.

180. **Hadeishi, T. and Kimura, H.,** Direct measurements of concentration of trace elements in gallium arsenide crystals by Zeeman atomic absorption spectroscopy, *J. Electrochem. Soc.,* 126, 1988, 1979.

181. **Korenaga, T.,** Atomic absorption spectroscopic determination of arsenic in antimony compounds using solvent extraction and arsine generation, *Mikrochim. Acta,* 1, 435, 1979.

182. **Nakashima, S.,** Atomic absorption determination of arsenic in sea water by hydride generation after separating by flotation, *Bunseki Kagaku,* 28, 561, 1979.

183. **Odanaka, Y., Matano, O., and Goto, S.,** Determination of inorganic and methylated arsenicals in environmental materials by graphite furnace atomic absorption spectrometry. Enhancing and depressing effects of various coexisting reagents, *Bunseki Kagaku,* 28, 517, 1979.

184. **Peats, S.,** Determination of arsenic in seaweed and related products by atomic absorption spectrophotometry using the MHS-1 hydride generation system, *At. Absorpt. Newsl.,* 18, 118, 1979.

185. **Sefzik, E.,** Determination of arsenic, lead, cadmium, chromium and selenium with respect to the requirements of drinking water regulations by means of flameless atomic absorption spectroscopy, *Vom Wasser,* 50, 285, 1979 (in German).

186. **Welz, B. and Melcher, M.,** Use of a new antifoaming agent for the determination of arsenic in urine with the hydride AA technique, *At. Absorpt. Newsl.,* 18, 121, 1979.

187. **Boyer, K. W., Capar, S. G., Jones, J. W., Suddendorf, R. F., and Forwalter, J.,** Multielement/trace analysis identifies 48 elements in foods. Simultaneous element analysis and data reduction reaches parts per billion, *Food Process.,* 40, 72, 1979.

188. **Carelli, G., Iannaccone, A., LaBua, R., and Rimatori, V.,** Interference effect in the atomic absorption spectrometric determination of arsenic in filter-collected air samples, *Anal. Chim. Acta,* 111, 287, 1979.

189. **Flanjak, J. and Lee, H. Y.,** Trace metal content of livers and kidneys of cattle, *J. Sci. Food Agric.,* 30, 503, 1979.

190. **Iverson, D. G., Anderson, M. A., Holm, T. R., and Stanforth, R. R.,** An evaluation of column chromatography and flameless atomic absorption spectrophotometry for arsenic speciation as applied to aquatic systems, *Environ. Sci. Technol.,* 13, 1491, 1979.

191. **Robert, R. and Balaes, G.,** Atomic absorption spectrophotometry with hydride generation: an improved technique, *Natl. Inst. Metall. Repub. S. Afr. Rep.,* No. 2023, 1979.

192. **Stockton, R. A. and Irgolic, K. J.,** The Hitachi graphite furnace-Zeeman atomic absorption spectrometer as an automated, element-specific detector for high pressure liquid chromatography: the separation of arsenobetaine, arsenocholine and arsenite/arsenate, *J. Environ. Anal. Chem.,* 6, 313, 1979.

193. **Fonds, A. W., Kempf, T., Minderhoud, A., and Sonneborn, M.,** Heavy metal content of various kinds of water, *Reinhalt. Wassers,* 78, 1979.

194. **Fricke, F. L., Robbins, W. B., and Caruso, J. A.,** Trace element analysis of food and beverages by atomic absorption spectrometry, *Anal. At. Spectrosc.,* 2, 185, 1979.

195. **Reichert, J. K. and Gruber, H.,** Determination of arsenic and selenium by atomic absorption spectrophotometry. Comparison between two types of quartz cuvettes, *Vom Wasser,* 52, 289, 1979.

196. **Welz, B. and Melcher, M.,** Hydride atomic absorption spectroscopic technique in trace analysis, *Labor Praxis,* 3, 41, 1979 (in German).

197. **El-Enany, F. F., Mahmoud, K. F., and Varma, M. M.,** Single organic extraction for determination of ten heavy metals in sea water, *J. Water Pollut. Control. Fed.,* 51, 2545, 1979.

198. **Iadevaia, R., Aharonson, N., and Woolson, E. A.,** Extraction and cleanup of soil arsenic residues for analysis by high pressure liquid chromatographic-graphite furnace atomic absorption, *J. Assoc. Off. Anal. Chem.,* 63, 742, 1980.

199. **Inhat, M. and Thompson, B. K.,** Acid digestion, hydride evolution atomic absorption spectrophotometric method for determining arsenic and selenium, *J. Assoc. Off. Anal. Chem.,* 63, 1814, 1980.

200. **Kuldvere, A.,** Atomic absorption determination of arsenic in seaweed by arsine generation using the Perkin-Elmer MHS-1, *At. Spectrosc.,* 52, 929, 1980.

201. **Nadkarni, R. A.,** Multitechnique multielemental analysis of coal and fly ash, *Anal. Chem.,* 52, 929, 1980.

202. **Slovak, Z. and Docekal, B.,** Sorption of arsenic, antimony and bismuth on glycol methacrylate gels with bound thiol groups for direct sampling in electrothermal atomic absorption spectrometry, *Anal. Chim. Acta,* 117, 293, 1980.

203. **Stein, V. B., Canelli, E., and Richards, A. H.,** Determination of arsenic in potable, trash and estuarine water by flameless atomic absorption, *At. Spectrosc.,* 1, 133, 1980.

204. **Dupire, S. and Hoenig, M.,** Effect of complex matrixes in the determination of trace elements by flameless atomic absorption spectrometry. II. Determination of arsenic and antimony in plants, *Analusis,* 8, 153, 1980.

205. **Hahn, M. H., Mulligan, K. J., Jackson, M. E., and Caruso, J. A.,** Sequential determination of arsenic, selenium, germanium and tin as their hydrides by gas-solid chromatography with an atomic absorption detector, *Anal. Chim. Acta,* 118, 115, 1980.

206. **Hinners, T. A.,** Arsenic speciation: limitations with direct hydride analysis, *Analyst (London),* 105, 751, 1980.

207. **Holak, W.,** Analysis of foods for lead, cadmium, copper, zinc, arsenic and selenium using closed system sample digestion: collaborative study, *J. Assoc. Off. Anal. Chem.,* 63, 485, 1980.

208. **Hubert, J., Candelaria, R. M., and Applegate, H. G.,** Determination of lead, zinc, cadmium and arsenic in environmental samples, *At. Spectrosc.,* 1, 90, 1980.

209. **Robert, R. V. D., Balaes, G., and Steele, T. W.,** Study of the measurement by electrothermal atomization and atomic absorption spectrophotometry of hydride forming elements, *Natl. Inst. Metall. Repub. S. Afr. Rep.,* No. 2053, 1980.

210. **Yanagi, K.,** A new procedure for determining arsenic in natural waters by means of atomic absorption spectrophotometry combined with the techniques of arsine evolution by sodium borohydride and of its collection in a trap of liquid nitrogen, *Bunseki Kagaku,* 29, 194, 1980.

211. **Agemian, H. and Thompson, R.,** Simple semiautomated atomic absorption spectrometric method for the determination of arsenic and selenium in fish tissue, *Analyst (London),* 105, 902, 1980.

212. **Arbab-Zavar, M. H. and Howard, A. G.,** Automated procedure for the determination of soluble arsenic using hydride generation atomic absorption spectroscopy, *Analyst (London),* 105, 744, 1980.

213. **Chakrabarti, D., DeJonghe, W., and Adams, F.,** Determination of arsenic by electrothermal atomic absorption spectrometry with a graphite furnace. I. Difficulties in the direct determination, *Anal. Chim. Acta,* 119, 331, 1980.

214. **Clark, P. J., Zingaro, R. A., Irgolic, K. J., and McGinley, A. N.,** Arsenic and selenium in Texas lignite, *Int. J. Environ. Anal. Chem.,* 7, 295, 1980.

215. **Cox, D. H.,** Arsine evolution-electrothermal atomic absorption method for the determination of nanogram levels of total arsenic in urine and water, *J. Anal. Toxicol.,* 4, 207, 1980.

216. **Brinckman, F. E., Jewett, K. L., Iverson, W. P., Irgolic, K. J., Erhardt, K. C., and Stockton, R. A.,** Graphite furnace atomic absorption spectrophotometers as automated element-specific detectors for high pressure liquid chromatography. The determination of arsenite, arsenate, methyl arsonic acid and dimethyl arsinic acid, *J. Chromatogr.*, 191, 31, 1980.

217. **Dekersabiec, A. M.,** Determination of arsenic, selenium and bismuth in rocks and soils by atomic absorption spectrophotometry, *Analusis*, 8, 97, 1980.

218. **Hon, P.K., Lau, D.W., Cheung, W. C., and Wong, M. C.,** The atomic absorption spectrometric determination of As, Bi, Pb, Sb, Se and Sn with a flame-heated silica T-tube after hydride generation, *Anal. Chim. Acta*, 115, 355, 1980.

219. **Lomashevich, S. A., Antonova, A.I., and Dvinina, A. P.,** Atomic absorption determination of arsenic in metallurgical products, *Zavod. Lab.*, 46, 230, 1980.

220. **Manning, D. C., Ediger, R. D., and Hoult, D. W.,** Determination of arsenic in synthetic fuel process waters, *At. Spectrosc.*, 1, 52, 1980.

221. **Barudi, W. and Beilig, H. J.,** Heavy metal content of vegetables which grow above ground and fruits, *Z. Lebensm. Unters. Forsch.*, 170, 254, 1980.

222. **Jonsen, J., Helgeland, K., and Steinnes, E.,** Trace elements in human serum. Regional distribution in Norway, in *Geomedical Aspects of Present and Future Research*, Laag, J., Ed., Global Book Resources Ltd., London, 1980, 89.

223. **Rombach, N., Apel, R., and Tschochner, F.,** Trace determination of arsenic, barium and mercury in plastics with flameless atomic absorption spectroscopy and ICP emission spectral analysis, *GIT Fachz. Lab.*, 24(12), 1165, 1980.

224. **Motuzova, G. V. and Obukhov, A. L.,** Effect of conditions for soil decomposition on the results of the determination of trace nutrient content on it, *Biol. Nauk (Moscow)*, 11, 87, 1980.

225. **Knight, M. J.,** Comparison of Four Digestion Procedures Not Requiring Perchloric Acid for the Trace Element Analysis of Plant Material, Rep. ANL/LRP-TM-18, Argonne National Laboratory, Argonne, Ill., 1980, 31; available from the NTIS, Springfield, Va.

226. **Agemian, H. and Bedek, E.,** Semi-automated method for the determination of total arsenic and selenium in soils and sediments, *Anal. Chim. Acta*, 119, 323, 1980.

227. **Bansho, K., Aoki, T., and Umezaki, Y.,** Determination of arsenic (III) and arsenic (V) by atomic absorption spectrometry, *Kogai*, 15, 272, 1980 (in Japanese).

228. **Buchet, J. P., Lauwerys, R., and Roels, H.,** Comparison of several methods for the determination of arsenic compounds in water and urine, their application for the study of arsenic metabolism and monitoring workers exposed to arsenic, *Int. Arch. Occup. Environ. Health*, 46, 11, 1980.

229. **Chakrabarti, D., DeJonghe, W., and Adams, F.,** Determination of arsenic by electrothermal atomic absorption spectrometry with a graphite furnace. II. Determination of arsenic (III) and arsenic (V) after extraction, *Anal. Chim. Acta*, 120, 121, 1980.

230. **Del Monte Tomba, M. G. and Luperi, N.,** Determination of traces of arsenic, antimony, bismuth and vanadium in steel and cast iron by graphite furnace atomic absorption spectrometry, *Metall. Lab.*, 72, 353, 1980 (in Italian and English).

231. **Drasch, G., Meyer, L. V., and Kauret, G.,** Application of the graphite furnace for detection of arsenic in biological samples by the hydride-atomic absorption spectroscopic technique, *Z. Anal. Chem.*, 304, 141, 1980.

232. **Freeman, G. H., Gutred, M., and Morris, L. R.,** Line profile study of the 193.76 nm arsenic emission line from lamps used in atomic absorption spectroscopy, *Spectrochim. Acta*, 35b, 687, 1980.

233. **Iverson, D. G., Anderson, M. A., Holm, T. R., and Stanforth, R. R.,** Column chromatography and flameless atomic absorption methods for the arsenic speciation in sediments, *Contam. Sediments*, 2, 29, 1980.

234. **Liddle, J. R., Brooks, R. R., and Reaves, R. D.,** Some parameters affecting hydride generation from arsenic (V) for atomic absorption spectrophotometry, *J. Assoc. Off. Anal. Chem.*, 63, 1175, 1980.

235. **Lindh, U., Brune, D., Nordberg, G., and Wester, P. O.,** Levels of antimony, arsenic, cadmium, copper, lead, mercury, selenium, silver, tin and zinc in bone tissue of industrially exposed workers, *Sci. Total Environ.*, 16, 109, 1980.

236. **Matsubara, K., Ota, N., Taniguchi, M., and Narita, K.,** Atomic absorption spectrophotometric determination of small amounts of arsenic, tin and aluminum in steel by the hydride generation method and the injection method, *Trans. Iron Steel Inst. Jpn.*, 20, 13406, 1980.

237. **Norman, E. A., Orlova, E. S., and Shestakova, A. I.,** Determination of arsenic, tin and antimony impurities in steel by atomic absorption with electrothermal atomization, *Zavod. Lab.*, 46, 1108, 1980.

238. **Tam, G. K. H. and Lacroix, G.,** Determination of arsenic in urine and feces by dry ashing-atomic absorption spectrometry, *Int. J. Environ. Anal. Chem.*, 8, 283, 1980.

239. **Van Willis, W., El-Ahraf, A., Vinjamoori, D. V., and Aref, K.,** Analysis of animal feed ingredients and soil amendment products from beef cattle manure for selected trace metals using atomic absorption spectrophotometry, *J. Food Prot.,* 43, 834, 1980.

240. **Webster, J.,** Estimation of arsenic concentrations in sewage sludges from the Lothian region, *Water Pollut. Control,* 79, 405, 1980.

241. **Thiex, N.,** Solvent extraction and flameless atomic absorption determination of arsenic in biological materials, *J. Assoc. Off. Anal. Chem.,* 63, 496, 1980.

242. **Weigert, P.,** Heavy metal content of chicken eggs, *Z. Lebensm. Unters. Forsch.,* 171, 118, 1980.

243. **Woolson, E. A. and Aharonson, N.,** Separation and detection of arsenical pesticide residues and some of their metabolites by high pressure liquid chromatography-graphite furnace atomic absorption spectrometry, *J. Assoc. Off. Anal. Chem.,* 63, 523, 1980.

244. **Howard, A. G. and Arbab-Zavar, M. H.,** Determination of organic arsenic (III) and arsenic (V) methyl arsenic and dimethyl arsenic species by selective hydride evolution-atomic absorption spectroscopy, *Analyst (London),* 106, 213, 1981.

245. **Korenaga, T.,** Atomic absorption spectrophotometric determination of trace amounts of arsenic in acrylic fibers containing antimony oxide with solvent extraction and arsine generation, *Analyst (London),* 106, 40 981.

246. **Ricci, G. R., Shepard, L. S., Colovos, G., and Hester, N. E.,** Ion chromatography with atomic absorption spectrometric detection for determination of organic and inorganic arsenic species, *Anal. Chem.,* 53, 610, 1981.

247. **Sinemus, H. W., Melchov, M., and Welz, B.,** Influence of valence state on the determination of antimony, arsenic, bismuth, selenium and tellurium in lake water using the hydride AA technique, *At. Spectrosc.,* 2, 81, 1981.

248. **Subramanian, K. S. and Meranger, J. C.,** Determination of As (III), As(V), Sb(III), Sb(V), Se(IV) and Se(VI) by extraction with ammonium pyrrolidine dithiocarbamatemethyl-isobutyl ketone and electrothermal atomic absorption spectrometry, *Anal. Chim. Acta,* 124, 131, 1981.

249. **Aksyuk, A. F., Merzlyakova, N. M., and Tarkhova, L. P.,** Determination of trace elements in desalinated seawater by an atomic absorption method, *Gig. Sanit.,* 5, 72, 1981 (in Russian).

250. **Cowgill, U. M.,** The chemical composition of bananas. Market basket values, *Biol. Trace Elem. Res.,* 3(1), 33, 1981.

251. **Xie, Y.,** Simultaneous determination of trace arsenic in water by hydride generation spectrophotometry, *Fen Hsi Hua Hsueh,* 9(1), 85, 1981 (in Chinese).

252. **Clark, J. R.,** Multielement extraction system for the determination of 18 trace elements in geochemical samples, *Anal. Chem.,* 53, 6, 1981.

253. **Clark, J. R. and Viets, J., G.,** Back–extraction system for the determination of 18 trace elements in geochemical samples, *Anal. Chem.,* 53, 65, 1981.

254. **Yanagi, K. and Ambe, M.,** Determination of arsenic in biological, environmental and geological materials by arsine evolution-flameless atomic absorption spectrophotometry, *Bunseki Kagaku,* 30, 209, 1981.

255. **Brooks, R. R., Ryan, D. E., and Zhang, H. F.,** Use of a tantalum-coated graphite furnace tube for the determination of arsenic by flameless atomic absorption spectrometry, *At. Spectrosc.,* 2(5), 161, 1981.

256. **Uthus, E. O., Collings, M. C., Cornatzer, W. E., and Nielson, F. H.,** Determination of total arsenic in biological samples by arsine generation and atomic absorption spectrometry, *Anal. Chem.,* 53, 2221, 1981.

257. **Yudelevich, I. G., Buyanova, L. M., Beisel, N. F., and Kozhanova, L. A.,** Atomic absorption determination of concentration profiles of major components of $A_{III}B_V$ type semiconductor films, *At. Spectrosc.,* 2, 165, 1981.

258. **Guo, X. and Wang, S.,** A new type of double capillary nebulizer used in atomic absorption analysis, *Fen Hsi Hua Hsueh,* 9, 258, 1981 (in Chinese).

259. **Brooks, R. R., Ryan, D. E., and Zhang, H.,** Atomic absorption spectrometry and other instrumental methods for quantitative measurements of arsenic, *Anal. Chim. Acta,* 131, 1, 1981.

260. **Koreckova, J., Frech, W., Lundberg, E., Persson, J. A., and Cedergren, A.,** Investigations of reactions involved in electrothermal atomic absorption procedures. X. Factors influencing the determination of arsenic, *Anal. Chim. Acta,* 130, 267, 1981.

261. **Welz, B., Grobenski, Z., and Melcher, M.,** Determination of antimony, arsenic, selenium, tellurium, bismuth and tin using the hydride-AAS technique, *Spektrometertagung (Votr.),* 13, 337, 1981 (in German).

262. **Mueller, H. and Siepe, V.,** Quantitative determination of arsenic, lead, cadmium, mercury and selenium in foods by flameless atomic absorption spectrophotometry, *Dtsch. Lebensm. Rundsch.,* 77, 392, 1981.

263. **Crock, J. G. and Lichte, F. E.,** The determination of trace-level arsenic in geological materials by semiautomated hydride generation-atomic absorption spectroscopy, *U.S. Geol. Surv. Open File Rep.,* 81, 20, 1981.

264. **Fukushima, I.,** A Study on Monitoring Methodology for Trace Elements in the Environment Through Chemical Analysis of Hair. Part of a Coordinated Program on Nuclear Methods in Health-Related Monitoring of Trace Element Pollutants, Rep. IAEA-R-2480/F, International Atomic Energy Agency, Vienna, 1981; INIS Atomindex 12, Abstract No. 626438, 1981.

265. **Blakemore, W. M. and Billedeau, S. M.,** Analysis of laboratory animal feed for toxic and essential elements by atomic absorption and inductively coupled argon plasma emission spectrometry, *J. Assoc. Off. Anal. Chem.,* 64, 1284, 1981.

266. **Malykh, V. D., Erkovich, G. E., Mukhamedshin, F. Y., and Kudryavina, A. K.,** Atomic absorption determination of an arsenic micro impurity in gold and silver, *Zavod. Lab.,* 47, 36, 1981.

267. **Adelman, H., Jenniss, S. W., and Katz, S. A.,** Interlaboratory analysis of sewage sludge, *Am. Lab.,* 31, 1981.

268. **Kahn, H. L., Cristiano, L. C., Dulude, G. R., and Sotera, J. J.,** Automated hydride analysis, *Am. Lab.,* 13, 136, 138, 141, 1981.

269. **Britske, M. E. and Slabodenyuk, I. V.,** Determination of arsenic in products of nonferrous metallurgy by atomic absorption, *Nauch. Tr. NII Tsvet. Met.,* 48, 41, 1981 (in Russian).

270. **Yamashige, T., Yamamoto, M., and Yamamoto, Y.,** Sensitive method for determination of arsenic ambient particulates utilizing arsine generation followed by heated quartz cell atomic absorption spectrophotometry, *Bunseki Kagaku,* 30, 324, 1981.

271. **Yamamoto, M., Urata, K., Murashige, K., and Yamamoto, Y.,** Differential determination of As(III) and As(V), and Sb(III) and Sb(V) by hydride generation-atomic absorption spectrophotometry and its application to the determination of these species in seawater, *Spectrochim. Acta,* 36b, 671, 1981.

272. **Welz, B. and Melcher, M.,** Determination of antimony, arsenic, bismuth, selenium, tellurium and tin in metallurgical samples using the hydride AA technique, *Spectrochim. Acta,* 36b, 439, 1981.

273. **Tekula-Buxbaum, P.,** Indirect sequential determination of phosphorus and arsenic in high-purity tungsten and its compounds by atomic absorption spectrophotometry, *Mikrochim. Acta,* 2, 183, 1981.

274. **Subramanian, K. S.,** Rapid electrothermal atomic absorption method for arsenic and selenium in geological materials via hydride evolution, *Z. Anal. Chem.,* 305, 382, 1981.

275. **Shkinev, V. M., Khavezov, I., Spivakov, B. Y., Mareva, S., Ruseva, E., Zolotov, Y. A., and Iordanov, N.,** Solvent extraction of arsenic(V) by dialkyl tin dinitrates. Extraction-atomic absorption determination of arsenic with a flame and graphite furnace, *Zh. Anal. Khim.,* 36, 896, 1981.

276. **Sanzolone, R. F. and Chao, T. T.,** Matrix modification with silver for the electrothermal atomization of arsenic and selenium, *Anal. Chim. Acta,* 128, 225, 1981.

277. **Saeed, K. and Thomassen, Y.,** Spectral interferences from phosphate matrixes in the determination of arsenic antimony, selenium and tellurium by electrothermal atomic absorption spectrometry, *Anal. Chim. Acta,* 130, 281, 1981.

278. **Peacock, C. J. and Singh, S. C.,** Inexpensive, simple hydride generation system with minimum interference for the atomic absorption spectrophotometry of arsenic, *Analyst (London),* 106, 931, 1981.

279. **Maher, W. A.,** Determination of inorganic and methylated arsenic species in marine organisms and sediments, *Anal. Chim. Acta,* 126, 157, 1981.

280. **Jastrow, J. D., Zimmerman, C. A., Dvorak, A. J., and Hinchman, R. R.,** Plant growth and trace element uptake on acidic coal refuse amended with lime or fly ash, *J. Environ. Qual.,* 10, 154, 1981.

281. **Itsuki, K. and Ikeda, T.,** Atomic absorption spectrometry of antimony and arsenic using graphite furnace with the aid of lanthanum, *Bunseki Kagaku,* 30, 684, 1981.

282. **Inui, T., Terada, S., and Tamura, H.,** Determination of arsenic by arsine generation with reducing tube followed by graphite furnace atomic absorption spectrometry, *Z. Anal. Chem.,* 305, 189, 1981.

283. **Ikeda, M., Jiro, N., and Nakahara, T.,** Investigation of some fundamental conditions for the determination of arsenic, lead and tellurium with continuous hydride generation-atomic absorption spectrometry, *Bunseki Kagaku,* 30, 368, 1981.

284. **Hirayama, T., Sakagami, Y., Nohara, M., and Fukui, S.,** Wet digestion with nickel and graphite furnace atomic absorption spectrophotometry for the determination of total arsenic in food samples, *Bunseki Kagaku,* 30, 278, 1981.

285. **Grabinski, A. A.,** Determination of As(III), As(V), monomethylarsenate and dimethylarsenate by ion exchange chromatography with flameless atomic absorption spectrometric detection, *Anal. Chem.,* 53, 966, 1981.

286. **Dornemann, A. and Kleist, H.,** Hydride generation method for determining arsenic by atomic absorption spectrometry with atomization in a hot quartz tube. Elimination of metal ion interference by addition of chelating agents, *Z. Anal. Chem.,* 305, 379, 1981.

287. **DeCarlo, E. H., Zeitlin, H., and Fernando, Q.,** Simultaneous separation of trace levels of germanium, antimony, arsenic and selenium from an acid matrix by adsorbing colloid flotation, *Anal. Chem.,* 53, 1104, 1981.

288. **Brown, R. M., Fry, R. C., Moyers, J. L., Northway, S. J., Denton, M. B., and Wilson, G. S.,** Interference by volatile nitrogen oxides and transition-metal catalysts in the preconcentration of arsenic and selenium as hydrides, *Anal. Chem.*, 53, 1560, 1981.

289. **Brooks, R. R., Ryan, D. E., and Zhang, H. F.,** Use of a tantulum-coated graphite furnace tube for the determination of arsenic by flameless atomic absorption spectrometry, *At. Spectrosc.*, 2, 161, 1981.

290. **Berndt, H. and Messerschmidt, J.,** Microanalytical method of flame atomic absorption and emission spectrometry (platinum loop method). II. Determination of As, Bi, Cd, Hg, Pb, Sb, Se, Te, Tl, Zn, Li, Na, K, Cs, Sr, and Ba, *Spectrochim. Acta*, 36b, 809, 1981.

291. **Elson, C. M., Bem, E. M., and Ackman, R. G.,** Determination of heavy metals in a menhaden oil after refining and hydrogenation using several analytical methods, *J. Am. Oil Chem. Soc.*, 58, 1024, 1981.

292. **Kumagai, H. and Saeki, K.,** Variation pattern of heavy metal content of short-neck clam *Tapes japonica* with its growth, *Nippon Suisan Gakkaishi*, 47, 1511, 1981.

293. **Welz, B. and Melcher, M.,** Mutual interactions of elements in the hydride technique in atomic absorption spectrometry. I. Influence of selenium on arsenic determination, *Anal. Chim. Acta*, 131, 17, 1981.

294. **Khalighie, J., Ure, A. M., and West, T. S.,** Atom-trapping atomic absorption spectrometry of arsenic, cadmium, lead, selenium and zinc in air-acetylene and air-propane flames, *Anal. Chim. Acta*, 131, 27, 1981.

295. **Jackson, K. F., Benedik, J. E., and Birholz, F. A.,** Metals distribution in shale oil fractions, *Oil Shale Symp. Proc.*, 14, 75, 1981.

296. **Brodie, K. G. and Rowland, J. J.,** Trace analysis of arsenic, lead and tin, *Eur. Spectrosc. News*, 36, 41, 1981.

297. **Arafat, N. M. and Glooschenko, W. A.,** Method for the simultaneous determination of arsenic, aluminum, iron, zinc, chromium and copper in plant tissue without the use of perchloric acid, *Analyst (London)*, 106, 1174, 1981.

298. **Senesi, N. and Polemio, M.,** Trace element addition to soil by application of NPK fertilizers, *Fert. Res.*, 2, 289, 1981.

299. **Irgolic, K. J., Banks, C. H., Bottino, N. R., Chakraborti, D., Gennity, J. M., Hillman, D. C., O'Brien, D. H., Pyles, R. A., Stockton, R. A. et al.,** Analytical and biochemical aspects of the transformation of arsenic and selenium compounds into biomolecules, *NBS Spec. Publ.*, 618, 244, 1981.

300. **Jones, E. A. and Dixon, K.,** The separation of trace elements in manganese dioxide, *Natl. Inst. Metall. Repub. S. Afr. Rep.*, No. 2131, 1981.

301. **Pacey, G. E. and Ford, J. A.,** Arsenic speciation by ion-exchange separation and graphite furnace atomic absorption spectrophotometry, *Talanta*, 28, 935, 1981.

302. **Monteil, A. and Welte, B.,** Comparison of the different methods of attack during the determination of micropollutants in the sediments, Congr. Mediterr. Ing. Quim. (Actas) 2nd, C 20-1/C 20-13, F.O.I.M., Barcelona, Spain, 1981 (in French).

303. **Balaes, G. E. E. and Robert, R. V. D.,** Determination by atomic absorption spectrophotometry of impurities in manganese dioxide, *Natl. Inst. Metall. Repub. S. Afr. Rep.*, No. 2094, 1981.

304. **Brooks, P. J. and Evans, W. H.,** Determination of total inorganic arsenic in fish, shellfish and fish products, *Analyst (London)*, 106, 514, 1981.

305. **Christensen, E. R. and Nan-Kwang, C.,** Fluxes of arsenic, lead, zinc and cadmium to Green Bay and Lake Michigan sediments, *Environ. Sci. Technol.*, 15, 553, 1981.

306. **Massee, R. and Maessen, F. J. M. J.,** Losses of silver, arsenic, cadmium, selenium and zinc traces from distilled water and artificial seawater by sorption on various container surfaces, *Anal. Chim. Acta*, 127, 181, 1981.

307. **Taguchi, M., Takagi, H., Iwashima, K., and Yamagata, N.,** Metal content of shark muscle powder biological reference material, *J. Assoc. Off. Anal. Chem.*, 64, 260, 1981.

308. **Vajda, F.,** Determination of arsenic and antimony in steels by atomic absorption, *Banyasz. Kohasz. Lopak Kohasz.*, 114, 113, 1981 (in Hungarian).

309. **Tsalev, D. and Petrov, I.,** Selective determination of arsenite and arsenate in soil extracts using hydride generation atomic absorption spectrometry, *Dokl. Bolg. Akad. Nauk*, 34, 1413, 1981.

310. **Tarui, T. and Tokairin, H.,** Determination of trace metal elements in petroleum. IV. Determination of arsenic in petroleum by combustion in an oxygen bomb followed by graphite furnace atomic absorption spectrometry, *Bunseki Kagaku*, 31, T45, 1982 (in Japanese).

311. **Benard, H. and Pinta, M.,** Determination of arsenic in atmospheric aerosols by atomic absorption with electrothermal atomization, *At. Spectrosc.*, 3, 8, 1982.

312. **Ebdon, L., Wilkinson, J. R., and Jackson, K. W.,** Simple and sensitive continuous hydride generation system for the determination of arsenic and selenium by atomic absorption and atomic fluorescence spectrometry, *Anal. Chim. Acta*, 136, 191, 1982.

313. **Fish, R. H., Brinckman, F. E., Jewett, K. L.,** Finger printing inorganic arsenic and organoarsenic compounds *in situ* oil shale retort and process waters using a liquid chromatograph coupled with an atomic absorption spectrometer as a detector, *Environ. Sci. Technol.*, 16, 174, 1982.

314. **Haring, B. J. A., Van Delft, W., and Bom, C. M.,** Determination of arsenic and antimony in water and soil by hydride generation and atomic absorption spectroscopy, *Z. Anal. Chem.,* 310, 217, 1982.

315. **Kida, A.,** Matrix effects of metal salts for determination of arsenic by graphite furnace atomic absorption spectrophotometry and their application to water analysis, *Bunseki Kagaku,* 31, 1, 1982.

316. **Schierling, P., Oefele, C., and Schaller, K. H.,** Determination of arsenic and selenium in urine samples with the hydride AAS technique, *Aertz. Lab.,* 28, 21, 1982 (in German).

317. **Subramanian, K. S. and Meranger, J. C.,** Rapid hydride evolution-electrothermal atomization atomic absorption spectrophotometric method for determining arsenic and selenium in human kidney and liver, *Analyst (London),* 107, 157, 1982.

318. **Tam, G. K. H. and Lacroix, G.,** Dry ashing hydride generation atomic absorption spectrometric determination of arsenic and selenium in foods, *J. Assoc. Off. Anal. Chem.,* 65, 647, 1982.

319. **Nygaard, D. D. and Lowry, J. H.,** Sample digestion procedures for simultaneous determination of arsenic, antimony and selenium by inductively coupled argon plasma spectrometry with hydride generation, *Anal. Chem.,* 54, 803, 1982.

320. **Ondon, J. M. and Biermann, A. H.,** Effects of Particle-Control Devices on Atmospheric Emissions of Minor and Trace Elements from Coal Combustion, Rep. EPA-600/9-80-039d, U.S. Environmental Protection Agency, Office of Research and Development, Symp. Transfer Util. Part Control Technol. 2nd, PB81-122228, Corvallis, Ore., 1980, 454; CA 96,90897g, 1982.

321. **Nadkarni, R. A.,** Application of hydride generation-atomic absorption spectrometry to coal analysis, *Anal. Chim. Acta,* 135, 363, 1982.

322. **Heinrichs, H. and Keltsch, H.,** Determination of arsenic, bismuth, cadmium, selenium and thallium by atomic absorption spectrometry with a volatilization technique, *Anal. Chem.,* 54, 1211, 1982.

323. **Bodewig, F. G., Valenta, P., and Nuernberg, H. W.,** Trace determination of arsenic (III) and arsenic (V) in natural waters by differential pulse anodic stripping voltammetry, *Fresenius' Z. Anal. Chem.,* 311, 187, 1982.

324. **Costantini, S., Giordano, R., and Ravagnan, P.,** Arsenic determination in ground water and cistern water, *Ann. Ist. Super. Sanit.,* 16(2), 287, 1980 (in Italian); CA 96,186929v, 12, 1982.

325. **Gunn, A. M.,** The Determination of Arsenic and Selenium in Raw and Potable Waters by Hydride Generation/Atomic Absorption Spectrometry — A Review, Tech. Rep. 169, Water Res. Cent., Medmenham Laboratory, Medmenham, Marlow, Bucks., England, 1982.

326. **Ueta, T., Mori, K., Nishida, S., and Yoshihara, T.,** Hygienic chemical studies on dental treatment materials. I. Content of cadmium, lead and arsenic in dental casting alloys and acrylic denture resins, *Tokyo Toritsu Eisei Kenkyusho Kenkyu Nempo,* 32, 110, 1982.

327. **Subramanian, K. S., Leung, P. C., and Meranger, J. C.,** Determination of arsenic (III, V, total) in polluted waters by graphite furnace atomic absorption spectrometry and anodic stripping voltammetry, *Int. J. Environ. Anal. Chem.,* 11, 121, 1982.

328. **Woolson, E. A., Aharonson, N., and Iadevaia, R.,** Application of high-performance liquid chromatography flameless atomic absorption method to the study of alkyl arsenical herbicide metabolism in soil, *J. Agric. Food Chem.,* 30, 580, 1982.

329. **Mautner, G. S. and Zhidkova, L. B.,** Method for the atomic absorption determination of arsenic in seawater and its intercalibration with colorimetric and spectrophotometric methods, *Metadiki Anal. Mor. Vod. Tr. Sov-bolg. Sotrudmchestva, L.,* 72, 1981; from *Ref. Zh. Khim.,* Abstr. No. 76170, 1982.

330. **Austenfeld, F. A. and Berghoff, R. L.,** An improved method for the selective determination of trace quantities of arsenite and arsenate in plant material, *Plant Soil,* 64, 267, 1982.

331. **George, G. M., Frahm, L. J., and McDonnell, J. P.,** Graphite furnace atomic absorption spectrophotometric determination of 4-hydroxy-3-nitrobenzenearsonic acid other organic arsenicals and inorganic arsenic in finished animal feed, *J. Assoc. Off. Anal. Chem.,* 65, 711, 1982.

332. **Yamagata, N.,** Interlaboratory comparison study on the reliability of environmental analysis. Soil and sediment 1978-80, *Bunseki Kagaku,* 31, T1, 1982.

AUTHOR INDEX

SUBJECT INDEX (ARSENIC)

ANTIMONY, Sb (ATOMIC WEIGHT 121.75)

Antimony is found in nature as a sulfide and also occurs as a free element in a very small amount. It is mainly used in the manufacture of bearing and type metal alloys, vulcanized rubber, and paints. It has some biological uses.

Preparation of sample solution is very important due to the volatile nature of its halides and hydrolytic tendencies. Metallic antimony is insoluble in cold dilute acids, but is soluble in hydrochloric acid in the presence of an oxidant, such as nitric acid. Usually loss of antimony occurs during acid digestion or when an attempt is made to concentrate the acid solutions. Alloy samples are dissolved in hydrochloric-nitric acids mixture (5 mℓ HCl + 5 mℓ HNO$_3$ + 20 mℓ water). To prevent hydrolysis of the solution, complexing agents such as fluoride or tartaric acid should be added to the solution.

In the antimony determination using the AA method, spectral interferences occur at the 217.6-nm line when the copper concentration exceeds 10,000 mg/ℓ. This interference can be avoided by using the alternate wavelength 231.2 nm and matrix matching of the standards with the similar concentration of copper. Copper and nickel depress antimony absorbance in reducing air-acetylene flame. When antimony is determined in the presence of lead at 217.6 nm, lead also causes spectral interference. A narrow band pass will eliminate this interference or else alternate wavelengths can also be used. Samples which contain high solids and acid concentrations also interfere.

Standard Solution

To prepare 1000 mg/ℓ of antimony solution, dissolve 1.0000 g of metallic antimony in 100 mℓ of hydrochloric acid plus 2 mℓ of nitric acid and dilute to 1 ℓ with deionized water, or dissolve 2.743 g of potassium antimonyl tartrate hemihydrate (K(SbO)C$_4$H$_4$O$_6$1/2H$_2$O) in deionized water and dilute to 1 ℓ.

Instrumental Parameters

Wavelength	217.6 nm
Slit	0.2 nm
Light source	Hollow cathode lamp
Flame	Air-acetylene, oxidizing (lean, blue)
Sensitivity	0.2—0.55 mg/ℓ
Detection limit	0.07 mg/ℓ
Optimum range	30 mg/ℓ
Secondary wavelengths and their sensitivities	
206.8 nm	0.25—0.85 mg/ℓ
231.2 nm	0.4—1.3 mg/ℓ
212.7 nm	1.5—12.0 mg/ℓ

REFERENCES

1. **Willis, J. B.,** Determination of lead and other heavy metals in urine by atomic absorption spectroscopy, *Anal. Chem.,* 34, 614, 1962.
2. **Slavin, W., Sprague, S., and Manning, D. C.,** Detection limits in analytical atomic absorption spectrophotometry, *At. Absorpt. Newsl.,* 18, 1962.
3. **Sattur, T. W.,** Routine atomic absorption analyses on non-ferrous alloys and plant intermediates, *At. Absorpt. Newsl.,* 5, 37, 1966.
4. **Ivanov, N. P., Minervina, L. V., Baranov, S. V., Profralidi, L. G., and Olikov, I. I.,** Tubes without electrodes with high frequency excitation of the spectrum of In, Ga, Bi, Sb, Tl, Pb, Mg, Ca and Cu, as a radiation source in atomic absorption analysis, *Zh. Anal. Khim.,* 21, 1129, 1966 (in Russian).
5. **Roth, D. J.,** Ratio determination of antimony trisulfide and potassium chlorate by atomic absorption spectrophotometry, in *Developments in Applied Spectrophotometry,* Vol. 5, Plenum Press, New York, 1966, 403.
6. **Krumpack, J.,** Semiquantitative analysis by atomic absorption, *At. Absorpt. Newsl.,* 6, 20, 1967.
7. **Dagnall, R. M., Thompson, K. C., and West, T. S.,** The use of microwave excited electrodeless discharge tubes as a spectral sources in atomic absorption spectroscopy, *At. Absorpt. Newsl.,* 6, 117, 1967.
8. **Mostyn, R. A. and Cunningham, A. F.,** Determination of antimony by atomic absorption spectrometry, *Anal. Chem.,* 39, 433, 1967.
9. **Slavin, S. and Sattur, T. W.,** Spectral interference of lead on antimony, *At. Absorpt. Newsl.,* 7, 11, 1968.
10. **Meranger, J. C. and Somers, E.,** Determination of antimony in titanium dioxide by atomic absorption spectrophotometry, *Analyst (London),* 93, 799, 1968.
11. **Walker, C. R., Vita, C. A., and Sparks, R. W.,** Determination of antimony, iron and molybdenum in nickel or uranium by atomic absorption spectroscopy, *Anal. Chim. Acta,* 47, 1, 1969.
12. **Yanagisawa, M., Suzuki, M., and Takeuchi, T.,** Atomic absorption spectrophotometry of antimony, *Anal. Chim. Acta,* 47, 121, 1969.
13. **Burke, K. E.,** Study of the scavanger properties of Mn (IV) oxide with atomic absorption spectrometry. Determination of microgram quantities of antimony, bismuth, lead and tin in nickel, *Anal. Chem.,* 42, 1536, 1970.
14. **Ivanov, N. P. and Mikhel'son, D. M.,** Determination of antimony in lead by atomic absorption, *Tr. Nauchno Issled Proekt. Inst. Azot. Prom. Produkt. Org. Sintez,* 5, 66, 1970.
15. **Murthy, G. K., Rhea, V., and Peeler, J. T.,** Levels of antimony, cadmium, chromium, cobalt, manganese and zinc in institutional diets, *Environ. Sci. Technol.,* 5, 436, 1971.
16. **Nichols, D. J.,** Determination of antimony in geological materials by atomic absorption spectrophotometry with particular reference to soils, *Anal. Chim. Acta,* 55, 59, 1971.
17. **Headridge, J. B. and Smith, D. R.,** Determination of trace amounts of antimony in mild steels by solvent extraction followed by atomic absorption spectrophotometry, *Lab. Pract.,* 20, 312, 1971.
18. **Burke, K. E.,** Determination of microgram amounts of antimony, bismuth, lead and tin in aluminum, iron and nickel-base alloys by nonaqueous atomic absorption spectroscopy, *Analyst (London),* 97, 19, 1972.
19. **Goulin, J. U., Holt, J. L., and Miller, R. E.,** Determination of tin and antimony in type metal using atomic absorption spectrophotometry, *Anal. Chem.,* 44, 1042, 1972.
20. **Stresko, V. and Martiny, E.,** Determination of antimony in geological materials by atomic absorption spectrometry, *At. Absorpt. Newsl.,* 11, 4, 1972.
21. **Pollack, E. N. and West, S. J.,** The determination of antimony at submicrogram levels by atomic absorption spectrophotometry, *At. Absorpt. Newsl.,* 11, 104, 1972.
22. **Yamamoto, Y., Kumamaru, T., Hayashi, Y., and Tsujino, R.,** Enhancement of sensitivity for antimony determination in atomic absorption spectrophotometry by introducing stibine into an argon-hydrogen flame, *Anal. Lett.,* 5, 419, 1972.
23. **Kobayashi, S. and Otobe, S.,** Determination of small amounts of antimony by atomic absorption spectrometry combined with solvent extraction, *Bunseki Kagaku,* 21, 1648, 1972.
24. **Schreiber, B. E. and Frei, R. W.,** Determination of trace amounts of antimony by flameless atomic absorption spectroscopy, *Int. J. Environ. Anal. Chem.,* 2, 149, 1972.
25. **Ng, W. K.,** Determination of antimony in galena by atomic absorption spectrometry, *Anal. Chim. Acta,* 64, 292, 1973.
26. **Renshaw, G. D., Pounds, C. A., and Pearson, E. F.,** The quantitative estimation of lead, antimony and barium in gunshot residues by non-flame atomic absorption spectrophotometry, *At. Absorpt. Newsl.,* 12, 55, 1973.
27. **Schmidt, F. J. and Royer, J. L.,** Submicrogram determination of arsenic, selenium, antimony and bismuth by atomic absorption utilizing sodium borohydride reduction, *Anal. Lett.,* 6, 17, 1973.

28. **Yanagisawa, M., Takeuchi, T., and Suzuki, M.,** Flameless atomic absorption spectrometry of antimony, *Anal. Chim. Acta,* 64, 381, 1973.
29. **Tan, K. T.,** An automated method for the determination of arsenic and antimony, *Anal. Lett.,* 6, 603, 1973.
30. **Quarrell, T. M., Powell, R. J. W., and Cluley, H. J.,** Determination of tin and antimony in lead alloy for cable sheathing by atomic absorption spectroscopy, *Analyst (London),* 98, 443, 1973.
31. **Frech, W.,** Rapid determination of antimony in steel by flameless atomic absorption, *Talanta,* 21, 565, 1974.
32. **Goulden, P. D. and Brooksbank, P.,** Automated atomic absorption determination of arsenic, antimony and selenium in natural waters, *Anal. Chem.,* 46, 1431, 1974.
33. **Miksovsky, M. and Rubeska, I.,** Determination of antimony, indium and thallium by atomic absorption spectrometry, *Chem. Listy,* 68, 299, 1974 (in Czech).
34. **Norris, J. D. and West, T. S.,** Use of an argon-hydrogen flame for the atomic absorption and atomic fluorescence spectrometry of antimony, *Anal. Chim. Acta,* 71, 458, 1974.
35. **Terashima, S.,** Atomic absorption analysis of microamounts of arsenic and antimony in silicates by the arsine and stibine generation method, *Biokhimiya,* 39, 1331, 1974.
36. **Thompson, K. C. and Thomerson, D. R.,** Atomic absorption studies on the determination of antimony, arsenic, bismuth, germanium, lead, selenium, tellurium and tin by utilizing the generation of covalent hydrides, *Analyst (London),* 99, 595, 1974.
37. **Thornton, K. and Burke, K. E.,** Modification to the extraction-atomic absorption method for the determination of antimony, bismuth, lead and tin, *Analyst (London),* 99, 469, 1974.
38. **Van Loon, J. C. and Brooker, E. J.,** A simplified method for determining arsenic and antimony by the hydride evolution-atomic absorption method, *Anal. Lett.,* 7, 505, 1974.
39. **Yasuda, S. and Kakiyama, H.,** Determination of arsenic and antimony by flameless atomic absorption spectroscopy using a carbon tube atomizer, *Bunseki Kagaku,* 23, 620, 1974.
40. **Josephson, M. and Dixon, K.,** Determination of minor amounts of antimony in ores and concentrates by atomic absorption spectrophotometry, *Natl. Inst. Metall. Repub. S. Afr. Rep.,* No. 1665, 1974.
41. **Condylis, A. and Mejean, B.,** Recent applications of atomic absorption to metallurgical analysis. Determination of low concentrations of the elements, *Analusis,* 3, 94, 1975.
42. **Goleb, J. A. and Midkif, C. R.,** Determination of barium and antimony in gunshot residue by flameless atomic absorption spectroscopy using a tantalum strip atomizer, *Appl. Spectrosc.,* 29, 44, 1975.
43. **Harrington, D. E. and Bramstedt, W. R.,** The determination of tin, antimony and tantalum in the presence of precious metals by atomic absorption spectroscopy, *At. Absorpt. Newsl.,* 14, 36, 1975.
44. **Kamada, T., Kumamaru, T., and Yamamoto, Y.,** Rapid determination of trace amounts of arsenic (III, V), antimony (III,V)and selenium (IV) in water by atomic absorption spectrophotometry with a carbon tube atomizer, *Bunseki Kagaku,* 24, 89, 1975 (in Japanese).
45. **Ratcliffe, D. B., Byford, C. S., and Osman, P. B.,** The determination of arsenic, antimony and tin in steels by flameless atomic absorption spectrometry, *Anal. Chim. Acta,* 75, 457, 1975.
46. **Schmidt, F. J., Royer, J. L., and Muir, S. M.,** Automated determination of arsenic, selenium bismuth and tin by atomic absorption utilizing sodium borohydride reduction, *Anal. Lett.,* 8, 123, 1975.
47. **Oguro, H.,** Determination of antimony in fire-retardant polypropylene by atomic absorption spectrophotometry, *Bunseki Kagaku,* 24, 797, 1975.
48. **Welch, E. P. and Chao, T. T.,** Determination of trace amounts of antimony in geological materials by atomic absorption spectrometry, *Anal. Chim. Acta,* 76, 65, 1975.
49. **Belcher, R., Bogdanski, S. L., Henden, E., and Townshend, A.,** Elimination of interferences in the determination of arsenic and antimony using molecular emission cavity analysis (MECA), *Analyst (London),* 100, 522, 1975.
50. **Goto, T.,** Atomic absorption spectrophotometric analysis of antimony iodide complex extraction with a high molecular amine, *Bunseki Kagaku,* 24, 520, 1975.
51. **Yudelevich, I. G., Povleva, G. V., and Tokareva, A. G.,** Atomic absorption determination of antimony and bismuth in chemical reagents, *Izv. Sib. Otd. Akad. Nauk SSSR Ser. Khim. Nauk,* 5, 10, 1975.
52. **Barnett, W. B. and McLaughlin, E. A.,** The atomic absorption determination of antimony, arsenic, bismuth, cadmium, lead and tin in iron, copper and zinc alloys with the graphite furnace, *Anal. Chim. Acta,* 80, 285, 1975.
53. **Crecelius, E. A., Bothner, M. H., and Carpenter, R.,** Geochemistries of arsenic, antimony, mercury and related elements in sediments of Puget Sound, *Environ. Sci. Technol.,* 9, 325, 1975.
54. **Acatini, C., de Berman, S. N., Colombo, O., and Fondo, O.,** Determination of silver, copper, lead, tin, antimony, iron, calcium, zinc, magnesium, potassium and manganese in canned tomatoes by atomic absorption spectrophotometry, *Rev. Asoc. Bioquim. Argent.,* 40, 175, 1975 (in Spanish).
55. **Dixon, K., Mallet, R. C., and Kocaba, R.,** Determination by atomic absorption spectrophotometry of arsenic, selenium, tellurium, antimony and bismuth on ion-exchange resins, *Natl. Inst. Metall. Repub. S. Afr. Rep.,* No. 1689, 1975.

56. **Yurinov, Y. V., Ryabkova, O. D., Katkova, D. E., and Semavin, Y. N.,** Atomic absorption analysis for antimony solution, *Tr. Ural. Nauchno Issled. Proekt. Inst. Med. Prom.,* 18, 209, 1975 (in Russian).

57. **Shelton, B. J.,** Determination of silver, selenium, tellurium, antimony, tin, lead and arsenic in anode sludges, *Natl. Inst. Metall. Repub. S. Afr. Rep.,* No. 1771, 1975.

58. **Fiorino, J. A., Jones, J. W., and Capar, S. G.,** Sequential determination of arsenic, selenium, antimony and tellurium in foods via rapid hydride evolution and atomic absorption spectrometry, *Anal. Chem.,* 48, 120, 1976.

59. **Kunselman, G. C. and Huff, E. A.,** The determination of arsenic, antimony, selenium and tellurium in envrionmental water samples by flameless atomic absorption, *At. Absorpt. Newsl.,* 15, 29, 1976.

60. **Muradov, V. G., Muradova, O. N., and Truzin, G. G.,** Measurement of equilibrium partial pressure of antimony atoms by the atomic absorption spectroscopic method, *Zh. Prikl. Spektrosk.,* 24, 5, 1976.

61. **Yamamoto, Y. and Kumamaru, T.,** Comparative study of zinc tablet and sodium borohydride tablet reduction systems in the determination of arsenic, antimony and selenium by atomic absorption spectrophotometry via their hydrides, *Z. Anal. Chem.,* 281, 353, 1976.

62. **Chambers, J. C. and McClellan, B. E.,** Enhancement of atomic absorption sensitivity for copper, cadmium, antimony, arsenic and selenium by means of solvent extraction, *Anal. Chem.,* 48, 2061, 1976.

63. **Hamner, R. M., Lechak, D. L., and Greenberg, P.,** Determination of silver, arsenic, bismuth, antimony, selenium, and tellurium in chromium metal with flameless atomic absorption spectroscopy, *At. Absorpt. Newsl.,* 15, 122, 1976.

64. **Jedrzejewska, H. and Malusecka, M.,** Interference effects of some carriers on the atomic absorption of chromium, manganese, antimony, iron, bismuth and lead in an air-acetylene flame, *Chem. Anal.,* 21, 585, 1976.

65. **Aznarez Aldvan, J. and Castillo Suarez, J. R.,** Determination of antimony by atomic absorption spectrophotometry after coprecipitation with manganese dioxide, *Quim. Anal.,* 30, 207, 1976 (in Spanish).

66. **Aslin, G. E. M.,** Determination of arsenic and antimony in geological materials by flameless atomic absorption spectrometry, *J. Geochem. Explor.,* 6, 321, 1976.

67. **Konanor, N. K. and Van Loon, G. W.,** Determination of lead and antimony in firearm discharge residues on hands by anodic stripping voltammetry, *Talanta,* 24, 184, 1977.

68. **Michael, S. S.,** Determination of sulfide samples containing antimony by atomic absorption spectrophotometry, *Anal. Chem.,* 49, 451, 1977.

69. **Pandey, L. P., Ghose, A., and Dasgupta, P.,** Determination of zinc, silver, copper, iron and antimony in lead metal by atomic absorption spectrophotometry, *J. Inst. Chem.,* 49, 35, 1977.

70. **Husain, D. and Slater, N. K.,** Kinetic study of ground state antimony atoms Sb ($5^4S_{3/2}$) by atomic absorption spectrophotometry, *J. Photochem.,* 7, 59, 1977.

71. **Kamada, T. and Yamamoto, Y.,** Selective determination of antimony (III) and antimony (V) with ammonium pyrrolidine dithiocarbamate, sodium diethyl dithiocarbamate and dithizone by atomic absorption spectrometry with a carbon tube atomizer, *Talanta,* 24, 330, 1977.

72. **Kneip, T. J.,** Arsenic, selenium and antimony in urine and air. Analytical method by hydride generation and atomic absorption spectroscopy, *J. Health Lab. Sci.,* 14, 53, 1977.

73. **Nagafuchi, Y., Fukamachi, K., and Morimoto, M.,** Atomic absorption spectrophotometric determination of microamounts of antimony using solvent extraction with sodium dimethyl dithiocarbamate, *Bunseki Kagaku,* 26, 729, 1977.

74. **Valente, I. and Bowen, H. J. M.,** Method for the separation of antimony (III) from antimony (V) using polyurethane foam, *Analyst (London),* 102, 842, 1977.

75. **Clay, A. F.,** Rapid determination of antimony on lead-tin base materials by atomic absorption spectrophotometry, *Lab. Pract.,* 26, 690, 1977.

76. **Aruscavage, P.,** Determination of arsenic, antimony and selenium in coal by atomic absorption spectrometry with graphite tube atomizer, *J. Res. U.S. Geol. Surv.,* 5, 405, 1977.

77. **Kozusnikova, J. and Kolarova, A.,** Determination of antimony in steel by flameless atomic absorption spectrophotometry, *Hutn. Listy,* 32, 810, 1977.

78. **Dujmovic, M.,** Determination of Bi, Hg, Sb, Sn, Te and Pb in aqueous solutions by flameless atomic absorption spectroscopy, *G.I.T. Fachz. Lab.,* 21, 861, 1977.

79. **Haynes, B. W.,** Electrothermal atomic absorption determination of arsenic and antimony in combustible municipal solid waste, *At. Absorpt. Newsl.,* 17, 49, 1978.

80. **Chan, C. Y. and Vijan, P. N.,** Semiautomated determination of antimony in rocks, *Anal. Chim. Acta,* 101, 33, 1978.

81. **Collett, D. L., Fleming, D. E., and Taylor, G. A.,** Determination of antimony by stibine generation and atomic absorption spectrophotometry using a flame-heated silica furnace, *Analyst (London),* 103, 1074, 1978.

82. **Eller, P. M. and Haartz, J. C.,** Sampling and analytical methods for antimony and its compounds. A review, *J. Am. Ind. Hyg. Assoc.,* 39, 790, 1978.

83. **Yamada, H., Uchino, K., Koizumi, H., Toda, T., and Yasuda, K.,** Spectral interference in antimony analysis with high temperature furnace atomic absorption, *Anal. Lett.,* A11, 855, 1978.

84. **Yamashige, T. and Shigetomi, Y.,** Selective determination of antimony in atmospheric dusts by extraction-atomic absorption spectrophotometry, *Nippon Kagaku Kaishi,* 7, 972, 1978.

85. **Zakharchuk, N. F., Yudelevich, I. G., Alimova, E. V., Terenteva, L. A., and Beizel, N. F.,** Layer-by-layer determination of antimony in epitaxial silicon layers by inversion voltammetry and atomic absorption spectrophotometry, *Zh. Anal. Khim.,* 33, 1977, 1978.

86. **Hocquellet, P.,** Application of electrothermal atomization to the determination of arsenic, antimony, selenium and mercury by atomic absorption spectrometry, *Analusis,* 6, 426, 1978.

87. **Kubota, T. and Ueda, T.,** Atomic absorption spectrophotometric determination of antimony with use of borohydride solution as reductant, *Bunseki Kagaku,* 27, 692, 1978.

88. **Fricke, F. L., Robbins, W. B., and Caruso, J. A.,** Trace element analysis of food and beverages by atomic absorption spectrometry, *Prog. Anal. At. Spectrosc.,* 2, 185, 1979.

89. **Haynes, B. W.,** Arsenic, antimony, selenium and tellurium determination in high-purity copper by electrothermal atomization, *At. Absorpt. Newsl.,* 18, 46, 1979.

90. **Evans, W. H., Jackson, F. J., and Dellar, D.,** Evaluation of a method for determination of total antimony, arsenic and tin in foodstuffs using measurement by atomic absorption spectrophotometry with atomization in a silica tube using the hydride generation technique, *Analyst (London),* 104, 16, 1979.

91. **Jackwerth, E., Willmer, P. G., Hohn, R., and Berndt, H.,** A simple accessory for the determination of mercury and the hydride-forming elements (As, Bi, Sb, Se and Te) using flameless atomic absorption spectroscopy, *At. Absorpt. Newsl.,* 18, 66, 1979.

92. **Lundberg, E. and Frech, W.,** Direct determination of trace metals in solid samples by atomic absorption spectrometry with electrothermal atomizers. I. Investigations of homogeneity for lead and antimony in metallurgical materials, *Anal. Chim. Acta,* 104, 67, 1979.

93. **Milner, B. A., Whiteside, P. J., and Price, W. J.,** Determination of trace amounts of arsenic and antimony in zinc powder, *Analyst (London),* 104, 474, 1979.

94. **Nyagah, C. G. and Wandiga, S. O.,** Use of complexing ligands in the determination of antimony and tin by atomic absorption spectrometry, *Talanta,* 26, 333, 1979.

95. **Ohta, K. and Suzuki, M.,** Electrothermal atomic absorption spectrometry of antimony by use of a molybdenum microtube atomizer, *Talanta,* 26, 207, 1979.

96. **Palliere, M. and Gernez, G.,** Apparatus for the production of volatile hydrides for As, Sb, Sn, Ge, Se and Te determination by atomic absorption spectrometry, *Analusis,* 7, 46, 1979.

97. **Alevato, S. J. and Curtius, A. J.,** Determination of copper and antimony in lead alloy by atomic absorption spectrophotometry, *Mikrochim. Acta,* 1, 361, 1979.

98. **Aziz-Alrahman, A. M. and Headridge, J. B.,** Determination of antimony and other trace elements in iron and steel by atomic absorption spectrophotometry with introduction of solid samples into an induction furnace, *Analyst (London),* 104, 944, 1979.

99. **Cross, J. B.,** Determination of stibine in air with pyridine-silver diethyl dithiocarbamate scrubber and flameless atomic absorption spectrometry, *Anal. Chem.,* 51, 2033, 1979.

100. **Donaldson, E. M.,** Determination of antimony in concentrates, ores and nonferrous materials by atomic absorption spectrophotometry after iron-lanthanum collection, or by the iodide method after further xanthate extraction, *Talanta,* 26, 999, 1979.

101. **Forrester, J.E., Lehecka, V., Johnston, J. R., and Ott, W. L.,** Direct determination of trace quantities of antimony, arsenic, bismuth, cadmium, lead, selenium, silver, tellurium and thallium in high-purity nickel by electrothermal atomic absorption spectrometry, *At. Absorpt. Newsl.,* 18, 73, 1979.

102. **Hernandez, S. and Itturriaga, H.,** Study on the determination of antimony using extraction followed by atomic absorption. I. Inorganic ligand, *An. Quim.,* 75, 540, 1979.

103. **Hernandez, S. and Itturriaga, H.,** Study on the determination of antimony extraction followed by atomic absorption, I. Inorganic ligand, *An. Quim.,* 75, 545, 1979.

104. **Kramer, G. W.,** Determination of antimony in lead-antimony alloys by atomic absorption spectroscopy using indium as an internal standard, *Appl. Spectrosc.,* 33, 468, 1979.

105. **Tonini, C.,** Determination of antimony in polyester fibers by atomic absorption spectroscopy, *Tinctoria,* 76, 117, 1979.

106. **Nakahara, T., Kobayashi, S., and Musha, S.,** Determination of trace antimony in phosphoric acid by hydride generation and non-dispersive flame atomic fluorescence spectrometry, *Anal. Chem.,* 51, 1589, 1979.

107. **Vijan, P. N.,** Determination of antimony in environmental samples by AA, *Am. Lab.,* 11, 32, 1979.

108. **Baeckman, S. and Karlsson, R. W.,** Determination of lead, bismuth, zinc, silver and antimony in steel and nickel-base alloys by atomic absorption spectrophotometry using direct atomization of solid samples in a graphite furnace, *Analyst (London),* 104, 1017, 1979.

109. **Boyer, K. W., Capar, S. G., Jones, J. W., Suddendorf, R. F., and Forwalter, J.,** Multielement/trace analysis identifies 48 elements in foods. Simultaneous element analysis and data reduction reaches parts per billion, *Food process.,* 40, 72, 1979.

110. **Fishkova, N. L. and Petrakova, Z. A.,** Atomic absorption determination of antimony in geological materials, *Zh. Anal. Khim.,* 34, 2354, 1979.

111. **Ward, R. J., Black, C. D., and Watson, G. J.,** Determination of antimony in biological materials by electrothermal atomic absorption spectroscopy, *Clin. Chim. Acta,* 99, 143, 1979.

112. **Gennaro, C. D. and Muttoni, E.,** Determination of antimony in dry yeast by atomic absorption spectrometry, *Riv. Zootec. Vet.,* 2, 97, 1979 (in Italian).

113. **Welz, B. and Melcher, M.,** Hydride atomic absorption spectroscopic technique in trace analysis, *Labor Praxis,* 3, 41, 1979 (in German).

114. **Zakhariya, A. N., Olenovich, N. L., and Khtornoi, A. M.,** Atomic absorption determination of tin, lead, antimony and bismuth in copper products, *Ukr. Khim. Zh.,* 46, 421, 1980.

115. **Hon, P. K., Lau, O. W., Cheung, W. C., and Wong, M. C.,** The atomic absorption spectrometric determination of arsenic, bismuth, lead, antimony, selenium and tin with a flame-heated silica-tube after hydride generation, *Anal. Chim. Acta,* 115, 355, 1980.

116. **Dupire, S. and Hoenig, M.,** Effect of complex matrices in the determination of trace elements by flameless atomic absorption spectrometry. II. Determination of arsenic and antimony in plants, *Analusis,* 8, 153, 1980.

117. **Nadkarni, R. A.,** Multitechnique multielemental analysis of coal and fly ash, *Anal. Chem.,* 52, 929, 1980.

118. **Nakamura, S., Fudagawa, N., and Kawase, A.,** Determination of antimony in plant materials by Zeeman atomic absorption spectrophotometry after coprecipitation with manganese dioxide, *Bunseki Kagaku,* 29, 477, 1980.

119. **Nakashima, S.,** Selective determination of antimony (III) and antimony (V) by atomic absorption spectrophotometry after stibine generation, *Analyst (London),* 105, 732, 1980.

120. **Newbury, M. L.,** Analysis of gunshot residue for antimony and barium by flameless atomic absorption spectrophotometry, *J. Can. Soc. Forensic Sci.,* 13, 19, 1980.

121. **Robert, R. V. D., Balaes, G., and Steele, T. W.,** Study of the measurement by electrothermal atomization and atomic absorption spectrophotometry of hydride-forming elements, *Natl. Inst. Metall. Repub. S. Afr. Rep.,* No. 2053, 1980.

122. **Slovak, Z. and Docekal, B.,** Sorption of arsenic, antimony and bismuth on glycol methacrylate gels with bound thiol groups for direct sampling in electrothermal atomic absorption spectrometry, *Anal. Chim. Acta,* 117, 293, 1980.

123. **Welz. B. and Melcher, M.,** Influence of valence state on the determination of antimony in steel using the hydride AA technique, *At. Spectrosc.,* 1, 145, 1980.

124. **Woidich, H. and Pfannhauser, W.,** Determination of antimony in biological materials and environmental samples by atomic absorption spectrophotometry, *Nahrung,* 24, 367, 1980.

125. **Arpadjan, S. and Stefanova, V.,** Extraction of antimony (III) from strongly acid media and determination of antimony in copper, lead, zinc and cadmium by extraction-atomic absorption spectrophotometry, *Z. Anal. Chem.,* 303, 409, 1980.

126. **Del Monte Tamba, M. G. and Luperi, N.,** Determination of traces of arsenic, antimony, bismuth and vanadium in steel and cast iron by graphite furnace atomic absorption spectrometry, *Metall. Ital.,* 72, 253, 1980 (in Italian and English).

127. **Hughes, M. J.,** Analysis of Roman tin and pewter ingots, *Occas. Pap. British Museum,* 17, 41, 1980.

128. **Katskov, D. A., Grinshtein, I. L., and Kruglikova, L. P.,** Study of the evaporation of the metals In, Ga, Th, Ge, Sn, Pb, Sb, Bi, Se and Te from a graphite surface by the atomic absorption method, *Zh. Prikl. Spektrosk.,* 33, 804, 1980.

129. **Lindh, U., Brune, D., Nordberg, G., and Wester, P. O.,** Levels of antimony, arsenic, cadmium, copper, lead, mercury, selenium, silver, tin and zinc in bone tissue of industrially exposed workers, *Sci. Total Environ.,* 16, 109, 1980.

130. **Morita, K., Shimizu, M., Inoue, B., and Ishida, T.,** Graphite furnace atomic absorption spectrometry of antimony and selenium in whole blood, *Okayama-Ken Kankyo Hoken Senta Nempo,* 4, 94, 1980.

131. **Norman, E. A., Orlova, E. S., and Shestakova, A. I.,** Determination of arsenic, tin and antimony impurities in steel by atomic absorption with electrothermal atomization, *Zavod. Lab.,* 46, 1108, 1980.

132. **Clark, J. R. and Viets, J. G.,** Multielement extraction system for the determination of 18 trace elements in geochemical samples, *Anal. Chem.,* 53, 61, 1981.

133. **Clark, J. R. and Viets, J. G.,** Back extraction of trace elements from organometallic-halide extracts for determination by flameless atomic absorption spectrometry, *Anal. Chem.,* 53, 165, 1981.

134. **Kitagawa, K., Suzuki, M., Aoi, N., and Tsuge, S.,** Analytical and spectral features of atomic magneto-optical rotation spectroscopy (The Atomic Faraday Effect) of Sb, Bi, Ag and Cu with a hollow cathode lamp operated in a pulse mode, *Spectrochim. Acta,* 36b, 21, 1981.

135. **Nakashima, S.,** Determination of antimony in water by atomic absorption spectrophotometry following flotation separation, *Bull. Chem. Soc. Jpn.,* 54, 291, 1981.
136. **Sinemus, H. W., Melcher, M., and Welz, B.,** Influence of valence state on the determination of antimony, arsenic, bismuth, selenium and tellurium in lake water using the hydride AA technique, *At. Spectrosc.,* 2, 81, 1981.
137. **Subramanian, K. S. and Meranger, J. C.,** Determination of arsenic (III), arsenic (V), antimony (III), antimony (V), selenium (IV) and selenium (VI) by extraction with ammonium pyrrolidine dithiocarbamate-methyl isobutyl ketone and electrothermal atomic absorption spectrometry, *Anal. Chim. Acta,* 124, 131, 1981.
138. **Tsujii, K.,** Differential determination of antimony (III) and antimony (V) by nondispersive atomic fluorescence spectrometry using a sodium borohydride reduction technique, *Anal. Lett.,* 14, 181, 1981.
139. **Yamamoto, M., Urata, K., and Yamamoto, Y.,** Differential determination of antimony (III) and antimony (V) by hydride generation-atomic absorption spectrophotometry, *Anal. Lett.,* 14, 21, 1981.
140. **Aksyuk, A. F., Merzlyakova, N. M., and Tarkhova, L. P.,** Determination of trace elements in desalinated seawater by an atomic absorption method, *Gig. Sanit.,* 5, 72, 1981 (in Russian).
141. **Higashi, Y. and Masuko, E.,** Microdetermination of antimony by reduction with zinc followed by flameless atomic absorption spectrometry, *Kenkyu Hokoku-Nara-Ken Kogyo Shikengo,* 6, 31, 1980; CA 95, 21448w, 1981.
142. **Welz, B., Grobenski, Z., and Melcher, M.,** Determination of antimony, arsenic, selenium, tellurium, bismuth and tin using the hybride-AAS technique, *Spektrometertagung (Votr.),* 13, 377, 1981 (in German).
143. **Crock, J. G. and Lichte, F. E.,** The determination of trace level antimony in geological materials by semiautomated hydride generation-atomic absorption spectroscopy, *U.S. Geol. Surv. Open File Rep.,* 81, 1, 1981.
144. **Guo, X. and Wang, S.,** A new type of double capillary nebulizer used in atomic absorption analysis, *Fen Hsi Hua Hsueh,* 9, 258, 1981 (in Chinese).
145. **Balaes, G. E. and Robert, R. V. D.,** Determination by atomic absorption spectrophotometry of impurities in manganese dioxide, *Natl. Inst. Metall. Repub. S. Afr. Rep.,* No. 2034, 1981.
146. **Boyer, K. W., Jones, J. W., Linscott, D., Wright, S. K., Stroube, W., and Cunningham, W.,** Trace element levels in tissues from cattle fed on sewage sludge-amended diet, *J. Toxicol. Environ. Health,* 8, 281, 1981.
147. **Farley, K. R., Sheetz, K. W., and Whitehead, A. B.,** Determination of antimony in smelter flue dusts by atomic absorption spectrometry, *Rep. Invest. U.S. Bur. Mines,* RI 8510, 1981.
148. **Vajda, F.,** Determination of arsenic and antimony in steels by atomic absorption, *Banyasz Kihasz Lapok Kohasz,* 114, 113, 1981 (in Hungarian).
149. **Nomura, N.,** Trace analysis of antimony by atomic absorption spectroscopy, *Kochi Kogyo Koto Semmon Gakko Gakujutsu Kiyo,* 17, 65, 1981.
150. **Bencze, K.,** Determination of antimony in whole blood as antimony trihydride by atomic absorption spectroscopy (AAS), *Aerztl. Lab.,* 27, 347, 1981.
151. **Bansho, K., Aoki, T., and Umezaki, Y.,** Determination of antimony and selenium by atomic absorption spectrometry via hydride generation using sodium tetrahydroborate pellets, *Koyai Shigen Kenkuosho Iho,* 11, 37, 1981.
152. **Jones, E. A. and Dixon, K.,** The separation of trace elements in manganese dioxide, *Natl. Inst. Metall. Repub. S. Afr. Rep.,* No. 2131, 1981.
153. **Watson, C. A.,** Development and testing of hydride generation methods for antimony and selenium in organic matter. The work of the metallic impurities in organic matter subcommittee of the Analytical Methods Committee, *Anal. Proc. (London),* 18, 482, 1981.
154. **Ondov, J. M. and Biermann, A. H.,** Effects of Particle-Control Devices on Atmospheric Emissions of Minor and Trace Elements from Coal Combustion, EPA-600/9-80-039d, U.S. Envrionmental Protection Agency, Office of Research and Development, Symp. Transfer Util. Part Control. Technol. 2nd, PB 81-122228, Corvallis, Ore., 1980, 455-85; CA 96, 90897g, 1982.
155. **Miyakawa, H.,** Determination of antimony by thorium hydroxide coprecipitation flameless atomic absorption, *Nenpo-Fukui-Ken Kogyo Shikenjo,* 128, 1980; CA 96, 115109s, 1982.
156. **Cacho Palomar, J. and Nerin de la Puerta, C.,** Atomic absorption spectrophotometric determination of antimony following extraction of its trioctylamine bromo complex in methyl isobutyl ketone, *Afinidad,* 39, 51, 1982.
157. **Smith, B. M. and Griffiths, M. B.,** Determination of lead and antimony in urine by atomic absorption spectroscopy with electrothermal atomization, *Analyst (London),* 107, 253, 1982.
158. **Raverby, M.,** Analysis of long-range bullet entrance holes by atomic absorption spectrophotometry and scanning electron microscopy, *J. Forensic Sci.,* 27, 92, 1982.
159. **Haring, B. J., Van Delft, W., and Bom, C. M.,** Determination of arsenic and antimony in water and soil by hydride generation and atomic absorption spectroscopy, *Z. Anal. Chem.,* 310, 217, 1982.

AUTHOR INDEX

SUBJECT INDEX (ANTIMONY)

BISMUTH, Bi (ATOMIC WEIGHT 208.9804)

Bismuth is found in combined form as an oxide or a sulfide in nature. It occurs as a free element in a very small amount. The metal is useful for the preparation of alloys with low melting points, medicinal compounds and amalgams, etc.

Bismuth determination is required in the analyses of ores and evaluation of minerals as well as in residues from lead refining, alloys, electric fuses, amalgams, solders, antifriction metals, geomaterials, biomaterials, and other bismuth compounds.

A sample for the determination of bismuth is usually dissolved in nitric acid or in a mixture of hydrochloric and nitric acids. In the presence of silica, bismuth usually stays as an oxycompound and causes loss in bismuth values during the analysis.

The atomic absorption method (flame or nonflame) is one of the best analytical method for the determination of bismuth. This method combined with microsampling and hydride generation techniques enables one to determine bismuth at the 0.005 to 0.06 mg/ℓ level quantitatively. No chemical interferences have been reported in the air-acetylene flame. Use of an electrodeless discharge lamp provides the same sensitivity as obtained from the hollow cathode lamp.

Standard Solution

To prepare 1000 mg/ℓ of standard solution, dissolve 1.000 g of bismuth metal in a minimum volume of (1:1) nitric acid and dilute to 1 ℓ with 2% (v/v) nitric acid.

Instrumental Parameters

Wavelength	223.1 nm
Slit	0.2 nm
Light source	Hollow cathode lamp
Flame	Air-acetylene, oxidizing (lean, blue)
Sensitivity	0.2—0.45 mg/ℓ
Detection limit	0.04 mg/ℓ
Optimum range	15 mg/ℓ

Secondary wavelengths and their sensitivities

222.8 nm	0.5 mg/ℓ
306.8 nm	0.5—1.0 mg/ℓ
206.2 nm	1.7—3.7 mg/ℓ
227.7 nm	4.0—6.1 mg/ℓ

REFERENCES

1. **Willis, J. B.**, Determination of lead and other heavy metals in urine by atomic absorption spectroscopy, *Anal. Chem.*, 34, 614, 1962.
2. **Slavin, W., Sprague, S., and Manning, D. C.**, Detection limits in analytical atomic absorption spectrophotometry, *At. Absorpt. Newsl.*, 18, 1964.
3. **Sattur, T. W.**, Routine atomic absorption analysis on nonferrous alloys and plant intermediates, *At. Absorpt. Newsl.*, 5, 37, 1966.
4. **Vollmer, J.**, Bismuth hollow cathode lamp, *At. Absorpt. Newsl.*, 5, 12, 1966.
5. **Kinser, R. E.**, Determination of bismuth and tellurium in body tissues of animals by atomic absorption spectrophotometry, *Am. Ind. Hyg. Assoc. J.*, 27, 260, 1966.
6. **Ivanov, N. P., Minervina, L. V., Baranov, S. V., Pofralidi, L. G., and Olikov, I. I.**, Tubes without electrodes with high-frequency excitation of the spectrum of In, Ga, Bi, Sb, Tl, Pb, Mg, Ca, and Cu as a radiation source in atomic absorption radiation, *Zh. Anal. Khim.*, 21, 1129, 1966 (in Russian).
7. **Scott, T. C., Roberts, E. D., and Cain, D. A.**, Determination of minor constituents in ferrous materials by atomic absorption spectrophotometry, *At. Absorpt. Newsl.*, 6, 1, 1967.
8. **Venghiattis, A.**, A technique for the direct sampling of solids without prior dissolution, *At. Absorpt. Newsl.*, 6, 19, 1967.
9. **Manning, D. C., Vollmer, J., and Fernandez, F.**, Shielded bismuth hollow cathode lamps, *At. Absorbt. Newsl.*, 6, 17, 1967.
10. **Marshall, D. and Schrenk, W. G.**, Atomic absorption characteristics of bismuth using a turbulent flow total-consumption burner, *Appl. Spectrosc.*, 21, 246, 1967.
11. **Devoto, G.**, Determination of bismuth in urine by atomic absorption spectrophotometry, *Boll. Soc. Ital. Biol. Sper.*, 44, 1253, 1968 (in Italian).
12. **Bishop, C. T. and Harris, B. N.**, Tellurium interference in the determination of bismuth, chromium, and nickel by atomic absorption spectroscopy, *At. Absorpt. Newsl.*, 8, 110, 1969.
13. **Endo, V., Hata, T., and Nakahara, V.**, Determination of Ti, V, N, Cr, Pb and Bi in iron ore by atomic absorption method, *Bunseki Kagaku*, 18, 833, 1969.
14. **Poluektov, N. S., Zelyukova, Y. V., and Nikonova, M. P.**, Atomic absorption determination of bismuth by means of the exhaust gases of the flame, *Zavod. Lab.*, 35, 163, 1969 (in Russian).
15. **Husler, J.**, Atomic absorption determination of bismuth in soils, rocks and ores, *At. Absorpt. Newsl.*, 9, 31, 1970.
16. **Headridge, J. B. and Richardson, J.**, Determination of trace amounts of bismuth in ferrous alloys by solvent extraction followed by atomic absorption spectrophotometry, *Analyst (London)*, 95, 930, 1970.
17. **Burke, K. E.**, Study of the scavanger properties of manganese (IV) oxide with atomic absorption spectrometry. Determination of microgram quantities of Sb, Bi, Pb and Sn in nickel, *Anal. Chem.*, 42, 1536, 1970.
18. **Hermon, S. E. and Rennie, R. J.**, Rapid determination of lead and bismuth in aluminum alloys either separately or in the presence of each other using atomic absorption spectrophotometry, *Metallurgia*, 82, 201, 1970.
19. **Bohnstedt, U.**, Determination of traces of lead, bismuth, zinc and cadmium in alloys: an example of the use of combined analytical procedures, *DEW Tech. Ber.*, 11, 101, 1971 (in German).
20. **Kisfauldi, G. and Lenhof, M.**, Determination of bismuth in cast iron and steel by atomic absorption flame spectrometry, *Anal. Chim. Acta*, 55, 442, 1971 (in French).
21. **Newland, B. T. N. and Mostyn, R. A.**, The determination of trace amounts of bismuth in nickel alloys by atomic absorption spectrometry, *At. Absorpt. Newsl.*, 10, 89, 1971.
22. **Headridge, J. B. and Richardson, J.**, Determination of trace amounts of bismuth in ferrous alloys by solvent extraction followed by atomic absorption spectrophotometry, *B.I.S.R.A. Open Rep.*, Mg/D/421, British Iron and Steel Research Association, Scarborough, 1971.
23. **Burke, K. E.**, Determination of microgram amounts of antimony, bismuth, lead and tin in aluminum, iron and nickel-base alloys by nonaqueous atomic absorption spectroscopy, *Analyst (London)*, 97, 19, 1972.
24. **Hall, R. J. and Farber, T.**, The determination of bismuth in body tissues and fluids after administration of controlled doses, *J. Assoc. Off. Anal. Chem.*, 55, 639, 1972.
25. **Hoften, M. E. and Hubbard, D. P.**, Determination of trace amounts of bismuth in steels by solvent extraction and atomic absorption or atomic fluorescence spectrometry, *Anal. Chim. Acta*, 62, 311, 1972.
26. **Jackwerth, E., Hoehn, R., and Koos, K.**, Trace enrichment by partial dissolution of the matrix in presence of mercury. Determination of bismuth, copper, lead, nickel, silver, gold and palladium in high purity cadmium by atomic absorption spectrometry, *Z. Anal. Chem.*, 264, 1973 (in German).
27. **Nakahara, T., Munemori, M., and Musha, S.**, Atomic absorption spectrophotometric determination of bismuth in premixed inert gas (entrained air)-hydrogen flames, *Bull. Chem. Soc. Jpn.*, 46, 1166, 1973.

28. **Schmidt, F. J. and Royer, J. L.,** Submicrogram determination of arsenic, selenium, antimony and bismuth by atomic absorption utilizing sodium borohydride reduction, *Anal. Lett.,* 6, 17, 1973.

29. **Tsukuhara, I. and Yamamoto, T.,** Determination of micro amounts of bismuth in lead, copper, tin and nickel metals by solvent extraction and atomic absorption spectrometry, *Anal. Chim. Acta,* 63, 464, 1973.

30. **Blakeley, J. H., Manson, A., and Zatka, V. J.,** Improvements in the manganese dioxide collection of trace lead and bismuth in nickel, *Anal. Chem.,* 45, 1941, 1973.

31. **Langmyhr, F. J., Solberg, R., and Wold, L. T.,** Atomic absorption spectrometric determination of silver, bismuth and cadmium in sulfide ores by direct atomization from the solid state, *Anal. Chim. Acta,* 69, 267, 1974.

32. **Stahlavska, A., Prokopova, H., and Tuzar, M.,** Use of spectral-analytical methods in drug analysis. Determination of bismuth, aluminum, magnesium, calcium, titanium and silicon in antacids by atomic absorption spectrophotometry, *Pharmazie,* 29, 140A, 1074.

33. **White, J. A., Harper, W. L., Friedman, A. P., and Banas, V. E.,** Determination of bismuth in nickel base super alloys by atomic absorption spectrophotometry, *Appl. Spectrosc.,* 28, 192, 1974.

34. **Muradov, V. G., Muradova, O. N., and Yablochkov, E. U.,** Measurement of the equilibrium partial pressure of bismuth atoms in the gas phase by an atomic absorption spectroscopic method, *Zh. Prikl. Spektrosk.,* 20, 1076, 1974.

35. **Thompson, K. C. and Thomerson, D. R.,** Atomic absorption studies on the determination of antimony, arsenic, bismuth, germanium, lead, selenium, tellurium and tin by utilizing the generation of covalent hydrides, *Analyst (London),* 99, 595, 1974.

36. **Thornton, K. and Burke, K. E.,** Modification to the extraction-atomic absorption method for the determination of antimony, bismuth, lead and tin, *Analyst (London),* 99, 469, 1974.

37. **Welcher, G. G., Kriege, O. H., and Marks, J. Y.,** Direct determination of trace quantities of lead, bismuth, selenium, tellurium and thallium in high temperature alloys by non-flame atomic absorption spectrophotometry, *Anal. Chem.,* 46, 1227, 1974.

38. **Schmidt, F. J., Royer, J. L., and Muir, S. M.,** Automated determination of arsenic, selenium, antimony, bismuth, and tin by atomic absorption utilizing sodium borohydride reduction, *Anal. Lett.,* 8, 123, 1975.

39. **Allain, P.,** Determination of bismuth in blood, urine and cerebrospinal fluid by flameless atomic absorption spectrophotometry, *Clin. Chim. Acta,* 64, 281, 1975 (in French).

40. **Bedard, M. and Kerbyson, J. D.,** Determination of trace bismuth in copper by hydride evolution atomic absorption spectrophotometry, *Anal. Chem.,* 47, 1441, 1975.

41. **Frech, W.,** Rapid determination of bismuth in steels by flameless atomic absorption, *Z. Anal. Chem.,* 275, 353, 1975.

42. **Kirk, M., Perry, E. G., and Arritt, J. M.,** Separation and atomic absorption measurement of trace amounts of lead, silver, zinc, bismuth and cadmium in high nickel alloys, *Anal. Chim. Acta,* 80, 163, 1975.

43. **Kono, T. and Nemori, A.,** Extraction of copper, cadmium, lead, silver and bismuth with iodide methyl isobutyl ketone in atomic absorption spectrophotometric analysis, *Bunseki Kagaku,* 24, 419, 1975.

44. **Tarasevich, N. I., Kozyreva, G. V., and Portugal'skaya, Z. P.,** Extraction-atomic absorption determination of indium, bismuth and lead trace impurities in rocks and soils, *Vestn. Mosk. Univ. Khim.,* 16, 241, 1975.

45. **Yudelevich, I. G., Poleva, G. V., and Tokareva, A. G.,** Atomic absorption determination of antimony and bismuth in chemical reagents, *Izv. Sib. Otd. Nauk SSSR Ser. Khim. Nauk,* 5, 10, 1975.

46. **Barnett, W. B. and McLaughlin, E. A.,** The atomic absorption determination of antimony, arsenic, bismuth, cadmium, lead and tin in iron, copper and zinc alloys, *Anal. Chim. Acta,* 80, 285, 1975.

47. **Dixon, K., Mallet, R. C., and Kocaba, R.,** Determination of atomic absorption spectrophotometry of arsenic, selenium, tellurium, antimony, and bismuth on ion exchange resins, *Natl. Inst. Metall. Repub. S. Afr. Rep.,* No. 1689, 1975.

48. **Joshi, M. M. and Gopal, R.,** Spectroscopic investigations of flame test for tin and bismuth, *Indian J. Pure Appl. Phys.,* 14, 325, 1976.

49. **Drinkwater, J. E.,** Atomic absorption determination of bismuth in complex nickel-base alloys by generation of its covalent hydrides, *Analyst (London),* 101, 672, 1976.

50. **Hammer, R. M., Lechak, D. L., and Greenberg, P.,** Determination of silver, arsenic, bismuth, antimony, selenium and tellurium in chromium metal with flameless atomic absorption spectroscopy, *At. Absorpt. Newsl.,* 15, 122, 1976.

51. **Jedrzejewska, H. and Malusecka, M.,** Interference effects of some carriers on the atomic absorption of chromium, manganese, antimony, iron, bismuth and lead in an air-acetylene flame, *Chem. Anal.,* 21, 585, 1976.

52. **Wunderlich, E. and Hadeler, W.,** Determination of traces of zinc, cadmium and bismuth in pure copper with AAS, *Z. Anal. Chem.,* 281, 300, 1976 (in German).

53. **Kahn, H. L., Bancroft, M., and Emmel, R. H.,** Solving precision problems in flameless AA sampling, *Res. Dev.*, 27, 30, 1976.
54. **Rooney, R. C.,** Determination of bismuth in blood and urine, *Analyst (London)*, 101, 749, 1976.
55. **Ficklin, W. H. and Ward, F. N.,** Flameless atomic absorption determination of bismuth in soils and rocks, *J. Res. U.S. Geol. Surv.*, 4, 217, 1976.
56. **Marks, J. Y., Welcher, G. G., and Spellman, R. J.,** Atomic absorption determination of lead, bismuth, selenium, tellurium, thallium and tin in complex alloys using direct atomization from metal chips in the graphite furnace, *Appl. Spectrosc.*, 31, 9, 1977.
58. **Gladney, E. S.,** Matrix modification for the determination of bismuth by flameless atomic absorption, *At. Absorpt. Newsl.*, 16, 114, 1977.
59. **Andrews, D. G. and Headridge, J. B.,** Determination of bismuth in steels and cast irons by atomic absorption spectrophotometry with an induction furnace. Direct analysis of solid samples, *Analyst (London)*, 102, 436, 1977.
60. **Djudzman, R., Van den Eeckhout, E., and De Moerloose, P.,** Determination of bismuth by atomic absorption spectrophotometry with electrothermal atomization after low-temperature ashing, *Analyst (London)*, 102, 688, 1977.
61. **Pellerin, F., Goule, J. P., and Dumitrescu, D.,** Formation of bismuth complexes and soluble chelates with various drugs and foods: detection by atomic absorption, *Ann. Pharm. Fr.*, 35, 281, 1977.
62. **Sato, A. and Saitoh, N.,** Determination of beryllium and bismuth in spring water by atomic absorption spectroscopy using a carbon tube atomizer and co-precipitation with zirconium hydroxide, *Bunseki Kagaku*, 26, 747, 1977.
63. **Tsalev, D., Arpadzhyan, S., and Khavezov, I.,** Extraction atomic absorption determination of bismuth in tungsten trioxide, *Dokl. Bolg. Akad. Nauk*, 30, 1297, 1977.
64. **Dujmovic, M.,** Determination of bismuth, mercury, antimony, tin, tellurium and lead in aqueous solutions by flameless atomic absorption spectroscopy, *GIT Fachz. Lab.*, 21, 861, 1977.
65. **Kujirai, O., Kobayashi, T., and Sudo, E.,** Determination of trace quantities of lead and bismuth in heat-resisting alloys by atomic absorption spectrometry with heated graphite atomizer, *Trans. Jpn. Inst. Met.*, 18, 775, 1977.
66. **Viets, J. G.,** Determination of silver, bismuth, cadmium, copper, lead and zinc in geologic materials by atomic absorption spectrometry with tricaprylyl methyl ammonium chloride, *Anal. Chem.*, 50, 1097, 1978.
67. **Headridge, J. B. and Thompson, R.,** Determination of bismuth in nickel-base alloys by atomic absorption spectrometry with introduction of solid samples into an induction furnace, *Anal. Chim. Acta*, 102, 33, 1978.
68. **Ohta, K. and Suzuki, M.,** Atomic absorption spectrometry of bismuth with electrothermal atomization from metal atomizers, *Anal. Chim. Acta*, 96, 77, 1978.
69. **Watling, R. J.,** The use of a slotted tube for the determination of lead, zinc, cadmium, bismuth, cobalt, manganese and silver by atomic absorption spectrometry, *Anal. Chim. Acta*, 97, 395, 1978.
70. **Inui, S., Terada, S., Tamura, H., and Ichinose, N.,** Extraction and atomic absorption spectrometric determination of traces of bismuth with zinc debenzyl dithiocarbamate in aluminum alloys and solder alloys, *Z. Anal. Chem.*, 292, 282, 1978.
71. **Rombach, N. and Kock, K.,** Atomic absorption spectroscopic trace determination of bismuth in organic matrix using a hydride system, *Z. Anal. Chem.*, 292, 365, 1978.
72. **Heinrichs, H.,** Determinations of bismuth, cadmium and thallium in 33 international standard reference rocks by fractional distillation combined with flameless atomic absorption spectrometry, *Z. Anal. Chem.*, 294, 345, 1979.
73. **Jackwerth, E., Hoehn, R., and Musaick, K.,** Preconcentration of traces of silver, gold, bismuth, copper and palladium from pure lead by partial precipitation of the matrix with sodium borohydride as a reducing agent, *Z. Anal. Chem.*, 299, 362, 1979.
74. **Jackwerth, E., Willmer, P. G., Hohn, R., and Berndt, H.,** A simple accessory for the determination of mercury and the hydride-forming elements (As, Bi, Sb, Se and Te) using flameless atomic absorption spectroscopy, *At. Absorpt. Newsl.*, 18, 66, 1979.
75. **Kane, J. S.,** Determination of nanogram amounts of bismuth in rocks by atomic absorption spectrometry with electrothermal atomization, *Anal. Chim. Acta*, 106, 325, 1979.
76. **Palliere, M. and Gernez, G.,** Apparatus for the production of volatile hydrides for arsenic, antimony, bismuth, tin, germanium, selenium and tellurium determination by atomic absorption spectrometry, *Analusis*, 7, 46, 1979.
77. **Arpadjan, S. and Alexandrova, I.,** Determination of trace elements (lead, bismuth, copper, zinc and silver) in high-purity tin by flame atomic absorption, *Z. Anal. Chem.*, 298, 159, 1979.
78. **Forrester, J. E., Lehecka, V., Johnston, J. R., and Ott, W. L.,** Direct determination of trace quantities of antimony, arsenic, bismuth, cadmium, lead, selenium, silver, tellurium and thallium in high-purity nickel by electrothermal atomic absorption spectrometry, *At. Absorpt. Newsl.*, 18, 73, 1979.

79. **Tsalev, D. and Petrova, V.,** Hexamethylene ammonium hexamethylene dithiocarbamate-n-butyl acetate as an extractant of bismuth, cadmium, cobalt, mercury, indium, lead, and palladium from acidic media, *Dokl. Bolg. Akad. Nauk,* 32, 911, 1979.

80. **Baeckman, S. and Karlsson, R. W.,** Determination of lead, bismuth, zinc, silver and antimony in steel and nickel-base alloys by atomic absorption spectrophotometry using direct atomization of solid samples in a graphite furnace, *Analyst (London),* 104, 1017, 1979.

81. **Donaldson, E. M.,** Determination of bismuth in ores, concentrates and nonferrous alloys by atomic absorption spectrophotometry after separation by diethyl dithiocarbamate extraction or iron collection, *Talanta,* 26, 1119, 1979.

82. **Inui, T., Fudagawa, N., and Kawase, A.,** Extraction and atomic absorption spectrometric determination of bismuth with electrothermal atomization, *Z. Anal. Chem.,* 299, 190, 1979.

83. **Rezchikov, V. G. and Usvatov, V. A.,** Atomic absorption determination of copper, lead and bismuth in silver nitrate, *Zavod. Lab.,* 45, 1112, 1979.

84. **Welz, B. and Melcher, M.,** Hydride atomic absorption spectroscopic technique in trace analysis, *Labor Prixis,* 3, 41, 1979 (in German).

85. **Dekersabiec, A. M.,** Determination of arsenic, selenium and bismuth in rocks and soils by atomic absorption spectrophotometry, *Analusis,* 8, 97, 1980.

86. **Hon, P. K., Lau, O. W., Cheung, W. C., and Wong, M. C.,** The atomic absorption spectrometric determination of arsenic, bismuth, lead, antimony, selenium and tin with a flame-heated silica T-tube after hydride generation, *Anal. Chim. Acta,* 115, 355, 1980.

87. **Zakhariya, A. N., Olenovich, N. L., and Khutornoi, A. M.,** Atomic absorption determination of tin, lead, antimony and bismuth in copper blisters, *Ukr. Khim. Zh.,* 46, 421, 1980.

88. **Aihara, M. and Kiboku, M.,** Determination of bismuth and thallium by atomic absorption spectrophotometry after extraction with potassium xanthate-methyl isobutyl ketone, *Bunseki Kagaku,* 29, 243, 1980.

89. **Roberts, R. V. D., Balaes, G., and Steele, T. W.,** Study of the measurement by electrothermal atomization and atomic absorption spectrophotometry of hydride-forming elements, *Natl. Inst. Metall. Repub. S. Afr. Rep.,* No. 2053, 1980.

90. **Slovak, Z. and Docekal, B.,** Samples of arsenic, antimony and bismuth in glycol methacrylate gels with bound thiol groups for direct sampling in electrothermal atomic absorption spectrometry, *Anal. Chim. Acta,* 117, 293, 1980.

91. **Del Monte Tamba, M. G. and Lupiri, N.,** Determination of trace of arsenic, antimony, bismuth and vanadium in steel and cast iron by graphite furnace atomic absorption spectrometry, *Metall. Ital.,* 72, 253, 1980 (in Italian and English).

92. **Goda, A., Moriyama, K., and Harimaya, S.,** Solid sample technique for the determination of trace bismuth in iron and steel by atomic absorption spectrometry with a graphite tube atomizer, *Trans. Iron Steel Inst. Jpn.,* 20, B405, 1980 (in English).

93. **Katskov, B. A., Grinshtein, I. L., and Kruglikova, L. P.,** Study of the evaporation of the metals indium, gallium, thallium, germanium, tin, lead, antimony, bismuth, selenium and tellurium from a graphite surface by the atomic absorption method, *Zh. Prikl. Spektrosk.,* 33, 804, 1980.

94. **Nakashima, S.,** Determination of submicrogram amounts of bismuth in water by atomic absorption spectrophotometry following flotation separation, *Z. Anal. Chem.,* 303, 10, 1980.

95. **Baker, A. A. and Headridge, J. B.,** Determination of bismuth, lead, and tellurium in copper by atomic absorption spectrometry with introduction of solid sample into an induction furnace, *Anal. Chim. Acta,* 125, 93, 1981.

96. **Clark, J. R. and Viets, J. G.,** Multielement extraction system for the determination of 18 trace elements in geochemical samples, *Anal. Chem.,* 53, 61, 1981.

97. **Clark, J. R. and Viets, J. G.,** Back-extraction of trace elements from organometallic-halide extracts for determination by flameless atomic absorption spectrometry, *Anal. Chem.,* 53, 65, 1981.

98. **Sinemus, H. W., Melcher, M., and Welz, B.,** Influence of valence state on the determination of antimony, arsenic, bismuth, selenium and tellurium in lake water using the hydride AA technique, *At. Spectrosc.,* 2, 81, 1981.

99. **Shevchuk, I. A., Dovzhenko, N. P., and Kravtsova, Z. N.,** Atomic absorption determination of lead and bismuth in steels using ion-exchange chromatography, *Ukr. Khim. Zh.,* 47(7), 773, 1981 (in Russian).

100. **Carleer, R., Francois, J. P., and Van Poucke, L. C.,** Determination of the main, minor and trace elements in lead/tin based solder by atomic absorption spectrophotometry, *Bull. Soc. Chim. Belg.,* 90(4), 357, 1981.

101. **Arpadjan, S. and Aslanova, N.,** Concentration by extraction and determination of bismuth in chemical lead by flame atomic absorption spectroscopy, *Z. Chem.,* 21(5), 192, 1981 (in German).

102. **Kimura, M. and Kawanami, K.,** Separation and preconcentration of trace amounts of several metals in sodium perchlorate using activated carbon as a collector, *Nippon Kagaku Kaishi,* 1, 1, 1981 (in Japanese).

103. **Berndt, H. and Messerschmidt, J.,** *o,o*-Diethyl dithiophosphate for trace enrichment on activated carbon. I. Analysis of high purity gallium-aluminum determination of element traces by flame AAS (injection method and loop AAS), *Fresenius' Z. Anal. Chem.,* 308, 104, 1981 (in German).

104. **Berndt, H. and Messerschmidt, J.,** Microanalytical method of flame atomic absorption and emission spectrometry (platinum loop method). II. Determination of As, Bi, Cd, Hg, Pb, Sb, Se, Tl, Zn, Li, Na, K, Cs, Sr and Ba, *Spectrochim. Acta,* 36b, 809, 1981.

105. **Welz, B., Grobenski, Z., and Melcher, M.,** Determination of antimony, arsenic, selenium, tellurium, bismuth and tin using the hydride-AAS technique, *Spektrometertagung (Votr.),* 13, 337, 1981 (in German).

106. **Gusinski, M. N. and Pashadzhanov, A. M.,** Dependence of the sensitivity of the atomic absorption determination of bismuth on the temperature and the flame type, *Zh. Anal. Khim.,* 36, 1755, 1981.

107. **Pashadzhanov, A. M., Moreiskaya, L. V., and Nemodruk, A. A.,** Effect of various substances on bismuth atomic absorption in atomic absorption spectrometry, *Zh. Anal. Khim.,* 36, 1343, 1981.

108. **Guo, X. and Wang, S.,** A new type of double capillary nebulizer used in atomic absorption analysis, *Fen Hsi Hua Hsueh,* 9, 258, 1981.

109. **Shaburova, V. P., Ruchkin, E. D., Yudelevich, I. G., and Kruglov, L. M.,** Atomic absorption method of analysis for piezoceramic materials of the lead zirconate titanate system, *Izv. Sib. Otd. Akad. Nauk SSSR Ser. Khim. Nauk,* 5, 95, 1981.

110. **Ikeda, M., Nishibe, J., and Nakahara, T.,** Study of some fundamental conditions for the determination of antimony, tin, selenium and bismuth by continuous hydride generation-flameless atomic absorption spectrometry, *Bunseki Kagaku,* 30, 545, 1981.

111. **Burba, P. and Schaefer, W.,** Atomic absorption spectrometric determination (flame-AAS) of heavy metals in bacterial leach liquors (Jarosite) after their analytical separation on cellulose, *Erzmetall,* 34, 582, 1981 (in German).

112. **Frech, W., Lundberg, E., and Barbooti, M. H.,** Direct determination of trace metals in solid samples by atomic absorption spectrometry with electrothermal atomizers. IV. Interference effects in the determination of lead and bismuth in steels, *Anal. Chim. Acta,* 131, 45, 1981.

113. **Tunstall, M., Berndt, H., Sommer, D., and Ohls, K.,** Direct determination of trace elements in aluminum with flame atomic absorption spectrometry and ICP emission spectrometry — a comparison, *Erzmetall,* 34, 588, 1981.

114. **Senesi, N. and Polemio, M.,** Trace element addition to soil by application of NPK fertilizers, *Fert. Res.,* 2, 289, 1981.

115. **Alexander, P. W. and Joseph, J. P.,** Selective potentiometric determination of bismuth (III) with a coated-wire electrode, *Talanta,* 28, 931, 1981.

116. **Gusinskii, M. N., Pashadzhanov, A. M., and Nemodruk, A.,** Study of interference in the atomic absorption determination of bismuth, *Zh. Anal. Khim.,* 37, 35, 1982.

117. **Gusinskii, M. N., Pashadzhanov, A. M., and Nemodruk, A. A.,** Use of water-miscible organic solvents in the extraction atomic absorption determination of the trace amounts of bismuth in rocks, ores and lead-zinc concentrates, *Zh. Anal. Khim.,* 37, 42, 1982 (in Russian).

118. **Nadkarni, R. A.,** Application of hydride generation atomic absorption spectrometry to coal analysis, *Anal. Chim. Acta,* 135, 363, 1982.

119. **Heinrichs, H. and Keltsch, H.,** Determination of arsenic, bismuth, cadmium, selenium and thallium by atomic absorption spectrometry with a volatilization technique, *Anal. Chem.,* 54, 1211, 1982.

120. **Matsumoto, K., Nishio, M., Yukari, M., and Terada, K.,** Decomposition of tin(IV) oxide, antimony(III) oxide and bismuth(III) oxide by fusion with ammonium iodide and its application for analysis of environmental samples, *Bunseki Kagaku,* 31, 141, 1982.

121. **Lee, D. S.,** Determination of bismuth in environmental samples by flameless atomic absorption spectrometry with hydride generation, *Anal. Chem.,* 54, 1682, 1982.

122. **Jones, E. A.,** The separation and determination of trace elements in chromic oxide, *Rep. MINTEK,* M15, Council for Mineral Technology, Randburg, South Africa, 1982.

123. **Howell, D. J. and Dohnt, B. R.,** The accurate determination of bismuth in lead concentrates and other non-ferrous materials by atomic absorption spectrometry after separation and preconcentration of the bismuth with mercaptoacetic acid, *Talanta,* 29, 391, 1982.

124. **Jackwerth, E. and Salewski, S.,** Contribution to the multielement preconcentration from pure cadmium, *Z. Anal. Chem.,* 30, 108, 1982 (in German).

125. **Nemodruk, A. A. and Pashadzhanov, A. M.,** Atomic absorption determination of trace bismuth in metals and alloys, *Zh. Anal. Khim.,* 37, 442, 1982.

126. **Astrom, O.,** Flow injection analysis for the determination of bismuth by atomic absorption spectrometry with hydride generation, *Anal. Chem.,* 54, 190, 1982.

AUTHOR INDEX

SUBJECT INDEX (BISMUTH)

Group VIA Elements — S, Se, and Te

SULFUR, S (ATOMIC WEIGHT 32.06)

Sulfur occurs free in nature near volcanoes and hot springs. It is also found in meteorites, rocks, ores, and minerals. It is one of the earliest known elements used in medicine. Sulfur is mostly used in the manufacture of dyes, black gunpowder, matches, sulfuric acid, sulfite paper, fungicides, insecticides, and fumigants. It is also used in vulcanization of rubber, bleaching of dried fruits, electrical insulators, and art and science products.

Sulfur determination is required in clay, alkaline soils, water, air, gases, wool, stock medicants, iron and steel, cinders, plating baths, environmental pollution, coal, fertilizers, and other agricultural products.

Sample preparation depends on the nature of the sample. Usually fusion with sodium carbonate or sodium carbonate and potassium nitrate or zinc oxide is used. Direct dissolution of the sample with the mineral acids will result in the loss of sulfur as sulfide. Most of the atomic absorption methods are the indirect procedures for the sulfur determination.

Standard Solution

To prepare 2000 mg/ℓ sulfur solution, dissolve 8.243 g of ammonium sulfate $(NH_4)_2SO_4$ in deionized water and dilute to 1 ℓ.

REFERENCES

1. **Coedo, A. G. and Jiminez Seco, J. L.,** Determination of Fe, Mn, Ca, Mg, Al, Y and S in minerals by atomic absorption, *Rev. Metal. Cenim.,* 4, 58, 1968 (in Spanish).
2. **Syty, A. and Dean, J. A.,** Determination of phosphorus and sulfur in fuel rich air-hydrogen flames, *Appl. Spectrosc.,* 7, 1331, 1968.
3. **Dunk, R., Mostyn, R. A., and Hoare, H. C.,** The determination of sulfate by indirect atomic absorption spectroscopy, *At. Absorpt. Newsl.,* 8, 79, 1969.
4. **Fuwa, K. and Vallee, B. L.,** Molecular flame absorption spectrometry for sulfur, *Anal. Chem.,* 41, 188, 1969.
5. **Jungreis, E. and Anavi, Z.,** Determination of sulfite ion (or sulfur dioxide) by atomic absorption spectroscopy, *Anal. Chim. Acta,* 45, 190, 1969.
6. **L'vov, B. V. and Khartsyzov, A. D.,** Atomic absorption determination of sulfur, phosphorus, iodine and mercury from resonance lines in the vacuum ultraviolet spectral region, *Zh. Prikl. Spektrosk.,* 11, 413, 1969.
7. **Magyar, B. and Santos, F.,** Indirect sulfate determination in the presence of phosphate by atomic absorption spectrometry, *Helv. Chim. Acta,* 52, 820, 1969.
8. **Galindo, G. G., Appelt, H., and Schalscha, E. B.,** Sulfur determination in soil extracts by an indirect atomic absorption spectrophotometric method, *Soil Sci. Soc. Am. Proc.,* 33, 974, 1969.
9. **Pleskach, L. I. and Chirkova, G. D.,** Flame photometric determination of sulfate ions in natural waters, *Zavod. Lab.,* 35, 115, 1969.
10. **Rose, S. A. and Blotz, D. F.,** Indirect determination of sulfur dioxide by atomic absorption spectrometry after precipitation of lead sulfate, *Anal. Chim. Acta,* 44, 239, 1969.
11. **Sinha, R. C. P., Singhal, K. C., and Banerjee, B. K.,** Estimation of chloride and sulfate in ammonium bicarbonate by atomic absorption spectrophotometry, *Technology,* 6, 198, 1969.
12. **Wollin, A.,** Microdetermination of total sulfur by atomic absorption spectrophotometry, *At. Absorpt. Newsl.,* 9, 43, 1970.
13. **Aldous, K. M., Dagnall, R. M., and West, T. S.,** Flame spectroscopic determination of sulfur and phosphorus in organic and aqueous matrixes by using a simple filter photometer, *Analyst (London),* 95, 417, 1970.
14. **Borden, F. Y. and McCormick, L. H.,** An indirect method for the measurement of sulfate by barium absorption spectrophotometry, *Soil Sci. Soc. Am. Proc.,* 34, 705, 1970.
15. **Varley, J. A. and Chin, P. Y.,** Determination of water soluble sulfate in acidic sulfate soils by atomic absorption spectroscopy, *Analyst (London),* 95, 592, 1970.
16. **Taylor, H. E., Gibson, J. H., and Skogerboe, R. K.,** Determination of trace amounts of sulfur by atomic absorption and emission spectrometry, *Anal. Chem.,* 42, 1569, 1970.
17. **Liteanu, C. and Manoliu, C.,** Indirect determination of sulfur in pyrites by flame photometric titration, *Rev. Roum. Chim.,* 16, 411, 1971.
18. **Looyenga, R. W. and Huber, C. O.,** Determination of sulfate by atomic absorption inhibition titration, *Anal. Chim. Acta,* 55, 179, 1971.
19. **Pleskach, L. I. and Chirkova, G. D.,** Flame photometric determination of small concentrations of sulfate ion in natural waters, *Zavod. Lab.,* 37, 168, 1971.
20. **Sevcik, J.,** Selective detection of sulfur, chlorine and nitrogen with the help of the combination of flame ionization and flame photometric detectors, *Chromatographia,* 4, 195, 1971.
21. **Kirkbright, G. F. and Marshall, M.,** Direct determination of sulfur by atomic absorption spectrometry in a separated nitrous oxide-acetylene flame, *Anal. Chem.,* 44, 1288, 1972.
22. **Kirkbright, G. F., Marshall, M., and West, T. S.,** Direct determination of sulfur in oils by atomic absorption spectrometry using an inert gas shielded nitrous oxide-acetylene flame, *Anal. Chem.,* 44, 2379, 1972.
23. **Lautenbacher, H. W. and Baker, H. W.,** Pollution control: how much sulfur in fuel oils?, *Am. Lab.,* 4, 45, 1972.
24. **Lin, C. I. and Huber, C. O.,** Determination of phosphate silicate and sulfate in natural and waste water by atomic absorption inhibition titration, *Anal. Chem.,* 44, 2200, 1972.
25. **Monteil, A.,** Indirect determination of sulfate ions (in potable water) by atomic absorption spectrophotometry, *Trib. CEBEDEAU,* 25, 292, 1972 (in French).
26. **Veillon, C. and Park, J. Y.,** Use of the salet phenomena in the determination of sulfur and phosphorus in aqueous and organic samples, *Anal. Chim. Acta,* 60, 293, 1972.
27. **Baker, D. E.,** A new approach to soil testing. I. Ionic equilibria involving H, K, Ca, Mg, Mn, Fe, Cu, Zn, Na, P and S, *Soil Sci. Soc. Am. Proc.,* 27, 537, 1973.

28. **Niim, Y. and Endo, J.,** Analysis for sulfur in steel by atomic absorption spectrophotometric acid combustion methods. Indirect determination of sulfur in steel by atomic absorption spectrophotometry, *Tetsu To Hagane,* 59, 223, 1973.

29. **Everett, G. L., West, T. S., and Williams, R. W.,** Trace analysis for sulfur and phosphorus in aqueous solution by a carbon filament atom reservoir technique, *Anal. Chim. Acta,* 68, 387, 1974.

30. **Kunishi, M. and Ohno, S.,** Indirect determination of sulfate ion in rhodium plating solution by atomic absorption spectrophotometry, *At. Absorpt. Newsl.,* 13, 29, 1974.

31. **Leskovar, R. and Weidmann, G.,** Determination of free sulfate in serum by flame photometry, *Z. Klin. Chem. Klin. Biochem.,* 12, 98, 1974.

32. **Weidmann, G. and Leskovar, R.,** Determination of free sulfate in urine by flame photometry, *Z. Klin. Chem. Klin. Biochem.,* 12, 103, 1974.

33. **Ametani, K.,** Atomic absorption spectrochemical analysis of sulfur in the nickel arsenide type sulfides containing iron and titanium and in the sulfo spinels containing chromium, *Bunseki Kagaku,* 23, 745, 1974.

34. **Kirkbright, G. F. and Wilson, P. J.,** Application of a demountable hollow cathode lamp as a source for the direct determination of sulfur, iodine, arsenic, selenium and mercury by atomic absorption flame spectrometry, *Anal. Chem.,* 46, 1414, 1974.

35. **Stanton, R. E.,** Determination of calcium, cobalt, iron, molybdenum, phosphorus, selenium and sulfur in plant material, *Lab. Pract.,* 23, 233, 1974.

36. **Galle, O. K. and Hathaway, L. R.,** Indirect determination of sulfate in water by atomic absorption, *Appl. Spectrosc.,* 29, 518, 1975.

37. **Meyer, J. L. and Rundquist, R. T.,** Determination of inorganic sulfate in urine by atomic absorption spectroscopy, *Biochem. Med.,* 12, 398, 1975.

38. **Pleskach, L. I. and Chirkova, G. D.,** Flame photometric determination of sulfate sulfur in super phosphates, *Zh. Anal. Khim.,* 30, 729, 1975.

39. **Adams, M. J. and Kirkbright, G. F.,** The direct determination of sulfur by atomic absorption spectrometry using a graphite furnace electrothermal atomizer, *Can. J. Spectrosc.,* 21, 127, 1976.

40. **Bataglia, O. C.,** Indirect determination of sulfur in plants by atomic absorption spectrophotometry, *Cien. Cult. (Sao Paulo),* 28, 672, 1976 (in Portuguese).

41. **Childs, A. H. and Schrenk, W. G.,** Some characteristics of low pressure sulfur, microwave-excited electrodeless discharge lamps, *Appl. Spectrosc.,* 30, 507, 1976.

42. **Goldshtein, M. M. and Yudelevich, I. G.,** Indirect atomic absorption methods of determining fluorine, chlorine, iodine, phosphorus and sulfur, *Zh. Anal. Khim.,* 31, 801, 1976.

43. **Sand, J. R., Liu, J. H., and Huber, C. O.,** Determination of silicate, phosphate and sulfate by calcium atomization inhibition titration, *Anal. Chim. Acta,* 87, 79, 1976.

44. **Siemer, D. D., Woodriff, R., and Robinson, J.,** Sulfate determination in natural waters by nonresonance line furnace atomic absorption, *Appl. Spectrosc.,* 31, 168, 1977.

45. **Yoshida, Z. and Takahashi, M.,** Indirect determination of submicrogram amounts of sulfide by flameless atomic absorption spectrometry of mercury, *Mikrochim. Acta,* 1, 459, 1977.

46. **Pleskach, L. I.,** Use of an atomic absorption spectrophotometer for determining nitrogen and sulfur from the molecular absorption of ammonia, hydrogen sulfide and sulfur dioxide, *Zh. Anal. Khim.,* 34, 600, 1979.

47. **Eklund, R. H. and Holcombe, J. A.,** Signal depression in electrothermal atomic absorption spectrometry by nitrate and sulfate ions, *Anal. Chim. Acta,* 109, 97, 1979.

48. **Hassan, S. S. and Eldesouki, M. H.,** Microdetermination of divalent sulfur in aliphatic compounds by visual titrimetry, potentiometry and atomic absorption spectrometry, *Mikrochim. Acta,* 2, 27, 1979.

49. **Syty, A.,** Determination of sulfide by evolution of hydrogen sulfide and absorption spectrometry in the gas phase, *Anal. Chem.,* 51, 911, 1979.

50. **Iosof, V. and Calinescu, E.,** Atomic absorption spectrophotometric determination of sulfides in fresh water, *Rev. Chim.,* 30, 958, 1979.

51. **Yoshimura, C. and Noda, Y.,** Deoxidation effect of sulfur on flame atomic absorption spectrometry of metals by direct atomization of their oxides, *Nippon Kagaku Kaishi,* 8, 1098, 1979.

52. **Oeien, A.,** Determination of easily soluble sulfate and total sulfur in soil by indirect atomic absorption, *Acta Agric. Scand.,* 29, 71, 1979.

53. **Couto, M. I. and Curtius, A. J.,** Indirect analyses of sulfate in brines by atomic absorption, *Appl. Spectrosc.,* 34, 228, 1980.

54. **Clegg, J. B., Grainger, F., and Gale, I. G.,** Quantitative measurements of impurities in gallium arsenide, *J. Mater. Sci.,* 15, 747, 1980.

55. **Michalk, D. and Manz, F.,** Determination of inorganic plasma sulfate by indirect atomic absorption spectrophotometry, *Clin. Chim. Acta,* 107, 43, 1980.

56. **Ray, R. C., Nayar, P. K., Misra, A. K., and Sethunathan, N.,** Determination of sulfide in flooded acid-sulfate soils by an indirect atomic absorption spectrophotometric method, *Analyst (London),* 105, 984, 1980.

57. **Sasaki, H. and Nagumo, S.,** Determination of sulfate ion by atomic absorption spectrometry after pre-treatment with barium chromate-acidic suspension method, *Bunseki Kagaku,* 29, 95t, 1980.
58. **Reim, R. E. and Hawn, D. D.,** Determination of total sulfur in hydrocarbons by reductive pyrolysis with polarographic detection, *Anal. Chem.,* 53, 1088, 1981.
59. **Hassan, S. S. and Eldesouki, M. H.,** Potentiometric and atomic absorption spectrometric determination of sulfon amides in pharmaceutical preparations, *J. Assoc. Off. Anal. Chem.,* 64, 1158, 1981.
60. **Koizumi, H., Hadeishi, T., and McLaughlin, R. D.,** New technique for the determination of sulfur dioxide by a Zeeman shifted atomic line, *Spectrochim. Acta,* 36b, 483, 1981.
61. **Marshall, G. and Midgley, D.,** Mercury displacement detection for the determination of picogram amounts of sulfite ion or sulfur dioxide by atomic spectrometry, *Anal. Chem.,* 53, 1760, 1981.
62. **Wieckowska, J.,** The determination of trace elements in oils, *Mikrochim. Acta,* 1, 195, 1982.

AUTHOR INDEX

SUBJECT INDEX

SELENIUM, Se (ATOMIC WEIGHT 78.96)

Selenium is found associated with tellurium and in copper, iron, meteorite iron, and minerals. It is used in photoelectric cells, solar cells, electrical rectifiers, the electronic and solid state industries, xerography, the glass and ceramic industry, as a photographic toner, and as a stabilizer of lubricating oils, etc. Selenium metal is not toxic, but its compounds are known for toxicity and produce serious health effects on animals and humans.

Selenium determination is required in the analyses of copper concentrates, flue dusts, smelter slags, feed, food, water, petroleum products, and biological samples. Sample preparation is very important; loss may occur due to coprecipitation or volatilization. The loss due to volatilization is severe in the presence of halides and in higher than 6 N hydrochloric acid concentration. Sulfuric-perchloric treatment of organic samples at 100°C and covered dissolution below 100°C is recommended.

In air-acetylene flame, nonatomic species absorb strongly at 196.0 and 204.0 nm. Background correction should be used to improve signal-to-noise ratio. An increase in sensitivity may be achieved in argon-hydrogen flame with a high solids single slot burner. This flame is not suitable for use with organic solvents or in the presence of other chemical interferences. Nitrous oxide-acetylene flame is recommended to avoid large chemical interference, but with less sensitivity. Better results can be obtained with the hydride generation method.

Standard Solution

To prepare 1000 mg/ℓ solution of selenium, dissolve 1.000 g of selenium metal in 100 mℓ of 1:1 nitric acid, heat gently, cool, and dilute to 1 ℓ with deionized water.

Instrumental Parameters

Wavelength	196.0 nm
Slit	2.0 nm
Light source	Hollow cathode lamp
Flame	Air-acetylene, oxidizing (lean, blue)
Sensitivity	0.59 mg/ℓ
Detection limit	0.13 mg/ℓ
Optimum range	50 mg/ℓ
Secondary wavelengths and their sensitivities	
204.0 nm	3.0—6.0 mg/ℓ
206.3 nm	12.0—22.0 mg/ℓ
207.5 nm	40.0—75.0 mg/ℓ

REFERENCES

1. **Allan, J. E.,** Absorption flame photometry below 2000 Å, *4th Australian Spectroscopy Conf.*, Australian Academy of Science, Canberra, August 1963.
2. **Rann, C. S. and Hambly, A. N.,** The determination of selenium by atomic absorption spectrophotometry, *Anal. Chim. Acta*, 32, 346, 1965.
3. **Mulford, C. E.,** Solvent extraction techniques for atomic absorption spectroscopy, *At. Absorpt. Newsl.*, 5, 88, 1966.
4. **Mulford, C. E.,** Low-temperature ashing for determination of volatile metals by atomic absorption spectroscopy, *At. Absorpt. Newsl.*, 5, 135, 1966.
5. **Dagnall, R. M., Thompson, K. C., and West, T. S.,** The use of microwave excited electrodeless discharge tubes as spectral sources in atomic absorption spectroscopy, *At. Absorpt. Newsl.*, 6, 117, 1967.
6. **Schroeder, H. A.,** Effects of selenate, selenite and tellurite as the growth and early survival of mice and rats, *J. Nutr.*, 92, 334, 1967.
7. **Chakrabarti, C. L.,** The atomic absorption spectroscopy of selenium, *Anal. Chim. Acta*, 42, 379, 1968.
8. **Kahn, H. L. and Schallis, J. E.,** Improvement of detection limits for As, Se and other elements with an argon-hydrogen flame, *At. Absorpt. Newsl.*, 7, 5, 1968.
9. **Kirkbright, G. F., Sargent, M., and West, T. S.,** The determination of arsenic and selenium by atomic absorption spectroscopy in a nitrogen separated air-acetylene flame, *At. Absorpt. Newsl.*, 8, 34, 1969.
10. **Peterson, E. A.,** Determination of selenium in stainless steels by atomic absorption spectrophotometry, *At. Absorpt. Newsl.*, 9, 129, 1970.
11. **Rudnevskii, N. K., Demarin, V. T., and Nyrkova, A.,** Use of atomic absorption spectra for determining excess quantities of selenium and zinc in zinc selenide, *Zh. Prikl. Spektrosk.*, 12, 156, 1970.
12. **Zook, E., Greene, F., and Morris, E.,** Nutrient composition of selected wheats and wheat products. Distribution of manganese, copper, nickel, zinc, magnesium, lead, tin, cadmium, chromium and selenium as determined by atomic absorption spectroscopy and colorimetry, *Cereal Chem.*, 47, 720, 1970.
13. **Johns, P.,** The atomic absorption of arsenic and selenium, *Spectrovision*, 24, 6, 1970.
14. **Nakahara, T., Munemori, M., and Musha, S.,** Determination of selenium in sulfur by atomic absorption spectrophotometry, *Anal. Chim. Acta*, 50, 51, 1970.
15. **Kirkbright, G. F. and Ranson, L.,** Use of the nitrous oxide-acetylene flame for determination of arsenic and selenium by atomic absorption spectrometry, *Anal. Chem.*, 43, 1238, 1971.
16. **Lau, H. K. Y. and Lott, P. F.,** Indirect atomic absorption method for the determination of selenium, *Talanta*, 18, 303, 1971.
17. **Manning, D. C.,** A high sensitivity arsenic-selenium sampling system for atomic absorption spectroscopy, *At. Absorpt. Newsl.*, 10, 123, 1971.
18. **Reichel, W.,** Differential reduction and atomic absorption determination of selenium, *Anal. Chem.*, 43, 1501, 1971.
19. **Kirkbright, G. F., Ranson, L., and West, T. S.,** Use of a triple pass optical arrangement for the determination of arsenic and selenium by atomic absorption spectroscopy in an inert gas shielded nitrous oxide-acetylene flame, *Spectrosc. Lett.*, 5, 25, 1972.
20. **Severne, B. C. and Brooks, R. R.,** Rapid determination of selenium by atomic absorption spectrophotometry, *Anal. Chim. Acta*, 58, 216, 1972.
21. **Baird, R. B., Pourian, S., and Gabrielian, S. M.,** Determination of trace amounts of selenium in waste waters by carbon rod atomization, *Anal. Chem.*, 44, 1887, 1972.
22. **Yamamoto, Y., Kumamaru, T., Hayashi, Y., and Manke, M.,** Enhancement of sensitivity for selenium determination in atomic absorption spectrophotometry by introducing selenide into an argon-hydrogen flame, *Anal. Lett.*, 5, 717, 1972.
23. **Severne, B. C. and Brooks, R. R.,** Rapid determination of selenium and tellurium by atomic absorption spectrophotometry, *Talanta*, 19, 1467, 1972.
24. **Schmidt, F. J. and Royer, J. L.,** Submicrogram determination of arsenic, selenium, antimony and bismuth by atomic absorption utilizing sodium borohydride reduction, *Anal. Lett.*, 6, 17, 1973.
25. **Balandina, N. S., Yudelevich, I. G., Khrapai, V. P., and Abakumov, D. N.,** Atomic absorption determination of selenium and tellurium in pure silver, *Izv. Sib. Otd. Akad. Nauk SSSR, Ser. Khim. Nauk*, 5, 93, 1973.
26. **Nakahara, T., Nishino, H., Munemori, M., and Musha, S.,** Atomic absorption spectrophotometric determination of arsenic and selenium in premixed inert gas (entrained air)-hydrogen flame with a multiflame burner, *Bull. Chem. Soc. Jpn.*, 46, 1706, 1973.
27. **Balandina, N. S., Yudelevich, I. G., Khrapai, V. P., and Abakumov, D. V.,** Atomic absorption determination of selenium and tellurium in pure gold, *Zh. Anal. Khim.*, 28, 2054, 1973.

28. **Caldwell, J. S., Lishka, R. J., and McFarren, E. F.,** Evaluation of a low-cost arsenic and selenium determination at microgram per liter levels, *J. Am. Water Works Assoc.,* 65, 731, 1973.

29. **Baird, R. B. and Gabrielian, S. M.,** A tantalum foil lined graphite tube for the analysis of arsenic and selenium by atomic absorption spectrometry, *Appl. Spectrosc.,* 28, 273, 1974.

30. **Freeman, H. C. and Uthe, J. F.,** An improved hydride generation apparatus for determining arsenic and selenium by atomic absorption spectroscopy, *At. Absorpt. Newsl.,* 13, 75, 1974.

31. **Supp, G. R. and Gibbs, I.,** The determination of selenium and tellurium in organic accelerators used in vulcanization process for natural and synthetic rubbers by atomic absorption spectrometry, *At. Absorpt. Newsl.,* 13, 71, 1974.

32. **Goulden, P. D. and Brooksbank, P.,** Automated atomic absorption determination of arsenic, antimony and selenium in natural waters, *Anal. Chem.,* 46, 1431, 1974.

33. **Inhat, M. and Westerby, R. J.,** Application of flameless atomization to the atomic absorption determination of selenium in biological samples, *Anal. Lett.,* 7, 257, 1974.

34. **Kirkbright, G. F. and Wilson, P. J.,** Application of a demountable hollow cathode lamp as a source for the direct determination of sulfur, iodine, arsenic, selenium and mercury by atomic absorption spectrometry, *Anal. Chem.,* 46, 1414, 1974.

35. **Lansford, M., McPherson, E. M., and Fishman, M. J.,** Determination of selenium in water, *At. Absorpt. Newsl.,* 13, 103, 1974.

36. **Ng, S., Munroe, M., and McSharry, W.,** Determination of selenium in animal feed premix by atomic absorption spectroscopy, *J. Assoc. Off. Anal. Chem.,* 57, 1260, 1974.

37. **Thompson, K. C. and Thomerson, D. R.,** Atomic absorption studies on the determination of antimony, arsenic, bismuth, germanium, lead, selenium, tellurium and tin by utilizing the generation of covalent hydrides, *Analyst (London),* 99, 595, 1974.

38. **Welcher, G. G., Kriege, O. H., and Marks, J. Y.,** Direct determination of trace quantities of lead, bismuth, selenium, tellurium and thallium in high temperature alloys by nonflame atomic absorption spectrophotometry, *Anal. Chem.,* 46, 1227, 1974.

39. **Stanton, R. E.,** Determination of calcium, cobalt, copper, iron, molybdenum, phosphorus, selenium and sulfur in plant material, *Lab. Pract.,* 23, 233, 1974.

40. **King, H. G. and Morrow, R. W.,** Determination of Arsenic and Selenium in Surface Water by Atomic Absorption to Support Environmental Monitoring Programs, Rep. Y-1956, U.S. Atomic Energy Commission, Oak Ridge, Tenn., 1974.

41. **Shelton, B. J.,** Determination of silver, selenium, tellurium, antimony, tin, lead and arsenic in anode sludges, *Natl. Inst. Metall. Repub. S. Afr. Rep.,* No. 1771, 1975.

42. **Henn, E. L.,** Determination of selenium in water and industrial effluents by flameless atomic absorption, *Anal. Chem.,* 47, 428, 1975.

43. **Kamada, T., Kumamaru, T., and Yamamato, Y.,** Rapid determination of trace amounts of arsenic (III, V), antimony (III, V) and selenium (IV) in water by atomic absorption spectrophotometry with a carbon rod atomizer, *Bunseki Kagaku,* 24, 89, 1975 (in Japanese).

44. **Lavrakas, V., Barry, E., and Golembeski, T.,** Determination of submicrogram amounts of selenium in rocks by atomic absorption spectroscopy, *Talanta,* 22, 547, 1975.

45. **Schmidt, F. J., Royer, J. L., and Muir, S. M.,** Automated determination of arsenic, selenium, antimony, bismuth and tin by atomic absorption utilizing sodium borohydride reduction, *Anal. Lett.,* 8, 123, 1975.

46. **Shendrikar, A. D. and West, P. W.,** The rate of loss of selenium from aqueous solution stored in various containers, *Anal. Chim. Acta,* 74, 189, 1975.

47. **Siemer, D. D. and Hagemann, L.,** An improved hydride generation-atomic absorption apparatus for selenium determination, *Anal. Lett.,* 8, 323, 1975.

48. **Burch, R. E., Williams, R. V., Kahn, H. K. J., Jetton, M. M., and Sullivan, J. F.,** Tissue trace element and enzyme content in pigs fed a low manganese and selenium, *J. Lab. Clin. Med.,* 86, 132, 1975.

49. **Chau, Y. K., Wong, P. T., and Gouldon, P. D.,** Gas chromatography-atomic absorption method for the determination of dimethyl selenide and dimethyl diselenide, *Anal. Chem.,* 47, 2279, 1975.

50. **Golembeski, T.,** Determination of submicrogram amounts of selenium in rocks by atomic absorption spectroscopy, *Talanta,* 22, 547, 1975.

51. **Martin, T. D., Kopp, J. F., and Ediger, R. D.,** Determination of selenium in water, waste water, sediment and sludge by flameless atomic absorption spectroscopy, *At. Absorpt. Newsl.,* 14, 109, 1975.

52. **Mesman, B. B. and Thomas, T. C.,** Two atomic absorption methods for the determination of subnanogram amounts of arsenic and selenium, *Anal. Lett.,* 8, 449, 1975.

53. **Ng, S. and McSharry, W.,** Atomic absorption spectrophotometric determination of selenium in animal feed premix using the vapor generation technique, *J. Assoc. Off. Anal. Chem.,* 58, 987, 1975.

54. **Nishigaki, N.,** Methylmercury and selenium in umbilical cords of inhabitants of Minimata area, *Nature (London),* 258, 324, 1975.

55. **Ohta, K. and Suzuki, M.,** Determination of selenium in metallurgical samples by flameless atomic absorption spectrometry, *Anal. Chim. Acta,* 77, 288, 1975.

56. **Vogel, C. H. and Etten, N.,** Analytical and preparative method for the separation of selenium from trace contaminants, *Z. Anal. Chem.,* 275, 349, 1975 (in German).

57. **Takeda, M., Inamasu, Y., and Tomida, T.,** On mercury and selenium contained in tuna fish tissues, *J. Shimonaseki Univ. Fish.,* 23, 145, 1975 (in Japanese).

58. **Dixon, K., Mallet, R. C., and Kocaba, R.,** Determination by atomic absorption spectrophotometry of arsenic, selenium, tellurium, antimony and bismuth on ion exchange resins, *Natl. Inst. Metall. Rep. S. Afr. Rep.,* No. 1689, 1975.

59. **Rudnevskii, N. K., Demarin, V. T., Sklemina, L. V., and Tumanova, A. N.,** Influence of some organic solvents on sensitivity of atomic absorption determination of iron, cadmium, copper, magnesium, sodium, vanadium, selenium and zinc, *Tr. Khim. Khim. Tekhnol.,* 1, 106, 1975 (in Russian).

60. **Hamner, R. M., Lechak, D. L., and Greenberg, P.,** Determination of silver, arsenic, bismuth, antimony, selenium and tellurium in chromium metal with flameless atomic absorption spectroscopy, *At. Absorpt. Newsl.,* 15, 122, 1976.

61. **Inhat, M.,** Selenium in foods: evaluation of atomic absorption spectrometric techniques involving hydrogen selenide generation and carbon furnace atomization, *J. Assoc. Off. Anal. Chem.,* 59, 911, 1976.

62. **Danchik, R. S., Thompson, D. E., and Hillegass, H. F.,** Determination of trace amounts of tellurium and selenium in gallium metal, *Anal. Lett.,* 9, 687, 1976.

63. **McDaniel, M., Shendrikar, A. D., Reiszner, K. D., and West, P. W.,** Concentrations and determination of selenium from environmental samples, *Anal. Chem.,* 48, 2240, 1976.

64. **Radziuk, B. and Van Loon, J. C.,** Atomic absorption spectroscopy as a detector for the gas chromatographic study of volatile selenium alkanes from *astragalus racemosus, Sci. Total Environ.,* 6, 251, 1976.

65. **Walker, H. H., Runnels, J. H., and Merryfield, R.,** Determination of trace quantities of selenium in petroleum and petroleum products by atomic absorption spectrometry, *Anal. Chem.,* 48, 2056, 1976.

66. **Fiorino, J. A., Jones, J. W., and Capar, S. G.,** Sequential determination of arsenic, selenium, antimony and tellurium in foods via rapid hydride evolution and atomic absorption spectrometry, *Anal. Chem.,* 48, 120, 1976.

67. **Inhat, M.,** Atomic absorption spectrometric determination of selenium with carbon surface atomization, *Anal. Chim. Acta,* 82, 293, 1976.

68. **Johnson, C. A. and Lewin, J. F.,** The determination of some "toxic" metals in human liver as a guide to normal levels in New Zealand. II. Arsenic, mercury and selenium, *Anal. Chim. Acta,* 82, 79, 1976.

69. **Kunselman, G. C. and Huft, E. A.,** The determination of arsenic, antimony, selenium and tellurium in environmental water samples by flameless atomic absorption, *At. Absorpt. Newsl.,* 15, 29, 1976.

70. **Pierce, F. D. and Brown, H. R.,** Inorganic interference study of automated arsenic and selenium determination with atomic absorption spectrometry, *Anal. Chem.,* 48, 693, 1976.

71. **Pierce, F. D., Lamoreaux, T. C., Brown, H. R., and Fraser, R. S.,** An automated technique for the submicrogram determination of selenium and arsenic in surface waters by atomic absorption spectroscopy, *Appl. Spectrosc.,* 30, 38, 1976.

72. **Vijan, P. N. and Wood, G. R.,** An automated submicrogram determination of selenium in vegetation by quartz tube furnace atomic absorption spectrophotometry, *Talanta,* 23, 89, 1976.

73. **Chambers, J. C. and McClellan, B. E.,** Enhancement of atomic absorption sensitivity for copper, cadmium, antimony, arsenic and selenium by means of solvent extraction, *Anal. Chem.,* 48, 2061, 1976.

74. **Bedard, M. and Kerbyson, J. D.,** Determination of arsenic, selenium, tellurium and tin in copper by hydride evolution atomic absorption spectrophotometry, *Can. J. Spectrosc.,* 21, 64, 1976.

75. **Corbin, D. R. and Barnard, W. M.,** Atomic absorption spectrophotometric determination of arsenic and selenium in water by hydride generation, *At. Absorpt. Newsl.,* 15, 116, 1976.

76. **Yamamoto, Y. and Kumamaru, T.,** Comparative study of zinc tablet and sodium borohydride tablet reduction systems in the determination of arsenic, antimony, and selenium by atomic absorption spectrophotometry via their hydrides, *Z. Anal. Chem.,* 281, 353, 1976.

77. **Clinton, O. E.,** Determination of selenium in blood and plant material by hydride generation and atomic absorption spectroscopy, *Analyst (London),* 102, 187, 1977.

78. **Hermann, R.,** A simple and rapid decomposition technique for the atomic absorption spectrophotometric determination of selenium in glass by hydride generation, *At. Absorpt. Newsl.,* 16, 44, 1977.

79. **Ishizaki, M.,** Determination of selenium in biological materials by flameless atomic absorption spectrometry using a carbon tube atomizer, *Bunseki Kagaku,* 26, 206,

80. **Marks, J. Y., Welcher, G. G., and Spellman, R. J.,** Atomic absorption determination of lead, bismuth, selenium, tellurium, thallium and tin in complex alloys using direct atomization from metal chips in the graphite furnace, *Appl. Spectrosc.,* 31, 9, 1977.

81. **Szydlowski, F. J.,** Comparative study of the determination of selenium in high carbohydrate nutrient carriers using flameless atomic absorption spectroscopy and fluorimetry, *At. Absorpt. Newsl.,* 16, 60, 1977.

82. **Condylis, A. and Hocquaux, H.,** Atomic absorption spectrophotometric determination of arsenic, selenium and tin in metallurgical matrixes, *Analusis,* 5, 228, 1977.
83. **Inhat, M. and Miller, H. J.,** Analysis of foods for arsenic and selenium by acid digestion, hydride evolution atomic absorption spectrophotometry, *J. Assoc. Off. Anal. Chem.,* 60, 813, 1977.
84. **Inhat, M. and Miller, H. J.,** Acid digestion hydride evolution atomic absorption spectrophotometric method for determining arsenic and selenium in foods: collaborative study. I, *J. Assoc. Off. Anal. Chem.,* 60, 1414, 1977.
85. **Kneip, T. J.,** Arsenic, selenium and antimony in urine and air, Analytical method by hydride generation and atomic absorption spectroscopy, *J. Health Lab. Sci.,* 14, 53, 1977.
86. **Neve, J. and Hanocq, M.,** The determination of selenium traces after extraction with 4-chloro-1,2-diaminobenzene by flameless atomic absorption spectrophotometry. Application to biological samples, *Anal. Chim. Acta,* 93, 85, 1977.
87. **Pierce, F. D. and Brown, H. R.,** Comparison of inorganic interferences in atomic absorption spectrometric determination of arsenic and selenium, *Anal. Chem.,* 49, 1417, 1977.
88. **Shum, G. T., Freeman, H. C., and Uthe, J. F.,** Flameless atomic absorption spectrophotometry of selenium in fish and food products, *J. Assoc. Off. Anal. Chem.,* 60, 1010, 1977.
89. **Siemer, D. D. and Koteel, P. K.,** Comparison of methods of hydride generation atomic absorption spectrometric arsenic and selenium determination, *Anal. Chem.,* 49, 1096, 1977.
90. **Yasuda, K., Taguchi, M., Tamura, S., and Toda, S.,** Determination of selenium in biological samples by solvent extraction-graphite furnace atomic absorption spectrometry, *Bunseki Kagaku,* 26, 442, 1977.
91. **Brodie, K. G.,** A comparative study: determining arsenic and selenium by atomic absorption spectrophotometry, *Int. Lab.,* 65, 68, 74, 1977.
92. **Aruscavage, P.,** Determination of arsenic, antimony and selenium in coal by atomic absorption spectrometry with graphite tube atomizer, *J. Res. U.S. Geol. Surv.,* 5, 405, 1977.
93. **Cutter, G. A.,** Species determination of selenium in natural waters, *Anal. Chim. Acta,* 98, 59, 1978.
94. **Kamada, T., Shiraishi, T., and Yamamoto, Y.,** Differential determination of selenium (IV) and selenium (VI) with sodium diethyldithiocarbamate, ammonium pyrrolidinedithiocarbamate and dithizone by atomic absorption spectrophotometry with a carbon tube atomizer, *Talanta,* 25, 15, 1978.
95. **Pyen, G. and Fishman, M.,** Automated determination of selenium in water, *At. Absorpt. Newsl.,* 17, 47, 1978.
96. **Wauchope, R. D.,** Selenium and arsenic levels in soybean from different production regions of the United States, *J. Agric. Food Chem.,* 26, 226, 1978.
97. **Egaas, E. and Julshamn, K.,** A method for the determination of selenium and mercury in fish products using the same digestion procedures, *At. Absorpt. Newsl.,* 17, 135, 1978.
98. **Flanjak, J.,** Atomic absorption spectrometric determination of arsenic and selenium in offal and fish by hydride generation, *J. Assoc. Off. Anal. Chem.,* 61, 1299, 1978.
99. **Ishizaki, M.,** Simple method for determination of selenium in biological materials by flameless atomic absorption spectrometry using a carbon tube atomizer, *Talanta,* 25, 167, 1978.
100. **Manning, D. C.,** Spectral interferences in graphite furnace atomic absorption spectroscopy. I. The determination of selenium in an iron matrix, *At. Absorpt. Newsl.,* 17, 107, 1978.
101. **Meyer, A., Hofer, C., and Toelg, G.,** Determination of selenium in the ppb range by furnace atomic absorption spectrometry in copper, copper alloys, silver, gold, lead and bismuth after volatilization in an oxygen stream, *Z. Anal. Chem.,* 290, 292, 1978.
102. **Reichert, J. K. and Gruber, H.,** Application of technically simplified hydride system for arsenic and selenium determination by atomic absorption spectrophotometry, *Vom Wasser,* 51, 191, 1978.
103. **Yamashige, T., Ohmoto, Y., and Shigetomi, Y.,** Ion exchange separation and determination of selenium in ambient particulates by heated quartz cell atomic absorption spectrophotometry, *Bunseki Kagaku,* 27, 607, 1978.
104. **Hocquellet, P.,** Application of electrothermal atomization to the determination of arsenic, antimony, selenium and mercury by atomic absorption spectrometry, *Analusis,* 6, 426, 1978.
105. **Papanastasiu, G. and Bronsch, K.,** Determination of selenium in feeding stuffs by flameless atomic absorption spectroscopy, *Z. Therphysiol. Tierernaehr. Futtermittelkd.,* 40, 325, 1978 (in German).
106. **Balusov, V. A., Solomatin, V. T., Gubin, S. P., Rysev, A. P., and Tikhonov, G. P.,** Spectrophotometric and atomic absorption determination of selenium (IV) using a ferrocenium cation, *Izv. Akad. Nauk SSSR Ser. Khim.,* 4, 921, 1979.
107. **Bower, N. W. and Ingle, J. D.,** Precision of flame atomic absorption spectrometric measurements of aluminum, chromium, cobalt, europium, lead, manganese, nickel, potassium, selenium, silicon, titanium and vanadium, *Anal. Chem.,* 51, 72, 1979.
108. **Haynes, B. W.,** Arsenic, antimony, selenium and tellurium determination in high-purity copper by electrothermal atomization, *At. Absorpt. Newsl.,* 18, 46, 1979.

109. **Jackwerth, E., Willmer, P. G., Hohn, R., and Berndt, H.,** A simple accessory for the determination of mercury and the hydride-forming elements (As, Bi, Sb, Se and Te) using flameless atomic absorption spectroscopy, *At. Absorpt. Newsl.,* 18, 66, 1979.

110. **Kirkbright, G. F. and Taddia, M.,** Use of tellurium (IV) to reduce interferences from some metal ions in the determination of selenium by hydride generation and atomic absorption spectrometry, *At. Absorpt. Newsl.,* 18, 68, 1979.

111. **Nakashima, S.,** Flotation separation and atomic absorption spectrometric determination of selenium (IV) in water, *Anal. Chem.,* 51, 654, 1979.

112. **Ivanov, G. and Koen, E.,** Flameless atomic absorption method for the determination of selenium microamounts in blood and organs, *Khig. Zdraveopaz.,* 22, 524, 1979.

113. **Palliere, M. and Gernez, G.,** Apparatus for the production of volatile hydrides for arsenic, antimony, bismuth, tin, germanium, selenium and tellurium determination by atomic absorption spectrometry, *Analusis,* 7, 46, 1979.

114. **Raie, R. M. and Smith, H.,** The determination of selenium in biological material by thermal neutron activation analysis and atomic absorption spectrometry, *J. Radioanal. Chem.,* 48, 185, 1979.

115. **Elson, C. M. and MacDonald, A. S.,** Determination of selenium in pyrite by an ion exchange-electrothermal atomic absorption spectrometric method, *Anal. Chim. Acta,* 110, 153, 1979.

116. **Forrester, J. E., Lehecka, V., Johnston, J. R., and Ott, W. L.,** Direct determination of trace quantities of antimony, arsenic, bismuth, cadmium, lead, selenium, silver, tellurium and thallium in high-purity nickel by electrothermal atomic absorption spectrometry, *At. Absorpt. Newsl.,* 18, 73, 1979.

117. **Reichert, J. K. and Gruber, H.,** Determination of arsenic and selenium by atomic absorption spectrophotometry. Comparison between two types of quartz cuvettes, *Vom Wasser,* 52, 289, 1979.

118. **El-Shaarawy, M. I., Maes, R. A., and Otten, J.,** Selenium determination in urine by atomic absorption spectrophotometry, *Indian J. Hosp. Pharm.,* 16, 136, 1979.

119. **Lund, W. and Bye, R.,** Flame atomic absorption analysis for selenium after electrothermal preconcentration, *Anal. Chim. Acta,* 110, 279, 1979.

120. **Meyer, A., Hofer, C., Tolg, G., Raptis, S., and Knapp, G.,** Cross interferences by elements in the determination of traces of selenium by the hydride-AAS procedure, *Z. Anal. Chem.,* 296, 337, 1979.

121. **Saeed, K., Thomassen, Y., and Langmyhr, F. J.,** Direct electrothermal atomic absorption spectrometric determination of selenium in serum, *Anal. Chim. Acta,* 110, 285, 1979.

122. **Sefzik, E.,** Determination of As, Pb, Cd, Cr and Se with respect to the requirements of drinking water regulations by means of flameless atomic absorption spectroscopy, *Vom Wasser,* 50, 285, 1979 (in German).

123. **Verlinden, M. and Deelstra, H.,** Study of the effects of elements that form volatile hydrides on the determination of selenium by hydride generation atomic absorption spectrometry, *Z. Anal. Chem.,* 296, 253, 1979.

124. **Fricke, F. L., Robbins, W. B., and Caruso, J. A.,** Trace element analysis of food and beverages by atomic absorption spectrometry, *Prog. Anal. At. Spectrosc.,* 2, 185, 1979.

125. **Arikawa, Y., Hirai, S., and Ozawa, T.,** Simultaneous determination of selenium and tellurium in native sulfur by atomic absorption spectrophotometry, *Bunseki Kagaku,* 28, 653, 1979.

126. **Boyer, K. W., Caper, S. G., Jones, J. W., Suddendorf, R. F., and Forwalter, J.,** Multielement/trace analysis identifies 48 elements on foods. Simultaneous element analysis and data reduction reaches parts per billion, *Food Process.,* 40, 72, 1979.

127. **Rubenstein, D. A. and Soares, J. H.,** The effect of selenium on the biliary excretion and tissue deposition of two forms of mercury in the broiler chick, *Poult. Sci.,* 58, 1289, 1979.

128. **Flanjack, J. and Lee, H. Y.,** Trace metal content of livers and kidneys of cattle, *J. Sci. Food Agric.,* 30, 503, 1979.

129. **Welz, B. and Melcher, M.,** Hydride atomic absorption spectroscopic technique in trace analysis, *Labor Prixis,* 3, 41, 1979 (in German).

130. **Fonds, A. W., Kempf, T., Minderhoud, A., and Sonneborn, M.,** Heavy metal content of various kinds of water, *Reinhalt Wasser,* 76, 1979.

131. **Hon, P. K., Lau, O. W., Cheung, W. C., and Wong, M. C.,** The atomic absorption spectrometric determination of arsenic, bismuth, lead, antimony, selenium and tin with a flame-heated silica T-tube after hydride generation, *Anal. Chim. Acta,* 115, 355, 1980.

132. **Neve, J., Hanocq, M., and Molle, L.,** Critical study of the flameless atomic absorption determination of selenium in complex media after extraction with aromatic o-diamines and addition of nickel (II), *Anal. Chim. Acta,* 115, 133, 1980.

133. **Belskii, N. K. and Ochertyanova, L. I.,** Atomic absorption determination of the deviation of cadmium chromium selenide ($CDCr_2Se_4$) crystals from the stoichiometric composition, *Z. Anal. Khim.,* 35, 604, 1980.

134. **Dekersabiec, A. M.,** Determination of arsenic, selenium and bismuth in rocks and soils by atomic absorption spectrophotometry, *Analusis,* 8, 97, 1980.

135. **Stein, V. B., Canelli, E., and Richards, A. H.,** Determination of dissolved selenium in fresh and estuarine waters by flameless atomic absorption, *At. Spectrosc.,* 1, 61, 1980.

136. **Szydlowski, F. J. and Vianzon, F. R.,** Further studies on the determination of selenium using graphite furnace atomic absorption spectroscopy, *At. Spectrosc.,* 1, 39, 1980.

137. **Agemian, H. and Thomson, R.,** Simple semiautomated atomic absorption spectrometric method for the determination of arsenic and selenium in fish tissue, *Analyst (London),* 105, 902, 1980.

138. **Clark, P. J., Zingaro, R. A., Irgolic, K. J., and McGinley, A. N.,** Arsenic and selenium in Texas lignite, *Int. J. Environ. Anal. Chem.,* 7, 295, 1980.

139. **Erzinger, J. and Puchelt, H.,** Determination of selenium in geochemical reference samples using flameless atomic absorption spectrometry, *Geostandards Newsl.,* 4, 13, 1980.

140. **Kamada, T. and Yamamoto, Y.,** Use of transition elements to enhance sensitivity for selenium determination by graphite furnace atomic absorption spectrophotometry combined with a solvent extraction with the APDC-MIBK system, *Talanta,* 27, 473, 1980.

141. **Kirkbright, G. F., Shan, H. C., and Snook, R. D.,** Evaluation of some matrix modification procedures for use in the determination of mercury and selenium by atomic absorption spectroscopy with a graphite tube electrothermal atomizer, *At. Spectrosc.,* 1, 85, 1980.

142. **Nadkarni, R. A.,** Multitechnique multielemental analysis of coal and fly ash, *Anal. Chem.,* 52, 929, 1980.

143. **Ohta, K. and Suzuki, M.,** Determination of selenium in water by electrothermal atomic absorption spectrometry, *Z. Anal. Chem.,* 302, 177, 1980.

144. **Roberts, R. V. D., Balaes, G., and Steele, T. W.,** Study of the measurement by electrothermal atomization and atomic absorption spectrophotometry of hydride forming elements, *Natl. Inst. Metall. Repub. S. Afr. Rep.,* No. 2053, 1980.

145. **Kahn, M. H., Mulligan, K. J., Jackson, M. E., and Caruso, J. A.,** Sequential determination of arsenic, selenium, germanium and tin as their hydrides by gas-solid chromatography with an atomic absorption detector, *Anal. Chim. Acta,* 118, 115, 1980.

146. **Hocquellet, P.,** Determination of selenium in animal feeds by atomic absorption spectrometry with electrothermal atomization, *Ann. Falsif. Expert. Chim.,* 73, 129, 1980.

147. **Holak, W.,** Analysis of foods for lead, cadmium, copper, zinc, arsenic and selenium using closed system sample digestion collaborative study, *J. Assoc. Off. Anal. Chem.,* 63, 485, 1980.

148. **Inhat, M. and Thompson, B. K.,** Acid digestion, hydride evolution atomic absorption spectrophotometric method for determining arsenic and selenium in foods. II. Assessment of collaborative study, *J. Assoc. Off. Anal. Chem.,* 63, 814, 1980.

149. **Scheiba, M. and Wobig, D.,** Determination of the miscibility gap of liquid potassium selenium alloys, *Z. Phys. Chem.,* 119, 23, 1980.

150. **Tada, Y., Yonemoto, T., Iwasa, A., and Nakagawa, K.,** Graphite furnace atomic absorption spectrophotometry of selenium in blood by application of enhancement effect of rhodium, *Bunseki Kagaku,* 29, 248, 1980.

151. **Verlinden, M., Baart, J., and Deelstra, H.,** Optimization of the determination of selenium by atomic absorption spectrometry. Comparison of two hydride generation systems, *Talanta,* 27, 633, 1980.

152. **Weigert, P.,** Heavy metal content of chicken eggs, *Z. Lebensm. Unters. Forsch.,* 171, 18, 1980.

153. **Agemian, H. and Bedek, E.,** Semi-automated method for the determination of total arsenic and selenium in soils and sediments, *Anal. Chim. Acta,* 119, 323, 1980.

154. **Alexander, J., Saeed, K., and Thomassen, Y.,** Thermal stabilization of inorganic and organoselenium compounds for direct electrothermal atomic absorption spectrometry, *Anal. Chim. Acta,* 120, 377, 1980.

155. **Vickrey, T. M. and Buren, M. S.,** Factors affecting selenium atomization efficiency in graphite furnace atomic absorption, *Anal. Lett.,* 13, 1465, 1980.

156. **Vijan, P. and Leung, D.,** Reduction of chemical interference and speciation studies in the hydride generation atomic absorption method for selenium, *Anal. Chim. Acta,* 120, 141, 1980.

157. **Workman, S. M. and Soltanpour, P. N.,** Importance of prereducing selenium (VI) to selenium (IV) and decomposing organic matter in soil extracts prior to determination of selenium using hydride generation, *Soil Sci. Soc. Am. J.,* 44, 1331, 1980.

158. **El-Shaarawy, M. I.,** Method for selenium determination using atomic absorption spectrophotometry, *Clin. Lab. (Rome),* 4, 206, 1980 (in Italian).

159. **Hofsommer, H. J. and Bielig, H. J.,** Significance of selenium and its analytical detection in fruits and vegetables, *Dtsch. Lebensm. Rundsch.,* 76, 419, 1980.

160. **Katskov, D. A., Grinshtein, I. L., and Kruglikova, L. P.,** Study of the evaporation of the metals indium, gallium, thallium, germanium, tin, lead, antimony, bismuth, selenium and tellurium from a graphite surface by the atomic absorption method, *Zh. Prikl. Spektrosk.,* 33, 804, 1980.

161. **Lindh, U., Brune, D., Nordberg, G., and Wester, P. O.,** Levels of antimony, arsenic, cadmium, copper, lead, mercury, selenium, silver, tin and zinc in bone tissue of industrially exposed workers, *Sci. Total Environ.,* 16, 109, 1980.

162. **Morita, K., Shimizu, M., Inoue, B., and Ishida, T.,** Graphite furnace atomic absorption spectrometry of antimony and selenium in whole blood, *Okayama-Ken Kankyo Hoken Senta Nempo*, 4, 94, 1980.

163. **Neve, J. and Hanocq, M.,** Application of a new method for selenium determination in various biological samples, *J. Pharm. Belg.*, 35, 345, 1980 (in French).

164. **Rail, C. D., Kidd, D. E., and Hadley, W. M.,** Determination of selenium in tissues, serum and blood of wild rodents by graphite furnace atomic absorption spectrophotometry, *Int. J. Environ. Anal. Chem.*, 8, 79, 1980.

165. **Jonsen, J., Helgeland, K., and Steinnes, E.,** Trace elements in human serum. Regional distribution in Norway, in *Geomed. Aspects Present, Future Res., 1978*, Laag, J., Ed., 1980, 189.

166. **Capdevila, C. and Alduan, F. A.,** Determination of Trace Amounts of Selenium in Minerals and Rocks by Flameless Atomic Absorption Spectrometry, Rep. JEN-472, 1980 (Spanish); INIS Atomindex 12, 3, Abstr. No. 577161, 1981.

167. **Clark, J. R. and Viets, J. G.,** Multielement extraction system for the determination of 18 trace elements in geochemical samples, *Anal. Chem.*, 53, 61, 1981.

168. **Clark, J. R. and Viets, J. G.,** Back-extraction of trace elements from organometallic-halide extracts for determination by flameless atomic absorption spectrometry, *Anal. Chem.*, 53, 65, 1981.

169. **Cox, D. H. and Bibb, A. E.,** Hydrogen selenide evolution-electrothermal atomic absorption method for determining nanogram levels of total selenium, *J. Assoc. Off. Anal. Chem.*, 64, 265, 1981.

170. **Fishkova, N. L., Nazarenko, I. I., Vilenkin, V. A., and Petrokova, Z. A.,** Extraction-atomic absorption determination of selenium geological samples, *Zh. Anal. Khim.*, 36, 115, 1981 (in Russian).

171. **Reamer, D. C., Veillon, C., and Tokousbalides, P. T.,** Radiotracer techniques for evaluation of selenium hydride generation systems, *Anal. Chem.*, 53, 245, 1981.

172. **Sighinolfi, G. P. and Gorgoni, C.,** Atomic absorption spectrochemical analysis for ultratrace elements in geological materials by hydride forming techniques: selenium, *Talanta*, 28, 169, 1981.

173. **Sinemus, H. W., Melcher, M., and Welz, B.,** Influence of valence state on the determination of antimony, arsenic, bismuth, selenium and tellurium in lake water using the hydride AA technique, *At. Spectrosc.*, 2, 81, 1981.

174. **Subramanian, K. S. and Meranger, J. C.,** Determination of arsenic (III), arsenic (V), antimony (III), antimony (V), selenium (IV) and selenium (VI) by extraction with ammonium pyrrolidine dithiocarbamate methyl isobutyl ketoneketone and electrothermal atomic absorption spectrometry, *Anal. Chim. Acta*, 124, 131, 1981.

175. **Thompson, D. D. and Allen, R. J.,** Rapid determination of selenium in nutritional supplements by flameless atomic absorption technique using a novel sample preparation, *At. Spectrosc.*, 2, 53, 1981.

176. **Hahn, M. H., Kuennen, R. W., Caruso, J. A., and Fricke, F. L.,** Determination of trace amounts of selenium in corn, lettuce, potatoes, soybeans and wheat by hydride generationd/condensation and flame atomic absorption spectrometry, *J. Agric. Food Chem.*, 29, 792, 1981.

177. **Han, H., Kaiser, G., and Toelg, G.,** Decomposition of biological materials, rocks and soils in pure oxygen under dynamic conditions for the determination of selenium at trace levels, *Anal. Chim. Acta*, 128, 9, 1981.

178. **Meyer, A., Hofer, C., Knapp, G., and Toelg, G.,** Determination of selenium in the μg/g and ng/g range in inorganic and organic matrixes by AAS after volatilization in a dynamic system, *Fresenius' Z. Anal. Chem.*, 305, 1, 1981.

179. **Xie, Y. Y.,** Simultaneous determination of trace arsenic and selenium in water by hydride generation-spectrophotometry, *Fen Hsi Hua Hsueh*, 9, 85, 1981 (in Chinese).

180. **Capel, I. D., Pinnock, M. H., Dorrell, H. M., Williams, D. C., and Grant, E. C. G.,** Comparison of concentrations of some trace, bulk and toxic metals in the hair of normal and dyslexic children, *Clin. Chem. (Winston-Salem, N.C.)*, 27, 879, 1981.

181. **Krivan, V., Geiger, H., and Franz, H. E.,** Determination of Fe, Co, Zn, Se, Rb, and Cs in NBS bovine liver, blood plasma and erythrocytes by INAA and AAS, *Fresenius' Z. Anal. Chem.*, 305, 399, 1981.

182. **Verlinden, M., Deelstra, H., and Adriaenssens, E.,** The determination of selenium by atomic absorption spectrometry. A review, *Talanta*, 28, 637, 1981.

183. **Suzuki, M., Ohta, K., Yamakita, T., and Katsuno, T.,** Electrothermal atomization with a metal microtube in atomic absorption spectrometry, *Spectrochim. Acta*, 36b, 679, 1981.

184. **Khalighie, J., Ure, A. M., and West, T. S.,** Atom-trapping atomic absorption spectrometry of arsenic, cadmium, lead, selenium and zinc in air-acetylene and air-propane flames, *Anal. Chim. Acta*, 131, 27, 1981.

185. **Welz, B., Grobenski, Z., and Melcher, M.,** Determination of antimony, arsenic, selenium, tellurium, bismuth and tin using the hydride-AAS technique, *Spektrometertagung (Votr.)*, 13, 337, 1981.

186. **Mueller, H. and Siepe, V.,** Quantitative determination of arsenic, lead, cadmium, mercury and selenium in foods by flameless atomic absorption spectrophotometry, *Dtsch. Lebensm. Rundsch.*, 77, 392, 1981.

187. **Holen, B., Bye, R., and Lund, W.,** Atomic absorption spectrometry of selenium in an argon-hydrogen flame after electrothermal preconcentration, *Anal. Chim. Acta,* 130, 257, 1981.

188. **Guo, X. and Wang, S.,** A new type of double capillary nebulizer used in atomic absorption analysis, *Fen Hsi Hua Hsueh,* 9(3), 258, 1981 (in Chinese).

189. **Aldeman, H., Jeniss, S. W., and Katz, S. A.,** Interlaboratory analysis of sewage sludge, *Am. Lab.,* 31, Dec. 1981.

190. **Kahn, H. L., Cristiano, L. G., Dulude, G. R., and Sotera, J. J.,** Automated hydride analysis, *Am. Lab.,* 136, 138, 141, 1981.

191. **Blakemore, W. M. and Billedeau, S. M.,** Analysis of laboratory animal feed for toxic and essential elements by atomic absorption and inductively coupled argon plasma emission spectrometry, *J. Assoc. Off. Anal. Chem.,* 64, 1284, 1981.

192. **Neve, J., Hanocq, M., and Molle, L.,** Atomic absorption spectrophotometric determination of ultramicro amounts of selenium (IV) and total selenium in sulfuric acid, *Fresenius' Z. Anal. Chem.,* 308, 448, 1981.

193. **Fukushima, I.,** A Study on Monitoring Methodology for Trace Elements in the Environment Through Chemical Analysis of Hair. Part of a Coordinated Program on Nuclear Methods in Health-Related Monitoring of Trace Element Pollutants, Rep. IAEA-R-2480/F, International Atomic Energy Agency, Vienna, 1981; INIS Atomindex 12, Abstr. No. 626438, 1981.

194. **Holen, B., Bye, R., and Lund, W.,** Determination of selenium in technical sulfuric acid by flame atomic absorption spectrometry after electrothermal preconcentration, *Anal. Chim. Acta,* 131, 37, 1981.

195. **Berndt, H. and Messerschmidt, J.,** Microanalytical method of flame atomic absorption and emission spectrometry (platinum loop method). II. Determination of arsenic, bismuth, cadmium, mercury, lead, antimony, selenium, tellurium, thallium, zinc, lithium, sodium, potassium, cesium, strontium and barium, *Spectrochim. Acta,* 36b, 809, 1981.

196. **Brown, R. M., Fry, R. C., Moyers, J. L., Northway, S. J., Denton, M. B., and Wilson, G. S.,** Interference by volatile nitrogen oxides and transition-metal catalysts in the preconcentration of arsenic and selenium as hydrides, *Anal. Chem.,* 53, 1560, 1981.

197. **DeCarlo, E. H., Zeitlin, H., and Fernando, Q.,** Simultaneous separation of trace level of germanium, antimony, arsenic and selenium from an acid matrix by absorbing colloid flotation, *Anal. Chem.,* 53, 1104, 1981.

198. **Futekov, L., Parichkova, R., and Specker, H.,** Methylmethacrylate as extractant and solvent in flame AAS. Determination of selenium in industrial products, *Z. Anal. Chem.,* 306, 378, 1981.

199. **Pleban, P. A., Kerkay, J., and Pearson, K. H.,** Cadmium, copper, lead, manganese and selenium levels and glutathione peroxidase activity in human kidney cortex, *Anal. Lett.,* 14b, 1089, 1981.

200. **Ikeda, M., Nishibe, J., and Nakahara, T.,** Study of some fundamental conditions for the determination of antimony, tin, selenium and bismuth by continuous hydride generation-flameless atomic absorption spectrometry, *Bunseki Kagaku,* 30, 545, 1981.

201. **Jastrow, J. D., Zimmerman, C. A., Dvorak, A. J., and Hinchman, R. R.,** Plant growth and trace element uptake on acidic coal refuse amended with lime or fly ash, *Environ. Qual.,* 10, 154, 1981.

202. **Sanzolone, R. F. and Chao, T. T.,** Matrix modification with silver for the electrothermal atomization of arsenic and selenium, *Anal. Chim. Acta,* 128, 225, 1981.

203. **Saeed, K. and Thomassen, Y.,** Spectral interferences from phosphate matrixes in the determination of arsenic, antimony, selenium and tellurium by electrothermal atomic absorption spectrometry, *Anal. Chim. Acta,* 130, 281, 1981.

204. **Sanzolone, R. F. and Chao, T. T.,** Determination of submicrogram amounts of selenium in geological materials by atomic absorption spectrophotometry with electrothermal atomization after solvent extraction, *Analyst (London),* 106, 647, 1981.

205. **Subramanian, K. S.,** Rapid electrothermal atomic absorption method for arsenic and selenium in geological materials via hydride evolution, *Z. Anal. Chem.,* 305, 382, 1981.

206. **Welz, B. and Melcher, M.,** Determination of antimony, arsenic, bismuth, selenium, tellurium and tin in metallurgical samples using the hydride AA technique. I. Analysis of low alloy steels, *Spectrochim. Acta,* 36b, 439, 1981.

207. **Young, R. S.,** Analysis of nickel refinery slimes and residues, *Talanta,* 28, 25, 1981.

208. **Irgolic, K. J., Banks, C. H., Bottino, N. R., Chakraborti, D., Gennity, J. M., Hillman, D. C., O'Brien, D. H., Pyles, R. A., Stockton, R. A., et al.,** Analytical and biochemical aspects of the transformation of arsenic and selenium compounds into biomolecules, *NBS Spec. Publ.,* 618, 244, 1981.

209. **Bansho, K., Aoki, T., and Umezaki, Y.,** Determination of antimony, and selenium by atomic absorption spectrometry via hydride generation using sodium tetrahydroborate pellets, *Kogai Shigen Kenkyusho Iho,* 11, 37, 1981 (in Japanese).

210. **Berndt, H. and Messerschmidt, J.,** A microanalytical method for flame atomic absorption and emission spectrometry (platinum loop method). II. Determination of arsenic, bismuth, cadmium, mercury, lead, antimony, selenium, tellurium, thallium, zinc, lithium, sodium, potassium, cesium, strontium and barium, *Spectrochim. Acta,* 36b, 809, 1981.

211. **Watson, C. A.,** Development and testing of hydride generation methods for antimony and selenium in organic matter. The work of the metallic impurities in organic matter subcommittee of the Analytical Method Committee, *Anal. Proc. (London),* 18, 482, 1981.

213. **Karring, M., Pohjanvirta, R., Rahko, T., and Korpela, H.,** The influence of dietary molybdenum and copper supplementation on the contents of serum uric acid and some trace elements in cocks, *Acta Vet. Scand.,* 22, 289, 1981.

214. **Schweiger, E.,** Current status of selenium determination according to different methods, *Hydrochem. Hydrogeol. Mitt.,* 4, 247, 1981.

215. **Gunn, A. M.,** The Determination of Arsenic and Selenium in Raw and Potable Waters by Hydride Generation/Atomic Absorption Spectrometry — a Review, Tech. Rep. 169, Res. Cent. Medmenham Laboratory, Medmenham, Marlow, Bucks, England, 1981.

216. **Ondov, J. M. and Biermann, A. H.,** Effects of Particle-Control Devices on Atmospheric Emissions of Minor and Trace Elements from Coal Combustion, EPA-600/9-80-039d, U.S. Environmental Protection Agency, Office of Research and Development, 1980, Symp. Transfer Util. Part. Control. Technol. 2nd., PB81-122228, Corvallis, Ore., 1981, 454.

217. **Boyer, K. W., Jones, J. W., Linscott, D., Wright, S. K., Stroube, W., and Cunningham, W.,** Trace element levels in tissues from cattle fed a sewage sludge-amended diet, *J. Toxicol. Environ. Health,* 8, 281, 1981.

218. **Massee, R. and Maeesen, F. J. M. J.,** Losses of silver, arsenic, cadmium, selenium and zinc traces from distilled water and artificial seawater by sorption on various container surfaces, *Anal. Chim. Acta,* 127, 181, 1981.

219. **Schnitger, H. and Lieck, G.,** Determination of selenium in forage samples by hydride generation atomic absorption spectrometry, *Landwirtsch. Forsch.,* 34, 1, 1981 (in German).

220. **Varo, P. and Koivistoinen, P.,** Intercalibration of selenium analysis, *Kem. Kemi,* 8, 238, 1981.

221. **Vijayakumar, M., Ramakrishna, T. V., and Aravamudan, G.,** Determination of trace quantities of selenium by indirect atomic absorption spectrophotometry, *Talanta,* 29, 61, 1982.

222. **Subramanian, K. S. and Meranger, J. C.,** Rapid hydride evolution-electrothermal atomization atomic absorption spectrophotometric method for determining arsenic and selenium in human kidney and liver, *Analyst (London),* 107, 157, 1982.

223. **Xiao-quan, S., Long-zhu, J., and Zhe-ming, N.,** Determination of selenium in soil digests by graphite furnace atomic absorption spectrometry after extraction with 1,2-diamino-4-nitrobenzene, *At. Spectrosc.,* 3, 41, 1982.

224. **Schierling, P., Oefele, C., and Schaller, K. H.,** Determination of arsenic and selenium in urine samples by hydride AAS technique, *Aerztl. Lab.,* 28, 21, 1982 (in German).

225. **Roden, D. R. and Tallman, D. E.,** Determination of inorganic selenium species in groundwaters containing organic interferences by ion chromatography and hydride generation/atomic absorption spectrometry, *Anal. Chem.,* 54, 307, 1982.

226. **Pleban, P. A., Munyani, A., and Beachum, J.,** Determination of selenium concentrations and glutathione peroxidase activity in plasma and erythrocytes, *Clin. Chem. (Winston-Salem, N.C.),* 28, 311, 1982.

227. **Oyamada, N. and Ishizuki, M.,** Determination of trimethyl selenium ion and total selenium in human urine by graphite furnace atomic absorption spectrometry, *Bunseki Kagaku,* 31, 17, 1982.

228. **Jiang, S., De Jonghe, W., and Adams, F.,** Determination of alkyl selenide compounds in air by gas chromatography-atomic absorption spectrometry, *Anal. Chim. Acta,* 136, 183, 1982.

229. **Ebdon, L., Wilkinson, J. R., and Jackson, K. W.,** Simple and sensitive continuous hydride generation system for the determination of arsenic and selenium by atomic absorption and atomic fluorescence spectrometry, *Anal. Chim. Acta,* 136, 191, 1982.

230. **Dillon, L. J., Hilderbrand, D. C., and Groon, K. C.,** Flameless atomic absorption determination of selenium in human blood, *At. Spectrosc.,* 3, 5, 1982.

231. **Khalighie, J., Ure, A. M., and West, T. S.,** Atom-trapping absorption spectrometry with water-cooled metal collector tubes, *Anal. Chim. Acta,* 134, 271, 1982.

232. **Nadkarni, R. A.,** Application of hydride generation-atomic absorption spectrometry to coal analysis, *Anal. Chim. Acta,* 135, 363, 1982.

233. **Heinrichs, H. and Keltsch, H.,** Determination of arsenic, bismuth, cadmium, selenium and thallium by atomic absorption spectrometry with a volatilization technique, *Anal. Chem.,* 54, 1211, 1982.

234. **Tam, G. K. H. and Lacroix, G.,** Dry ashing, hydride generation atomic absorption spectrometric determination of arsenic and selenium in foods, *J. Assoc. Off. Anal. Chem.,* 65, 647, 1982.

235. **Abdulla, M., Norden, A., Schersten, B., Svenson, S., Thulin, T., and Oeckerman, P. A.,** The intake and urinary excretion of electrolytes and trace elements, in *Trace Element Metab. Man Anim.,* Proc. 4th Int. Symp., 1981, Gawthorne, J. M., Howell, J. M., and White, C. L., Eds., Springer Verlag, Berlin, 1982, 81.

AUTHOR INDEX

SUBJECT INDEX(SELENIUM)

TELLURIUM, Te (ATOMIC WEIGHT 127.06)

Tellurium is found in native form and is more often found as the telluride of gold and other metals. It is used in the glass and ceramic industry, as a secondary vulcanizing agent for rubber, in stainless steel to improve machinibility, in lead to decrease sulfuric acid corrosion, in alloys, and in semiconductors.

Tellurium determination is required in metallurgy, geological samples, biological samples, coal, environmental samples, and other industrial products. It is toxic and is considered to be a health hazard. Proper precautions should be taken to handle tellurium and its compounds.

Sample preparation for chemical analysis requires extra care due to volatilization loss. Covered acid digestion at low temperature will dissolve the sample except in the case of refractory material, where fusion is required.

A very large excess of calcium, copper, silicon, sodium, zinc, and zirconium affect the tellurium signal, otherwise no severe interferences have been reported. At low tellurium concentrations, dissolved acids can depress the tellurium signal. Use of background correction will reduce this effect. Matrix matching is recommended for best results. Improved sensitivity can be achieved with the hydride generation method.

Standard Solution

To prepare 1000 mg/ℓ of tellurium solution, dissolve 1.000 g of tellurium metal in 1:1:1 nitric acid, hydrochloric acid, and water mixture and dilute it to 1 ℓ with deionized water.

Instrumental Parameters

Wavelength	214.3 nm
Slit	0.4 nm
Bandpass	0.5 nm
Light source	Hollow cathode lamp
Flame	Air-acetylene, oxidizing (lean, blue)
Sensitivity	0.2—0.4 mg/ℓ
Detection limit	0.02 mg/ℓ
Optimum range	25.0 mg/ℓ

Secondary wavelengths and their sensitivities

225.9 nm	3.0—4.0 mg/ℓ
238.6 nm	18.0—28.0 mg/ℓ

REFERENCES

1. **Sprague, S., Manning, D. C., and Slavin, W.,** Determination of selenium and tellurium in copper by atomic absorption spectrophotometry, *At. Absorpt. Newsl.,* 20, 1964.
2. **Mulford, C. E.,** Solvent extraction techniques for atomic absorption spectroscopy, *At. Absorpt. Newsl.,* 5, 88, 1966.
3. **Sattur, T. W.,** Routine atomic absorption analyses on non-ferrous alloys and plant intermediates, *At. Absorpt. Newsl.,* 5, 37, 1966.
4. **Chakrabarti, C. L.,** The atomic absorption spectroscopy of tellurium, *Anal. Chim. Acta,* 39, 293, 1967.
5. **Marcec, M. V., Kinson, K., and Belcher, C. B.,** The determination of minor amounts of tellurium in iron and steel by atomic absorption spectrophotometry, *Anal. Chim. Acta,* 41, 447, 1968.
6. **Roth, D. J., Bohl, D. R., and Sellers, D. E.,** Determination of trace quantities of tellurium in bismuth by atomic absorption spectrophotometry, *At. Absorpt. Newsl.,* 7, 87, 1968.
7. **Schroeder, H. A.,** Effect of selenate, selenite and tellurite on the growth and early survival of mice and rats, *J. Nutr.,* 92, 334, 1968.
8. **Wu, J. Y. L., Droll, H. A., and Loft, P. F.,** Determination of tellurium by atomic absorption spectrophotometry, *At. Absorpt. Newsl.,* 7, 90, 1968.
9. **Musha, S., Munemori, M., and Nakahara, T.,** Atomic absorption spectrophotometric determination of tellurium, *J. Chem. Soc. Jpn. Pure Chem. Sect.,* 89, 495, 1968.
10. **Nakagawa, H. M. and Thompson, C. E.,** Atomic absorption determination of tellurium, *U.S. Geol. Surv.,* 600b, 123, 1968.
11. **Barnett, W. B. and Kahn, H. L.,** The determination of tellurium in steel by atomic absorption spectroscopy with the deuterium background corrector, *At. Absorpt. Newsl.,* 8, 21, 1969.
12. **Goecke, R. F.,** The determination of tellurium in copper ores, *At. Absorpt. Newsl.,* 8, 106, 1969.
13. **Davis, C. E. S., Ewers, W. E., and Fletcher, A. B.,** Determination of traces of tellurium in rocks by atomic absorption spectroscopy after volatilization with synthetic pyrite, *Proc. Australas. Inst. Min. Metall.,* 232, 67, 1969.
14. **Uny, G., Tardif, J. P., and Spitz, J.,** Determination of tellurium by atomic absorption spectrometry in antimony-gallium semiconductor alloys, *Anal. Chim. Acta,* 54, 91, 1971.
15. **Workman, E. J.,** Determination of mercury, indium and tellurium in mercury indium tellurium, *Analyst (London),* 97, 703, 1972.
16. **Severne, B. C. and Brooks, R. R.,** Rapid determination of selenium and tellurium by atomic absorption spectrophotometry, *Talanta,* 19, 1467, 1972.
17. **Beaty, R. D.,** Atomic absorption determination of nanogram quantities of tellurium using the sampling boat technique, *Anal. Chem.,* 45, 234, 1973.
18. **Jedrezjowska, H.,** Determination of tellurium in copper from vacuum refining and wire quality by atomic absorption spectrophotometry, *Chem. Anal. (Warsaw),* 18, 117, 1973.
19. **Thavornyutikarn, P.,** Tellurium analysis in organo-tellurium compounds by atomic absorption spectroscopy, *J. Organometal. Chem.,* 51, 237, 1973.
20. **Balandina, N. S., Yudelevich, I. G., Khrapai, V. P., and Abakumov, D. N.,** Atomic absorption determination of selenium and tellurium in pure silver, *Izv. Sib. Otd. Akad. Nauk SSSR Ser. Khim. Nauk,* 5, 93, 1973.
21. **Balandina, N. S., Yudelevich, I. G., Khrapai, V. D., and Abakumov, D. N.,** Atomic absorption determination of selenium and tellurium in pure gold, *Zh. Anal. Khim.,* 28, 2054, 1973.
22. **Beaty, R. D.,** Determination of tellurium in rocks by graphite furnace atomic absorption, *At. Absorpt. Newsl.,* 13, 38, 1974.
23. **Supp, G. R. and Gibbs, I.,** The determination of selenium and tellurium in organic accelerators used in the vulcanization process for natural and synthetic rubbers by atomic absorption spectrometry, *At. Absorpt. Newsl.,* 13, 71, 1974.
24. **Thompson, K. C. and Thomerson, D. R.,** Atomic absorption studies on the determination of antimony, arsenic, bismuth, germanium, lead, selenium, tellurium and tin by utilizing the generation of covalent hydrides, *Analyst (London),* 99, 595, 1974.
25. **Van Montfort, F. F. E.,** Trace analysis by microwave excitation of sealed samples. III. Determination of 0.005—25 ng of Te and 0.25—25 ng of Se in 0.5 mℓ of aqueous solution, *Talanta,* 21, 660, 1974.
26. **Welcher, G. G., Kriege, O. H., and Marks, J. Y.,** Direct determination of trace quantities of lead, bismuth, selenium, tellurium and thallium in high temperature alloys by non flame atomic absorption spectrophotometry, *Anal. Chem.,* 46, 1227, 1974.
27. **Dixon, K., Mallet, R. C., and Kocaba, R.,** Determination by atomic absorption spectrophotometry of arsenic, selenium, tellurium, antimony and bismuth in ion exchange resins, *Natl. Inst. Metall. Repub. S. Afr. Rep.,* No. 1689, 1975.

28. **Lockwood, T. H. and Limtiaco, L. P.,** Determination of beryllium, cadmium and tellurium in animal tissues using electronically excited oxygen and atomic absorption spectrophotometry, *Am. Ind. Hyg. Assoc. J.,* 36, 57, 1975.

29. **Watterson, J. R. and Neuerburg, G. J.,** Analysis for tellurium in rocks in concentrations down to 5 parts per billion, *J. Res. U.S. Geol. Surv.,* 3, 191, 1975.

30. **Shelton, B. J.,** Determination of silver, selenium, tellurium, antimony, tin, lead and arsenic in anode sludges, *Natl. Inst. Metall. Repub. S. Afr. Rep.,* No. 1771, 1975.

31. **Cobb, W. D., Foster, W. W., and Harrison, T. S.,** Determination of tellurium in leaded free–cutting steels by atomic absorption spectrometry, *Analyst (London),* 101, 39, 1976.

32. **Kunselman, G. C. and Huff, C. A.,** The determination of arsenic, antimony, selenium and tellurium in environmental water samples by flameless atomic absorption, *At. Absorpt. Newsl.,* 15, 29, 1976.

33. **Nazarenko, I. I., Kalenchuk, G. E., and Kislova, I. V.,** Atomic absorption determination of tellurium in ores after extraction with methyl isobutyl ketone, *Zh. Anal. Khim.,* 31, 498, 1976.

34. **Bedard, M. and Kerbyson, J. D.,** Determination of arsenic, selenium, tellurium and tin in copper by hydride evolution atomic absorption spectrophotometry, *Can. J. Spectrosc.,* 21, 64, 1976.

35. **Danchik, R. S., Thompson, D. E., and Hillegass, H. F.,** Determination of trace amounts of tellurium and selenium in gallium metal, *Anal. Lett.,* 9, 687, 1976.

36. **Hamner, R. M., Lechak, D. L., and Greenberg, P.,** Determination of silver, arsenic, bismuth, antimony, selenium and tellurium in chromium metal with flameless atomic absorption spectroscopy, *At. Absorpt. Newsl.,* 15, 122, 1976.

37. **Sverdlina, O. A., Kovykova, N. V., and Bokova, T. A.,** Determination of tellurium in arsenic (III) chloride by atomic absorption with use of graphite cell, *Zavod. Lab.,* 42, 429, 1976 (in Russian).

38. **Greenland, L. P. and Campbell, E. Y.,** Rapid determination of nanogram amounts of tellurium in silicate rocks, *Anal. Chim. Acta,* 87, 323, 1976.

39. **Marks, J. Y., Welcher, G. G., and Spellman, R. J.,** Atomic absorption determination of lead, bismuth, selenium, tellurium, thallium and tin in complex alloys using direct atomization from metal chips in the graphite furnace, *Appl. Spectrosc.,* 31, 9, 1977.

40. **Smith, R. G., Van Loon, J. C., Knechtel, J. R., Fraser, J. L., Pitts, A. E., and Hodges, A. E.,** A simple and rapid hydride generation atomic absorption method for the determination of arsenic in biological environmental and geological samples, *Anal. Chim. Acta,* 93, 61, 1977.

41. **Dujmovic, M.,** Determination of bismuth, mercury, antimony, tin, tellurium and lead in aqueous solutions by flameless atomic absorption spectroscopy, *GIT Fachz. Lab.,* 21, 861, 1977.

42. **Britske, M. E. and Sedykh, E. M.,** Nonflame atomic absorption determination of tellurium in metallurgical products, *Zh. Anal. Khim.,* 33, 321, 1978.

43. **Chao, T. T., Sanzolone, R. F., and Hubert, A. E.,** Flame and flameless atomic absorption determination of tellurium in geological materials, *Anal. Chim. Acta,* 96, 251, 1978.

44. **Bedrossian, M.,** Determination of microgram amounts of tellurium in steels by atomic absorption spectrometry, *Anal. Chem.,* 50, 1898, 1978.

45. **Haynes, B. W.,** Arsenic, antimony, selenium and tellurium determinations in high-purity copper by electrothermal atomization, *At. Absorpt. Newsl.,* 18, 46, 1979.

46. **Jackwerth, E., Willmer, P. G., Hohn, R., and Berndt, H.,** A simple accessory for the determination of mercury and the hydride forming elements (As, Bi, Sb, Se and Te) using flameless atomic absorption spectroscopy, *At. Absorpt. Newsl.,* 18, 66, 1979.

47. **Kamada, T., Sugita, N., and Yamamoto, Y.,** Differential determination of tellurium (IV) and tellurium (VI) with sodium dithiocarbamate, ammonium pyrrolidinedithiocarbamate and dithizone by atomic absorption spectrophotometry with a carbon-tube atomizer, *Talanta,* 26, 337, 1979.

48. **Palliere, M. and Gernez, G.,** Apparatus for the production of volatile hydrides for arsenic, antimony, bismuth, tin, germanium, selenium and tellurium determination by atomic absorption spectrometry, *Analusis,* 7, 46, 1979.

49. **Sighinolfi, G. P., Santos, A. M., and Martinelli, G.,** Determination of tellurium in geochemical materials by flameless atomic absorption spectroscopy, *Talanta,* 26, 143, 1979.

50. **Epting, M. A., Sweigart, J. R., and Nixon, E. R.,** Absorption and emission spectra of matrix-isolated germanium telluride, *J. Mol. Spectrosc.,* 78, 277, 1979.

51. **Forrester, J. E., Lehecka, V., Johnston, J. R., and Ott, W. L.,** Direct determination of trace quantities of antimony, arsenic, bismuth, cadmium, lead, selenium, silver, tellurium and thallium in high-purity nickel by electrothermal atomic absorption spectrometry, *At. Absorpt. Newsl.,* 18, 73, 1979.

52. **Grainger, F. and Gale, I. G.,** Direct analysis of solid cadmium mercury telluride by flameless atomic absorption using interactive computer processing, *J. Mater. Sci.,* 14, 1370, 1979.

53. **Jin, K., Taga, M., Yoshida, H., and Kime, S. H.,** Differential determination of tellurium (IV) and tellurium (VI) by atomic absorption spectrophotometry after hydride generation. Combined use of titanium (III) chloride as a prereductant and sodium borohydride solution, *Bull. Chem. Soc. Jpn.,* 52, 2276, 1979.

54. **Kamada, T., Sugita, N., and Yamamoto, Y.,** Differential determination of tellurium (IV) and tellurium (VI) with sodium diethyl dithiocarbamate, ammonium pyrrolidine dithiocarbamate and dithizone by atomic absorption spectrophotometry with a carbon tube atomizer, *Talanta,* 26, 337, 1979.

55. **Ohta, K. and Suzuki, M.,** Atomic absorption spectrometry of tellurium with electrothermal atomization in a molybdenum microtube, *Anal. Chim. Acta,* 110, 49, 1979.

56. **Arikawa, Y., Hirai, S., and Ozawa, T.,** Simultaneous determination of selenium and tellurium in native sulfur by atomic absorption spectrophotometry, *Bunseki Kagaku,* 28, 653, 1979.

57. **Boyer, K. W., Capar, S. G., Jones, J. W., Suddendorf, R. F., and Forwalter, J.,** Multielement/trace analysis identifies 48 elements in foods. Simultaneous element analysis and data reduction reaches parts per billion, *Food Process.,* 40, 72, 1979.

58. **Nakahara, T. and Musha, S.,** Determination of tellurium in copper refinery slimes by atomic absorption spectrometry with premixed inert gas (entrained air)-hydrogen flames, *Can. J. Spectrosc.,* 24, 138, 1979.

59. **Welz, B. and Melcher, M.,** Hydride atomic absorption spectroscopic technique in trace analysis, *Labor Praxis,* 3, 41, 1979 (in German).

60. **Price, W. J.,** New techniques of atomic absorption in clinical and biochemical analysis, *Sci. Ind.,* 12, 22, 1979.

61. **Sedykh, E. M., Belyaev, Y. I., and Sorokina, E. V.,** Matrix effect during electrothermal atomic absorption determination of silver, tellurium, lead, cobalt, nickel in materials of complex composition, *Zh. Anal. Khim.,* 36, 2162, 1980 (in Russian).

62. **Sedykh, E. M., Belyaev, Y. I., and Sorokina, E. V.,** Elimination of matrix effects in electrothermal atomic absorption determination of Ag, Pb, Co, Ni and Te in samples of complicated composition, *Zh. Anal. Khim.,* 35, 2348, 1980 (in Russian).

63. **Robert, R. V. D., Balaes, G. E., and Steele, T. W.,** Study of the measurement by electrothermal atomization and atomic absorption spectrophotometry of hydride forming elements, *Natl. Inst. Metall. Repub. S. Afr. Rep.,* No. 2053, 1980.

64. **Tews, H.,** Zeeman splitting of ground and excited acceptor states in zinc telluride, *Solid State Commun.,* 34, 611, 1980.

65. **Katskov, D. A., Grinshtein, I. L., and Kruglikova, L. P.,** Study of the evaporation of the metals indium, gallium, thallium, germanium, tin, lead, antimony, bismuth, selenium and tellurium from a graphite surface by the atomic absorption method, *Zh. Prikl. Spektrosk.,* 33, 804, 1980.

66. **Sinemus, H. W., Melcher, M., and Welz, B.,** Influence of valence state on the determination of antimony, arsenic, bismuth, selenium and tellurium in lake water using the hydride AA technique, *At. Spectrosc.,* 2, 81, 1981.

67. **Baker, A. A. and Headridge, J. B.,** Determination of bismuth, lead and tellurium in copper by atomic absorption spectrometry with introduction of solid samples into an induction furnace, *Anal. Chim. Acta,* 125, 93, 1981.

68. **Clark, J. R. and Viets, J. G.,** Multielement extraction system for the determination of 18 trace elements in geochemical samples, *Anal. Chem.,* 53, 61, 1981.

69. **Clark, J. R. and Viets, J. G.,** Back-extraction of trace elements from organometallic-halide extracts for determination by flameless atomic absorption spectrometry, *Anal. Chem.,* 53, 65, 1981.

70. **Welz, B., Grobenski, Z., and Melcher, M.,** Determination of antimony, arsenic, selenium, tellurium, bismuth and tin using the hydride-AAS technique, *Spektrometertagung (Votr.),* 13, 337, 1981.

71. **Guo, Y. and Wong, S.,** A new type of double capillary nebulizer used in atomic absorption analysis, *Fen Hsi Hua Hsueh,* 9(3), 258, 1981.

72. **Shan, X. and Ni, Z.,** Matrix modification for the differential determination of the tellurium (IV) and tellurium (VI) in water samples by graphite furnace atomic absorption spectrometry, *Huanjing Kexue Xuebao,* 1(1), 74, 1981.

73. **Tsukahara, I. and Yamamoto, T.,** Determination of tellurium in copper, lead and selenium by atomic absorption spectrometry after extraction of the trioctylmethylammonium-tellurium bromide complex, *Talanta,* 28, 585, 1981.

74. **Suzuki, M., Ohta, K., Yamakita, T., and Katsuno, T.,** Electrothermal atomization with a metal microtube in atomic absorption spectrometry, *Spectrochim. Acta,* 36b, 679, 1981.

75. **Aihara, M. and Kiboku, M.,** Extraction and atomic absorption spectrophotometric determination of indium and tellurium by using potassium xanthates, *Bunseki Kagaku,* 30, 295, 1981.

76. **Berndt, H. and Messerschmidt, J.,** Microanalytical method of flame atomic absorption and emission spectrometry (platinum loop method). II. Determination of As, Bi, Cd, Hg, Pb, Sb, Se, Te, Tl, Zn, Li, Na, K, Ce, Sr and Ba, *Spectrochim. Acta,* 36b, 809, 1981.

77. **Ikeda, M., Jiro, N., and Nakahara, T.,** Investigation of some fundamental conditions for the determination of arsenic, lead and tellurium with continuous hydride-generation atomic absorption spectrometry, *Bunseki Kagaku,* 30, 368, 1981.

78. **Saeed, K. and Thomassen, Y.,** Spectral interferences from phosphate matrixes in the determination of arsenic, antimony, selenium and tellurium by electrothermal atomic absorption spectrometry, *Anal. Chim. Acta,* 130, 281, 1981.
79. **Weibust, G., Langmyhr, F. J., and Thomassen, Y.,** Thermal stabilization of inorganic and organically bound tellurium for electrothermal atomic absorption spectrometry, *Anal. Chim. Acta,* 128, 23, 1981.
80. **Welz, B. and Melcher, M.,** Determination of antimony, arsenic, bismuth, selenium, tellurium and tin in metallurgical samples using the hydride AA technique. I. Analysis of low alloy steels, *Spectrochim. Acta,* 36b, 439, 1981.
81. **Young, R. S.,** Analysis of nickel refinery slimes and residues, *Talanta,* 28, 25, 1981.
82. **Mohr, F. and Luft, B.,** Determination of low tellurium contents in steels by using hydride atomic absorption, *Nene Huette,* 26, 431, 1981 (in German).
83. **Xiao-Wei, G. and Sheng-Zhang, W.,** New type of double capillary nebulizer for atomic absorption spectrometry (for determining hydride-forming elements), *Fen Hsi Hua Hsueh,* 9, 258, 1981 (in Chinese).
84. **Xiao-quan, S. and Zho-ming, N.,** Matrix modification for the differential determination of tellurium (IV) and tellurium (VI) in water samples by graphite furnace atomic absorption spectrometry, *Huanjing Kexue Xuebao,* 1, 74, 1981 (in Chinese).
85. **Nadkarni, R. A.,** Application of hydride generation-atomic absorption spectrometry to coal analysis, *Anal. Chim. Acta,* 135, 363, 1982.
86. **Kujirai, O., Kobayashi, T., Ide, K., and Sudo, E.,** Determination of traces of tellurium in heat-resisting alloys by graphite furnace atomic absorption spectrometry after coprecipitation with arsenic, *Talanta,* 29, 27, 1982.

AUTHOR INDEX

SUBJECT INDEX (TELLURIUM)

Group VIIA Elements — F, Cl, Br, and I

HALOGENS

FLUORINE, F (ATOMIC WEIGHT 18.9984)

Fluorine usually occurs in combined form in rocks and minerals. It is the most electronegative and reactive of all elements. Fluorine and its compounds are used in the production of uranium and other fluorochemicals. Its compounds are extensively used as metal cleaning agents in electrodeposition, in the etching of glass, insecticides, air conditioning, refrigeration, corrosion-resistant fluids and components, and in aerosol propellants. It is also used as a rocket propellant. Fluorine is used in drinking water supplies and toothpastes to prevent tooth decay. Its evaluation is required in minerals for the production of hydrofluoric acid, fluoridation of water supplies, plants, bones, and teeth, etc.

Solution preparation is very important due to the volatility of silicon tetrafluoride, loss of hydrofluoric acid in evaporation, and in the presence of calcium, iron, and aluminum due to precipitation as the fluorides. Some of the fluorides are water soluble and for hard-to-dissolve samples potassium or sodium carbonate fusion is used. Atomic absorption method is used as an indirect technique to determine fluorine as a fluoride.

CHLORINE, Cl (ATOMIC WEIGHT 35.453)

Chlorine occurs in nature, usually combined with sodium, potassium, and magnesium in rocks, minerals, quartz-bearing ores, sea water, mineral springs, and ground water. It is extensively used as a bleach, catalyst, in the extraction of gold from ores, chemical industries, manufacture of hydrochloric acid and other compounds, paper products, dyestuffs, textiles, petroleum products, medicine, antiseptics, germicides, insecticides, foodstuff, solvents, paints, and plastics, etc.

Most of the chlorides are easily soluble in water except lead, silver, and mercurous chlorides. The insoluble chlorides can be dissolved by alkali fusion. Atomic absorption spectroscopy is an indirect method to determine chlorine as its chloride.

BROMINE, Br (ATOMIC WEIGHT 79.904)

Bromine occurs generally as bromides of alkali and alkaline earth elements. It is the only liquid nonmetallic element. It is used in flame-proofing agents, water purification compounds, dyes, medicine, photography, sanitizers, fumigants, and antiknocking gasoline fluid. It is also used as an oxidizing agent. Its evaluation is required in the above-mentioned industries. Atomic absorption is a successful technique as an indirect method of analysis of bromine.

IODINE, I (ATOMIC WEIGHT 126.9045)

Iodine is found in a few rare minerals and sea water. Its compounds are important in organic chemistry. It is frequently employed in biological investigations. Its radioisotope is used for the treatment of thyroid glands. Iodine is used in medicine and photography, as well as in chemical analysis of other elements as an analytical reagent.

Direct determination of iodine has been attempted by some workers using an electrodeless discharge lamp. The principle resonance line of iodine lies at a wavelength less than 200 nm. Nonflame atomic absorption method has been more successful than the flame method. No significant chemical or physical interferences have been observed in the presence of a 5-times excess of Cu, Co, Al, Ni, Mg, Mo, K, Na, and Zn and 100-fold excess of fluoride, chloride, bromide, nitrate, phosphate, and sulfate ions. At 183.0 nm with a nitrogen-purged nitrous oxide-acetylene flame, a detection limit of 25 mg/ℓ has been reported. With the use of a graphite furnace, a detection limit of 0.4 mg/ℓ has been achieved for iodine determination.

REFERENCES

1. **Westerland-Helmerson, U.**, The determination of chloride as silver chloride by atomic absorption spectroscopy, *At. Absorpt. Newsl.*, 5, 97, 1966.
2. **Bartels, H.**, An indirect determination of serum chloride by atomic absorption methods, *At. Absorpt. Newsl.*, 6, 132, 1967.
3. **Ezell, J. B., Jr.**, Atomic absorption analysis of chloride in plant liquors, *At. Absorpt. Newsl.*, 6, 84, 1967.
4. **Bond, A. M. and O'Donnell, T. A.**, Determination of fluoride by atomic absorption spectrometry, *Anal. Chem.*, 40, 560, 1968.
5. **Dahms, H., Rock, R., and Seligson, D.**, Ionic activities of sodium, potassium and chloride in human serum, *Clin. Chem. (Winston-Salem, N.C.)*, 14, 859, 1968.
6. **Reichel, W. and Acs, L.**, Determination of chlorine in selenium by a distillation-atomic absorption procedure, *Anal. Chem.*, 41, 1886, 1969.
7. **Sinha, R. C. P., Singhal, K. C., and Banerjee, B. K.**, Estimation of chloride and sulfate in ammonium bicarbonate by atomic absorption spectrophotometry, *Technology*, 6, 198, 1969.
8. **Kumamaru, T.**, Determination of a small amount of iodide by atomic absorption spectrophotometry using a cadmium hollow cathode lamp, *Bull. Chem. Soc. Jpn.*, 42, 956, 1969.
9. **L'vov, B. V. and Khartsyzov, A. D.**, Atomic absorption determination of iodine in a graphite cuvet, *Zh. Anal. Khim.*, 24, 799, 1969.
10. **L'vov, B. V. and Khartyszov, A. D.**, Atomic absorption determination of sulfur, phosphorus, iodine, and mercury from resonance lines in the vacuum ultraviolet spectral region, *Zh. Prikl. Spektrosk.*, 11, 413, 1969.
11. **Thompson, K. C.**, The feasibility of determining iodine directly by atomic absorption and atomic fluorescence spectroscopy, *Spectrosc. Lett.*, 3, 59, 1970.
12. **Fujinuma, H., Kasama, K., Takeuchi, K., and Hirano, S.**, Indirect determination of traces of chloride by atomic absorption spectroscopy, *Bunseki Kagaku*, 19, 1487, 1970.
13. **Truscott, E. D.**, Determination of chlorine in a polyvinyl chloride matrix using the Schoeniger oxygen flask and absorption spectrometry, *Anal. Chem.*, 42, 1657, 1970.
14. **Deakin, J. J., Husain, D., and Wiesenfeld, J. R.**, Kinetic study of electronically excited iodine atoms, atomic iodine (5P [5] [2] P [1/2]), by time resolved atomic absorption atomic iodine (5P [4] 6 S [2] P [3/2]) from 5 P (5) ([2] P [1/2] [0] at lambda = 206.23 nm = 146.50, *Chem. Phys. Lett.*, 10, 146, 1971.
15. **Garrido, M. D., Llagund, C., and Garrido, J.**, Determination of chloride in wine by atomic absorption spectrophotometry, *Am. J. Enol. Viticult.*, 22, 44, 1971.
16. **Gutsche, B. and Herrmann, R.**, Determination of iodine in urine by flameless spectrometry, *Naunyn-Schmiedebergs Arch. Pharmacol.*, 270, 94, 1971.
17. **Tofuku, Y. and Hirand, S.**, Microdetermination of chloride by atomic absorption spectrophotometric titration, *Bunseki Kagaku*, 20, 142, 1971 (in Japanese).
18. **Yamamoto, Y., Kumamaru, T., Hayashi, Y., and Matsushita, S.**, An indirect atomic absorption spectrophotometric determination of iodide combined with solvent extraction of TRIS (1,10 phananthroline)-iron (II) cations, *J. Sci. Hiroshima Univ. Sec. A*, 35, 141, 1971.
19. **Sevcik, J.**, Selective detection of sulfur, chlorine and nitrogen with help of the combination of flame ionization and flame photometric detectors, *Chromatographia*, 4, 195, 1971.
20. **Gutsche, B. and Herrmann, R.**, Flame spectra of fluorine compounds, *Z. Anal. Chem.*, 258, 277, 1972.
21. **Jowsey, J., Johnson, W., Taves, D. R., and Kelly, P. J.**, Effects of dialysate calcium and fluoride on bone disease during regular hemodialysis, *J. Lab. Clin. Med.*, 79, 204, 1972.
22. **Kirkbright, G. F., West, T. S., and Wilson, P. J.**, The direct determination of iodine by atomic absorption spectroscopy in a nitrogen separated nitrous oxide-acetylene flame, *At. Absorpt. Newsl.*, 11, 53, 1972.
23. **Gambrell, J. W.**, The use of an atomic absorption spectrophotometer for end-point determination: application to chloride in waters, *At. Absorpt. Newsl.*, 11, 125, 1972.
24. **Kirkbright, G. F., West, T. S., and Wilson, P. J.**, The application of the Leipert amplification to increase sensitivity in the direct determination of iodine by atomic absorption spectrometry, *At. Absorpt. Newsl.*, 11, 113, 1972.
25. **Talmi, Y. and Morrison, G. H.**, Use of induction furnace source for spectrometric determination of iodine, *Anal. Chem.*, 1467, 1972.
26. **Tomkins, D. F. and Frank, C. W.**, Atomic absorption determination of chloride utilizing the Beilstein test, *Anal. Chem.*, 44, 1451, 1972.
27. **Belcher, R., Nadjafi, A., Rodriguez-Vazquez, J. A., and Stephen, W. I.**, The determination of chloride by atomic absorption spectrophotometry, *Analyst (London)*, 97, 993, 1972.
28. **Sadykhov, I. D. and Zeinalov, A. Y.**, Use of a flame photometric method for automatic determination of chloride salt content in petroleum, *Neftepererab. Neftekhim. (Moscow)*, 1972, 12, 1972.

29. **Smith, R. V. and Nessen, M. A.,** Analysis of chloride in amine hydrochlorides and quarternary ammonium chlorides by atomic absorption spectroscopy, *Microchem. J.,* 17, 638, 1972.

30. **de Alnedia, M. A. T. M., de Moraes, S., and Barberio, J. C.,** Indirect determination of iodine in organic compounds by atomic absorption spectrophotometry, *Relat. Inst. Energia Atom.,* IEA-285, 1973 (in Portuguese).

31. **Ametani, K.,** Estimation of chlorine content in single crystals of chromium chalcogenides spinels by atomic absorption spectrophotometry, *Bull. Chem. Soc. Jpn.,* 47, 242, 1974.

32. **Gutsche, B. and Herrmann, R.,** Flame spectrometric fluorine analysis by the strontium fluoride bands, *Z. Anal. Chem.,* 269, 260, 1974.

33. **Kirkbright, G. F., West, T. S., and Wilson, P. J.,** Determination of iodine by atomic absorption and emission spectrometry with a cathode sputtering cell, *Anal. Chim. Acta,* 68, 462, 1974.

34. **Adams, M. J., Kirkbright, G. F., and West, T. S.,** Direct determination of nanogram amounts of iodine by atomic absorption spectroscopy using a graphite tube atomizer, *Talanta,* 21, 573, 1974.

35. **Kirkbright, G. F. and Wilson, P. J.,** Application of a demountable hollow cathode lamp as a source for the direct determination of sulfur, iodine, arsenic, selenium and mercury by atomic absorption flame spectrometry, *Anal. Chem.,* 46, 1414, 1974.

36. **Kirkbright, G. F. and Wilson, P. J.,** The direct determination of iodine by atomic absorption spectrometry with the graphite furnace, *At. Absorpt. Newsl.,* 13, 140, 1974.

37. **Mansfield, J. M., West, T. S., and Dagnall, R. M.,** Determination of iodine by atomic absorption spectrometry using the platinum loop technique, *Talanta,* 21, 787, 1974.

38. **Gutsche, B., Kleinoeder, H., and Herrmann, R.,** Device for trace analysis for fluorine in reaction tubes by atomic absorption spectroscopy, *Analyst (London),* 100, 192, 1975.

39. **Tsukui, H. and Togashi, Y.,** Indirect determination of fluorine in a coating on low-hydrogen type arc welding electrodes and raw materials by separating precipitated calcium fluoride and atomic absorption spectrophotometry, *Tetsu To Hagane,* 61, 388, 1975.

40. **Becker-Ross, H. and Falk, H.,** Atomic absorption spectroscopy in the vacuum-UV range, Determination of bromine. *Spectrochim. Acta,* 30b, 253, 1975.

41. **Gutsche, B., Rudiger, K., and Herrmann, R.,** A method for the determination of fluoride concentration with the help of atomic absorption, *Spectrochim. Acta,* 30b, 441, 1975.

42. **Mosescu, N., Kalmutchi, G., and Badea, S.,** Iodine determination by flame photometry in some catalyst types, *Rev. Chim.,* 26, 677, 1975.

43. **Gil de la Pena, M. L. and Garrido, M. D.,** Determination of chloride and sodium in vinegar by atomic absorption spectrophotometry, *An. Bromatol.,* 27, 269, 1975.

44. **Kidani, Y. and Ito, E.,** Indirect determination of fluorine in organic compounds by atomic absorption spectrometry, *Bunseki Kagaku,* 25, 57, 1976.

45. **Tellinghuisen, J. and Clyne, M. A.,** Role of hyperfine structure in atomic absorption. Oscillator strengths in bromine and iodine atoms, *J. Chem. Soc. Faraday Trans.,* 2, 783, 1976.

46. **Fong, C. C. and Huber, C. O.,** Fluoride determination by magnesium atomic absorption inhibition-release effects, *Spectrochim. Acta,* 31b, 113, 1976.

47. **Gold'shtein, M. M. and Yudelevich, I. G.,** Indirect atomic absorption methods of determining fluorine, chlorine, iodine, phosphorus and sulfur, *Zh. Anal. Khim.,* 31, 801, 1976.

48. **Gray, D. R., Baker, H. J., and King, T. A.,** Atomic absorption of thermally dissociated iodine for laser applications, *J. Phys. D,* 10, 169, 1977.

49. **Tsunoda, K., Fujiwara, K., and Fuwa, K.,** Subnanogram fluorine determination by aluminum monofluoride molecular absorption spectrometry, *Anal. Chem.,* 49, 2035, 1977.

50. **Yoshimura, E., Tanaka, Y., Tsunoda, K., Toda, S., and Fuwa, K.,** Determination of chlorine by the indium monochloride absorption spectrum in carbon tube atomizer, *Bunseki Kagaku,* 26, 647, 1977.

51. **McHugh, J. B. and Turner, J. H.,** Indirect determination of chloride in plants by atomic absorption spectrophotometry, *U.S. Geol. Surv. Pap.,* 1129-A-1, F1, 1980.

52. **Baroccio, A. and Moauro, A.,** Determination of Some Elements in Vegetal Samples by Instrumental Neutron Activation. Comparison with the Data Obtained from Atomic Absorption, CNEN-RT/CHI 8015, Comitato Nazionale Energia Nucleare, Casaccio Nuclear Studies Center, Rome, Italy, 1980 (in Italian).

53. **Benos, D. J.,** Intracellular analysis of sodium, potassium and chloride in mouse erythrocytes, *J. Cell. Physiol.,* 105, 185, 1980.

54. **Wandiga, S. O.,** The concentrations of Zn, Cu, Pb, Mn, Ni and fluoride in rivers and lakes of Kenya, *Sinet,* 3, 67, 1981.

55. **Chuchalina, L. S., Yudelevich, I. G., and Chinenkova, A. A.,** Indirect atomic absorption method for determination of microgram amounts of halide ions. Determination of chloride, bromide and iodide ions in solution when one halide is present, *Zh. Anal. Khim.,* 36, 920, 1981.

56. **Nomura, T. and Karasawa, I.,** Determination of micromolar concentrations of iodide by electrothermal atomic absorption spectrometry, *Anal. Chim. Acta,* 126, 241, 1981.

57. **Boyer, K. W., Jones, J. W., Linscott, D., Wright, S. K., Stroube, W., and Cunningham, W.,** Trace element levels in tissues from cattle fed a sewage sludge-amended diet, *J. Toxicol. Environ. Health,* 8, 281, 1981.

58. **Lowe, M. D., Sutton, M. M., and Clinton, O. E.,** Capillary discharge lamp for the direct determination of iodine by flame atomic absorption spectroscopy, *Appl. Spectrosc.,* 36, 22, 1982.

59. **Ondon, J. M. and Biermann, A. H.,** Effects of Particle-Control Devices on Atmospheric Emissions of Minor and Trace Elements from Coal Combustion, 600/9-80-039d, U.S. Environmental Protection Agency, Office of Research and Development, Symp. Transfer Util. Part. Control. Technol. 2nd, PB 81-122228, Corvallis, Ore., 454, 1980; CA 96,90897g, 1982.

.AUTHOR INDEX

SUBJECT INDEX

Zero Group Elements — He, Ne, Ar, Kr, and Xe

ZERO GROUP OR INERT GASES OR NOBLE GASES

The distribution of inert gases or so-called noble gases in nature is determined entirely by nuclear reactions and their physical properties. These gases, helium, neon, argon, krypton, xenon, and radon, completely lack the chemical reactivity.

A number of binary mixtures of noble gases have been analyzed by the atomic absorption techniques. A detailed attempt to identify spectral lines for all the noble gases has been made by Goleb[1-3] (see reference at the end of this chapter). Determination of neon and argon by an indirect method has also been described by the same author. A few references are available in relation to the atomic absorption method applied to the noble analyses.

HELIUM, He (ATOMIC WEIGHT, 4.003)

Helium is found in natural gas from some wells, volcanic gases, etc. It is second to hydrogen in occurrence in the universe, but it is rare on earth. It is widely used in artificial air mixtures for deep-sea divers and others working under pressure. Helium is best used in weather balloons, cryogenic research, arc welding, as a protective gas in growing silicon and germanium crystals, in titanium and zirconium production, as a cooling medium for nuclear reactors, and as a gas for supersonic wind tunnels. Helium is the only liquid which does not solidify by the decrease in temperature.

NEON, Ne (ATOMIC WEIGHT 20.183)

Neon is present in the atmosphere to the extent of 0.002%. A very small amount is found as occluded gas in rocks. Its largest use is in making neon advertising signs, high-voltage indicators, lightning arrestors, wave meter tubes, television tubes, and in cryogenics as an economical refrigerant.

ARGON, Ar (ATOMIC WEIGHT 39.948)

Argon is the most abundant of the noble gases in the atmosphere, comprising approximately 0.94% by volume of dry air. It is used in electric light bulbs, fluorescent tubes, photo tubes, glow tubes, as an inert gas shield for arc welding and cutting, as a blanket for the production of titanium and other reactive elements, and in growing silicon and germanium crystals.

KRYPTON, Kr (ATOMIC WEIGHT 83.80)

Krypton is present in the atmosphere to the extent of about one part per million. It is used as a gas fill for small bright lamps, bactericidal lamp-starter tube, and special electron tubes. It is also used in electron etching and ion engines.

XENON, Xe (ATOMIC WEIGHT 131.30)

Xenon is obtained from the atmosphere which contains 1 part to 20 million. It is also found in gases evolved from mineral springs. It is used to prepare stroboscopic lamps, bactericidal lamps, and lamps used to excite lasers. It is also used in bubble chambers and probes in the atomic energy field and as a gas for ion engines. Xenon is also used as an experimental surgical anesthetic. Some of its compounds are used as strong oxidizing agents in the laboratory.

REFERENCES

1. **Goleb, J. A.,** Near ultraviolet-visible atomic absorption spectra of the noble gases, *Anal. Chem.,* 38, 1059, 1966.
2. **Goleb, J. A.,** The determination of neon and argon in helium by atomic absorption spectrophotometry, *Anal. Chim. Acta,* 41, 229, 1968.
3. **Goleb, J. A.,** Use of an alternating electromagnetic field to modulate light in an electrodeless tube for the determination of neon and argon in helium by atomic absorption, *Anal. Chim. Acta,* 51, 343, 1970.
4. **Bernat, M., Bieri, R. H., Koide, M., Goldberg, E. D., and Griffin, J. J.,** Uranium, thorium, potassium and argon in marine phillipsites, *Geochim. Cosmochim. Acta,* 34, 1053, 1970.
5. **McElfresh, P. H. and Parsons, M. L.,** Some new observations on the atomic absorption of argon and neon in helium by atomic absorption, *At. Absorpt. Newsl.,* 11, 69, 1972.

AUTHOR INDEX

SUBJECT INDEX (ZERO GROUP)

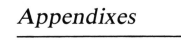

Appendixes

Appendix I
GLOSSARY OF ATOMIC ABSORPTION TERMS AND DEFINITIONS

Absorbance, A
Logarithm to base 10 of the reciprocal of the transmittance

$$A = \log_{10}(1/T)$$

Absorbance scale expansion
The ratio of the magnitude of the scale reading to the actual absorbance
Absorption cells
A simple tube system in which the flame from a total consumption burner is directed in at
   one end and the combustion products emerge from the other
Absorption coefficient, $k\nu$
The absorption coefficient at a discrete frequency, ν, is defined by

$$I_\nu = I_\nu^0 \, e^{-k\nu L}$$

where I_ν^0 and I_ν are the initial and final intensities of radiation of frequency ν passed through
   an absorption cell of length L
Absorptivity, a
Absorbance divided by the product of concentration (c, in g/ℓ) times the path length (b in
   cm)

$$a = A/bc$$

Note: Absorptivity is a constant to normalize the units and is experimentally determined.
Accuracy
The measure of the agreement between a measured value and the value expected as "true"
Angstrom, Å
Unit of length equal to 1/6438.4694 of the wavelength of the red line of cadmium (for
   practical purposes, it is considered equal to 10^{-8} cm)
Atomic absorption spectroscopy, AAS or AA
An analytical method for the determination of elements, based on the absorption of radiation
   by free atoms
Atomic vapor
A vapor that contains free atoms of the analysis element
Atomization
The process that converts the analysis element, or its compounds, to an atomic vapor
Atomizer
The device, usually a flame, used to produce and stabilize or maintain, a population of free
   atoms
Background absorption
It is an interference due to light scattering by particles in the flame and molecular absorption
   of light from the lamp by molecules in the flame
Bandpass
The width of the observed peak in wavelength units between the two points where the
   transmitted intensity is equal to half that at the maximum
Beer's law
Absorptivity of a substance is constant with respect to changes in concentration

$$I = I_o 10^{-abc}$$

where,

I = transmitted radiation power

I_o = incident radiant power

a = absorptivity

b = path length

c = concentration of the absorbing species in the analyzing beam

Blank

That which contains no analyte element

Boat sampling

Boat sampling is based on the principal of evaporating the sample completely and quickly in the flame in order to record a high narrow absorption peak

Bohr's equation

Transition between two quantized states corresponds to the absorption or emission of energy in the form of electromagnetic radiation, the frequency v of which is determined by

$$\Delta E = E_1 - E_2 = hv$$

where E_1 and E_2 are the energies in the initial and final states, respectively, and h is Planck's constant

Boosted output lamp

A lamp in which a secondary discharge is used to increase the emission of characteristics absorbed radiation (this increase is normally both absolute and relative to other radiation from the lamp)

Burner angle

The acute angle between the plane of the flame produced by a long path burner and the optical axis of the monochromator

Calibration

The relationship between the absorption indicated by the instrument and the concentration of the element that produces it is established in the calibration curve

Carrier gas

The gas used to convey the sample mist to the atomizer

Centrifugal nebulizer

A nebulizer in which the sample solution is brought into contact with a rapidly spinning disk

Characteristic absorbed radiation

Radiation that is specifically absorbed by free atoms of the analysis element

Chemical interference

If the sample being analyzed contains a thermally stable compound with the analyte that is not totally decomposed by the energy of the flame, a chemical interference exists and as a result the number of atoms in the flame capable of absorbing light is reduced.

Concentration

The quantity of a substance contained in a unit quantity of sample (the units should always be stated explicitly, usually $\mu g/m\ell$ or mg/ℓ)

Continuum spectra

The energy is distributed in an uninterrupted manner between all wavelengths within a given domain

Continuum source

A continuum source emits light over a broad spectrum of wavelengths instead of at specific lines, e.g., deuterium arc lamp in UV or tungsten-iodide lamp for visible wavelengths

Curvature corrector

A device or factor that alters the relationship between the signal and the readout as a function of signal, such that there is a linear response between the readout and the desired physical measurement, e.g., concentration

Depressions

An interference that causes a decreased instrument response

Detection limit

The detection limit is defined as the concentration of an element that will produce an absorbance equal to twice the SD of a series of measurement of a solution, the concentration of which is distinctly detectable above the base line

$$\text{Detection limit} = \frac{\text{standard concentration} \times 2 \text{ standard deviations}}{\text{mean}}$$

Deviation from Beer's law

The deviation from Beer's law is largely due to the result of unabsorbed and unabsorbable light (stray light, nonhomogeneities of temperature and space in the absorption cell, line broadening, and absorption at nearby lines) reaching the detector

Diffraction grating

Light is deflected by a pattern of fine slits at different angles according to its wavelength

Direct injection burner

A burner in which liquid is nebulized directly into the flame (the flame obtained with such a burner is normally turbulent)

Double-beam spectrometer

The light from the source lamp is divided into a sample beam, which is focused through the sample cell, and a reference beam, which is directed around the sample cell, and the readout represents the ratio of the sample and the reference beams

Double-channel spectrometers

These allow simultaneous measurements at two wavelengths either in absorption or emission and require two separate optical systems — primary source, monochromator, and detector

Electrical damping

It reduces the noise shown by a meter or on a recorder trace. Degree of damping is usually 0.2 to 2 sec with a meter and 1 to 4 sec with a recorder

Electrodeless discharge tube (EDT or EDL)

A tube containing the element to be determined in a readily vaporized form and constructed so as to enable a discharge to be induced in the vapor (this discharge can be used as a source of characteristic radiation)

Emission interference

At high analyte concentrations, the atomic absorption analysis for highly emissive elements sometimes exhibits poor analytical precision, if the emission signal falls within the spectral bandpass being used.

Enhancement

An interference that causes an increased instrument response

Flame noise

It is caused by the refractive index variations in the region between the hot parts of the flame and the cold surrounding atmosphere and from small variations in effective path length of the flame cell

Flow spoiler

A device, in a spray chamber, for removing large droplets from a mist

Fluctuational concentration limit

Same as ''detection limit''

Fluctuational sensitivity

Same as "detection limit"

Fuel

The substance, usually a gas, that is burned to provide the atomizing flame

Grating

See "diffraction grating".

Hollow cathode lamp (HCL)

A discharge lamp with a hollow cathode, usually cylindrical, used in atomic spectroscopy
to provide characteristic radiation

Interference

A general term for an effect that modifies the instrumental response to a particular concentration of the analysis element

Ionization buffer

A spectroscopic buffer used to minimize or stabilize the ionization of free atoms of the
analysis element

Ionization interference

It occurs when the flame temperature has enough energy to cause the removal of an electron
from the atom, creating an ion. As these electronic rearrangements deplete the number
of ground state atoms, absorption is reduced.

Laminar flow burner

In laminar flow burner the direction of flow is generally perpendicular to the flame front.
It sustains a highly stable flame and the proportion of the sample mist normally introduced
has little or no influence over the flame characteristics except when organic solvents are
employed or the volume ratio of liquid sample and combustion gases exceeds a critical
ratio of about 1:5000.

Limit of detection

The minimum concentration or amount of an element that can be detected with 95% certainty,
assuming a normal distribution of errors.

Long path burner

A burner constructed to produce a flame that is extended in one direction at right angles to
the direction of movement of the flame gases

Long tube device

A device in which an atomizing flame is directed into a tube lying along the optical axis of
an atomic absorption spectrometer

Matrix effect

An interference caused by differences between the sample and a standard containing only
the analysis element and, where appropriate, a solvent

Matrix matching

To match the standards and samples with respect to the materials that are present in excess
of 1% in the total solution

Mean, $\overline{X}$

A statistical term for the central tendency of a group or series of data (the mean is defined
as the sum of all of the individual test results divided by the number of data)

$$\overline{X} = (X_i)/N$$

Mean error

An accuracy reference that is defined as the difference between the true or accepted value
and mean average value resulting from a series of determinations

Micrometer (formerly micron)

Unit of length equal to 10^{-6} m

Multislot burner

A burner head that contains several parallel slots

Nanometer (formerly millimicron)

Unit of length equal to 10^{-9} m, almost, but not exactly, equal to 10 Å

Nebulization

The process that converts a liquid to a mist

Nebulization efficiency

The ratio of the amount of sample reaching the atomizer to the total amount of sample entering the nebulizer

Nebulizer

A device for the nebulization of a liquid

Noise level

The noise level of an atomic absorption spectrometer is defined as that concentration of the analysis element that would give a signal equal to 1/50 of the sum of 20 measurements. The 20 measurements are made as follows. The output of an atomic absorption spectrometer operating on an a blank solution is recorded for ten time periods each of ten times the constant of the instrument. The maximum displacements that occur to both sides of the median line in each of the ten periods are measured.

Nonabsorbing line

A line from the conventional hollow-cathode lamp that lies closer to the resonance line, but which is not absorbed by the sample atom.

Nonspecific absorption

This effect is due to the presence of dried and semidried salt particles in the flame that scatter or absorb the incident radiation from the source.

Observation height

The vertical distance between the optical axis of the monochromator and the top of the burner

Oxidant

The substance, usually a gas, used to oxidize the fuel in a flame.

Oxidizing flame

Fuel-lean flame (it is the hottest flame and contains enough oxidant to obtain a clear blue flame)

Pneumatic nebulizer

A type of nebulizer in which the liquid sample is introduced into a high-velocity jet of the oxidant (or inert gas) employed in the combustible mixture used to form the flame

Precision

The measure of the agreement among test results as measured in terms of the SD or relative SD

Premix system

A sampling unit in which the fuel, oxidant gas, and sample mist are mixed in a spray chamber before entering the flame. Flames obtained using this system are normally laminar.

Radiation generator

Equipment for producing characteristic absorbed radiation (the equipment normally consists of a lamp and power supply unit)

Radiation scattering

An interference effect caused by scattering of radiation from drops or particles associated with the atomic vapor

Range

A statistical term signifying the difference between the highest and lowest values in a series of test results

Reciprocal linear dispersion (RLD)

The spectral range per unit distance at the focal plane of the spectral dispersing instrument (RLD is preferably designated as nm/mm)

Reducing flame

Fuel-rich flame (it is coolest in temperature)

Relative error

The mean error expressed as a percentage of the true result

Relative standard deviation (RSD)

A statistical term defined as the SD of a series of test results expressed as a percentage of the mean of this series

Releasing agent

A spectroscopic buffer used to reduce interferences attributable to the formation of involatile compounds in the atomizer

Repeatability

A statistical term to measure the index of precision of a single aspect of a complete analytical scheme (normally, repeatability is the SD of a series of test results without varying the operator or apparatus)

Resonance radiation

Characteristic absorbed radiation that corresponds to the transfer of an electron from the ground state level to a higher energy level in the atom

Sampling unit

The part of an atomic absorption spectrometer that accepts the sample solution and prepares it for atomization

Scale expansion

In atomic absorption instruments, scale expansion simply involves an electrical expansion of the presented signal by a chosen factor, e.g., 2, 5, 10, or 20. Scale expansion is valuable when the noise level is 0.5% absorption (0.002 absorbance units) or less and facilitates the reading of small scale deflexions.

Scatter

Scatter is the result of the presence of small solid particles in the resonance beam. These solid particles may be caused by the inability of the flame to vaporize a high dissolved solid content of the sample solution or may be due to the formation of particles, e.g., carbon in the flame.

Semipremix system

A sampling unit in which either fuel or oxidant gas is added to the sample mist after the spray chamber

Sensitivity

It is defined as that concentration in solution of the analysis element which will produce a change, compared to pure solvent, of 0.0044 absorbance units (i.e., 1% absorption) in the optical transmission of the atomic vapor at the wavelength of the radiation used.

$$\text{Sensitivity} = \frac{\text{concentration of standard} \times 0.0044}{\text{measured absorbance}}$$

Sensitivity check

The sensitivity check (in mg/ℓ) value is the concentration of an element that will produce a signal of approximately 0.2 absorbance units under optimum conditions at the wavelength listed.

Separated flame

A flame in which the diffusion combustion zone is so separated from the primary combustion zone as to enable the two zones to be observed independently (separation may be effected either mechanically or by an inert gas shield)

Spectral bandwidth, $\Delta\lambda$

The wavelength or frequency interval of radiation leaving the exit slit of a monochromator between limits set at a radiant power level half way between the continuous background and the peak of an emission line or an absorption band of negligible intrinsic width

Spectral interference

When an absorbing wavelength of an element present in the sample, but not being determined falls within the bandwidth of the absorption line of the element of interest, causes the spectral interference

Spectral resolution

The ratio, $\lambda/\Delta\lambda$, where λ is the wavelength of the region being examined and $\Delta\lambda$ is the spectral bandwidth (resolution can also be defined as v/v, where v and v refer to the wave number scale)

Spectrometer

An optical instrument with an entrance slit, a dispersing device, and with one or more exit slits with which measurements are made at selected wavelengths within the spectral range or by scanning over the range (The quantity detected is a function of radiant power.)

Spectrophotometer

A spectrometer with associated equipment. It furnishes the ratio, or a function of the ratio, of two radiant power measurements as a function of spectral wavelength. These two measurements may be separated in time, space, or both.

Spectroscopic buffer

A substance that is part of, or is added to, a sample and which reduces interference effects

Spray chamber

The vessel wherein a mist is generated prior to transfer to an atomizer

Standard deviation, σ

A statistical term indicating the distribution of test results in a Gaussian distribution of a series of results (it is defined as the square root of the variance)

Stoichiometric flame

It is characterized by tinges of yellow in the flame.

Stray radiation energy

All radiant energy that reaches and is sensed by the detector at wavelengths that do not correspond to the spectral energy under consideration

Time constant

An index designating the period ($t_1 - t_0$) required for a damped, dynamic system to decay to $1/e$ of the original value, i.e., at time t_0

Transmittance, T

The ratio of the radiant power transmitted by a sample to the radiant power incident on the sample

Variance, σ^2

The variance is defined by the equation,

$$\sigma^2 = \frac{\Sigma[\overline{X} - X_i]^2}{N}$$

$$s^2 = \frac{\Sigma[\overline{X} - X_i]^2}{N - 1}$$

where $\overline{X}$ = mean

X_i = value of ith measurement

N = number of measurements

Wavelength

The distance, measured along the line of propagation, between two points that are in phase on adjacent waves

Appendix II
ABSTRACTS AND REVIEWS

1. **Milazzo, G.,** Atomic absorption (a review), *Chim. Ind. 44*, 493, 1962 (in Italian).
2. **Johansson, A.,** Atomic absorption spectrophotometry (a review), *Svensk. Kem. Tidskr.*, 74, 415, 1962 (in Swedish).
3. **Gilbert, P. T.,** Atomic absorption spectroscopy. A review of recent developments, in Proc. 6th Conf. Anal. Chem. Nuclear Reactor Technology, AECTID-7655, Gatlinburg, October 1962, 1963.
4. **Slavin, W.,** Atomic absorption instrumentation and technique. A review, in *Analysis Instrumentation,* Fowler, L., Rose, D. K., and Harmon, R. G., Eds., Plenum Press, New York, 1964.
5. **Slavin, W.,** Atomic absorption spectroscopy — a critical review, *Appl. Spectrosc.*, 20, 281, 1966.
6. **West, T. S.,** Atomic analysis in flames, *Endeavour,* 44, 1967.
7. **Reynolds, R. J.,** Atomic absorption spectroscopy, its principles, applications and future development, *Lab. Equip. Dig.*, 6, 79, 1968.
8. **Slavin, W. and Slavin, S.,** Recent trends in analytical atomic absorption spectroscopy, *Appl. Spectrosc.*, 23, 421, 1969.
9. **West, T. S.,** Atomic flame spectroscopy in trace analysis. I. Atomic absorption spectroscopy, *Miner. Sci. Eng.*, 1, 3, 1969.
10. **Kirkbright, G. F.,** Application of non-flame atom cells in atomic absorption and atomic fluorescence spectroscopy. A review, *Analyst (London),* 96, 609, 1971.
11. **Platt, P.,** *Annual Reviews in Analytical Chemistry,* SAC, London, 1971.
12. **Amos, M. D.,** Nonflame atomization in AAS — a current review, *Am. Lab.*, 4, 57, 1972.
13. **Stone, R. G. and Warren, J.,** Atomic absorption spectrometry: a review of modern instrumentation and techniques, *Lab. Equip. Dig.*, 12, 49, 1973.
14. **Woodriff, R.,** Atomization chambers for atomic absorption spectrochemical analysis. Review, *Appl. Spectrosc.*, 28, 413, 1974.
15. **Kharlamov, I. P. and Eremina, G. V.,** Use of an atomic absorption method for analyzing steels and alloys. Review of research published in 1966- 1971, *Zavod. Lab.*, 40, 385, 1974.
16. **Winefordner, J. D., Fitzgerald, J. J., and Omenetto, N.,** Review of multielement atomic spectroscopic methods, *Appl. Spectrosc.*, 29, 369, 1975.
17. **Aidarov, T. K.,** Status of developments and some trends in developing atomic absorption spectrometer instrumentation. A review, *Zh. Prikl. Spektrosk.*, 26, 779, 1977.
18. **Willis, J. B.,** Review: analytical atomic spectroscopy at the CSIRO, Division of Chemical Physics, *Anal. Chim. Acta,* 106, 175, 1979.
19. *Atomic Absorption Newsletter,* Perkin-Elmer Corporation, Norwalk, Conn., up to 1979.
20. **Blake, C. J.,** Sample preparation methods for the analysis of metals in foods by atomic absorption spectrometry. A literature review, *Sci. Tech. Surv. Br. Food Manuf. Ind. Res. Assoc.*, 122, 57, 1980.
21. **Young, E. F.,** A review of the spectrophotometer, *Opt. Spectra,* 14, 44, 1980.
22. **Kirkbright, G. F.,** Current status and future needs in atomic absorption instrumentation, *Anal. Chem.*, 52, 736a, 1980.
23. **Marcus, L. H.,** Atomic Absorption Spectroscopy, Rep. NERACUSGNT 0 56; Order No. PB 80-859614, 1980; available from the NTIS, Springfield, Va.; *Govt. Rep. Announce. Index (U.S.),* 80(26), 5637, 1980.
24. **Berndt, H. and Messerschmidt, J.,** Loop atomic absorption/emission spectrometry. A brief review, *Aerztl. Lab.*, 28, 133, 1981.
25. **Cresser, M. S. and Sharp, B. L.,** Annual reports on analytical atomic spectroscopy, *Reviewing,* 10, 1, 1981.
26. *Atomic Spectroscopy,* bibliography published twice a year, Perkin-Elmer Corporation, Norwalk, Conn., up to December 1982.
27. Chemical Abstracts, American Chemical Society, up to 1982.
28. CA Selects, Atomic Spectroscopy and Trace Element Analysis, American Chemical Society, up to 1982.
29. *Analytical Chemistry,* annual reviews, American Chemical Society.
30. *Analytical Abstracts,* Society for Analytical Chemistry, London, monthly publication.
31. **Masek, P. R., Sutherland, I., and Grivell, S.,** *Atomic Absorption* and *Flame Emission Spectroscopy Abstracts,* Sci. & Tech. Agency, London, bimonthly publications.
32. Annual Reports on Analytical Atomic Spectroscopy, S.A.C. London, from 1972 forward.
33. **Horlick, G.,** Atomic absorption, atomic fluorescence and flame spectrometry, *Anal. Chem.*, 54, 276r, 1982.
34. **Baranov, S. V., Baranova, I. V., and Ivanov, N. P.,** Spectral lamps for atomic absorption spectrometry. Review, *Zh. Prikl. Spektrosk.*, 36, 357, 1982.

35. **Komarek, J. and Sommer, L.,** Organic complexing agents in atomic absorption spectrometry. A review, *Talanta,* 29, 159, 1982.
36. **Koirtyohann, S. R. and Kaiser, M. L.,** Furnace atomic absorption — a method approaching maturity, *Anal. Chem.,* 54, 1515a, 1982.

Appendix III
REVIEWS

1. **Menzies, A. C.,** Trends in automatic spectrochemical analysis, *Acta Suppl.,* 106, 1957.
2. **Malmstadt, H. V.,** Atomic absorption spectrochemical analysis, in *Encyclopedia of Spectroscopy,* Clark, G. L., Ed., Reinhold, New York, 1960.
3. **David, D. J.,** The application of atomic absorption to chemical analysis (a review), *Analyst (London),* 85, 779, 1960.
4. **Menzies, A. C.,** A study of atomic absorption spectroscopy, *Anal. Chem.,* 32, 898, 1960.
5. **Robinson, J. W.,** Atomic absorption spectroscopy, *Anal. Chem.,* 32, 17a, 1960.
6. **Poluektov, N. S.,** Atomic absorption flame photometry, *Zavod. Lab.,* 7, 830, 1961.
7. **Robinson, J. W.,** Recent advances in atomic absorption spectroscopy, *Anal. Chem.,* 33, 1067, 1961.
8. **Walsh, A.,** Application of atomic absorption spectra to chemical analysis, in *Advances in Spectroscopy,* Vol. 2, Thompson, H. W., Ed., Interscience, New York, 1961.
9. **Robinson, J. W.,** Flame photometry and atomic absorption spectroscopy, in *Analytical Chemistry,* Crouthamel, C. E., Ed., Pergamon Press, 1961.
10. **Leithe, W.,** Absorption flame photometry in analytical chemistry, *Angew. Chem.,* 73, 488, 1961.
11. **Butler, L. R. P.,** Atomic absorption spectroscopy, *S. Afr. Ind. Chem.,* 15, 162, 1961.
12. **David, D. J.,** Emission and atomic absorption spectrochemical methods, in *Modern Methods of Plant Analysis,* Vol. 5, Eds. Peach and Tracy, Springer-Verlag, Berlin, 1962.
13. **David, D. J.,** Atomic absorption spectrochemical analysis with particular reference to plant analysis, *Rev. Univ. Ind. Santander,* 4, 207, 1962 (in Spanish).
14. **Franswa, C. E. M.,** Atomic absorption spectrophotometry, *Chem. Weekbl.,* 58, 177, 189, 1962 (in Dutch).
15. **Johansson, A.,** Atomic absorption spectrophotometry, *Svensk. Kem. Tidskr.,* 74, 415, 1962.
16. **L'vov, B. V.,** Theory and method of atomic absorption analysis, *Zavod. Lab.,* 28, 931, 1962 (in Russian).
17. **Milazzo, G.,** Atomic absorption, *Chim. Ind. (Milan),* 44, 493, 1962 (in Italian).
18. **Allan, J. E.,** A review of recent work in atomic absorption spectroscopy, *Spectrochim. Acta,* 18, 605, 1962.
19. **Kahn, H. L. and Slavin, W.,** Atomic absorption analysis, *Int. Sci. Technol.,* 1962.
20. **Robinson, J. W.,** Atomic absorption spectroscopy, *Ind. Chem.,* 38, 226, 362, 1962.
21. **Tabeling, R. W. and Devany, J. J.,** Factors influencing sensitivity in atomic absorption, in *Developments in Applied Spectroscopy,* Vol. 1, Ashby, W. D., Ed., Plenum Press, New York, 1962.
22. **Willis, J. B.,** Atomic absorption spectroscopy, in Proc. Royal Australian Chemical Institute, July-September 1962.
23. **Walsh, A.,** Atomic absorption spectroscopy, in *Proc. 10th Coll. Spectroscopium Internationale,* Lippincott, E. R. and Margoshes, M., Eds., Spartan Books, Rochelle Park, N.J., 1963.
24. **Willis, J. B.,** Analysis of biological materials by atomic absorption spectroscopy, in *Methods of Biochemical Analysis,* Vol. 11, Glick, D., Ed., Interscience, New York, 1963.
25. **Gilbert, P. T.,** Atomic absorption spectroscopy: a review of recent developments, in Proc. 6th Conf. Anal. Chem. Nuclear Reactor Technology, AEC TID-7655, Gatlinburg, 1963.
26. **David, D. J.,** Recent developments in atomic absorption analysis, *Spectrochim. Acta,* 20, 1185, 1964.
27. **Slavin, W.,** Quantitative metal analysis by atomic absorption spectrophotometry, *Chim. Ind.,* 46, 60, 1964 (review in Italian).
28. **Slavin, W.,** Atomic absorption instrumentation and technique. A review, in *Analysis Instrumentation,* Fowler, L., Roe, D. K., and Harmon, R. G., Eds., Plenum Press, 1964; *At. Absorpt. Newsl.,* No. 24, September 1964.
29. **Lockyer, R.,** Atomic absorption spectroscopy, in *Advances in Analytical Chemistry and Instrumentation,* Reilley, C. N., Ed., Interscience, New York, 1964.
30. **Poluektov, N. S. and Zelyukova, Y. V.,** Absorption flame photometry, *Zavod. Lab.,* 30, 33, 1964.
31. **Robinson, J. W.,** The future of atomic absorption spectroscopy, in *Developments in Applied Spectroscopy,* Vol. 4, Plenum Press, New York, 1965.
32. **Ulrich, W. F. and Shifrin, N.,** Atomic absorption spectrophotometry, *Analyzer,* 6, 10, 1965.
33. **Slavin, W.,** The application of atomic absorption spectroscopy to analytical, biochemistry and toxicology, *Occup. Health Rev.,* 17, 9, 1965 (in French and English).
34. **Slavin, W.,** The application of atomic absorption spectroscopy to geochemical prospecting and mining, *At. Absorpt. Newsl.,* 4, 243, 1965.
35. **Zettner, A.,** Principles and applications of atomic absorption spectroscopy, in *Advances in Clinical Chemistry,* Vol. 7, Sobotka, H., Ed., Academic Press, New York, 1965.
36. **Herrmann, R.,** Principles and applications of atomic absorption spectroscopy in flames, *Z. Klin. Chem.,* 6, 178, 1965 (in German).
37. **Rubeska, I. and Velicka, I.,** Atomic absorption spectrophotometry, *Chem. Listy,* 59, 769, 1965.

38. **Schleser, F. H.,** Atomic absorption spectrophotometry, *Z. Instrumentenkd.,* 73, 25, 1965 (in German).

39. **Adams, P. B.,** Flame and atomic absorption spectrometry, in *Standard Methods of Chemical Analysis, Instrumental Analysis,* Vol. 3B, 6th ed., Welcher, F. J., Ed., D Van Nostrand, Princeton, N.J., 1966.

40. **Slavin, W.,** Recent developments in analytical atomic absorption spectroscopy ·*At. Absorpt. Newsl.,* 5, 42, 1966.

41. **Walsh, A.,** Some recent advances in atomic absorption spectroscopy, *J. N.Z. Inst. Chem.,* 30, 7, 1966.

42. **Walsh, A. and Willis, J. B.,** Atomic absorption spectrometry, in *Standard Methods of Instrumental Methods,* Vol. 3, Part A, Welcher, F. J., Ed., D Van Nostrand, Princeton, N.J., 1966.

43. **Suzuki, M. and Takeuchi, T.,** Recent advances in atomic absorption spectroscopic analysis, *Bunseki Kagaku,* 15, 1003, 1966 (in Japanese).

44. **West, T. S.,** Inorganic trace analysis, *Chem. Ind. (London),* 25, 1005, 1966.

45. **Weberling, R. P. and Cosgrove, J. F.,** Atomic absorption spectroscopy, in *Trace Analysis,* Morrison, G. H., Ed., Interscience, New York, 1966.

46. **Koirtyohann, S. R.,** Recent developments in atomic absorption and flame emission spectroscopy, *At. Absorpt. Newsl.,* 6, 77, 1967.

47. **Slavin, W.,** Atomic absorption: recent developments and applications, *Chim. Ind.,* 49, 60, 1967 (in Italian).

48. **West, T. S.,** Atomic analysis in flames, *Endeavor,* 26, 44, 1967.

49. **Allaire, R. F., Brachett, F. P., and Shafer, J. T.,** Analytical benefits of atomic absorption spectrophotometric methods, *J. Soc. Motion Pict. Television Eng.,* Oct. 1967.

50. **Angino, E. E. and Billings, G. K.,** *Atomic Absorption Spectrometry in Geology,* Elsevier, Amsterdam, 1967.

51. **Walsh, A.,** Atomic absorption spectroscopy, *Aust. Phys.,* 4, 185, 1967.

52. **Willis, J. B.,** Recent advances in atomic absorption spectroscopy, *Rev. Pure Appl. Chem.,* 17, 111, 1967.

53. **Panday, V. K. and Ganguly, A. K.,** A short review on the determination of elements by atomic absorption spectrophotometry, *Trans. Bose Res. Inst. Calcutta,* 30, 131, 1967.

54. **Bellod, R. P.,** Atomic absorption flame photometry, *Inform. Quim. Anal. (Madrid),* 14, 137, 1967.

55. **Kahn, H. L.,** Principles and practice of atomic absorption, in *Trace Inorganics in Water,* Baker, R. A., Ed., Advances in Chemistry Series 73, American Chemical Society, Washington, D.C., 1968.

56. **Lewis, L. L.,** Atomic absorption spectrometry — applications and problems, *Anal. Chem.,* 40, 28a, 1968.

57. **Scholes, P. H.,** The application of atomic absorption spectrophotometry to the analysis of iron and steel, *Analyst (London),* 93, 197, 1968.

58. **Walsh, A.,** Atomic absorption spectroscopy: a foreword, *Appl. Opt.,* 7, 1259, 1968.

59. **Beamish, F. E., Lewis, C. L., and Van Loon, J. C.,** A critical review of atomic absorption spectrochemical and X-ray fluorescence methods for the determination of noble metals. II, *Talanta,* 16, 1, 1969.

60. **Slavin, W. and Slavin, S.,** Recent trends in analytical atomic absorption spectroscopy, *Appl. Spectrosc.,* 23, 421, 1969.

61. **Walsh, A.,** Physical aspects of atomic absorption, *ASTM STP,* 443, 3, 1969.

62. **Britske, M. E.,** Atomic absorption spectrophotometric analysis, *Zavod. Lab.,* 35, 1329, 1969.

63. **West, T. S.,** Flame emission and atomic absorption spectrometry, *Chem. Ind.,* 21, 387, 1970.

64. **L'vov, B. V.,** Progress in atomic absorption spectroscopy employing flame and graphite cuvette techniques, *Pure Appl. Chem.,* 23, 11, 1970.

65. **Pinta, M.,** New prospects in atomic absorption, *Meth. Physe. Analyse,* 6, 368, 1970 (in French).

66. **Mallett, R. C.,** Review of techniques for the determination of gold and silver by atomic absorption spectroscopy, *Miner. Sci. Eng.,* 2, 28, 1970.

67. **Winefordner, J. D., Svoboda, V., and Cline, L. J.,** A critical comparison of atomic emission, atomic absorption and atomic fluorescence flame spectrometry, *CRC Crit. Rev. Anal. Chem.,* 12, 233, 1970.

68. **Ando, A.,** Flame and atomic absorption spectrochemical analysis, *Bunseki Kagaku,* 20, 112, 1971.

69. **L'vov, B. V.,** Modern state and main problems of atomic absorption analysis, *Zh. Anal. Khim.,* 26, 590, 1971.

70. **Yudelevich, I. G., Shelpakova, I. R., Brusentsev, F. A., and Zayakina, S. B.,** Dispersion analysis of chemical spectrographic flame photometric and atomic absorption method for determining trace impurities, *Zh. Anal. Khim.,* 26, 2075, 1971.

71. **Kirkbright, G. F.,** Application of non-flame atom cells in atomic absorption and atomic fluorescence spectroscopy. A review, *Analyst (London),* 96, 609, 1971.

72. **Slavin, S.,** An atomic absorption bibliography for 1971, *At. Absorpt. Newsl.,* 11, 7, 1972.

73. **Levine, S. L.,** Atomic absorption spectrophotometry, *Chem. Technol.,* 110, 1972.

74. **Amos, M. D.,** Nonflame atomization in AAS. A current review, *Am. Lab.,* 4, 57, 1972.

75. **DeWaele, M. and Droeven, G.,** Atomic absorption spectrometry, *Ind. Chim. Belge,* 37, 322, 1972.

76. **Slavin, S.,** An atomic absorption bibliography for January-June 1972, *At. Absorpt. Newsl.,* 11, 74, 1972.

77. **Slavin, S.,** An atomic absorption bibliography for July-December 1972, *At. Absorpt. Newsl.,* 12, 9, 1973.

78. **Belcher, R.,** New methods for the determination of elements in trace amounts, *Z. Anal. Chem.,* 263, 257, 1973.

79. **Busch, K. W. and Morrison, G. H.,** Multielement flame spectroscopy, *Anal. Chem.,* 45, 712a, 1973.
80. **Buttgereit, G.,** Modified atomic absorption spectroscopic methods for trace metal analyses, *Z. Anal. Chem.,* 267, 81, 1973.
81. **Barnard, A. J., Jr. and Dudley, R. W.,** Tracing the elements, *Ind. Res.,* March 1973.
82. **Stone, R. G. and Warren, J.,** Atomic absorption spectrometry: a critical review of modern instrumentation and techniques, *Lab. Equip. Dig.,* 12, 49, 1973.
83. **Horncastle, D. C. J.,** Atomic absorption spectrophotometry, *Med. Sci. Law,* 13, 3, 1973.
84. **Sen Gupta, J. G.,** A review of the methods for the determination of the platinum group metals, silver and gold by atomic absorption spectroscopy, *Miner. Sci. Eng.,* 5, 207, 1973.
85. **Slavin, S.,** An atomic absorption bibliography for January-June 1973, *At. Absorpt. Newsl.,* 12, 77, 1973.
86. **Lagesson, V.,** Trace metal determination by atomic absorption spectrometry, *Kem. Tidskr.,* 85, 66, 1973.
87. **Reif, I., Fassel, V. A., and Kniseley, R. N.,** Spectroscopic flame temperature measurements and their physical significance. I. Theoretical concepts — a critical review, *Spectrochim. Acta,* 28b, 105, 1973.
88. **Slavin, S.,** An atomic absorption bibliography for July-December 1973, *At. Absorpt. Newsl.,* 13, 11, 1974.
89. **Thomerson, D. R. and Thompson, K. C.,** Recent developments in atomic absorption spectrometry, *Am. Lab.,* 6, 53, 1974.
90. **Winefordner, J. D. and Vickers, T. J.,** Flame spectrometry, *Anal. Chem.,* 46, 192r, 1974.
91. **Lisk, D. J.,** Recent developments in the analysis of toxic elements, *Science,* 184, 1137, 1974.
92. **Slavin, S.,** An atomic absorption bibliography for January-June 1974, *At. Absorpt. Newsl.,* 13, 84, 1974.
93. **Thomerson, D. R. and Price, W. J.,** Atomic absorption behavior of some of the rare earth elements, *Anal. Chim. Acta,* 72, 188, 1974.
94. **Walsh, A.,** Atomic absorption spectroscopy — stagnant or pregnant, *Anal. Chem.,* 46, 698a, 1974.
95. **Woodriff, R.,** Atomization chambers for atomic absorption spectrochemical analysis. Review, *Appl. Spectrosc.,* 28, 413, 1974.
96. **West, T. S.,** Atomic fluorescence and atomic absorption spectrometry for chemical analysis, *Analyst (London),* 99, 886, 1974.
97. **Hurlbut, J. A.,** History, Uses, Occurrences, Analytical Chemistry and Biochemistry of Beryllium. A Review, Rep. 2152, U.S. Atomic Energy Commission, 1974.
98. **Price, W. J.,** Atomic absorption: an essential tool in modern metallurgy, *Metals Mater.,* 8, 485, 1974.
99. **Reinhold, J. G.,** Review: trace elements — a selective survey, *Clin. Chem. (Winston-Salem, N.C.),* 21, 476, 1975.
100. **Slavin, S. and Lawrence, D. M.,** An atomic absorption bibliography for July-December 1974, *At. Absorpt. Newsl.,* 14, 1, 1975.
101. **Sunshine, I.,** Analytical toxicology, *Anal. Chem.,* 47, 212A, 1975.
102. **Ure, A. M.,** The determination of mercury by non-flame atomic absorption and fluorescence spectrometry — a review, *Anal. Chim. Acta,* 76, 1, 1975.
103. **Slavin, S. and Lawrence, D. M.,** An atomic absorption bibliography for January-June 1975, *At. Absorpt. Newsl.,* 14, 81, 1975.
104. **Winefordner, J. D., Fitzgerald, J. J., and Omenetto, N.,** Review of multielement atomic spectroscopic methods, *Appl. Spectrosc.,* 29, 369, 1975.
105. **Walker, G. W.,** Forensic science. Atomic absorption spectroscopy, *Chem. Br.,* 11, 440, 1975.
106. **Wondzinski, W.,** Atomic absorption and atomic fluorescence spectrophotometry, *GIT Fachz. Lab.,* 19, 671, 1975 (in German).
107. **Hieftje, G. M., Copeland, T. R., and De-Olivaren, D. R.,** Flame emission, atomic absorption and atomic fluorescence spectrometry, *Anal. Chem.,* 48, 142r, 1976.
108. **L'vov, B. V.,** Trace characterization of powders by atomic absorption spectrometry. The state of the art, *Talanta,* 23, 109, 1976.
109. **L'vov, B. V., Katskov, D. A., Kruglikova, L. P., and Polzik, L. K.,** Absolute analysis by flame atomic absorption spectroscopy: present status and some problems, *Spectrochim. Acta,* 31b, 49, 1976.
110. **Goldstein, S. A. and Walters, J. P.,** A review of considerations for high-fidelity imaging of laboratory spectroscopic sources. I, *Spectrochim. Acta,* 31b, 201, 1976.
111. **Goldstein, S. A. and Walters, J. P.,** A review of considerations of high-fidelity imaging of laboratory spectroscopic sources. II, *Spectrochim. Acta,* 31b, 295, 1976.
112. **Slavin, S. and Lawrence, D. M.,** An atomic absorption bibliography for January-June 1976, *At. Absorpt. Newsl.,* 15, 77, 1976.
113. **Kharlamov, I. P., Eremina, G. V., and Neimark, V. Y.,** Atomic absorption determination of harmful impurities of certain non-ferrous metals. A review, *Zavod. Lab.,* 42, 1320, 1976.
114. **O'Laughlin, J. W., Hemphill, D. D., and Pierce, J. O.,** Analytical Methodology for Cadmium in Biological Matter — A Critical Review, International Lead Zinc Research Organization, New York, 1976.
115. **Schuller, P. L. and Egan, H.,** Cadmium, Lead, Mercury and Methyl Mercury Compounds: Review of Methods of Trace Analysis and Sampling with Special Reference to Food, Rep. No. 92-5-1-00094-M-84, Food and Agriculture Organization, Rome, 1976.

116. **Slavin, S. and Lawrence, D. M.,** An atomic absorption bibliography for January-June 1977, *At. Absorpt. Newsl.,* 16, 89, 1977.

117. **Walsh, A.,** Atomic absorption spectroscopy and its applications, old and new, *Pure Appl. Chem.,* 49, 1621, 1977.

118. **Olwin, J. H.,** Metals in the life of man, *J. Anal. Toxicol.,* 1, 245, 1977.

119. **Kalman, S. M.,** The pathophysiology of lead poisoning: a review and a case report, *J. Anal. Toxicol.,* 1, 277, 1977.

120. **Fernandez, F. J.,** Metal speciation using atomic absorption as a chromatography detector. A review, *At. Absorpt. Newsl.,* 16, 33, 1977.

121. **Slavin, S. and Lawrence, D. M.,** An atomic absorption bibliography for July-December 1976, *At. Absorpt. Newsl.,* 16, 4, 1977.

122. **Aidarov, T. K.,** Status of developments and some trends in developing atomic absorption spectrometer instrumentation (review), *Zh. Prikl. Spektrosk.,* 26, 779, 1977.

123. **Jenkins, R.,** Recent developments in wavelength and energy dispersive spectrometry, *Pure Appl. Chem.,* 49, 1583, 1977.

124. **Langmyhr, F. J.,** Direct atomic absorption spectrometric analysis of geological materials. A review, *Talanta,* 24, 277, 1977.

125. **Peterson, G. E.,** The application of atomic absorption spectrophotometry to the analysis of non-ferrous alloys, *At. Absorpt. Newsl.,* 16, 133, 1977.

126. **Harrison, R. M. and Laxen, D. P. H.,** Comparative study of methods for analysis of total lead in soils, *Water Air Soil Pollut.,* 8, 2387, 1977.

127. **Kavanova, M. A.,** Atomic absorption spectrophotometry: basis for standardization of methods for determination of elements in natural waters, *Probl. Sovrem. Anal. Khim.,* 2, 23, 1977 (in Russian).

128. **Slavin, S. and Lawrence, D. M.,** An atomic absorption bibliography for July-December 1977, *At. Absorpt. Newsl.,* 17, 7, 1978.

129. **Berndt, H. and Slavin, W.,** Automated trace analysis of small samples using the "injection method" of flame atomic absorption spectrometry. A review, *At. Absorpt. Newsl.,* 17, 109, 1978.

130. **Eller, P. M. and Haartz, J. C.,** Sampling and analytical methods for antimony and its compounds — a review, *J. Am. Ind. Hyg. Assoc.,* 39, 790, 1978.

131. **Slavin, S. and Lawrence, D. M.,** An atomic absorption bibliography for January-June 1978, *At. Absorpt. Newsl.,* 17, 73, 1978.

132. **Walsh, A.,** Atomic spectroscopy — What next?, *At. Absorpt. Newsl.,* 17, 97, 1978.

133. **Weiss, E. B. and Rosenthal, I. M.,** Clinical use of laboratory determinations for metals, anti-convulsive agents and cardiovascular drugs, *J. Anal. Toxicol.,* 2, 166, 1978.

134. **Pinta, M.,** Present tendencies in atomic spectroscopy, *Analusis,* 6, 227, 1978 (in French).

135. **Jones, E. A. and Dixon, K.,** Review of the literature on the separation and determination of rare earth elements, *Natl. Inst. Metall. Repub. S. Afr. Rep.,* No. 1943, 1978.

136. **Slavin, S. and Lawrence, D. M.,** An atomic absorption bibliography for July-December 1978, *At. Absorpt. Newsl.,* 18, 18, 1979.

137. **Willis, J. B.,** Review: analytical atomic spectroscopy at the CSIRO Division of Chemical Physics, *Anal. Chim. Acta,* 106, 175, 1979.

138. **Wilson, D. L.,** Separation and concentration techniques for atomic absorption: a guide to the literature, *At. Absorpt. Newsl.,* 18, 13, 1979.

139. **Zingaro, R. Z.,** How certain trace elements behave, *Environ. Sci. Technol.,* 13, 282, 1979.

140. **Gladney, E. S., Perrin, D. R., Owens, J. W., and Knab, D.,** Elemental concentrations in the United States Geological Survey's geochemical exploration reference samples — a review, *Anal. Chem.,* 51, 1557, 1979.

141. **Lawrence, D. M.,** An atomic absorption bibliography for January-June 1979, *At. Absorpt. Newsl.,* 18, 77, 1979.

142. **Razumov, V. A. and Zvyagintsev, A. M.,** Nonselective weakening of light in atomic absorption and atomic fluorescence analysis. Review, *Zh. Prikl. Spektrosk.,* 31, 381, 1979.

143. **El-Shaarawy, M. I.,** Selected features of atomic absorption spectrophotometry and their pertinence to oil analysis, *Arabian J. Sci. Eng.,* 4, 75, 1979.

144. **Fan, P. Y.,** Atomic absorption spectroscopy and its application in medical examination, *Chung-Hua I Hsueh Chien Yen Tsa Chih,* 2, 175, 1979.

145. **Kinnunen, J.,** Determination of precious metals in mining and metallurgical products. A review, *Kem. Kemi,* 6, 617, 1979.

146. **Price, W. J.,** New techniques of atomic absorption in clinical and biochemical analysis, *Sci. Ind.,* 12, 22, 1979.

147. **Wang, P. L., Wu, T. C., Yau, W. C., Liu, C. L., and Ma, K. C.,** Thirty years development of atomic absorption analysis in China, *Fen Hsi Hua Hsueh,* 7, 378, 1979.

148. **Marcus, L. H.,** Atomic Absorption Spectroscopy (June 1970 - June 1980), Report 1980 NERACUSGNT 0456; Order No. PB80-859614, 1980; available from the NTIS, Springfield, Va.; *Gov. Rep. Announce. Index (U.S.),* 80, 5637, 1980.

149. **Horlick, G.,** Flame emission, atomic absorption and fluorescence spectrometry, *Anal. Chem.,* 52, 290, 1980.

150. **Johansson, A.,** Atomic absorption spectrometry, *Kem. Tidskr.,* 92, 26, 1980.

151. **Johansson, A.,** More study on chemical literature. II. Atomic absorption spectrometry, *Kem. Tidskr.,* 92, 48, 1980.

152. **Lawrence, D. M.,** An atomic spectroscopy bibliography for July-December 1979, *At. Spectrosc.,* 1, 8, 1980.

153. **Owens, J. W., Gladney, E. S., and Purtymun, W. D.,** Modification of trace element concentrations in natural waters by various field sampling techniques, *Anal. Lett.,* 13a, 253, 1980.

154. **Ali, S. L.,** Atomic absorption spectrophotometry — state of the art after 25 years, *Pharm. Ztg.,* 125, 450, 1980.

155. **De Galan, L. and Van Dalen, J. P. J.,** Atomic absorption spectrometry, *Pharm. Weekbl.,* 115, 689, 1980.

156. **Gilbert, R. K. and Platt, R.,** Measurement of calcium and potassium in clinical laboratories in the United States, 1971-1978, *Am. J. Clin. Pathol.,* 74, 508, 1980.

157. **Hughes, H.,** Analysis-survey. I. Oxide materials, *Iron Steel Int.,* 53, 13, 1980.

158. **Lawrence, D. M.,** An atomic spectroscopy bibliography for January-June 1980, *At. Spectrosc.,* 21, 94, 1980.

159. **L'vov, B. V.,** Twenty five years of analytical atomic absorption spectrometry, *Zh. Anal. Khim.,* 35, 1575, 1980.

160. **Walsh, A.,** The birth of modern atomic absorption spectroscopy, *Chimia,* 34, 427, 1980.

161. **Godden, R. G. and Thomerson, D. R.,** Generation of covalent hydrides in atomic absorption spectroscopy. A review, *Analyst (London),* 105, 1137, 1980.

162. **Grobenski, Z. and Schulze, H.,** Modern atomic spectroscopy techniques for trace metal determination in foods, *GIT Fachz. Lab.,* 24, 1156, 1980 (in German).

163. **Koirtyohann, S. R.,** A history of atomic absorption spectroscopy, *Spectrochim. Acta,* 35b, 663, 1980.

164. **Magyar, B. and Aeschbach, F.,** Why not ICP as atom reservoir for AAS?, *Spectrochim. Acta,* 35b, 839, 1980.

165. **Tonini, C.,** Atomic absorption spectroscopy in determining metals in textiles, *Tinctoria,* 77, 358, 1980 (in Italian).

166. **Willis, J. B.,** Some memories of the early days of atomic absorption spectroscopy, *Spectrochim. Acta,* 35b, 653, 1980.

167. **Walsh, A.,** Atomic absorption spectroscopy. Some personal recollections and speculations, *Spectrochim. Acta,* 35b, 639, 1980.

168. **Yoneda, S.,** Ppm and ppb. Detection limitation of elements, *Kagaku To Seibutsu,* 18, 312, 1980 (in Japanese).

169. **Lawrence, D. M.,** An atomic spectroscopy bibliography for July-December 1980, *At. Spectrosc.,* 2, 22, 1981.

170. **Lawrence, D. M.,** An atomic spectroscopy bibliography for January-June 1981, *At. Spectrosc.,* 2, 101, 1981.

171. **Verlinden, M., Deelstra, H., and Adriaenssens, E.,** Determination of selenium by atomic absorption spectrometry. A review, *Talanta,* 28, 637, 1981.

172. **Snook, R. D.,** A critical appraisal of the hydride generation method, *Anal. Proc. (London),* 18, 342, 1981.

173. **Van Loon, J. C.,** Review of methods for elemental speciation using atomic spectrometry detectors for chromatography, *Can. J. Spectrosc.,* 26, 22a, 1981.

174. **Miller, H. C., James, R. H., Dickson, W. R., Neptune, M. D., and Carter, M. H.,** Evaluation of methodology for survey analysis of solid wastes, *ASTM STP,* 760, 240, 1981.

175. **Varma, A.,** A Guide to Atomic Absorption Spectrophotometric Analysis of Alkali Metals, Instrumentation Laboratories Inc., Waltham, Mass., 1982.

176. **Varma, A.,** A Reference Guide for Antimony Using Atomic Absorption, Instrumentation Laboratories Inc., Waltham, Mass., 1982.

177. **Varma, A.,** A Literature Guide for Atomic Absorption Analysis of Boron, Aluminum, Gallium, Indium and Thallium, Instrumentation Laboratories Inc., Waltham, Mass., 1982.

178. **Varma, A.,** A Reference Guide for Calcium Using Atomic Absorption, Instrumentation Laboratories Inc., Waltham, Mass., 1982.

179. **Horlick, G.,** Atomic absorption, atomic fluorescence and flame spectrometry, *Anal. Chem.,* 54, 276r, 1982.

180. **Komarek, J. and Sommer, L.,** Organic complexing agents in atomic absorption spectrometry. A review, *Talanta,* 29, 159, 1982.

181. **Lawrence, D. M.,** An atomic spectroscopy bibliography for July-December 1981, *At. Spectrosc.,* 3, 13, 1982.

182. **Eaton, A., Oelker, G., and Leong, L.,** Comparison of AAS and ICP for analysis of natural waters, *At. Spectrosc.,* 3, 152, 1982.

183. **Falk, H.,** Limiting factors for intensity and line profile of radiation sources for atomic absorption spectrometry, *Prog. Anal. At. Spectrosc.,* 5, 205, 1982.

184. **Kirkbright, G. F.,** Some recent studies in optical emission and absorption spectroscopy for trace analysis, *Pure Appl. Chem.,* 54, 769, 1982.

185. **Lawrence, D. M.,** An atomic spectroscopy bibliography for January-June 1982, *At. Spectrosc.,* 3, 93, 1982.

186. **Schaller, K. H.,** Methods of quantitative determination of mercury in human biological materials, *Staub Reinhalt. Luft,* 42, 142, 1982.

187. **Werbicki, J. J., Jr.,** How much gold?, *Prod. Finish. (Cincinnati),* 46, 42, 1982.

188. **Yudelevich, I. G., Startseva, E. A., and Gordeev, G. A.,** Atomic absorption determination of platinum-group metals, *Zavod. Lab.,* 48, 23, 1982.

189. **Lawrence, D. M.,** An atomic spectroscopy bibliography for July-December 1982, *At. Spectrosc.,* 4, 10, 1983.

AUTHOR INDEX

Appendix IV
BOOKS

1. **Smith, E. A.,** *The Sampling and Assay of the Precious Metals,* 2nd ed., Griffin, London, 1947.
2. **Sandell, E. B.,** *Colorimetric Determination of Traces of Metals,* 3rd ed., Wiley Interscience, New York, 1959.
3. **Gibson, J. H., Grossman, W. E. L., and Cooks, W. D.,** The use of continuous source in atomic absorption spectroscopy, in *Anal. Chem. Proc. Feigl Anniversary Symposium,* Elsevier Amsterdam, 1962.
4. **David, D. J.,** Emission and atomic absorption spectrochemical methods, in *Modern Methods of Plant Analysis,* Vol. 5, Peach & Tracey, Eds., Springer-Verlag, Berlin, 1962.
5. **Roth, D. J.,** *Developments in Applied Spectroscopy,* Vol. 4, Plenum Press, New York, 1965.
6. **Walsh, A.,** Some recent advances in atomic absorption spectroscopy, in *12th Coll. Spectro. Intern. Exeter, 1965,* Hilger & Watts, London, 1965.
7. **Zettner, A.,** Principles and applications of atomic absorption spectroscopy, in *Advances in Clinical Chemistry,* Vol. 7, Sobotka, H., Ed., Academic Press, New York, 1965.
8. **Beamish, F. E.,** *The Analytical Chemistry of the Noble Metals,* Pergamon Press, Oxford, 1966.
9. **Elwell, W. T. and Gidley, J. A. F.,** *Atomic Absorption Spectrophotometry,* 1st ed., 1961, 2nd ed., Pergamon Press, Oxford, 1966.
10. **Movrodineau, R.,** Atomic absorption spectrophotometry, in *Encyclopedia of Industrial Chemical Analysis,* Snell, F. D. and Hilton, C. L., Eds., Interscience, New York, 1966.
11. **Robinson, J. W.,** *Atomic Absorption Spectroscopy,* Arnold, London, 1966.
12. **Rousselet, F.,** *Spectrophotométrie par Absorption Atomique Appliquée à la Biologie,* Sedes, Paris, 1966.
13. **Weberling, R. P. and Gosgrove, J. F.,** *Trace Analysis,* Morrison, G. H., Ed., Interscience, New York, 1966.
14. **Angino, E. E. and Billings, G. K.,** *Atomic Absorption Spectrometry in Geology,* Elsevier, Amsterdam, 1967.
15. **Ramirez-Munoz, J.,** *Atomic Absorption Spectroscopy,* Elsevier, New York, 1968.
16. **Slavin, W.,** *Atomic Absorption Spectroscopy,* Vol. 25, Elving, P. J. and Kolthoff, I. M., Eds., Interscience, New York, 1968.
17. **Dean, J. A. and Rains, T. C.,** *Flame Emission and Atomic Absorption Spectrometry,* Vol. 1, *Theory,* Marcel Dekker, New York, 1969.
18. **Price, W. J.,** *Chapters on Atomic Absorption and Fluorescence in Spectroscopy,* Browning, D. R., Ed., McGraw-Hill, London, 1969.
19. **Rubeska, I. and Moldan, B.,** *Atomic Absorption Spectrophotometry,* SNTL, Prague, 1967, Engl. ed., Iliffe Books, London, 1969.
20. **Christian, G. D. and Feldman, J. J.,** *Atomic Absorption Spectroscopy: Applications in Agriculture, Biology and Medicine,* Wiley Interscience, New York, 1970.
21. **Eardley, R. P. and Mountford, A. H.,** Application of atomic absorption spectroscopy in the analysis of ceramic materials, in *Automation in Production, Sampling and Testing of Silicate Materials, Symposium Proc.,* Society of Chemical Industry Publications, London, 1970.
22. **L'vov, B. V.,** *Atomic Absorption Spectrochemical Analysis,* transl. from the Russian by Divon, J. H., Adam Hilger, London, 1970.
23. **Reynolds, R. J., Aidoua, K., and Thompson, K. C.,** *Atomic Absorption Spectroscopy,* Griffin, London, 1970.
24. **Beamish, F. E. and Van Loon, J. C.,** *Recent Advances in the Analytical Chemistry of Noble Metals,* Pergamon Press, Oxford, 1972.
25. **Roelandt, I. and Guillaume, M.,** Comparative study on atomic absorption spectrometry and neutron activation: determination of rubidium in geochemistry, in *Proc. 3rd Int. Congr. on Atomic Absorption and Atomic Flame Spectrometry,* Adam Hilger, London, 1972.
26. **Ediger, R. D. and Coleman, R. L.,** An evaluation of anodic stripping voltammetry and nonflame atomic absorption as routine analytical tools, in *Trace Substances in Environmental Health,* Vol. 6, Hemphill, D. D., Ed., University of Missouri, Columbia, 1973.
27. **Green, H. C.,** *Atomic Absorption Spectroscopy in Metallurgical Research,* Publ. No. A.I.D.D. G97, Department of Scientific and Industrial Research, Auckland, New Zealand, 1973.
28. **Kirkbright, G. F. and Sargent, M.,** *Atomic Absorption and Fluorescence Spectroscopy,* Academic Press, New York, 1974.
29. **Price, W. J.,** *Analytical Atomic Absorption Spectrometry,* 2nd ed., Heyden & Son, London, 1974.
30. **Willis, J. B.,** Atomic absorption, atomic fluorescence and flame emission spectroscopy, in *Handbook of Spectroscopy,* Robinson, J. W., Ed., CRC Press, Cleveland, 1974, 799.
31. **Beamish, F. E. and Van Loon, J. C.,** *Analysis of Noble Metals,* Academic Press, New York, 1977.

32. **Reeves, R. D. and Brooks, R. R.,** *Trace Element Analysis of Geological Materials,* Vol. 51, Elving, P. J. and Winefordner, J. D., Eds., John Wiley & Sons, New York, 1978, 160.
33. **Van Loon, J. C.,** *Analytical Atomic Absorption Spectroscopy,* Academic Press, New York, 1980.
34. **Klein, A. A.,** Analysis of low-alloy steel using a sequential atomic absorption spectrophotometer equipped with an autosample, in *New Analytical Techniques for Trace Constituents of Metallic and Metal-bearing Ores,* Javier-Son, A., Ed., American Society for Testing and Materials, Philadelphia, 1981, 29.
35. **Rains, T. C.,** Determination of aluminum, barium, calcium, lead, magnesium and silver in ferrous alloys by atomic absorption spectrometry, in *New Analytical Techniques for Trace Constituents of Metallic and Metal-bearing Ores,* Javier-Son, A., Ed., American Society for Testing and Materials, Philadelphia, 1981.
36. **Magyar, B.,** Guidelines to planning of atomic absorption analysis, in *Studies in Analytical Chemistry,* Vol. 4, Elsevier, Amsterdam, 1982, 274.

Appendix VA
LIST OF HOLLOW CATHODE LAMPS
(SINGLE ELEMENT)

Element	Gas fill	Exit window	Wavelengths (nm)
Aluminum	Neon	Pyrex®	309.2, 396.2
Antimony	Neon	Quartz	217.6, 231.1
Arsenic	Argon	Quartz	193.7, 197.2
Barium	Neon	Pyrex®	350.1, 553.6
Beryllium	Neon	Quartz	234.8
Bismuth	Neon	Quartz	223.1, 306.8
Boron	Argon	Quartz	249.7
Cadmium	Neon	Quartz	228.8, 326.1
Calcium	Neon	Pyrex®	422.7
Cerium	Neon	Quartz	520.0, 569.7
Cesium	Neon	Pyrex®	455.6, 852.1
Chromium	Neon	Quartz/Pyrex®	357.9, 425.4
Cobalt	Neon	Quartz	240.7, 345.4
Copper	Neon	Pyrex®	324.7, 327.4
Dysprosium	Neon	Quartz	404.6, 418.7
Erbium	Neon	Quartz	386.3, 400.8
Europium	Neon	Quartz	459.4, 462.7
Gadolinium	Neon	Quartz	368.4, 407.9
Gallium	Neon	Quartz	287.4, 417.2
Germanium	Neon	Quartz	259.2, 265.1
Gold	Neon	Quartz	242.8, 267.6
Hafnium	Neon	Quartz	286.6, 307.2
Holmium	Neon	Quartz	405.4, 410.4
Indium	Neon	Quartz	304.0, 410.1
Iridium	Neon	Quartz	264.0, 285.0
Iron	Neon	Quartz	248.3, 372.0
Lanthanum	Neon	Quartz	392.8, 550.1
Lead	Neon	Quartz	217.0, 283.3
Lithium-natural	Neon	Pyrex®	670.8
Lithium-6	Neon	Pyrex®	670.8
Lithium-7	Neon	Pyrex®	670.8
Lutetium	Neon	Quartz	335.9, 337.6
Magnesium	Neon	Quartz	202.5, 285.2
Manganese	Neon	Quartz	279.5, 280.1
Mercury	Argon	Quartz	253.7
Molybdenum	Neon	Quartz	313.3, 317.0
Neodymium	Neon	Quartz/Pyrex®	463.4, 492.5
Nickel	Neon	Quartz	232.0, 341.5
Niobium	Neon	Quartz/Pyrex®	405.9, 408.0
Osmium	Argon	Quartz	290.9, 301.8
Palladium	Neon	Quartz	244.8, 247.6
Phosphorus	Neon	Quartz	213.6, 214.9
Platinum	Neon	Quartz	265.9, 299.8
Potassium	Neon	Pyrex®	404.4, 766.5
Praseodymium	Neon	Quartz	495.1, 513.3
Rhenium	Neon	Quartz/Pyrex®	346.0, 346.5
Rhodium	Neon	Quartz/Pyrex®	343.5, 350.7
Rubidium	Neon	Pyrex®	420.1, 780.0
Ruthenium	Neon	Quartz/Pyrex®	349.9, 392.5
Samarium	Neon	Quartz/Pyrex®	429.6, 476.0
Scandium	Neon	Quartz/Pyrex®	390.7, 391.2
Selenium	Neon	Quartz	196.0, 204.0
Silicon	Neon	Quartz	251.6, 288.1
Silver	Argon	Quartz/Pyrex®	328.1, 338.3

Appendix VA (continued)
LIST OF HOLLOW CATHODE LAMPS
(SINGLE ELEMENT)

Element	Gas fill	Exit window	Wavelengths (nm)
Sodium	Neon	Pyrex®	330.2, 589.0
Strontium	Neon	Pyrex®	460.7
Sulfur	Neon	Quartz	—
Tantalum	Argon	Quartz	271.4, 277.5
Tellurium	Neon	Quartz	214.3, 238.6
Terbium	Neon	Quartz/Pyrex®	431.9, 432.6
Thallium	Neon	Quartz	258.0, 276.7
Thorium	Neon	Quartz	—
Thulium	Neon	Quartz	371.7, 409.4
Tin	Neon	Quartz	224.6, 286.3
Titanium	Neon	Quartz/Pyrex®	364.3, 399.8
Tungsten	Neon	Quartz	255.1, 400.9
Uranium	Neon	Quartz/Pyrex®	348.9, 351.5
Vanadium	Neon	Quartz	318.4, 385.5
Ytterbium	Argon	Quartz/Pyrex®	346.4, 398.8
Yttrium	Neon	Quartz	407.7, 410.2
Zinc	Neon	Quartz	213.9, 307.6
Zirconium	Argon	Quartz/Pyrex®	351.9, 360.1

Appendix VB
LIST OF HOLLOW CATHODE LAMPS (MULTIELEMENTS)

Elements	Gas fill	Exit window
Aluminum-Calcium-Copper-Iron-Magnesium-Silicon-Zinc	Neon	Quartz
Aluminum-Calcium-Iron-Magnesium	Neon	Quartz
Aluminum-Calcium-Lithium-Magnesium	Neon	Quartz
Aluminum-Calcium-Iron-Titanium	Neon	Quartz
Aluminum-Calcium-Magnesium	Neon	Quartz
Aluminum-Iron-Magnesium	Neon	Quartz
Antimony-Arsenic-Bismuth	Neon	Quartz
Arsenic-Nickel	Neon	Quartz
Arsenic-Selenium-Tellurium	Neon	Quartz
Barium-Calcium-Strontium	Neon	Pyrex®
Barium-Calcium-Strontium-Magnesium	Neon	Quartz
Cadmium-Lead-Silver-Zinc	Neon	Quartz
Calcium-Aluminum-Magnesium	Neon	Quartz
Calcium-Iron-Aluminum-Magnesium	Neon	Quartz
Calcium-Magnesium	Neon	Quartz
Calcium-Magnesium-Aluminum-Lithium	Neon	Quartz
Calcium-Magnesium-Zinc	Neon	Quartz
Calcium-Zinc	Neon	Quartz
Chromium-Cobalt-Copper-Iron-Manganese-Nickel	Neon	Quartz
Chromium-Cobalt-Copper-Manganese-Nickel	Neon	Quartz
Chromium-Copper	Neon	Quartz
Chromium-Copper-Iron-Nickel-Silver	Neon	Quartz
Chromium-Copper-Nickel-Silver	Neon	Quartz
Chromium-Iron-Manganese-Nickel	Neon	Quartz
Cobalt-Chromium-Copper-Iron-Nickel-Manganese	Neon	Quartz
Cobalt-Copper-Iron-Manganese-Molybdenum	Neon	Quartz
Copper-Cadmium-Lead-Zinc	Neon	Quartz
Copper-Cobalt	Neon	Quartz

Appendix VB (continued)
LIST OF HOLLOW CATHODE LAMPS (MULTIELEMENTS)

Element	Gas fill	Exit window
Copper-Gallium	Neon	Quartz
Copper-Iron	Neon	Quartz
Copper-Iron-Lead-Nickel-Zinc	Neon	Quartz
Copper-Iron-Manganese-Zinc	Neon	Quartz
Copper-Iron-Nickel	Neon	Quartz
Copper-Lead-Tin-Zinc	Neon	Quartz
Copper-Lead-Zinc-Silver	Neon	Quartz
Copper-Manganese	Neon	Quartz
Copper-Zinc-Molybdenum	Neon	Quartz
Copper-Zinc-Molybdenum-Cobalt	Neon	Quartz
Gold-Copper-Iron-Nickel	Neon	Quartz
Gold-Nickel	Neon	Quartz
Gold-Silver	Neon	Quartz
Iron-Copper-Manganese	Neon	Quartz
Magnesium-Zinc	Neon	Quartz
Molybdenum-Copper-Iron	Neon	Quartz
Sodium-Potassium	Argon	Pyrex®
Zinc-Silver-Lead-Cadmium	Neon	Quartz

Appendix VC
LIST OF
ELECTRODELESS
DISCHARGE LAMPS

Aluminum	Phosphorus
Antimony	Potassium
Arsenic	Rubidium
Bismuth	Selenium
Cadmium	Tellurium
Cesium	Thallium
Germanium	Tin
Lead	Titanium
Mercury	Zinc

Appendix VIA
LIST OF MANUFACTURERS FOR ATOMIC ABSORPTION
SPECTROPHOTOMETERS AND ACCESSORIES

1. Analabs/Foxboro Analytical, a unit
   of Foxboro Analytical
   80 Republic Drive
   North Haven, Conn. 06473, U.S.
   Phone (203) 288-8463
   Telex 96-3518, TWX 710-4650-267

2. Anaspec Ltd.
   P.O. Box 25
   Newbury, Berks., RG14 5LL,
   England, U.K.
   Phone (0635)44329
   Telex 849266

3. Baird Corporation
   125 Middlesex Turnpike
   Bedford, Mass. 01730, U.S.
   Phone (617) 276-6000
   Telex 923491

4. Barnes Analytical Division,
   Spectra-Tech. Inc.
   652 Glenbrook Road
   Stamford, Conn. 06906, U.S.
   Phone (800) 243-9186, (203) 357-
   7055

5. Bausch & Lomb Inc. Instruments &
   Systems Division
   42 East Avenue
   Rochester, N.Y. 14692, U.S.
   Phone (716) 338-6000
   Telex 646748

6. Beckman Instruments Inc. Scientific
   Instruments Division
   Campus Drive and Jamboree
   Boulevard
   Irvine, Calif. 92713, U.S.
   Phone (714) 833-0751
   Telex 06-78413, TWX 910-592-
   1260

7. Beckman Instruments Inc.
   2500 Harbor Boulevard
   Fullerton, Calif. 92634, U.S.
   Phone (714) 871-4848
   Telex 06-78413, TWX 910-592-1260

8. Beckman Instruments International
   SA, Analytical Instruments
   Division
   17 Rue des Pierres-du Niton
   Geneva 1211 Switzerland
   Phone (022) 36 20 70
   Telex 22 596

9. Buck Scientific
   58 Fort Point Street
   East Norwalk, Conn. 06855, U.S.
   Phone (203) 853-9444
   Telex 643589

10. Burkard Instruments AG
    Buckhauserstrasse 26, CH-8048
    Zurich, Switzerland
    Phone (01) 491-5000
    Telex 56025 buin ch

11. Cecil Instruments Ltd.
    Milton Industrial Estate
    Milton, Cambridge, CB4 4A2,
    England, U.K.
    Phone (0223) 66821
    Telex 817479

12. CHEMetrics, Inc.
    Rt. 28
    Calverton, Va. 22016, U.S.
    Phone (703) 788-9026

13. Elico Pvt. Ltd.
    Instruments B-17
    Sanathnagar Industrial Estate
    Hyderabad 500 018, Andhra
    Pradesh, India
    Phone (0842) 26 0285
    Telex 0155-714 PH HD IN

14. Fisher Scientific Co.
    711 Forbes Avenue
    Pittsburgh, Pa. 15219, U.S.
    Phone (412) 562-8300

15. Foss Electric (Ireland) Ltd.
    Sandyford Industrial Estate
    Foxrock, Dublin, 18, Ireland
    Phone 953301
    Telex 24316

16. Gilford Instrument Laboratories,
    Inc.
    132 Artino St.
    Oberlin, Ohio 44074, U.S.
    Phone (216) 774-1041
    Telex 980-456 (GILLABSLS)

17. Harrick Scientific Corp.
    88 Broadway, Box 351
    Ossining, N.Y. 10562, U.S.
    Phone (914) 762-0020

18. Hi-Tech Scientific Ltd.
    Brunel Road
    Salisbury, Wilts. SP2 7PU,
    England, U.K.
    Phone (0722) 20322
    Telex 477019

19. Hilger Analytical Ltd.
    Westwood Industrial Estate
    Ramsgate, Margate, Kent CT9 4JL,
    England, U.K.
    Phone 0843-25131 or 24261
    Telex 96252

20. Hitachi Scientific Instruments,
    Analytical Instruments Department
    460 East Middlefield Road
    Mountain View, Calif. 94043, U.S.
    Phone (415) 969-1100
    Telex 171429

21. Instrumentation Laboratory Inc.
    113 Hartwell Avenue
    Lexington, Mass. 02173, U.S.
    Phone (800) 225-4040
    Telex 92-3440

22. Instrumentation Laboratory Inc.
    Analytical Instrument Division
    Jonspin Road
    Wilmington, Mass. 01887, U.S.
    Phone (617) 658-5125
    TWX 710-347-1274

23. Instrumentation Laboratory Ltd.
    European Headquarters
    Kelvin Close, Birchwood Science
    Park
    Warrington, Cheshire, England,
    U.K.
    Phone (0925) 810141
    Telex 627713

24. Jobin-Yvon
    Longjumeau, France

25. Kontron AG, Anal. Division
    Bernerstrasse-Sued 169,8048
    Zurich, Switzerland
    Phone (01) 435 41 11
    Telex 822191 kon

26. LKB Produkter, Inc.
    Box 305
    S-161 26 Bromma, Sweden
    Phone (08) 98 00 40
    Telex 10 492

27. LP Italiana SpA
    via Carlo Riale
    1514,20157, Milan, Italy
    Phone (02)376-4646 or 3641
    Telex 330896 LP I

28. Optical S. P. A.
    Milan, Italy

29. Oxford Instruments Ltd.
    Osney Mead
    Oxford, OXZ ONX, England, U.K.
    Phone Oxford 41456
    Telex 83413 OXINST G

30. Perkin Elmer Corporation,
    Instrument Group
    Main Avenue
    Norwalk, Conn. 06856, U.S.
    Phone (203) 762-1000
    Telex 00096-5954, TWX 710-468-
    3213

31. Perkin Elmer & Co. GmbH
    Bodenseewerk
    P.O. Box 1120
    D-7770 Uberlingen, Federal
    Republic of Germany
    Phone 07551-811
    Telex 0733902

32. Perkin Elmer Ltd.
    Post Office Lane
    Beaconsfield, Bucks., HP9 IQA,
    England, U.K.
    Phone (04-946) 6161
    Telex 83257

33. Preiser Scientific Inc.
    900 MacCorkle Avenue S.W.
    Charleston, W.Va. 25322, U.S.
    Phone (304) 344-4031
    TWX 710-938-1634

34. Pye Unicam Ltd., A Scientific &
    Industrial Co. of Philips
    York Street
    Cambridge, CBI 2PX, England,
    U.K.
    Phone 44 223 358866
    Telex 817331

35. Sargent-Welch Scientific Co.
    7300 N. Linder Avenue
    Skokie, Ill. 60077, U.S.
    Phone (312) 677-0600
    Telex 710-938-1634

36. Shandon Southern Instrument Inc.
    515 Broad Street
    Sewickley, Pa. 15143, U.S.
    Phone (412) 741-8400
    Telex 86-6200

37. Shandon Southern Ltd.
    Camberley, England, U.K.

38. Shimadzu (Europa) GmbH, Import/
    Export
    Ackerstrasse 111,4000
    Dusseldorf 1, Federal Republic of
    Germany
    Phone 0211/66 63 71
    Telex 8 586 839

39. Shimadzu International Marketing
    Division
    Shinjuku Mitsui Building
    No. I-I Nishi-Shinjuku 2-chome
    Shinjuku-ku, Tokyo, 160, Japan

40. Shimadzu Scientific Instruments
    9147-H Red Branch Road
    Columbia, Md. 21045, U.S.
    Phone (301) 997-1227
    Telex 87-959

41. Sixes and Sevens
    Sumner, Wash. 98390, U.S.

42. Spectrametrics Inc.,
    204 Andover Street
    Andover, Mass. 01810, U.S.
    Phone (617) 475-7015
    Telex 94-7134

43. Spectro Products Inc.
    North Haven, Conn. 06473, U.S.

44. Tennelco, Inc.
    601 Oak Ridge Turnpike
    Oak Ridge, Tenn. 37830, U.S.
    Phone (615) 483-8405
    TWX 810-572-18

45. Varian Associates Inc., Instrument
    Group
    611 Hansen Way
    Palo Alto, Calif. 94303, U.S.
    Phone (415) 493-4000
    Telex 34-8476, TWX 910-373-
    1731

46. Varian AG
    Steinhauserstrasse
    CH-6300 Zug, Switzerland
    Phone (042) 23 25 75
    Telex 868841

47. Varian Canada, Inc.
    45 River Drive
    Georgetown, Ont. L7G 2J4, Canada
    Phone (416) 457-4130
    Telex 069-7502

48. Varian SA
    Francisco Petrarca 326
    Mexico D.F., Mexico 5
    Phone (905) 545-4077
    Telex 177-4410

49. Varian Industria a Comercio Ltda.
    Avenida Drive, Cardoso de Melo
    1644
    CEP 04548 Sao Paulo, Brazil
    Phone 240-1622
    Telex 24314

50. Varian, Analytica Corp.
    3rd Matsuda Bldg., 2-2-6 Ohkubo
    Shinjuku-ku, Tokyo 160, Japan
    Phone 03-204-1211
    Telex J26471

51. Varian GmbH
    Alsfelderstrasse 6, D-6100
    Darmstadt, Federal Republic of
    Germany
    Phone (06151) 7031
    Telex 419429

52. Varian Associates Ltd.
    28 Manor Road
    Walton-on-Thames/Surrey,
    England, U.K.
    Phone (09322) 43 741
    Telex 928070

Appendix VIB
MANUFACTURERS OF HOLLOW CATHODE LAMPS

1. Atomic Spectral Lamps Pty. Ltd.
   23-31 Islington Street
   Melbourne, Victoria, Australia

2. Cathodeon Ltd.
   Nuffield Road
   Cambridge, CB4 ITF, England,
   U.K.
   Phone 44 (0223) 65222
   Telex 81685

3. FIVRE
   Via Panciatichi 70
   Firenze-Castello, Italy

4. Hitachi Ltd.
   1-4 Marunouchi
   Chiyoba-ku, Tokyo 160, Japan

5. Dr. Kern und Sprenger GmbH
   Post Box 751
   Florenz Sartorius Strasse 5
   Gottingen, Federal Republic of
   Germany

6. Perkin Elmer Corporation,
   Instrument Division
   Norwalk, Conn. 06852, U.S.
   Phone (203) 762-1000
   Telex 00096-5954, TWX 710-468-
   3213

7. Quartzlampen GmbH
   Hoehensonnen Strasse Hanau/Main
   Federal Republic of Germany

8. Rank Precision Industries Ltd.
   98 St. Pancras Way
   London, England, U.K.

9. Westinghouse Electric Corporation,
   Industrial and Government Tube
   Division
   Westinghouse Circle
   Horseheads, N.Y. 14845, U.S.
   Phone (607) 796-3211
   TWX 5102521588

Appendix VIC
LIST OF CHEMICAL SUPPLIERS (STANDARD SOLUTIONS, COMPLEXING AGENTS, ORGANOMETALLICS, AND CERTIFIED STANDARDS)

1. Aldrich Chemical Co.
   2371 North 30th Street
   Milwaukee, Wis. 53210, U.S.

2. Aldrich Chemical Co. Inc.
   940 West St. Paul Avenue
   Milwaukee, Wis. 53233, U.S.
   Phone (414) 273-3850

3. Alfa Inorganics
   8 Congress Street
   Beverly, Mass. 01915, U.S.

4. Alfa Products Thiokol/Ventron
   Division
   P.O. Box 299
   152 Andover Street
   Danvers, Mass. 01923, U.S.
   Phone (617) 777-1970
   Telex 951360 ALFA PROD DARS

5. Alpha Analytical Laboratories
   Division of Alpha Metals Inc.
   Jersey City, N.J. 07604, U.S.

6. Analytical Standards Lab.
    Box 21
    434 02 Kiungsbacka 2, Sweden

7. Analytical Standards Ltd.
    Fjallagatan 18
    41317 Goteborg, Sweden

8. Anderman & Co. Ltd.
    Battlebudge House
    87-95 Tooley Street
    London, SE1, England, U.K.

9. Angstrom Inc.
    P.O. Box 248
    Belleville, Mich. 48111, U.S.

10. Apache Chemicals Inc.
    Grant Street, P.O. Box 126
    Seward, Ill. 61077, U.S.
    Phone (815) 247-8491
    Corp. TWX 910 - 642 - 0601

11. Baird Corp.
    125 Middlesex Turnpike
    Bedford, Mass. 01730, U.S.
    Phone (617) 276-6000
    Telex 923491

12. J.T. Baker Chemical Co.
    222 Red School Lane
    Phillipsburg, N.J. 08865, U.S.
    Phone (201) 859-2151
    Telex 847480 (Domestics)
    Telex 831644 (Export)

13. Baker Chemikalien
    6080 Gross Gerau
    Postfach 1661
    Federal Republic of Germany
    Phone (06152) 710371
    Telex 0 419 1113

14. J.T. Baker Chemicals BV.
    Rijster borgher weg 20
    P.O. Box 1
    7400 AA Deventer, The
      Netherlands
    Phone (05700) 11341
    Telex 49072

15. J.T. Baker S.A. de C.V.
    Apartado Postal No. 75595
    Col. Lindvista. Deleg. Gustavo A.
    Madero 07300, Mexico D.F.
    Phone 569 - 1100
    Telex 1772336

16. J.T. Baker Chemical Co.
    Suite 1102, World Trade Center
    Telok Blangah Road
    Singapore 0409
    Phone 2735285
    Telex RS 39323 BACHEM

17. Barnes Engineering Co.
    30 Commerce Road
    Stamford, Conn. 06902, U.S.
    Phone (203) 348-5381

18. B D H Chemicals Ltd.
    Poole, Dorset BH12 4NN, England,
      U.K.

19. B D H
    Via Greda
    142 Milano, Italy 20126

20. Bie and Berntsen
    Sandbaekvej 7
    2610 Roedovre, Denmark

21. Bio-Rad Laboratories
    2200 Wright Avenue
    Richmond, Calif. 94804, U.S.
    Phone (415) 234-4130
    Telex 337-732

22. Bio-Rad Laboratories Ltd.
    Caxton Way
    Holywell Industrial Estate
    Watford, Herts. WD1 8RP,
      England, U.K.
    Phone (0923) Watford 40322
    Telex 8813192

23. Bracco
    Via Folli 50
    Milano, Italy 20134

24. Brammer Standard Co., Inc.
    5607 Fountainbridge Lane
    Houston, Tex. 77069, U.S.
    Phone (713) 440-9396
    Telex 775-376

25. Bundensanstalt fur Material
    Prufung
    Unter der Eichen 85
    1000 Berlin 45, Federal Republic of
    Germany

26. Burt & Harvey Ltd.
    Poole, Dorset BH12 4NN England,
    U.K.

27. Carlo Erba Farmatalia, Analytical
    Division
    Via C. Imbonati
    24 20159 Milano
    Italy
    Telex 330314

28. Conostan, Conoco, Inc.
    P.O. Box 1267
    Ponka City, Okla. 74601, U.S.

29. Curtis Mathis Scientific, Inc.
    9999 Stuebner - Airline
    Houston, Tex. 77001, U.S.
    Phone (713) 820-1661

30. Deutsch Vertretung
    Dipl-Met. G. Winopal
    Universal-Forschungsbedarf
    Echternfeld 25, Postfach 40
    3000 Hannover 51, Federal
    Republic of Germany

31. Durham Raw Materials Ltd.
    1-4 Great Tower Street
    London, EC3 R4AB, England,
    U.K.

32. Eastman Organic Chemicals,
    Eastman Kodak Co.
    343 State Street
    Rochester, N.Y. 14650, U.S.
    Phone (716) 724-4000

33. F & J Scientific
    79 Far Horizon Drive
    Monroe, Conn. 06468, U.S.

34. Firma Merck AG
    Postfach 4119
    6100 Darmstadt 2
    Federal Republic of Germany

35. Firma Riedel-de-Haen AG
    Wunstorfer Strasse
    3016 Seelze-Hannover, Federal
    Republic of Germany

36. Fisons Scientific Equipment
    Bishop Meadow Road
    Loughborough, Leics. LE11 ORG,
    England, U.K.
    Phone 0509 - 31166
    Telex 341110

37. Fluka AG
    Chemische Fabrik
    CH-9470 Buchs, Switzerland
    Phone 085 6 0275
    Telex 855282, 855283

38. Fluka Feinchemikalien GmbH
    Postfach 1346
    7910 New-Ulm, Federal Republic
    of Germany

39. Hopkins & Williams Ltd.
    P.O. Box 1
    Romford Essex, RM1 1HA,
    England

40. Instrumentation Laboratory Inc.
    113 Hartwell Avenue
    Lexington, Mass. 02173, U.S.
    Phone (800) 225-4040
    Telex 92 - 3440

41. Johnson Matthey Chemicals Ltd.
    74 Hatton Garden
    London, ECIP IAE, England, U.K.

42. A. Johnson and Co.
    Box 57
    201 20 Malmö
    Sweden

43. K & K Laboratories
    121 Express Street
    Plainville, N.Y. 11803, U.S.

44. Kebo-grave
    Domnarvag 4
    16391 Spaanga, Sweden

45. Koch Light Laboratories Ltd.
    37 Hollands Road
    Haverhill, Suffolk, England, U.K.
    Phone Haverhill 702436
    Telex 81504

46. Labassco ab
    Almedahlsv, Id,
    Box 5205
    A02 24 Gothenberg, Sweden

47. Mallinckrodt, Inc., Science
      Products Division
    675 McDonnell Boulevard
    P.O. Box 5840
    St. Louis, Mo., U.S.
    Phone (314) 895-2340

48. May & Baker Ltd.
    Dagenham,
    Essex, RM10 7XS, England, U.K.

49. M B H Analytical Ltd.
    Station House
    Potters Bar
    Herts EN6 IAL, England, U.K.

50. Merck
    65 Rue Victoire
    Paris, France

51. Merck Chemicals (Pty) Ltd.
    Wrench Road
    Isando, Republic of South Africa

52. E Merck
    Frankfurter Strasse 250,
    Postfach 4119
    D - 6100 Darmstadt, Federal
      Republic of Germany
    Phone 0615/721
    Telex 04 - 19 - 325 emd d, cable,
      emerck darmstadt

53. Molybdenum Corp. of America
    280 Park Avenue
    New York, N.Y. 10017, U.S.

54. National Bureau of Standards Office
      of Standard Reference Materials
    Room B311, Chemistry Building
    Washington, D.C. 20234, U.S.

55. National Spectrographic
      Laboratories Inc.
    19500 South Miles Road
    Cleveland, Ohio 44128, U.S.

56. pH-tamm Laboratoriet ab
    Liljeborgsv 12 752 36
    Uppsala, Sweden

57. Prolabo
    12 Rue Pelee
    Paris, France

58. Regine Brooks RBS
    Pariser Strasse 5
    5300 Bonn I, Federal Republic of
      Germany

59. Research Organic/Inorganic
      Chemical Corp.
    11686 Sheldon Street
    Sun Valley, Calif. 91352, U.S.

60. Riedel de Haen
    171 Av Jean Jaures
    Aubervilliers, France

61. Rielel de Haen AG
    Wunstorter Strasse 40
    D 3016 Seelze 1, Federal Republic
      of Germany

62. Spex Industries Inc.
    3880 Park Avenue
    Metuchen, N.J. 08840, U.S.
    Phone (201) 549 - 7144
    TWX 710-998-0531

63. Struers K/S
    Volhojs Alle 176
    DK - 2610 Rodovre
    Copenhagen, Denmark
    Phone (45) 1 70 80 90
    Telex 19625

64. U.S. Geological Survey National
     Center
     Reston, Va. 22092, U.S.

65. Ventron GmbH
     Postfach 6540
     7500 Karlsruhe I, Federal Republic
      of Germany
     Phone 0721 - 557061
     Telex 7826579

Index

INDEX

A

Agriculture and products
Aluminum, II: 191
Antimony, II: 339
Arsenic, II: 319
Barium, I: 263
Beryllium, I: 197
Bismuth, II: 351
Boron, II: 187
Cadmium, II: 117
Calcium, I: 227
Cesium, I: 191
Chlorine, II: 395
Chromium, I: 333
Cobalt, I: 435
Copper, II: 3
Gallium, II: 205
Gold, II: 61
Indium, II: 211
Iron, I: 403
Lead, II: 253
Lithium, I: 157
Magnesium, I: 203
Manganese, I: 371
Mercury, II: 155
Molybdenum, I: 355
Nickel, I: 451
Nitrogen, II: 309
Phosphorus, II: 311
Potassium, I: 177
Rubidium, I: 187
Selenium, II: 369
Silicon, II: 227
Silver, II: 47
Sodium, I: 157
Strontium, I: 255
Sulfur, II: 363
Tin, II: 241
Titanium, I: 297
Vanadium, I: 313
Zinc, II: 75

B

Biology and biological products
Aluminum, II: 191
Antimony, II: 339
Arsenic, II: 319
Barium, I: 263
Beryllium, I: 197
Bismuth, II: 351
Cadmium, II: 117
Calcium, I: 227
Chlorine, II: 395
Chromium, I: 333
Cobalt, I: 435

Copper, II: 3
Fluorine, II: 395
Gallium, II: 205
Gold, II: 61
Iodine, II: 395
Iron, I: 403
Lead, II: 253
Lithium, I: 157
Magnesium, I: 203
Manganese, I: 371
Mercury, II: 155
Nickel, I: 451
Palladium, I: 483
Phosphorus, II: 311
Platinum, I: 497
Potassium, I: 177
Rubidium, I: 187
Selenium, II: 369
Silicon, II: 227
Silver, II: 47
Sodium, I: 165
Strontium, I: 255
Sulfur, II: 363
Tellurium, II: 385
Thallium, II: 217
Vanadium, I: 313
Zinc, II: 75

C

Clinical, forensic, and toxicology
Aluminum, II: 191
Antimony, II: 339
Beryllium, I: 197
Cadmium, II: 117
Calcium, I: 227
Cobalt, I: 435
Iron, I: 403
Lead, II: 253
Lithium, I: 157
Magnesium, I: 203
Manganese, I: 371
Nickel, I: 451
Zinc, II: 75

E

Energy producing products
Aluminum, II: 191
Antimony, II: 339
Arsenic, II: 399
Barium, I: 263
Beryllium, I: 197
Boron, II: 187
Cadmium, II: 117
Calcium, I: 227

G

M

P

Pharmaceuticals
  Aluminum, II: 191
  Arsenic, II: 319
  Bismuth, II: 351
  Cadmium, II: 117
  Calcium, I: 227
  Chromium, I: 333
  Copper, II: 3
  Germanium, II: 237
  Iron, I: 403
  Lead, II: 253
  Magnesium, I: 203
  Manganese, I: 371
  Mercury, II: 155
  Nickel, I: 451
  Strontium, I: 255
  Sulfur, II: 363
  Zinc, II: 75

W

Waste products
  Aluminum, II: 191
  Antimony, II: 339
  Arsenic, II: 319
  Barium, I: 263
  Cadmium, II: 117
  Calcium, I: 227
  Chromium, I: 333
  Cobalt, I: 435
  Copper, II: 3
  Iron, I: 403
  Lead, II: 253
  Magnesium, I: 203
  Manganese, I: 371
  Mercury, II: 155
  Molybdenum, I: 355
  Nickel, I: 451
  Phosphorus, II: 311
  Selenium, II: 369
  Silicon, II: 227
  Silver, II: 47

Sulfur, II: 363
Tin, II: 241
Zinc, II: 75
Water
  Aluminum, II: 191
  Antimony, II: 339
  Arsenic, II: 319
  Barium, I: 263
  Beryllium, I: 197
  Bismuth, II: 351
  Boron, II: 187
  Cadmium, II: 117
  Calcium, I: 227
  Cesium, I: 191
  Chlorine, II: 395
  Chromium, I: 333
  Cobalt, I: 435
  Copper, II: 3
  Fluorine, II: 395
  Germanium, II: 237
  Gold, II: 61
  Indium, II: 211
  Iron, I: 403
  Lead, II: 253
  Lithium, I: 157
  Magnesium, I: 203
  Manganese, I: 371
  Mercury, II: 155
  Molybdenum, I: 355
  Nickel, I: 451
  Phosphorus, II: 311
  Potassium, I: 177
  Rubidium, I: 187
  Scandium, I: 273
  Selenium, II: 369
  Silver, II: 47
  Sodium, I: 165
  Strontium, I: 255
  Sulfur, II: 363
  Tellurium, II: 385
  Thallium, II: 217
  Tin, II: 241
  Tungsten, I: 365
  Uranium, I: 287
  Vanadium, I: 313
  Zinc, II: 75